AF361893

Advances in Structural and Mechanical Performances of Structures and Materials

Advances in Structural and Mechanical Performances of Structures and Materials

Guest Editors

Atsushi Suzuki
Dinil Pushpalal

Basel • Beijing • Wuhan • Barcelona • Belgrade • Novi Sad • Cluj • Manchester

Guest Editors

Atsushi Suzuki
Tohoku University
Sendai
Japan

Dinil Pushpalal
Tohoku University
Sendai
Japan

Editorial Office
MDPI AG
Grosspeteranlage 5
4052 Basel, Switzerland

This is a reprint of the Special Issue, published open access by the journal *Buildings* (ISSN 2075-5309), freely accessible at: https://www.mdpi.com/journal/buildings/special_issues/L276130F9P.

For citation purposes, cite each article independently as indicated on the article page online and as indicated below:

Lastname, A.A.; Lastname, B.B. Article Title. *Journal Name* **Year**, *Volume Number*, Page Range.

ISBN 978-3-7258-3107-4 (Hbk)
ISBN 978-3-7258-3108-1 (PDF)
https://doi.org/10.3390/books978-3-7258-3108-1

Cover image courtesy of Atsushi Suzuki

Contents

About the Editors

Atsushi Suzuki

Atsushi Suzuki, Ph.D., P.E. is an assistant professor in the Department of Architecture and Building Sciences at Tohoku University, Japan. He earned his Ph.D. in Structural Engineering in 2019. Following roles as a construction consultant in Japan and a research engineer in Mongolia, he embarked on his academic career as an assistant professor. Dr. Suzuki manages over 20 projects, addressing a wide range of topics ranging from the structural safety of steel structures to the decommissioning of the Fukushima nuclear power plant. His ongoing contributions to the academic community are reflected in his extensive publication record, comprising over 100 works, including 30 articles in top-tier journals, 25 international conference proceedings, and numerous domestic conference papers, both peer-reviewed and non-peer-reviewed. Dr. Suzuki's achievements have been recognized with prestigious awards from the Architectural Institute of Japan, the Japan Concrete Institute, and the Japan Society of Steel Construction.

Dinil Pushpalal

Dinil Pushpalal, Ph.D., was a professor at the Graduate School of International Cultural Studies, Tohoku University, Japan. He earned his Ph.D. in Material Science in 1997. After retirement from Tohoku University in 2024, Pushpalal is currently teaching and researching as a visiting professor at the Mongolian University of Science and Technology, Ulaanbaatar, Mongolia, by the invitation of the M-JEED (Mongolia Japan Engineering Education Development) Joint Research Program. He worked as a research engineer for Maeta Concrete Industry Ltd., Yamagata, Japan, from 1991 and engaged in cement and concrete research for more than ten years before joining Tohoku University in 2003. His debut research work dealt with developing a high-bending-strength cement–polymer composite, for which he has thirteen patents and many publications in renowned international journals.

Preface

The mechanical and structural performance of constructions is essential for ensuring the safety and well-being of society. Throughout their service lives, materials and structures are subjected to various demands, including dynamic loads such as earthquakes, wind, water, and soil pressure, as well as exposure to extreme conditions such as freeze–thaw cycles, corrosion, and heat. While these challenges vary by region, researchers and engineers worldwide continually strive to address the complex demands imposed by society.

Recent advancements in construction engineering have significantly matured within specialized fields. However, alongside these developments, new challenges have emerged. The construction industry continues to face numerous unresolved issues that require innovative solutions. To encourage progress in this field, an integrated approach that combines various aspects of engineering and materials science is essential. With this vision, this current Special Issue focuses on uniting achievements in both structural and material engineering as a foundational step toward further advancements.

This Special Issue showcases exceptional efforts aimed at addressing the aforementioned challenges. It includes 15 papers on topics spanning from materials science to structural engineering, covering a wide range of construction materials such as wood, concrete, and steel. The collective insights presented in this compilation provide valuable perspectives that can inspire future advancements for researchers and engineers worldwide.

Finally, the Guest Editors extend their heartfelt appreciation to all contributors for their invaluable work in shaping this Special Issue. Special thanks are due to the Managing Editor, Ms. Nina Tang, for her unwavering support throughout this process. Additionally, this Special Issue serves as part of a tribute project marking the remarkable career of Prof. Dr. Dinil Pushpalal at Tohoku University. I, Dr. Atsushi Suzuki, personally dedicate this Special Issue to him in gratitude for his guidance and cooperation as my supervisor for over a decade.

Atsushi Suzuki and Dinil Pushpalal
Guest Editors

Editorial

Guest Editorial on Advances in Structural and Mechanical Performances of Structures and Materials

Atsushi Suzuki [1,*] and Dinil Pushpalal [2]

1   Graduate School of Engineering, Tohoku University, Sendai 980-8579, Japan
2   School of Civil Engineering and Architecture, Mongolian University of Science and Technology, Ulaanbaatar 14191, Mongolia; dinil.pushpalal.b4@tohoku.ac.jp
*   Correspondence: atsushi.suzuki.c2@tohoku.ac.jp; Tel.: +81-22-795-7876

Received: 18 December 2024
Accepted: 21 December 2024
Published: 24 December 2024

Citation: Suzuki, A.; Pushpalal, D. Guest Editorial on Advances in Structural and Mechanical Performances of Structures and Materials. *Buildings* **2025**, *15*, 3. https://doi.org/10.3390/buildings15010003

1. Introduction

The mechanical and structural performance of structures is essential to ensuring the safety and well-being of society. Materials and structures are subjected to various demands throughout their service lives, including dynamic loads such as earthquakes, wind, water, and soil pressure and exposure to extreme conditions like freeze–thaw cycles, corrosion, and heat. While these challenges differ among regions, researchers and engineers worldwide continually strive to address the complex demands imposed by society.

Recent advancements in construction engineering have led to significant progress within specialized fields. However, alongside these developments, new challenges continue to emerge. The construction industry faces numerous unresolved issues that require innovative solutions. To advance this field further, an integrated approach combining various aspects of engineering and materials science is vital. In line with this vision, the current Special Issue aims to bring together structural and material engineering achievements as a foundational step toward new advancements.

This Special Issue highlights exceptional efforts to address the aforementioned challenges. It features 15 papers covering topics ranging from materials science to structural engineering, with a focus on a diverse array of construction materials, including wood, concrete, and steel. The authors who contributed to this issue are affiliated with countries across a large geographical distribution, including America, China, Cyprus, France, Germany, Greece, Japan, Poland, Romania, and Turkey. The insights compiled in this edition provide valuable perspectives that will inspire researchers and engineers worldwide to pursue future innovations.

2. Overview of Contributions

Suzuki et al. explored the application of piezoelectric sensors for structural damage detection [1]. The piezoelectric sensors were first used in cyclic loading tests by applying them to a component model of composite beam and frame subassemblies with a folded roof plate. The prospective positions of concrete cracking and buckling were predicted based on a robust finite element simulation. During the cyclic loading test, crack propagation in concrete and buckling of the roof plate were detected multiple times by the piezoelectric sensors. Hence, it was concluded that piezoelectric sensors can detect the structural damage of building components, demonstrating their potential use in inexpensive and stable monitoring systems.

Bellos and Konstantinidis introduced a robust approach to determining the yield and failure curvatures of reinforced concrete (RC) cross-sections while accounting for cracking effects [2]. Their proposed methodology also facilitates the calculation of required reinforcement or resistance moments at the yield and failure limits. This approach enables systematic strain calculations for various concrete and steel grades, accommodating both standard and non-standard cross-sections, thereby enhancing practical applications.

Ni and Aboutaha conducted a comprehensive finite element analysis (FEA) of the flexural strength of an AASHTO Type III girder-deck system strengthened to address debonding-damaged strands [3]. The proposed FEA model was validated using relevant experimental data. A parametric study was performed to investigate influential factors, and a total of 156 girder-deck systems were analyzed, considering variables such as the debonding level, span-to-depth ratio, strengthening type, and amount of reinforcing material. Finally, the study employed regression analysis to evaluate the relationship between ultimate strength and the amount of strengthening material.

Wang et al. conducted bending tests on simply supported frame-unit bamboo culm structures [4]. They developed a calculation model for the structure, introducing formulae to describe the stiffness balance between the joints and the overall structure. Furthermore, they proposed a simulation method to convert the actual structure into a beam element model, leveraging insights from their study. The predicted displacements obtained using the proposed method showed strong agreement with the experimentally observed results, demonstrating the adequacy and validity of the model.

Tanaka et al. applied an innovative non-destructive testing method utilizing subterahertz waves to evaluate the water content in concrete [5]. The study demonstrated that the reflectance of the applied frequencies increases with higher unit volume water content. Additionally, the data scatter was significantly reduced when higher frequency ranges were employed. Ultimately, an empirical equation was developed to classify the risk of reinforcement corrosion based on water content, paving the way for practical applications in diagnosing building integrity related to rebar corrosion.

Yoshino and Kimura explored the application of non-structural members as lateral supports for I-beams prone to lateral buckling [6]. Torsional tests were conducted on roof-folded plates to quantify their rotational stiffness and capacity with various connection methods between the plate and the beam. The experimental findings clarified the resistance mechanism of the roof plates, confirming their effectiveness as continuous bracing elements. Finally, the lateral buckling strength of the I-beam was evaluated based on Japanese and European design codes, incorporating the joint stiffness determined from the experiments.

Chen et al. addressed the challenges of track wear, fatigue cracks, and impact damage to the welded joints of azimuth tracks, a critical component of radio telescope wheel–rail systems [7]. This study included tensile tests using digital image correlation (DIC) technology and Vickers hardness tests on the base metal (BM), heat-affected zone (HAZ), modified layer, and weld zone (WZ). An FEA model was developed using a constitutive model to simulate rolling mechanical performance. The stress distribution under various pre-designed loads was analyzed, identifying the locations most susceptible to damage during regular operation, braking, and start-up conditions.

Based on experimental and database results, Kimura proposed equations for evaluating the ultimate strength ratio, rotational capacity, and energy dissipation capacity [8]. Cyclic loading tests were conducted on 11 specimens using different loading protocols, and an extensive database was constructed from prior studies. Using parameters such as the width-to-thickness ratio, shear span-to-depth ratio, and loading protocol, the proposed equations were validated against 65 specimens, demonstrating their accuracy in assessing structural performance.

Cavuslu and Dagli studied the time-dependent creep and earthquake performance of the historical Plaka stone bridge [9]. Two distinct 3D finite-difference models were developed for Khorasan mortar with varying additives, including egg white and eggshell, at mixture percentages ranging from 25% to 100%. Time-dependent creep analyses and simulations of seismic events in the region revealed that using a 50% eggshell-mixed Khorasan mortar positively influenced the creep and seismic behavior of restored and yet-to-be-restored bridges.

Narantogtokh et al. proposed a novel pin penetration test for determining the early-age compressive strength of concrete before demolding [10]. It is important to determine the early-age compressive strength at the actual construction site when concrete work is executed in cold weather conditions to prevent early-age frost damage. Most of the international norms and guidelines for cold weather concreting recommend obtaining at least 5 MPa strength before exposing concrete to early-age freezing. However, there are no suitable methods available for the measurement of low-strength concrete at a very early age, particularly before demolding at the construction site. The proposed non-destructive method was validated with penetration depth benchmarks of 8.0 mm and 6.7 mm, corresponding to compressive strengths of $5\,N/mm^2$ and $10\,N/mm^2$, respectively. The results provide practical guidance for determining demolding and initial curing timing.

Chen et al. investigated the bond-slip behavior between steel reinforcement and epoxy resin concrete through experimental research and FEA [11]. A total of 18 center-pullout tests and 14 FEA simulations revealed that the bond strength of epoxy resin concrete significantly exceeds that of ordinary concrete, being approximately 3.23 times higher for equivalent material strength levels. The study also highlighted post-peak behavior considerations for the practical application of epoxy resin concrete.

Erbasu et al. explored the mechanical behavior of wood-based composite materials at high temperatures, approaching those used in pyrolytic decomposition [12]. Experiments were conducted on glued laminated timber (glulam) elements, and the results were used to calibrate material models for numerical analysis. These findings enable future evaluations of various materials and combinations for constructing such composites.

Suzuki et al. examined the risk of overturning freestanding non-structural building contents in buckling-restrained braced frames (BRBFs) as part of business continuity planning (BCP) [13]. Various BRBFs were designed using different criteria, and incremental dynamic analyses (IDAs) were performed with artificially generated seismic waves. The results showed that peak floor accelerations could exceed thresholds in existing fragility curves, increasing the risk of overturning depending on regional hazard combinations.

Tapusi et al. developed a tool for generating topologically optimized geometries of glulam structural elements [14]. A neural network was trained to avoid overfitting, with parameters related to wood properties (type, density, modulus of elasticity) and model discretization. Future iterations will aim to improve the neural network's architecture for size and activation function, contributing to AI-driven optimal design of glulam members.

Pushpalal and Kashima conducted compressive strength tests on concrete incorporating five types of fly ash with high CaO and SO_3 levels [15]. Fly ash replacement of up to 70% of the total cementitious content of concrete was investigated in this study. This research aimed to clarify the influence of the naturally occurring anhydrite phase in fly ash on the compressive strength of concrete. X-ray diffraction Rietveld analysis determined the mineral composition and glass phase of the fly ash samples. The authors found when the CaO content in ash is high, the insoluble residue is low, showing the high cementitious properties of fly ash. Compressive strength results were compared between concrete with and without anhydrite in fly ash. The optimal amount of anhydrite was found to be $2 \pm 0.5\,kg/m^3$ of concrete, defining the applicable range for anhydrite-rich fly ash.

3. Conclusions

This Special Issue features 15 groundbreaking research studies on advancements in the structural and mechanical performance of materials and structures. The insights presented are expected to drive further progress and inspire future research in the fields of structural and materials engineering.

Funding: This research received no external funding.

Acknowledgments: We extend our gratitude to the authors who contributed their research to this Special Issue and to the reviewers for their meticulous evaluations. We also sincerely thank the editors for their dedication and perseverance in ensuring the success of this Special Issue.

Conflicts of Interest: The authors declare that they have no known competing financial interests or personal relationships that could have appeared to influence the work reported in this paper.

References

1. Suzuki, A.; Liao, W.; Shibata, D.; Yoshino, Y.; Kimura, Y.; Shimoi, N. Structural Damage Detection Technique of Secondary Building Components Using Piezoelectric Sensors. *Buildings* **2023**, *13*, 2368. [CrossRef]
2. Bellos, J.; Konstantinidis, A. Systematic Calculation of Yield and Failure Curvatures of Reinforced Concrete Cross-Sections. *Buildings* **2024**, *14*, 826. [CrossRef]
3. Ni, H.; Aboutaha, R. Bond-Damaged Prestressed AASHTO Type III Girder-Deck System with Retrofits: Parametric Study. *Buildings* **2024**, *14*, 902. [CrossRef]
4. Wang, G.; Zhuo, X.; Zhang, S.; Wu, J. Study on the Mechanical Properties and Design Method of Frame-Unit Bamboo Culm Members Based on Semi-Rigid Joints. *Buildings* **2024**, *14*, 991. [CrossRef]
5. Tanaka, A.; Arita, K.; Kobayashi, C.; Nishiwaki, T.; Tanabe, T.; Fujii, S. Fundamental Properties of Sub-THz Reflected Waves for Water Content Estimation of Reinforced Concrete Structures. *Buildings* **2024**, *14*, 1076. [CrossRef]
6. Yoshino, Y.; Kimura, Y. Rotational Stiffening Performance of Roof Folded Plates in Torsion Tests and the Stiffening Effect of Roof Folded Plates on the Lateral Buckling of H Beams in Steel Structures. *Buildings* **2024**, *14*, 1158. [CrossRef]
7. Chen, X.; Yin, R.; Yang, Z.; Lan, H.; Xu, Q. A Study on the Mechanical Characteristics and Wheel–Rail Contact Simulation of a Welded Joint for a Large Radio Telescope Azimuth Track. *Buildings* **2024**, *14*, 1300. [CrossRef]
8. Kimura, Y. Evaluation of Rotation Capacity and Bauschinger Effect Coefficient of I-Shaped Beams Considering Loading Protocol Influences. *Buildings* **2024**, *14*, 1376. [CrossRef]
9. Cavuslu, M.; Dagli, E. Egg White and Eggshell Mortar Reinforcing a Masonry Stone Bridge: Experiments on Mortar and 3D Full-Scale Bridge Discrete Simulations. *Buildings* **2024**, *14*, 1672. [CrossRef]
10. Narantogtokh, B.; Nishiwaki, T.; Takasugi, F.; Koyama, K.; Lehmann, T.; Jagiello, A.; Droin, F.; Ding, Y. A New Methodology to Estimate the Early-Age Compressive Strength of Concrete before Demolding. *Buildings* **2024**, *14*, 2099. [CrossRef]
11. Chen, P.; Li, Y.; Zhou, X.; Wang, H.; Li, J. Study on the Bond Performance of Epoxy Resin Concrete with Steel Reinforcement. *Buildings* **2024**, *14*, 2905. [CrossRef]
12. Erbasu, R.I.; Sabău, A.-D.; Tăpusi, D.; Teodorescu, I. Performance Assessment of Wood-Based Composite Materials Subjected to High Temperatures. *Buildings* **2024**, *14*, 3177. [CrossRef]
13. Suzuki, A.; Ohno, S.; Kimura, Y. Risk Assessment of Overturning of Freestanding Non-Structural Building Contents in Buckling-Restrained Braced Frames. *Buildings* **2024**, *14*, 3195. [CrossRef]
14. Tăpusi, D.; Sabău, A.-D.; Savu, A.-A.; Erbasu, R.-I.; Teodorescu, I. Numerical Methods for Topological Optimization of Wooden Structural Elements. *Buildings* **2024**, *14*, 3672. [CrossRef]
15. Pushpalal, D.; Kashima, H. Apparent Influence of Anhydrite in High-Calcium Fly Ash on Compressive Strength of Concrete. *Buildings* **2024**, *14*, 2899. [CrossRef]

Article

Structural Damage Detection Technique of Secondary Building Components Using Piezoelectric Sensors

Atsushi Suzuki [1,*], Wang Liao [2], Daiki Shibata [3], Yuki Yoshino [4], Yoshihiro Kimura [1] and Nobuhiro Shimoi [5]

[1] Graduate School of Engineering, Tohoku University, Sendai 980-8577, Japan; kimura@tohoku.ac.jp
[2] Shanghai Investigation, Design & Research Institute Co., Ltd., Shanghai 200434, China; liaowang@sidri.com
[3] School of Engineering, Tohoku University, Sendai 980-8577, Japan; shibata.daiki.p8@dc.tohoku.ac.jp
[4] National Institute of Technology, Sendai College, Sendai 015-0055, Japan; yoshinoy@sendai-nct.ac.jp
[5] Faculty of Systems Science and Technology, Akita Prefectural University, Yurihonjo 015-0055, Japan; shimoi@akita-pu.ac.jp
* Correspondence: atsushi.suzuki.c2@tohoku.ac.jp; Tel.: +81-22-795-7876

Abstract: With demand for the long-term continued use of existing building facilities, structural health monitoring and damage detection are attracting interest from society. Sensors of various types have been practically applied in the industry to satisfy this need. Among the sensors, piezoelectric sensors are an extremely promising technology by virtue of their cost advantages and durability. Although they have been used in aerospace and civil engineering, their application for building engineering remains limited. Remarkably, recent catastrophic seismic events have further reinforced the necessity of rapid damage detection and quick judgment about the safe use of facilities. Faced with these circumstances, this study was conducted to assess the applicability of piezoelectric sensors to detect damage to building components stemming from concrete cracks and local buckling. Specifically, this study emphasizes structural damage caused by earthquakes. After first applying them to cyclic loading tests to composite beam component specimens and steel frame subassemblies with a folded roof plate, the prospective damage positions were also found using finite element analysis. Crack propagation and buckling locations were predicted adequately. The piezoelectric sensors provided output when the concrete slab showed tensile cracks or when the folded roof plate experienced local buckling. Furthermore, damage expansion and progression were detected multiple times during loading tests. Results showed that the piezoelectric sensors can detect the structural damage of building components, demonstrating their potential for use in inexpensive and stable monitoring systems.

Keywords: concrete slab; damage detection; folded roof plate; piezoelectric sensor; structural health monitoring

Citation: Suzuki, A.; Liao, W.; Shibata, D.; Yoshino, Y.; Kimura, Y.; Shimoi, N. Structural Damage Detection Technique of Secondary Building Components Using Piezoelectric Sensors. *Buildings* **2023**, *13*, 2368. https://doi.org/10.3390/buildings13092368

Academic Editor: Huiyong Ban

Received: 20 August 2023
Revised: 4 September 2023
Accepted: 15 September 2023
Published: 17 September 2023

1. Introduction

In recent times, powerful earthquakes have struck cities worldwide, one after another. Since establishing the Sendai framework at the UN World Conference on Disaster Risk Reduction in 2015, the demand for resilience strengthening of structures is further attracting international researchers' interest. This situation certainly includes buildings of various types. Structural health monitoring fulfills an important role in the peri-disaster and post-disaster phases during and after disasters, respectively. Building collapse must be prevented even after an earthquake. A recent catastrophic event, the 2015 Gorka earthquake, caused the complete collapse of 500,000 buildings and the partial collapse of 250,000 buildings, according to the Japan International Cooperation Agency (JICA) [1]. Another recent seismic event, the 2023 Turkey–Syria earthquake, caused more than 164,000 buildings to be destroyed or severely damaged, as reported by the Turkish Ministry. Such damage derives from a lack of proper assessment of structural capacity and a lack of rapid seismic strengthening. Findings indicate that continuous efforts at damage prevention must be pursued locally and globally.

Most buildings in countries with strict seismic design criteria, such as those in Japan, can withstand strong ground motions. Nevertheless, several buildings have been reportedly unable to maintain their functions because of damage to secondary building components [2]. Eventually, some buildings became unsafe for use as evacuation shelters during post-disaster phases of recovery [3]. In addition, the durations required for emergent inspections have been raised as a primary concern. In spite of enormous efforts by engineers and public servants, damage inspections after the 2016 Kumamoto earthquake took 57 days to complete [4]. One reason underlying this long period is that there were few engineers able to complete on-site inspections at the municipality level. However, in contrast to restrictions on human resources, demands from society for the continuity of building use are skyrocketing.

Faced with these needs, structural health monitoring and damage detection are attracting interest among researchers and engineers. Generally, they are classifiable as global and local approaches [5]. It is noteworthy that global structural health monitoring uses the following representative methods: (1) natural frequency-based methods, (2) mode shape-based methods, (3) dynamically measured flexibility matrix-based methods, (4) neural network methods, and (5) generic algorithm methods [6]. Ji et al. [7] conducted full-scale shaking table tests as well as monitoring of building vibration. The results obtained by analyzing the shift of natural frequencies of building structures demonstrated the effectiveness of vibration-based damage diagnosis. Okada et al. reported the application of a three-dimensional structural monitoring system for a full-scale six-story RC building [8]. In addition, a cost-efficient method was established to interpolate responses from the limited recorded data. Moreover, Gislason et al. [9] proposed an automated structural health monitoring system based on time history analysis. Through rigorous numerical modeling, it was demonstrated that damage can be identified with story-level precision. The degree of damage can be quantified and accurately based on floor accelerations caused by ambient wind forces. In addition, Alampalli et al. [10] classically investigated the sensitivity of modal characteristics to damage in a laboratory-scaled bridge span. Through the comprehensive investigation, Alampalli et al. [10] concluded that the local damage does not necessarily change mode shapes more significantly at the damage location or near damage locations than in other areas.

Structural health monitoring and damage detection at the local level are also continuing their evolution internationally. For this purpose, sensors of various kinds, such as strain gauges, accelerometers, fiber optical sensors, displacement sensors, piezoelectric sensors, and Doppler vibrometers, have been developed to realize structural health monitoring [11]. The classical technique to detect local damage uses strain gauges. Recently, they have become widely available on the market. However, they are fragile and unsuitable for long-term monitoring. Consequently, they are commonly used for laboratory experiments. The recent development of image sensing has realized damage detection using digital images. Earlier achievements by Chida and Takahashi [4] enabled the detection and evaluation of quantitative damage at the ground level of timber houses using pre-post morphological processing combined with semantic segmentation by deep learning. From a simplified perspective, Kishiki et al. [12] attempted to visualize the residual strength of buckled steel members. The magnitude of buckling deformation was measured during cyclic loading tests. The strength and deformation magnitude were correlated. Ultimately, an evaluation equation was proposed for instant strength evaluation. The recent articles [13,14] revealed that the member performance was degraded by the environmental impact, proving the necessity of long-term monitoring to secure the safety of structures.

Recent efforts at structural health monitoring are being aimed at sensors of novel types, specifically piezoelectric sensors. Such sensors detect the applied force or displacement and then generate a voltage. Compared with other monitoring sensors and techniques, piezoelectric sensors provide numerous benefits such as small size, light weight, low cost, high sensitivity, and availability in various formats [11]. According to previous research [15], the system consisting of the microtremor, computer, and data logger requires 15,000 to 25,000 US dollars per measurement unit. In addition, another system comprised of the laser Doppler velocity meter, computer, and digital displacement gauge costs 45,000 to 60,000 US dollars per unit. The proposed method requires only piezoelectric sensors, computers, and circuits, resulting in less than 1000 US dollars. Therefore, the system established here has a cost advantage. By virtue of these benefits, piezoelectric sensors are applied practically for aerospace and civil engineering structures [16]. Earlier, Harada et al. [17] used piezoelectric sensors to detect crack propagation in steel specimens and RC beams. Conversions of the output voltage and strain are interrelated experimentally and theoretically. Furthermore, Harada et al. [17] reported that a charge amplifier with low energy consumption took stable measurements in a static condition. Therefore, Harada et al. [17] concluded that the piezoelectric sensor is effective, particularly for local and severe damage such as that associated with concrete crack expansion.

In terms of building applications, one of the authors enthusiastically investigated the application of piezoelectric sensors for building components. Earlier, the applicability of the sensor was studied for welded connections between a beam and column [18] because they are prone to being damaged in strong earthquakes. The result demonstrated that the piezoelectric sensor adequately detects structural damage in the inelastic phase. The piezoelectric sensors can resist a greater deformation than the strain gauges. Therefore, the sensors are reusable without replacement, even after a giant earthquake. The piezoelectric sensors, therefore, embrace the advantage of long-term structural health monitoring. Therefore, it was concluded that this sensor is promising for use in an inexpensive and durable health monitoring system.

Generally, buildings comprise main structural components (columns, beams, etc.) and secondary components (concrete slabs, folded roof plates, etc.). Earlier reports revealed that the secondary components function as a restraint on the primary structural members. Their restraint performance is generally represented as the spring stiffness or strength [19–25]. Steel beams are assembled with a concrete slab through shear connectors in an ordinary building with several floors (Figure 1a). It is widely recognized that a concrete slab demonstrates restraint performance along the in-plane direction and out-of-plane direction. Although cracks in the concrete slab originate during cyclically applied stress from the earthquake, enhancement of the buckling strength was confirmed by experimentation [26].

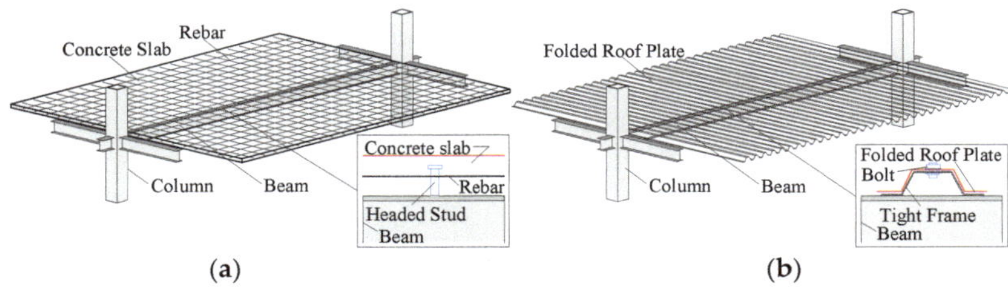

Figure 1. Installation of secondary structural components: (**a**) concrete slab; (**b**) folded roof plate.

By contrast, single-story buildings (gymnasiums, warehouses, etc.) on the top floor of multiple-story buildings can only have folded roof plates (Figure 1b). According to an earlier experiment, the folded roof plate demonstrates high restraint performance, but its rigidity is much lower than that of a concrete slab. Their buckling strength was derived theoretically and analytically in an earlier study [27]. In addition, the mechanical performance of the folded roof plate was evaluated at the component level [28]. As evaluation methods are becoming more sophisticated, as introduced above, the necessity of securing the designed restraint performance is being raised as an important concern. However, structural health monitoring and damage detection technology are usually intended for the global frame or for the primary structural components. Considering that the bracings are generally damaged before member buckling and subsequent strength deterioration, damage detection of secondary structural components is rather important.

Based on the discussion presented above, this study was conducted to detect concrete slabs and folded roof plate damage using inexpensive yet consistent and reliable measures. Specifically, this study applies piezoelectric sensors. For this purpose, this study applied cyclic loading tests to a component model of composite beam and steel frame subassembly with folded roof plates. Because the prospective damage position must be analyzed in advance, finite element analysis (FEA) is demonstrated for these assessments. The sensor output and the damage state were compared to investigate their adaptability to the damage detection of secondary building components.

The outcome of this research builds the foundation of the novel structural health monitoring system using the piezoelectric sensor. Specifically, this research focuses on the non-structural components more prone to deformation than the primary structural members. Therefore, the structural engineers and building owners can make a proper and prompt decision regarding the continuous use of the facilities.

2. Structural Damage Detection of a Concrete Slab Using Piezoelectric Sensors

2.1. Outline of Experiment

2.1.1. Specimen Configuration

Currently, headed stud shear connectors prevail as the shear connector, and the evaluation equations are available in the design provisions [29–31]. Their behavior had been investigated based on push-out specimens [32–34] and composite beams [35–37]. On the other hand, to demonstrate superior strength and stiffness, novel types of shear connectors are developed as alternatives: perfobond shear connectors [38–40], Y-type perfobond rib shear connectors [41–43], bar-ring shear connectors [44–46], puzzle-shaped shear connectors [47–49], and clothoid-shaped shear connectors [50–52]. Furthermore, recent research has revealed the innovative shear connectors' cyclic and fatigue behavior [53–57] and their mechanical performance with damaged concrete [58]. This research examines the clothoid-shaped shear connectors as a representative case of the novel shear connectors mentioned above.

Figure 2 portrays the configuration of the composite beam component model. The specimen was designed for earlier experiments [19–25]. The specimen comprised two concrete slabs, longitudinal and transversal reinforcements, four clothoid-shaped shear connectors, and two H-section steels. The specimen configuration was designed as symmetric relative to the z-axis to eliminate eccentricity. The clothoid-shaped shear connector thickness was 16 mm. The longitudinal diameter and transversal rebar were 10 mm.

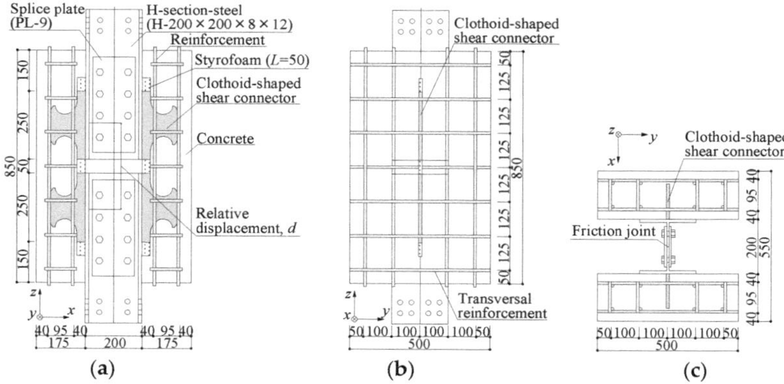

Figure 2. A component model of the composite beam (unit: mm): (**a**) side view; (**b**) front view; (**c**) cross-section view.

The casting orientation is widely acknowledged as influencing the mechanical performance of the shear connection. Therefore, the Japan Society of Steel Construction (JSSC) recommends replication of the structure concerned when engineers manufacture the shear connection component model [59]. For this study, we therefore poured concrete along the *x*-axis to model the composite beam. Two specimen parts were assembled using high-strength bolts and splice plates before loading tests.

Figure 3 presents an illustration of the loading frame. The specimen was placed widthwise on the footing beam. The single side of H-section steel was connected to the reaction member by high-strength bolts. Then, the imposed shear force was transferred to the reaction floor. The shear force was carried cyclically by a 1000 kN capacity hydraulic jack.

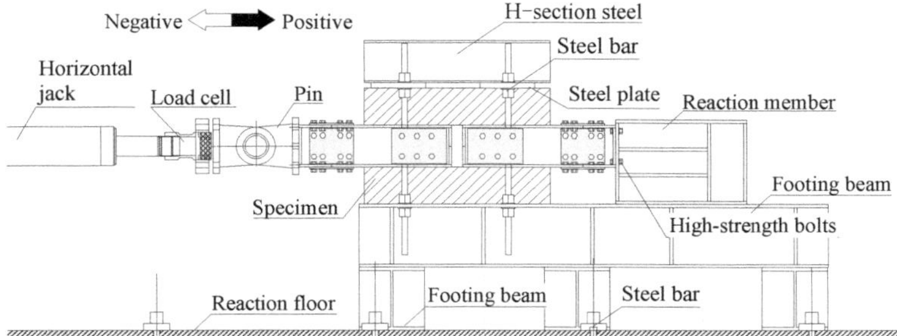

Figure 3. Loading frame.

2.1.2. Loading Protocols

For the loading test, the amplitude was controlled based on the relative displacement of two shear connectors (designated as connector relative displacement, *d*). The loading protocol was fully reversed loading. Compressive and tensile stresses were applied for the concrete slab in positive and negative side loading.

2.1.3. Material Properties

Table 1 shows the mix design of the concrete. The water/cement ratio (W/C) was 53.0%. The maximum aggregate size was 20 mm. Tables 2–4 present the material test results. The material tests of the concrete and steel members were conducted in accordance with JIS A 1108 [60] for the compressive strength test of the concrete, JIS A 1113 [61] for the splitting tensile strength of the concrete, and JIS Z 2241 [62] for the tensile strength test of the steel. The cylinder specimens were used for the material tests on concrete. The compressive and tensile strengths were, respectively, 26.4 N/mm^2 and 2.1 N/mm^2 (Table 2). The yield and ultimate strength of the shear connector, web, and flange plates were distributed, respectively, as 257–293 and 436–458 N/mm^2 (Table 3). The yield strength and the ultimate strength of the reinforcement were, respectively, 378 N/mm^2 and 509 N/mm^2 (Table 4).

Table 1. Mix proportions.

W/C [%]	s/a [%]	Unit Materials Content [kg/m^3]				
		Water	Cement	Sand	Gravel	Admixture
53.0	47.5	178	336	829	933	4.36

Table 2. Material properties (concrete).

Compressive Strength [N/mm^2]	Tensile Strength [N/mm^2]	Modulus of Elasticity [N/mm^2]
26.4	2.1	22,836

Table 3. Material properties (steel plate).

Part	Thickness [mm]	Yield Strength [N/mm^2]	Ultimate Strength [N/mm^2]	Elongation [%]
Connector	16	285	436	46
Web	8	293	458	37
Flange	12	257	440	43

Table 4. Material properties (steel bar).

Diameter [mm]	Yield Strength [N/mm^2]	Ultimate Strength [N/mm^2]	Elongation [%]
10	378	509	28

2.1.4. Curing Conditions

Specimens were demolded on the seventh day after concrete casting. They were air-cured up to the day of the loading test. The concrete specimens for the material tests were cured in the same room to give all of them an identical temperature history.

2.2. Outline of Finite Element Analysis

2.2.1. Configuration of Finite Element Analysis

An FEA model was demonstrated to determine the piezoelectric sensor positions. Figure 4 exhibits the FEA model configuration. Using the symmetricity of the specimen configuration, a quarter part of the specimen was extracted for the analysis. ABAQUS (ver. 2021) was used for this simulation. Embracing the advantage of the convergence, an explicit module was used to analyze the composite beam component model.

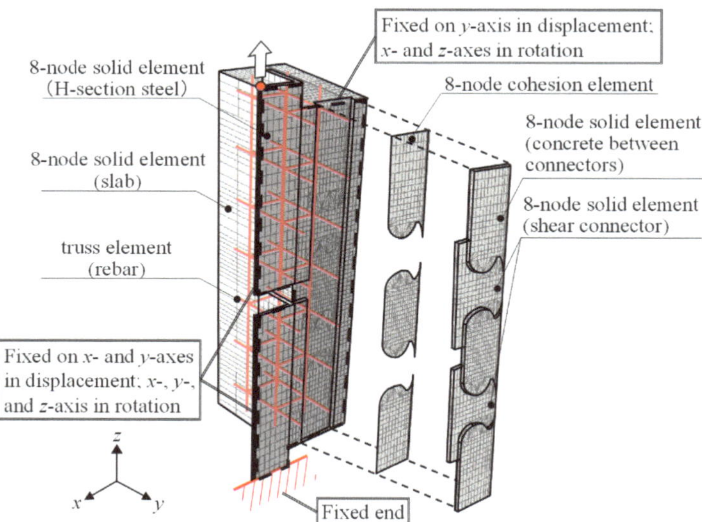

Figure 4. Finite element analysis model.

As exhibited in Figure 4, the H-section steel, shear connector, and concrete slab were built with the solid element. The reinforcements were modeled by the truss elements. The reinforcements were settled to resist the axial force only. Therefore, the analysis intended that the rebar yielded by means of the axial force, not by the combined stress of axial force and bending moment. Hence, the truss elements were adopted in the FEA. Also, the slippage was not observed in the preliminary analysis, justifying disregarding the bond fracture in the simulation. Separation through a tensile crack was represented by the 0.1-mm-thick cohesion element. The element was placed on the interface between the concrete block and the concrete between the shear connectors. The respective sides of the cohesion element were tied to the concrete block and the remaining concrete parts. When the deformation was imposed on the shear connector, the cohesion element deformed and carried the reaction force. According to the constitutive law described later, the cohesion element depleted its strength gradually and vanished in the ultimate state. The concrete blocks were therefore discretized after vanishing. The fine mesh was built near the shear connectors, where the stress transfer occurs prominently. Meanwhile, the mesh became coarse around the edge of the concrete slab to save computation time. The preliminary analysis was carried out, and the convergence was observed in the sensitivity study.

The clamped end constraint was given at the bottom of the H-section steel. The boundary condition to satisfy the symmetricity was provided to the web plate and concrete slab center. Loading was imposed at the top of the H-section steel.

The contact pair was defined between the shear connector-to-concrete and concrete-to-H-section steel. The friction and initial adhesion were not given because the shear connector surfaces were greased for the experiment. Furthermore, the embedded constraint was given to the reinforcements related to the concrete slab.

2.2.2. Constitutive Law of Concrete and Slab Separation

Figure 5a–c respectively presents the constitutive law of concrete under compressive, tensile, and cyclic stresses. The elastic limit was 40% of the maximum strength, as required by Eurocode-2 [63]. The parabolic function from the elastic limit to the maximum strength is presented in Equation (1). In addition, the strain at the peak strength can be computed using Equation (2) as follows:

$$\sigma_c = \left(\frac{k(\varepsilon_c/\varepsilon_{cm}) - (\varepsilon_c/\varepsilon_{cm})^2}{1 + (k-2)(\varepsilon_c/\varepsilon_{cm})} \right) \sigma_{cm} \tag{1}$$

$$\varepsilon_{cm} = 0.07\sigma_{cm}{}^{0.31} \leq 0.28 \, [\%] \tag{2}$$

where σ_{cm} denotes the compressive strength of concrete and ε_{cm} represents the strain at peak strength. An earlier study [64] showed that the strength dropped linearly to 85% of the ultimate strength at the ultimate strain $\sigma_{cu} = 0.01$.

The stress–strain relation in the tension side was defined by Equation (3) in accordance with an earlier study [65]. The tensile strength was computed using Equation (4) [66]. The fracture energy was calibrated using Equation (5) [67]. In the following equations, the Newton and millimeters are used to compute the respective physical quantities. One should note that the unit of output in Equation (5) is N/m, requiring the unit conversion to substitute Equation (3).

$$\sigma_t = \sigma_{ctm} \left(1 + 0.5\frac{\sigma_{ctm}}{G_F} w_t \right)^{-3} \tag{3}$$

$$\sigma_{ctm} = 0.291 \times \sigma_{cm}{}^{0.637} \tag{4}$$

$$G_F = 10 \times d_{max}{}^{1/3} \times \sigma_{cm}{}^{1/3} \tag{5}$$

In those equations, σ_t stands for the tensile strength of concrete, w_t expresses the crack width, σ_{ctm} denotes the ultimate tensile strength, G_F symbolizes the fracture energy, and d_{max} signifies the maximum size of the aggregate.

Figure 5c depicts the stiffness recovery of the unloading and reloading phases. The degree of stiffness degradation was determined by damage factors d_c and d_t. An earlier study [68] used the following equations for considering the constraint by the steel members. The present study therefore applies the identical equation.

$$d_t = 1.24\frac{k_t}{\sigma_{ctm}} w_t \ (\leq 0.99) \tag{6}$$

$$d_c = \begin{cases} \dfrac{k_{ci}\varepsilon_c}{\left[1+(\varepsilon_c/\varepsilon_0)^n\right]^{\frac{1}{n}}} & (\varepsilon_c \leq 0.0184) \\ 0.3485 & (\varepsilon_c > 0.0184) \end{cases} \tag{7}$$

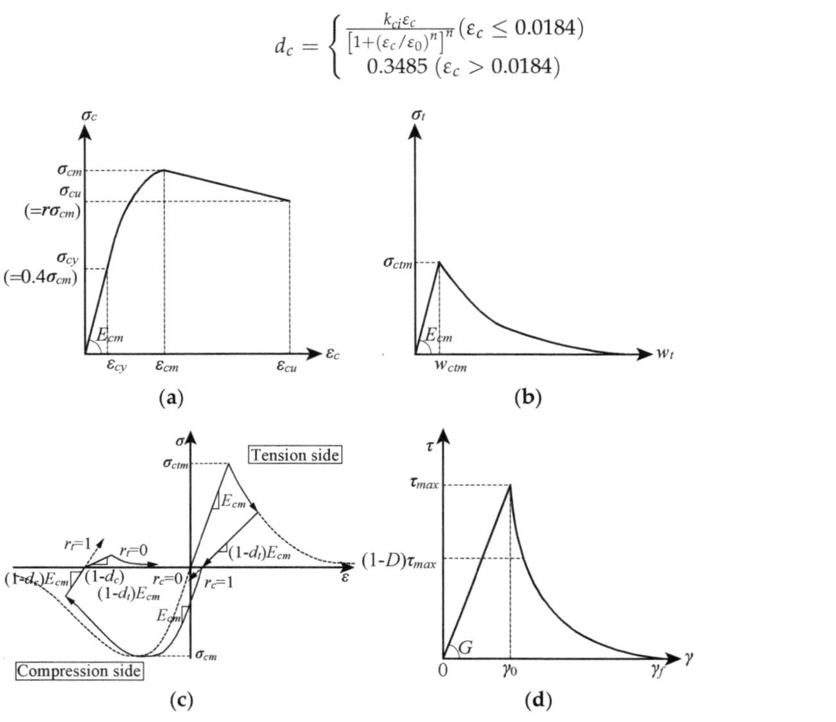

Figure 5. Constitutive law of concrete and separation: (**a**) compression; (**b**) tension; (**c**) cyclic; (**d**) separation.

In these equations, the constants k_{ci}, ε_0, n, and k_t are given, respectively, as 155, 0.0035, 1.08, and 386 N/mm^2/m in accordance with an earlier report [68].

2.2.3. Constitutive Law of Cohesion Element

Figure 5d presents the constitutive law of the cohesion element. The shear stress–strain relation was constructed based on earlier experimentally obtained results [69]. The shear stiffness G was calibrated using Equation (8). The damage function D governing strength deterioration was computed using Equation (9). The ultimate shear strength τ_{max} was calculated using Equation (10).

$$G = 43.0 \times (\omega/\omega_r) \times (\sigma_{cm}/\sigma_{cm,r})^{1/3} \tag{8}$$

$$D = \frac{\gamma - \gamma_0}{\gamma_f - \gamma_0} \tag{9}$$

$$\tau_{max} = \frac{\sigma_{cm}}{\sqrt{3}} \tag{10}$$

Therein, ω represents the concrete weight in unit volume, ω_r denotes the concrete weight in a unit volume of reference case (=24 kN/m^3), $\sigma_{cm,r}$ expresses the concrete strength of reference case (=25.9 N/mm^2), γ stands for the shear displacement, γ_0 signifies the effective displacement at damage initiation, and γ_f symbolizes the effective displacement at complete failure.

2.2.4. Constitutive Law of Shear Connector and Rebar

The stress–strain relation of the shear connector is determined by the combined hardening law with the three backstresses. The hardening parameters were set as identical to those reported from earlier research [70–75] because the same type of steel was used for this study.

2.2.5. Finite Element Analysis Results and Piezoelectric Sensor Positions

Figure 6 presents an illustration of the load–displacement relation and crack propagation. In Figure 6, the shear force once dropped around the connector relative displacement d of 1.9 mm. This reduction is derived from crack initiation at the embedded position of the shear connector. The plastic tensile strain distribution is presented in Figure 6b,d. The crack propagated toward the transversal direction. The most significant tensile stress concentrates near the shear connectors; hence, the critical position appears at the embedded place in the steel-concrete composite structure. However, the shear force recovered gradually using the stress transfer to the reinforcement. The damage spread with the increase in the loading amplitude (Figure 6c), after which the damage further propagated diagonally (Figure 6d). The observation presented in Figure 6 confirms that the damage should be detected aside from the embedded positions of the shear connectors.

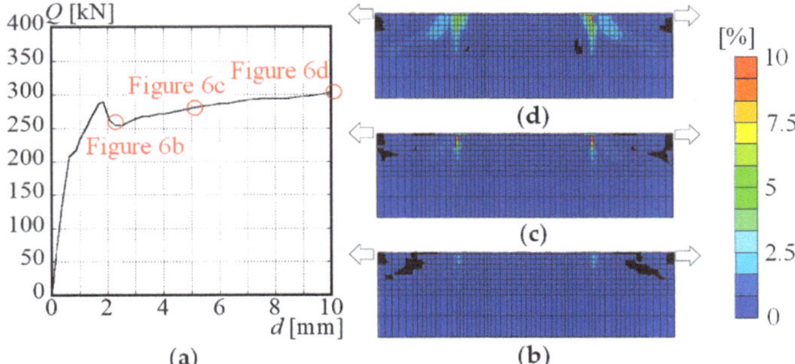

Figure 6. Distribution of tensile cracks: (**a**) load-displacement relation; (**b**) $d = 2.1$ mm; (**c**) $d = 5.0$ mm; (**d**) $d = 10.0$ mm.

Based on the observations presented above, this study determined the piezoelectric sensor installation position, as presented in Figure 7. The six piezoelectric sensors were attached to capture the tensile cracks. The piezoelectric sensor was expected to output the voltage through the movement by means of crack propagation. In addition, for crack width measurement, the PI gauges were set beside the piezoelectric sensors to ascertain the damage detection sensitivity. For the installation, the concrete slab surface was ground; the sensor was glued rigidly. During loading tests, the sensor was not peeled up to the final loading cycle.

2.3. Results of Experiments and Damage Detection

Figure 8 presents an illustration of the experimentally obtained result. Although the hysteresis was stable on the compression (positive) side, the strength deterioration and rapid displacement enlargement occurred several times. The first crack initiation was detected at $d = -1.4$ mm (point A). Figure 9 displays the fracture process. Although it is noticeable in the load–displacement relation, the cracks are not visible in Figure 9a. This finding corroborates that visual inspection alone presents some difficulty in identifying the concrete damage. The peak strength was performed at $d = -6.0$ mm (point B). The strength degraded gradually during subsequent loading cycles. The transversal damage at this stage is visible in Figure 9b. Furthermore, rapid strength degradation occurred at around $d = -10.0$ mm (point C) and $d = -13.5$ mm (point D).

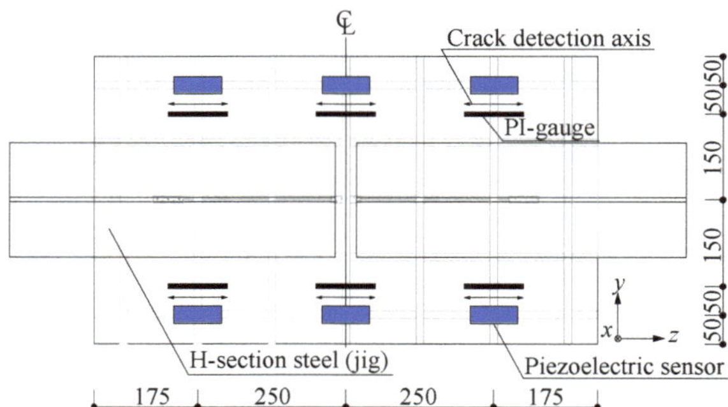

Figure 7. Positions of piezoelectric sensors.

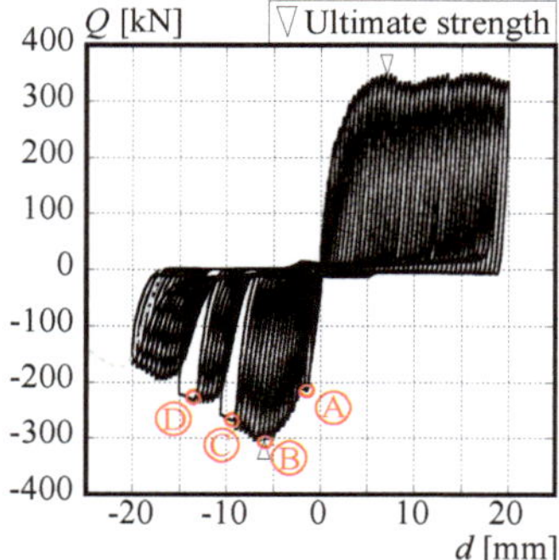

Figure 8. Cyclic behavior of the specimen.

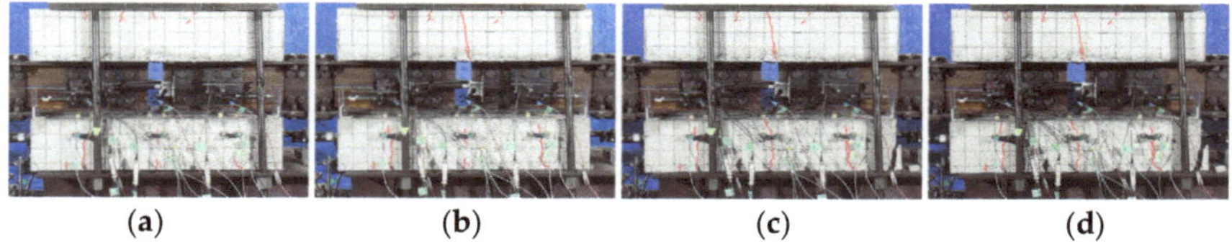

(a)   (b)   (c)   (d)

Figure 9. Fracture process: (**a**) $d = -2.0$ mm; (**b**) $d = -6.0$ mm; (**c**) $d = -10.0$ mm; (**d**) $d = -14.0$ mm.

Figure 10 shows the crack width w_c transition and the piezoelectric sensor output V_p. The horizontal axis is the absolute value of the loading amplitude. The skeleton part of the hysteresis curve on the negative side is extracted. In addition, Figure 10b demonstrates that the sensor position appropriately captures the tensile crack, thereby penetrating the sensor transversally. The piezoelectric sensor output is significant at $|d| = 1.6$ mm (point A), 10.0 mm (point C), and 14.5 mm (point D). In Figure 10, the crack width suddenly increases at point A and hits its peak near the ultimate shear strength (point B). During the subsequent loading cycle, the damage concentrates on the corresponding part. At point D, the position concerned becomes the origin of further damage. Consequently, the crack expands again, developing the prominent sensor output.

To deepen our understanding of piezoelectric sensor behavior, the relation between crack width and sensor output is presented directly for comparison in Figure 11. In Figure 11a, the reaction of the piezoelectric sensor was arranged by the crack width. As one might expect, the apparent relation is not visible in the figure. The noise around the origin was issued from the vibration during the unloading phase. Consequently, the piezoelectric sensor does not provide data related to the degree of concrete damage in the current measurement system. Instead, if the sensor output is arranged by the rate of crack expansion, the relation becomes much more apparent. This outcome corroborates that the piezoelectric sensor is helpful for detecting the damage origin. In future research, the method to convert the sensor output into the crack width will be calibrated, referring to an earlier investigation [8]. Therefore, this is a promising method of replacing the ordinary measurement scheme using strain or PI gauges.

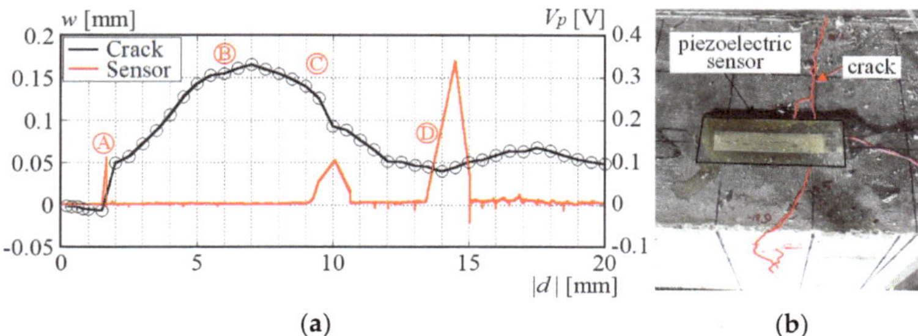

(a)   (b)

Figure 10. Transition of crack width and piezoelectric sensor output: (**a**) transition of crack width and sensor output; (**b**) crack and sensor positions.

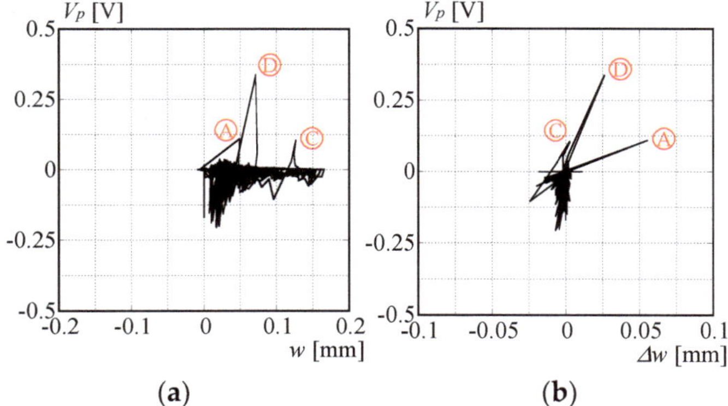

(a)   (b)

Figure 11. Relation between slab damage and piezoelectric sensor output: (**a**) crack width; (**b**) crack width velocity.

3. Damage Detection of a Folded Roof Plate Using Piezoelectric Sensor

3.1. Test Setup of Cyclic Loading Test on an I-Shaped Beam with a Folded Roof Plate

Figure 12 depicts the loading frame and specimen setup. The apparatus was designed in reference to earlier experimentation [76]. The force was imposed on the loading beam through the hydraulic jack. The specimen part, consisting of an I-shaped beam and the folded roof plate, was settled inside the frame. Consequently, the specimen was subjected to non-symmetric bending. In addition, both sides of the folded roof plate were pin-constrained with the slider (Figure 12b). Therefore, the folded roof plate does not carry the axial force.

The two specimens had different beam lengths. The beam heights were 250 mm and 200 mm for specimens No. 1 and No. 2. The respective beam widths were 80 mm and 100 mm. The beam lengths were 3675 mm and 3700 mm, with slenderness ratios of 222 and 166. The connection detail between the folded roof plate and the tight frame is presented in Figure 12c. In conformity with the actual buildings, the folded roof plate was bolted to the tight frame. The tight frame was fillet welded to the flange face. The tight frame had a width of 30 mm and a thickness of 3.2 mm. The material test results are presented in Table 5.

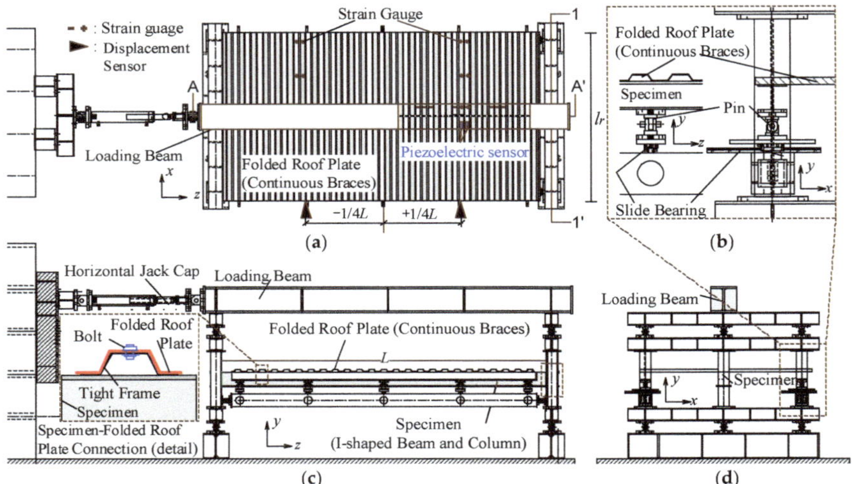

Figure 12. Loading frame: (**a**) floor plan; (**b**) slider detail; (**c**) A-A' section view; (**d**) 1-1' section view.

Table 5. Material test results.

Part	Specimen	Thickness [mm]	Yield Strength [N/mm²]	Ultimate Strength [N/mm²]
Flange	No. 1	6	313.9	465.9
	No. 2	6	294.8	443.5
Web	No. 1	9	368.3	475.1
	No. 2	8	334.1	456.2
Folded roof plate	No. 1	0.5	347.4	393.1
	No. 2	0.5	341.7	389.9

The loading amplitude was controlled by the column rotation. The amplitude was standardized by the full plastic rotation angle θ_p, which serves the beam to reach the full plastic moment $M_{p,b}$. The full plastic rotation angle θ_p is calculable as shown in Equation (11).

$$\theta_p = \frac{M_{p,b}L}{6E_bI_b}\left(1 + \frac{I_bH}{2I_cL}\right). \tag{11}$$

In this equation, E_b stands for the elastic modulus of steel, I_b represents the moment of inertia of beam along the strong axis, I_c denotes the moment of inertial of H-section column, and H stands for the column length. The loading protocol was the gradually increased loading. The target amplitude was enlarged to $\pm0.5\theta_p$, $\pm1.0\theta_p$, $\pm2.0\theta_p$, $\pm3.0\theta_p$, $\pm4.0\theta_p$, $\pm5.0\theta_p$, and $\pm6.0\theta_p$.

3.2. Preliminary Analysis of Folded Roof Plates and Piezoelectric Sensor Installation

This section describes the preliminary analysis of the folded roof plate that was conducted to ascertain the piezoelectric sensor position. A portion of the folded roof plate was extracted and subjected to the three-point bending test. The loading point was the middle part of the roof plate. The center part of the specimen was coupled with the reference point. The concentration force was given to the reference point. The loading modeled the flexural moment carried by beam rotation. Then, the force was applied monotonically downward.

Figure 13a shows the FEA model of a segment of the folded roof plate. The boundary conditions are also presented in Figure 13a. The analysis used ABAQUS (ver. 2021) with the standard solver package. The loading was force-controlled until strength deterioration, employing the originated local buckling. The finer mesh was built around the center part, where the most significant bending moment was carried, to reproduce the buckling behavior accurately. The constitutive law of the material was a simple bilinear model connecting the lower yield point and the ultimate strength of the material test result. The simulation did not consider fracture because the experiment did not observe it.

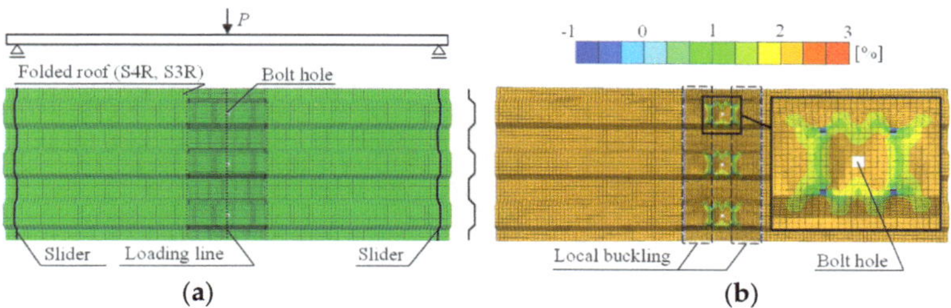

Figure 13. FEA model and buckling deformation: (**a**) FEA model; (**b**) buckling deformation.

Figure 13b presents the local buckling deformation. The buckling deformation occurred around the center part, issuing from the bolt hole. When lateral–torsional buckling originates in the I-shaped beam, the folded roof plate is subjected to the rotational moment transferred through the bolt. Ultimately, the folded roof plate starts to buckle. Furthermore, the magnitude of deformation is increased beside the hole. Based on these observations, the sensor location was determined as 100 mm distant from the bolt hole (Figure 12a).

3.3. Results of Cyclic Loading Tests and Piezoelectric Sensor Output

Figure 14a,b shows the cyclic loading test results. The horizontal axis is the column rotation. The vertical axis is the applied bending moment. The specimen demonstrates stable spindle-shaped hysteresis. However, with increasing loading amplitude, the beam failed because of lateral buckling, causing gradual strength deterioration. However, it is noteworthy that the structural performance is enhanced considerably compared to that of a bare steel beam. The superior structural performance stems from the rotational restraint provided by the folded roof plate. This trend reinforces the inference that the restraint from the folded roof plate improves the structural performance of beams that fail because of lateral buckling.

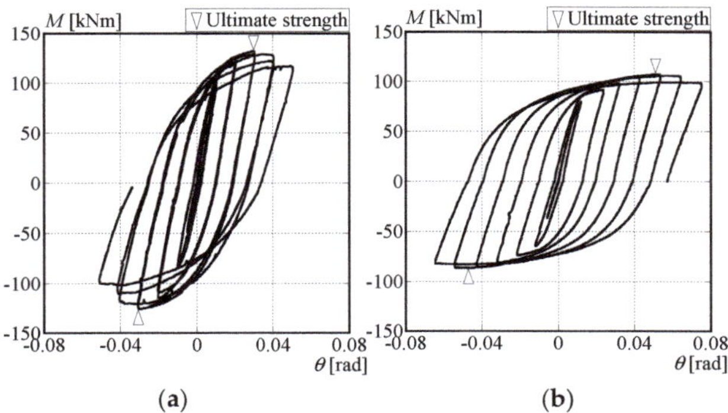

Figure 14. Cyclic behavior of an I-shaped beam assembled with a folded roof plate: (**a**) No. 1; (**b**) No. 2.

For a better understanding of sensor output, the cyclic hysteresis curve is converted to a skeleton curve. Figure 15 presents the procedure used to draw the skeleton curve. This procedure is identical to that used for earlier research [77]. Figure 15a portrays the hysteresis curve, Figure 15b is the cumulative hysteresis curve, and Figure 15c is the skeleton curve.

The hysteresis loop of a steel member can be decomposed into the skeleton part (drawn with bold lines), the Bauschinger part (drawn with dashed lines), and the elastic unloading part (drawn with dotted line). The skeleton part is defined as a part in which the steel member experiences stress for the first time. The Bauschinger parts are areas other than the elastic unloading part.

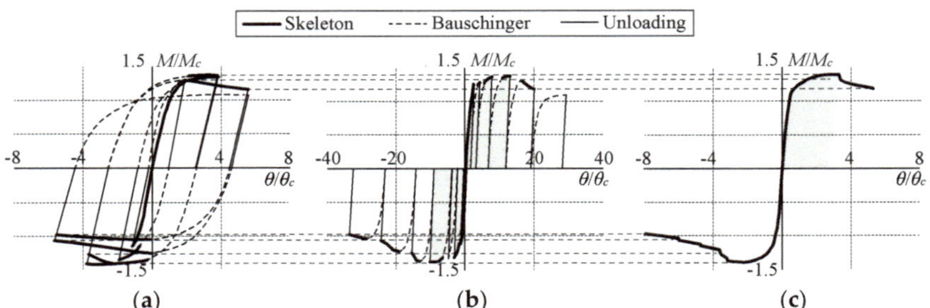

Figure 15. The procedure used to draw the skeleton curve: (**a**) hysteresis curve; (**b**) cumulative hysteresis curve; (**c**) skeleton curve.

Figure 16 presents the ultimate state of the folded steel plate attached to the specimen. As expected from the preliminary analysis, the buckling deformation became especially pronounced at the connection part. Therefore, the piezoelectric sensor captured the buckling deformation, as presented in Figure 16b.

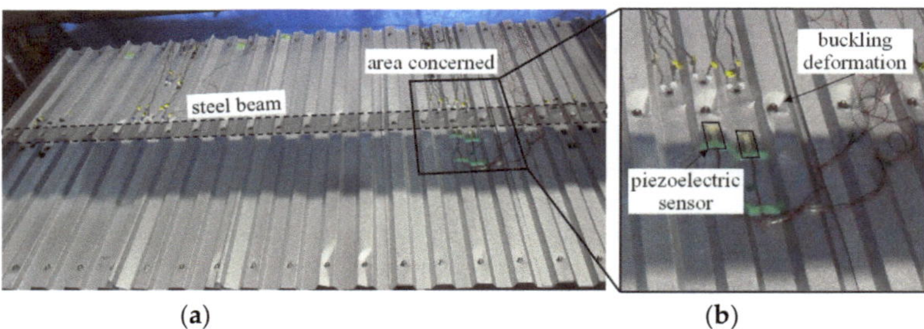

Figure 16. Ultimate state of the folded steel plate: (**a**) deformation on the whole part; (**b**) zoomed image of the sensor.

Figure 17 exhibits the skeleton curve and the sensor output at the respective measurement steps. In the initial phase, the significant sensor output was not visible because the out-of-plane deformation did not increase in the small loading amplitude. Instead, the piezoelectric sensor began to present a spike when the beam reached the ultimate state. It seems reasonable that the deformation of the folded steel plate increases after the sidesway of the beam. Therefore, damage detection of the piezoelectric sensor is primarily effective for verifying the possibility of buckling origination and the degradation of the restraint performance of folded roof plates.

3.4. Practical Applications and Future Research

The outcomes of this research emphasize the applicability of piezoelectric sensors for structural damage detection. Piezoelectric sensors are placed on the non-structural components, and the users can prepare a PC to detect the output voltage. In case the voltage exceeds the threshold, the alerting emails can be distributed to the building owners and users.

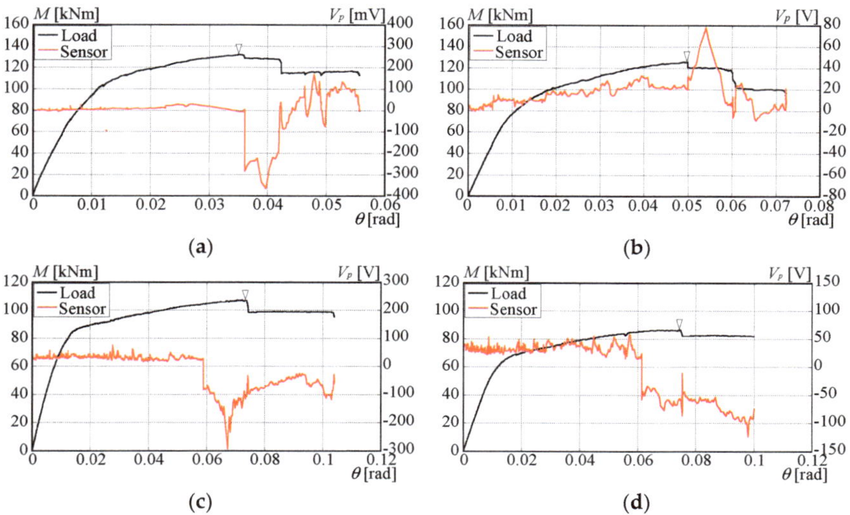

Figure 17. Skeleton curve and output of the piezoelectric sensor: (**a**) positive (No. 1); (**b**) negative (No. 1); (**c**) positive (No. 2); (**d**) negative (No. 2).

The configurations concerned in this research are limited to rectangular concrete slabs and thin steel plates. In contrast, the behavior of piezoelectric sensors attached to thick steel plates or circular rods has remained unclear. Further experimental study is necessary to broaden the application of this sensor.

All the loading tests reported in this paper were carried out statically. Therefore, a proper understanding of behavior under dynamic conditions is inevitable. Based on shaking table tests, a simplified system to monitor and alert building damage will be reported and verified for future research. Furthermore, the minimum degree of damage initiation will be quantified by comparing the measured physical quantities and voltage output.

4. Conclusions

This study was conducted to elucidate the applicability of a piezoelectric sensor for damage detection in secondary building components. The first half of the paper presented findings for a concrete slab, exhibiting rapid crack origination and expansion. The latter half described results for a folded roof plate, with bracing against torsional deformation of the I-shaped beam. A cyclic loading test and preliminary analysis were described for both components to elucidate the piezoelectric sensor behavior. A summary of the findings is presented below.

(1) The optimal sensor location for the concrete slab was determined to be beside the embedded position of the shear connector. That position in the case of the folded roof plate was found to be around the bolt connection.

(2) The piezoelectric sensor produced a prominent output when the concrete crack penetrated the sensor. Unlike standard strain gauges, the piezoelectric sensor can detect damage occurrence several times, which is a preferable characteristic for long-term monitoring.

(3) The piezoelectric sensor detected the forced deformation of a folded roof plate by beam torsion, thereby demonstrating the applicability of monitoring lateral buckling origination during the cyclical application of stress.

The application of piezoelectric sensors on the non-structural components possesses the potential to realize a novel scheme of structural health monitoring. The sensor output during the seismic event needs to be checked from the dynamic perspective through the experiment.

Author Contributions: Conceptualization, A.S. and N.S; methodology, A.S.; software, W.L.; validation, W.L. and Y.Y.; formal analysis, W.L.; investigation, D.S. and Y.Y.; resources, Y.K. and N.S.; data curation, A.S. and W.L.; writing—original draft preparation, A.S.; writing—review and editing, W.L., D.S. and Y.K.; visualization, A.S. and D.S.; supervision, Y.K. and N.S.; project administration, A.S., Y.K. and N.S.; funding acquisition, A.S., W.L.,Y.Y. and N.S. All authors have read and agreed to the published version of the manuscript.

Funding: This research was funded by JSPS KAKENHI (Grant Number 20H00290) (Principal Investigator: Dr. Nobuhiro Shimoi), JSPS KAKENHI (Grant Number 23K13392) (Principal Investigator: Dr. Atsushi Suzuki) and the Shanghai Pujiang Program (Grant No. 22PJD097) (Principal Investigator: Dr. Wang Liao), and JSPS KAKENHI (Grant Number 22K14369) (Principal Investigator: Dr. Yuki Yoshino). We express our deepest gratitude for their sincere support.

Data Availability Statement: The raw/processed data necessary to reproduce these findings cannot be shared at this time because the data also form part of an ongoing study.

Conflicts of Interest: The authors declare that they have no known competing financial interests or personal relationships that could have appeared to influence the work reported in this paper.

References

1. Japan International Cooperation Agency (JICA). *The Project for Assessment of Earthquake Disaster Risk for the Kathmandu Valley in Nepal*; Final Report, Volume 2 Main Report; JICA: Tokyo, Japan, 2018.
2. Matsumoto, Y.; Yamada, S.; Iyama, J.; Koyama, T.; Kishiki, S.; Shimada, Y.; Asada, H.; Ikenaga, M. Damage to steel educational facilities in the 2011 East Japan Earthquake: Part 1 Outline of the reconnaissance and damage to major structural components. In Proceedings of the 15th World Conference on Earthquake Engineering, Lisbon, Portugal, 24–28 September 2012.
3. Suzuki, A.; Fujita, T.; Kimura, Y. Identifying damage mechanisms of gymnasium structure damaged by the 2011 Tohoku earthquake based on biaxial excitation. *Structures* **2022**, *35*, 1321–1338. [CrossRef]
4. Chida, H.; Takahashi, N. Study on image diagnosis of timber houses damaged by earthquake using deep learning. *Jpn. Archit. Rev.* **2021**, *4*, 420–430. [CrossRef]
5. Chang, P.C.; Flatau, A.; Liu, S.C. Review paper: Health monitoring of civil infrastructure. *Struct. Health Monit.* **2003**, *2*, 257–267. [CrossRef]
6. Carden, E.P.; Fanning, P. Vibration based condition monitoring: A review. *Struct. Health Monit.* **2004**, *3*, 355–377. [CrossRef]
7. Ji, X.; Fenves, G.L.; Kajiwara, K.; Nakashima, M. Seismic damage detection of a full-scale shaking table test structure. *J. Struct. Eng.* **2011**, *137*, 14–21. [CrossRef]
8. Okada, K.; Shiroishi, R.; Morii, Y.; Sagawa, R. Method of structural health monitoring after earthquake using limited accelerometer—Case study of large-scale shaking table test on six story RC building. *Concr. J.* **2017**, *55*, 138–145. (In Japanese) [CrossRef]
9. Gislason, G.P.; Mei, Q.; Gul, M. Rapid and automated damage detection in buildings through ARMAX analysis of wind induced vibrations. *Front. Built Environ.* **2019**, *5*, 16. [CrossRef]
10. Alampalli, S.; Fu, G.; Dillon, E.W. Signal versus noise in damage detection by experimental modal analysis. *J. Struct. Eng.* **1997**, *123*, 237–245. [CrossRef]
11. Ju, M.; Dou, Z.; Li, J.; Qiu, X.; Shen, B.; Zhang, S.; Yao, F.; Gong, W.; Wang, K. Piezoelectric materials and sensors for structural health monitoring: Fundamental aspects, current status, and future perspectives. *Sensors* **2023**, *23*, 543. [CrossRef]
12. Kishiki, S.; Iwasaki, Y. Evaluation of residual strength based on local buckling deformation of steel column—Quick inspection method for steel structures based on the visible damage Part 3. *J. Struct. Constr. Eng. (Trans. AIJ)* **2017**, *82*, 735–743. (In Japanese) [CrossRef]
13. Gong, F.; Maekawa, K. Multi-scale simulation of freeze-thaw damage to RC column and its restoring force characteristics. *Eng. Struct.* **2018**, *156*, 522–536. [CrossRef]
14. Gong, F.; Wang, Z.; Xia, J.; Maekawa, K. Coupled thermos-hydro-mechanical analysis of reinforced concrete beams under the effect of frost damage and sustained load. *Struct. Concr.* **2021**, *22*, 3430–3445. [CrossRef]
15. Shimoi, N.; Nishida, T.; Obata, A.; Nakasho, K.; Cuadra, C. Comparison of displacement measurements in an exposed type column base using piezoelectric vibration sensors and piezoelectric limit sensors. *Akita Prefect. Univ. Web J. B* **2019**, *6*, 27–36.
16. Aabid, A.; Parveez, B.; Raheman, M.A.; Ibrahim, Y.E.; Anjum, A.; Hrairi, M.; Parveen, N.; Zayan, J.M. A review of piezoelectric material-based structural control and health monitoring techniques for engineering structures: Challenges and opportunities. *Actuators* **2021**, *10*, 101. [CrossRef]

17. Harada, T.; Yokoyama, K.; Tanabe, K. Study of crack detection of civil infrastructure using PVDF film sensor. *J. Struct. Eng.* **2013**, *59A*, 47–55.
18. Shimoi, N.; Cuadra, C.; Madokoro, H.; Nakasho, K. Comparison of displacement measurements and simulation on fillet weld of steel column base. *Int. J. Mech. Eng. Appl.* **2020**, *8*, 111–117. [CrossRef]
19. Suzuki, A.; Kimura, Y. Cyclic behavior of component model of composite beam subjected to fully reversed cyclic loading. *J. Struct. Eng.* **2019**, *145*, 04019015. [CrossRef]
20. Suzuki, A.; Abe, K.; Kimura, Y. Restraint performance of stud connection during lateral-torsional buckling under synchronized in-plane displacement and out-of-plane rotation. *J. Struct. Eng.* **2020**, *146*, 04020029. [CrossRef]
21. Suzuki, A.; Abe, K.; Suzuki, K.; Kimura, Y. Cyclic behavior of perfobond shear connectors subjected to fully reversed cyclic loading. *J. Struct. Eng.* **2021**, *147*, 04020355. [CrossRef]
22. Suzuki, A.; Suzuki, K.; Kimura, Y. Ultimate shear strength of perfobond shear connectors subjected to fully reversed cyclic loading. *Eng. Struct.* **2021**, *248*, 113240. [CrossRef]
23. Suzuki, A.; Kimura, Y. Mechanical performance of stud connection in steel–concrete composite beam under reversed stress. *Engineering Structures* **2021**, *249*, 113338. [CrossRef]
24. Suzuki, A.; Hiraga, K.; Kimura, Y. Cyclic behavior of steel-concrete composite dowel by clothoid-shaped shear connectors under fully reversed cyclic stress. *J. Adv. Concr. Technol.* **2023**, *21*, 76–91. [CrossRef]
25. Suzuki, A.; Hiraga, K.; Kimura, Y. Mechanical performance of puzzle-shaped shear connectors subjected to fully reversed cyclic stress. *J. Struct. Eng.* **2023**, *149*, 04023087. [CrossRef]
26. Liu, Y.; Sun, B. Experimental investigation and finite element analysis for mechanical behavior of steel–concrete composite beams under negative bending. In Proceedings of the International Conference on Mechanics and Civil Engineering, Wuhan, China, 13–14 December 2014.
27. Kimura, Y.; Sato, Y.; Suzuki, A. Effect of fork restraint of column on lateral buckling behavior for H-shaped steel beams with continuous braces under flexural moment gradient. *J. Struct. Constr. Eng. (Trans. AIJ)* **2022**, *87*, 316–327. (In Japanese) [CrossRef]
28. Liao, W.; Yoshino, Y.; Kimura, Y. Experimental study on the effect of H-shaped beam-folded roof plate joints on the rotational bracing stiffness of folded roof plates as continuous braces. *Steel Constr. Eng.* **2021**, *28*, 101–110.
29. *EN 1994-1-1*; Eurocode-4: Design of Composite Steel and Concrete Structures Part 1–1: Rules for Buildings. European Committee for Standardization: Brussels, Belgium, 2004.
30. Architectural Institute of Japan (AIJ). *Design Recommendations for Composite Constructions*; Maruzen Publishing Co. Ltd.: Tokyo, Japan, 2010.
31. American Institute of Steel Construction Inc. (AISC). *Specification for Structural Steel Buildings*; AISC: Chicago, IL, USA, 2016.
32. Hawkins, N.M.; Mitchell, D. Seismic response of composite shear connections. *J. Struct. Eng.* **1984**, *110*, 2120–2136. [CrossRef]
33. Oehlers, D.J. Deterioration in strength of stud connectors in composite bridge beams. *J. Struct. Eng.* **1990**, *116*, 3417–3431. [CrossRef]
34. Bursi, O.S.; Gramola, G. Behaviour of headed stud shear connectors under low-cycle high amplitude displacements. *Mater. Struct.* **1999**, *32*, 290–297. [CrossRef]
35. Lin, W.; Yoda, T.; Taniguchi, N. Fatigue tests on straight steel-concrete composite beams subjected to hogging moment. *J. Constr. Steel Res.* **2013**, *80*, 42–56. [CrossRef]
36. Lin, W.; Yoda, T.; Taniguchi, N.; Kasano, H.; He, J. Mechanical performance of steel-concrete composite beams subjected to a hogging moment. *J. Struct. Eng.* **2014**, *140*, 04013031. [CrossRef]
37. Song, A.; Wan, S.; Jiang, Z.; Xu, J. Residual deflection analysis in negative moment regions of steel-concrete composite beams under fatigue loading. *Constr. Build. Mater.* **2018**, *158*, 50–60. [CrossRef]
38. Japan Society of Civil Engineers (JSCE). *Standard Specification for Hybrid Structures–2014*; Maruzen Publishing Co. Ltd.: Tokyo, Japan, 2015.
39. Oguejiofor, E.C.; Hosain, M.U. Behavior of perfobond rib shear connectors in composite beams. *Can. J. Civ. Eng.* **1992**, *19*, 224–235. [CrossRef]
40. Oguejiofor, E.C.; Hosain, M.U. Numerical analysis of pushout specimens with perfobond rib connectors. *Comput. Struct.* **1997**, *62*, 617–624. [CrossRef]
41. Kim, S.H.; Kim, K.S.; Park, S.; Ahn, J.H.; Lee, M.K. Y-type perfobond rib shear connectors subjected to fatigue loading on highway bridges. *J. Constr. Steel Res.* **2016**, *122*, 445–454. [CrossRef]
42. Kim, S.H.; Kim, K.S.; Han, O.; Park, J.S. Influence of transverse rebar on shear behavior of Y-type perfobond rib shear connection. *Constr. Build. Mater.* **2018**, *180*, 254–264. [CrossRef]
43. Kim, S.H.; Batbold, T.; Shah, S.H.A.; Yoon, S.; Han, O. Development of shear resistance formula for the Y-type perfobond rib shear connector considering probabilistic characteristics. *Appl. Sci.* **2021**, *11*, 3877. [CrossRef]
44. Tanaka, T.; Sakai, J.; Kawano, A. Development of new mechanical shear connector using burring steel plate. *J. Struct. Constr. Eng. (Trans. AIJ)* **2013**, *78*, 2237–2245. [CrossRef]
45. Tanaka, T.; Sakai, J.; Kawano, A. Experimental study on elastic-plastic flexural behavior of composite beam by burring shear connector and perfobond-rib shear connector. *Steel Constr. Eng.* **2014**, *21*, 111–123.

46. Tanaka, T.; Norimatsu, K.; Sakai, J.; Kawano, A. Development of innovative shear connector in steel-concrete joint and establishment of rational design method. *Proc. Jpn. Concr. Inst.* **2013**, *35*, 1237–1242.
47. Lorenc, W.; Kozuch, M.; Rowinski, S. The behaviour of puzzle-shaped composite dowels—Part I: Experimental study. *J. Constr. Steel Res.* **2014**, *101*, 482–499. [CrossRef]
48. Lorenc, W.; Kozuch, M.; Rowinski, S. The behaviour of puzzle-shaped composite dowels—Part II: Theoretical investigations. *J. Constr. Steel Res.* **2014**, *101*, 500–518. [CrossRef]
49. Kopp, M.; Wolters, K.; Classen, M.; Hegger, J.; Gundel, M.; Gallwoszus, J.; Heinemeyer, S.; Feldmann, M. Composite dowels as shear connectors for composite beams—Background to the design concept for static loading. *J. Constr. Steel Res.* **2018**, *147*, 488–503. [CrossRef]
50. Lorenc, W. Concrete failure of composite dowels under cyclic loading during full-scale tests of beams for the "Wierna Rzeka" bridge. *Eng. Struct.* **2020**, *209*, 110199. [CrossRef]
51. Classen, M.; Hegger, J. Shear-slip behaviour and ductility of composite dowel connectors with pry-out failure. *Eng. Struct.* **2017**, *150*, 428–437. [CrossRef]
52. Lorenc, W.; Kurz, W.; Seidl, G. Hybrid steel-concrete sections for bridges: Definition and basis for design. *Eng. Struct.* **2022**, *270*, 114902. [CrossRef]
53. Christou, G.; Hegger, J.; Classen, M. Fatigue of clothoid shaped rib shear connectors. *J. Constr. Steel Res.* **2020**, *171*, 106133. [CrossRef]
54. Classen, M.; Gallwoszus, J. Concrete fatigue in composite dowels. *Struct. Concr.* **2016**, *17*, 63–73. [CrossRef]
55. Classen, M.; Gallwoszus, J.; Stark, A. Anchorage of composite dowels in UHPC under fatigue loading. *Struct. Concr.* **2016**, *17*, 183–193. [CrossRef]
56. Kim, H.Y.; Jeong, Y.J. Experimental investigation on behaviour of steel-concrete composite bridge decks with perfobond ribs. *J. Constr. Steel Res.* **2006**, *62*, 463–471. [CrossRef]
57. Christou, G.; Classen, M.; Wolters, K.; Broschart, Y. Fatigue of composite constructions with composite dowels. In Proceedings of the 14th Nordic Steel Construction Conference, Copenhagen, Denmark, 18–20 September 2019; pp. 277–282.
58. Kopp, M.; Christou, G.; Stark, A.; Hegger, J.; Feldmann, M. Integrated slab system for the steel and composite construction. *Bauingenieur* **2018**, *87*, 136–148.
59. Japan Society of Steel Construction (JSSC). *Guideline of Standard Push-out Tests of Headed Stud and Current Situation of Research on Stud Shear Connectors*; JSSC: Tokyo, Japan, 1996.
60. *JIS A 1108*; Method of Test for Compressive Strength of Concrete. Japan Industrial Standards (JIS): Tokyo, Japan, 2006.
61. *JIS A 1113*; Method of Test for Splitting Tensile Strength of Concrete. Japan Industrial Standards (JIS): Tokyo, Japan, 2006.
62. *JIS Z 2241*; Metallic Materials—Tensile Testing Method of Test at Room Temperature. Japan Industrial Standards (JIS): Tokyo, Japan, 2011.
63. *EN 1992-1-1*; Eurocode-2: Design of Concrete Structures Part 1–1: General Rules and Rules for Buildings. European Committee for Standardization: Brussels, Belgium, 2004.
64. Huu, T.N.; Kim, S.E. Finite element modeling of push-out tests for large stud shear connectors. *J. Constr. Steel Res.* **2009**, *65*, 1909–1920.
65. Hillerborg, A. Stability problems in fracture mechanics testing in fracture of concrete and rock. In Proceedings of the International Conference on Recent Developments in the Fracture of Concrete and Rocks, Houston, TX, USA, 20–24 March 1989; pp. 369–378.
66. Noguchi, T.; Tomosawa, F. Relationship between compressive strength and various mechanical properties of high strength concrete. *J. Struct. Constr. Eng. (Trans. AIJ)* **1995**, *472*, 11–16. [CrossRef]
67. Kitsutaka, Y.; Nakamura, S.; Mihashi, H. Simple evaluation method for the bilinear type tension softening constitutive law of concrete. *Proc. Jpn. Concr. Inst.* **1998**, *20*, 181–186.
68. Goto, Y.; Ghosh, K.P.; Kawanishi, N. Nonlinear finite-element analysis for hysteretic behavior of thin-walled circular steel columns with in-filled concrete. *J. Struct. Eng.* **2010**, *136*, 1413–1422. [CrossRef]
69. Doukan, Y.; Fujii, K.; Tamiya, Y.; Fujii, T. Consideration of load transfer mechanism of perforated rib shear connectors. *J. Struct. Eng.* **2014**, *60A*, 827–836.
70. Kimura, Y.; Suzuki, A.; Kasai, K. Estimation of plastic deformation capacity for I-shaped beams with local buckling under compressive and tensile forces. *J. Struct. Constr. Eng. (Trans. AIJ)* **2016**, *81*, 2133–2142. [CrossRef]
71. Suzuki, A.; Kimura, Y.; Kasai, K. Plastic deformation capacity of H-shaped beams collapsed with combined buckling under reversed axial forces. *J. Struct. Constr. Eng. (Trans. AIJ)* **2019**, *83*, 297–307. [CrossRef]
72. Suzuki, A.; Kimura, Y.; Kasai, K. Rotation capacity of I-shaped beams under alternating axial forces based on buckling-mode transitions. *J. Struct. Eng.* **2020**, *146*, 04020089. [CrossRef]
73. Suzuki, A.; Kimura, Y. Rotation capacity of I-shaped beam failed by local buckling in buckling-restrained braced frames with rigid beam-column connections. *J. Struct. Eng.* **2023**, *149*, 04022243. [CrossRef]
74. Fujak, S.M.; Kimura, Y.; Suzuki, A. Estimation of elastoplastic local buckling capacities and novel classification of I-beams based on beam's elastic local buckling strength. *Structures* **2022**, *39*, 765–781. [CrossRef]
75. Fujak, S.M.; Suzuki, A.; Kimura, Y. Estimation of ultimate capacities of Moment-Resisting Frame's subassemblies with Mid-Storey Pin connection based on elastoplastic local buckling. *Structures* **2023**, *48*, 410–426. [CrossRef]

76. Yoshino, Y.; Liao, W.; Kimura, Y. Restraint effect on lateral buckling load of continuous braced H-shaped beams based on partial frame loading tests. *J. Struct. Constr. Eng. (Trans. AIJ)* **2022**, *87*, 634–645. (In Japanese) [CrossRef]
77. Kimura, Y. Effect of loading hysteretic program on plastic deformation capacity and cumulative plastic deformation capacity for H-shaped beam with local buckling—Database of experimentally obtained results for cantilever with H-section. *J. Struct. Constr. Eng. (Trans. AIJ)* **2011**, *76*, 1143–1151. (In Japanese) [CrossRef]

 *buildings*

Article

Systematic Calculation of Yield and Failure Curvatures of Reinforced Concrete Cross-Sections

John Bellos [1,*] **and Apostolos Konstantinidis** [2]

[1] Department of Civil Engineering, Neapolis University Pafos, Paphos 8042, Cyprus
[2] BuildingHow PC, 11635 Athens, Greece; apl@pi.gr
* Correspondence: j.bellos@nup.ac.cy

Abstract: This paper examines and provides a robust solution to the problem of yield and failure curvatures of reinforced concrete (RC) cross-sections, taking into account cracking. At the same time, it calculates the corresponding necessary reinforcement or the moment of resistance in both yield and failure limit states. Computationally, the problem of determining the actual curvatures is reduced to the bending design problem of the cross-section in the yield and failure limit states. This study shows the researcher and the designer how to systematically calculate the strains for different concrete and steel grades and for standard or random cross-sections. This complex process is quite necessary to determine the respective curvatures. The main concept is presented with an emphasis on the "solution regions" as well as the critical cases of the "asymptotic regions", both in yield and failure limit states. Our wide-ranging research on RC element design under biaxial bending with axial force for both yield and failure limit states has been completed and validated via sophisticated algorithms and is available for publication.

Keywords: curvature; ductility; yield; failure; moment of resistance; reinforcement; bending; design

1. Introduction

The yield and failure curvatures, as well as the respective stiffnesses, depend on cross-section geometry, the amount and distribution of reinforcement, the reinforcement material properties, the concrete material properties, and the axial forces. The large relative displacements of the column heads in relation to their bases, and therefore those displacements of the associated diaphragms, significantly increase the characteristic periods of the building, thus resulting in low seismic accelerations. This fact is very important for the structural robustness of the building, especially when checking the resistance of an existing building structure to an earthquake.

Many researchers dealt with the specific problem in the past. Chen and Hsu [1] developed a semi-empirical formula for the curvature ductility of doubly reinforced beam sections, which, via performance-based design, takes into account the effect of reinforcement ratios as well as the reinforcement and concrete strengths. Hernández-Montes et al. [2] related the curvature ductility capacity of cross-sections designed with optimal reinforcement to those with symmetric reinforcement, for both unconfined and confined concrete cases, under varying axial loads, gross section area, and concrete strength. Chandrasekaran et al. [3] developed a closed form solution to estimate the curvature ductility of RC elements under service loads, considering the nonlinear characteristics of constitutive materials and the reinforcement ratios as required by Eurocodes. Arslan and Cihanli [4] produced a formula predicting the curvature ductility of reinforced high-strength concrete beams based on the parametric study of experimental results to evaluate the effects of various structural parameters. Lee [5] provided a prediction formula for the curvature ductility factor of doubly reinforced beam sections, taking into account the concrete strength, the tensile yield strength of steel, and the compressive ultimate strength of steel. Laterza et al. [6] performed an efficiency study of codal detailing rules for reinforcement

Citation: Bellos, J.; Konstantinidis, A. Systematic Calculation of Yield and Failure Curvatures of Reinforced Concrete Cross-Sections. *Buildings* **2024**, *14*, 826. https://doi.org/10.3390/buildings14030826

Academic Editors: Duc-Kien Thai, Atsushi Suzuki and Dinil Pushpalal

Received: 8 January 2024
Revised: 1 March 2024
Accepted: 14 March 2024
Published: 19 March 2024

design of primary columns and beams within the critical regions by comparing the codal design results to the measured curvature ductility. By examining the effects of spectral acceleration and a strong column factor, Zhou et al. [7] provided an empirical model in the form of a quantitative relationship between the curvature ductility demands of columns and the global displacement ductility demands of frame structures. Baji and Ronagh [8] developed a probabilistic method used to calculate curvature ductility by means of the central limit theorem, considering the specific behavior of the moment redistribution factor with respect to curvature ductility and plastic hinge length. Research on biaxial bending by Breccolotti et al. [9] produced a formula for the curvature ductility of reinforced short columns of varying section geometry, neutral axis direction, reinforcement ratios, and axial forces. Kollerathu [10] proposed an equation to evaluate and compare the curvature ductilities of reinforced masonry and RC walls, as a result of diagrams of flexural strength versus curvature. Recently, Foroughi and Yuksel [11] developed a predictive formula for the curvature ductility of doubly-reinforced beams by performing a numerical parametric study.

Finding the actual curvatures, both in yield and failure states, requires the calculation of concrete strain and steel strain under axial force and bending moment (or equivalently, reinforcement), an extremely complex computational problem with a wide range of solutions. Calculation tables were also used in the past, but they were available in failure states only and usually for specific materials. Nowadays, due to the variety of available materials and the demand for checking the actual strength and possibly retrofitting existing buildings, the design in limit states also becomes imperative. This is the reason for our extensive research to find a robust theoretical solution to the bending design problem, both in yield and failure limit states, a part of which is presented in this article.

2. Column Limit States

2.1. Column in Yield Limit State

Figure 1 presents an exaggerated model of a column in the yield limit state. Flexural cracks are perpendicular to the axis of the bar, while shear cracks have an inclination of 45° to 60° to the axis of the bar. Here, δ_1 is the displacement due to shear, which is linear and does not affect the curvature, and δ_2 is the displacement due to crack causing bending at the yield limit state.

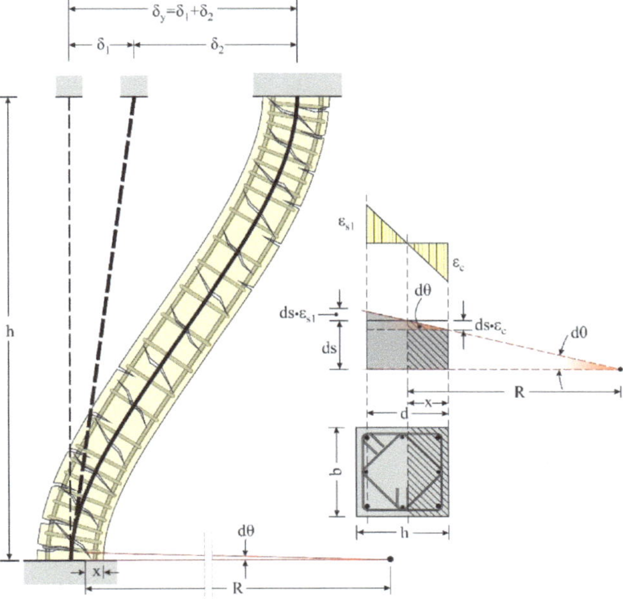

Figure 1. Column in yield limit state (not to scale).

Considering a differential length ds of the column at the side of its cross-section, the inner fiber compresses and shortens by $ds \cdot \varepsilon_c$, while the outer fiber stretches and expands by $ds \cdot \varepsilon_{s1}$. Then, the resulted differential central angle $d\theta$ is as follows:

$$d\theta = ds \cdot (\varepsilon_c + \varepsilon_{s1})/d \text{ and } d\theta = ds/R \tag{1}$$

Equation (1) yields the following:

$$ds/R = ds \cdot (\varepsilon_c + \varepsilon_{s1})/d \rightarrow \varphi_y = 1/R = (\varepsilon_c + \varepsilon_{s1})/d \tag{2}$$

where φ_y represents the actual yield curvature [12].

Note that in the yield state, it must be $\varepsilon_c \leq \varepsilon_{c2}$ and $\varepsilon_{s1} \leq \varepsilon_{yd}$, where the yield strain for at least one of the two materials, either ε_c of the concrete or ε_{s1} of the steel, has been reached.

2.2. Column in the Failure Limit State

The physical behavior of a column functioning in a failure limit state is represented in Figure 2 through the only possible observational method, which is the experimental one. The experiment, which was part of the "Anti-Seismic Thoraces" tests, took place in 1998 in the NTUA's Reinforced Concrete Laboratory under the auspices of Professor Theodosios Tassios. It is evident that the column failure takes place in relatively small regions at the ends, while the rest of the column is in a yield state (marginal yielding with cracking) [12,13].

Figure 2. The way to create a plastic joint at the two ends of a node due to strong alternating tension (Tests of "Anti-Seismic Thoraces"—Reinforced Concrete Laboratory, NTUA).

Figure 3 presents an exaggerated model of a column in the failure limit state. Flexural and shear cracks are apparent at the critical end regions of the column, while along the rest of the body, they remain similar to the yield limit state case.

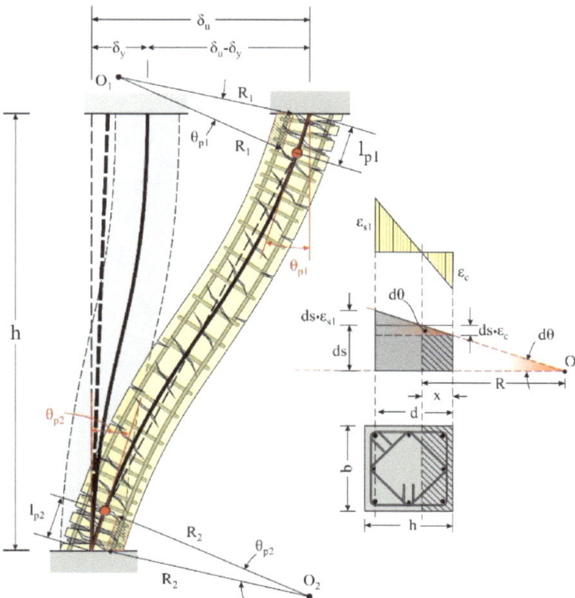

Figure 3. Column in the failure limit state with plastic joints at both ends (not to scale).

The failure curvatures φ_u are calculated in exactly the same way as the yield limit state, as follows:

$$\varphi_{u,j} = 1/R_j = (\varepsilon_c + \varepsilon_{s1})/d, \ j = 1,2 \ (1 = \text{head}, 2 = \text{base}) \tag{3}$$

When the reinforcement of the head and base is the same, as is generally the case, then $\varphi_{u,1} = \varphi_{u,2} = \varphi_u$. Note again that in the failure state, it must be $\varepsilon_c \leq \varepsilon_{cu2}$ and $\varepsilon_{s1} \leq \varepsilon_{ud}$, where the failure strain of at least one of the two materials, either ε_c of the concrete or ε_{s1} of the steel, has been reached.

2.3. Example: Calculation of Limit State Curvatures

Let us consider a fixed–fixed support column of height h = 3.0 m under axial force N_d = −800 kN (see Figure 1). The cross-section is 400 mm × 400 mm, f_{ck} = 30 MPa, γ_c = 1.50, f_{yk} = 500 MPa, γ_s = 1.15, ε_{su} = 20‰, and K = 1.0 with $d_1 = d_2 = 50$ mm. The applied reinforcement is 4Φ20 + 4Φ14 (=1860 mm^2, ρ = 1.16%). It is considered that 50% of the total reinforcements are placed at the corners, while the rest are distributed along the sides (see Figure 4).

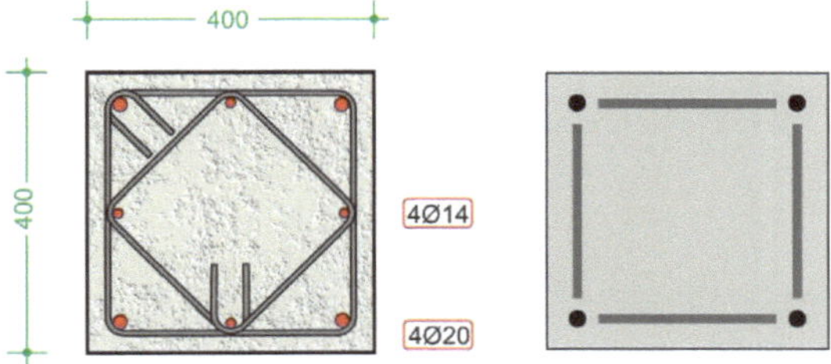

Figure 4. Column cross-section with its reinforcement and corresponding model for bending design.

Required: the yield curvatures $\varphi = 1/R$ and the moments of resistance M_{Rd} at the limit state:

(1) Yield limit y.
(2) Failure limit u.

2.3.1. Calculation of Yield Curvature

The uniaxial bending design of the cross-section yields $x = 187.5$ mm, $M_{Rd,y} = 185$ kN·m, $\varepsilon_c = 2.0‰$, and $\varepsilon_{s1} = 1.734‰$. Hence, Equation (2) provides the yield curvature as follows:

$$\varphi_y = (2.0 + 1.734) \times 10^{-3}/(0.40 - 0.05) = 10.67 \times 10^{-3}/\text{m} \rightarrow R_y = 94 \text{ m}$$

The respective elastic curvature (without cracks) at the base of the column is provided by the relation as follows:

$$\varphi_e = 1/R = M_{Rd,y}/(E \cdot I)$$

According to Eurocode 2 [14], §3.1.2(3), it is $f_{cm} = f_{ck} + 8 = 38$ MPa and $E = 22 \times (f_{cm}/10)^{0.3} \times 10^3 = 32.8$ GPa, so $E \cdot I = 32.8 \times (0.4 \times 0.4^3/12) = 70.0 \times 10^3$ kN·m^2. Substituting, we get:

$$\varphi_e = 185/\left(70.0 \times 10^3\right) = 2.64 \times 10^{-3}/\text{m} \rightarrow R_e = 379 \text{ m}$$

Therefore, it is:

$$\varphi_y/\varphi_e = 10.67 \times 10^{-3}/2.64 \times 10^{-3} = 4.04,$$

which is very important for determining the effective stiffness of a column according to Eurocode 8 [15], §4.3.1(6, 7).

2.3.2. Calculation of Failure Curvature

The uniaxial bending design of the cross-section yields $x = 156.1$ mm, $M_{Rd,u} = 219$ kN·m, $\varepsilon_c = 3.5‰$, and $\varepsilon_{s1} = 4.35‰$. Hence, from Equation (3), the failure curvature is as follows:

$$\varphi_u = (3.5 + 4.35) \times 10^{-3}/(0.40 - 0.05) = 22.42 \times 10^{-3}/\text{m} \rightarrow R_u = 45 \text{ m}$$

Similarly to the yield state case, the elastic curvature is as follows:

$$\varphi_e = 219/\left(70.0 \times 10^3\right) = 3.13 \times 10^{-3}/\text{m} \rightarrow R_e = 320 \text{ m}$$

Therefore, it is:

$$\varphi_u/\varphi_y = 22.42 \times 10^{-3}/10.67 \times 10^{-3} = 2.10 \text{ and } \varphi_u/\varphi_e = 22.42 \times 10^{-3}/3.13 \times 10^{-3} = 7.16.$$

3. Equilibrium of Internal and External Forces

The following relations are derived from Figure 5:

$$x = d \cdot \varepsilon_c/(\varepsilon_c + \varepsilon_s), \varepsilon_c = \varepsilon_s \cdot x/(d - x), \varepsilon_s = \varepsilon_c \cdot (d - x)/x, \varepsilon_{s2} = \varepsilon_c \cdot (x - d_2)/x \qquad (4)$$

$$k_F = \alpha_{cc} \cdot f_{cd} \cdot b, F_c = k_F \cdot x \cdot \alpha, z_c = x \cdot \kappa \qquad (5)$$

$$F_{s1} = A_{s1} \cdot \sigma_{s1}, F_{s2} = A_{s2} \cdot \sigma_{s2} \qquad (6)$$

There are two basic equations balancing the internal forces with the external forces of a cross-section under uniaxial bending with axial force (see Figure 5):

(a) Equilibrium equation of the forces F_{s1}, F_c, and F_{s2} with the axial force N_d

$$F_{s1} - F_c - F_{s2} = N_d \qquad (7)$$

(b) Equilibrium equation of the moments of the forces F_c and F_{s2} with the bending moment M_d in the cross-section center and the moment of axial force N_d

$$M_d - N_d \cdot z_{s1} = F_c \cdot (d - z_c) + F_{s2} \cdot (d - d_2) \tag{8}$$

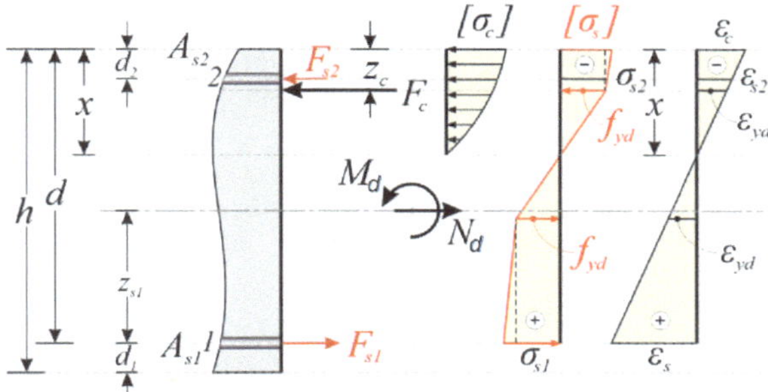

Figure 5. External and internal forces in a cross-section under uniaxial bending with axial force.

Using the Relations (4)–(6), Equations (7) and (8) transform into:

$$A_{s1} \cdot \sigma_{s1} = A_{s2} \cdot \sigma_{s2} + F_c + N_d \tag{9}$$

$$M_d = F_c \cdot (d - z_c) + A_{s2} \cdot \sigma_{s2} \cdot (d - d_2) + N_d \cdot z_{s1} \tag{10}$$

If $\rho_1 = A_{s1}/A_c$ and $\rho_2 = A_{s2}/A_c$, so $\rho_2/\rho_1 = A_{s2}/A_{s1}$ and $A_{s2} = A_{s1} \cdot \rho_2/\rho_1$, then Equations (9) and (10) can be written as follows:

$$A_{s1} = (F_c + N_d) / \left(\sigma_{s1} - \frac{\rho_2}{\rho_1} \cdot \sigma_{s2} \right) \tag{11}$$

$$M_d = F_c \cdot (d - z_c) + A_{s1} \cdot \frac{\rho_2}{\rho_1} \cdot \sigma_{s2} \cdot (d - d_2) + N_d \cdot z_{s1} \tag{12}$$

It is emphasized that the axial force N_d always has a given value, independent of the above relations.

The system of Equations (11) and (12) has three unknown variables in the corresponding problem, that is, ε_c, ε_s, and A_{s1} or M_d. Therefore, the system solution requires additional conditions to be set, as follows:

- First condition: the reinforcement ratio ρ_2/ρ_1 is provided.
- Second condition: either ε_c or ε_s should be in the limit state.

These two conditions, under certain assumptions, can replace the third equation. Nevertheless, the solution is rather difficult, especially in the failure state, due to numerous and complex combinations. The difficulty could be removed by using the trial solution method. However, this process would require the determination of the solution boundaries, which is also a quite complex problem.

4. Solution Regions in the Yield Limit State

The regions comprising possible solutions in the yield limit state are presented in Figure 6.

Let ρ_2/ρ_1 be the ratio of the compressive to the tensile reinforcement. For any given value of the ratio ρ_2/ρ_1, there is a characteristic case having compression zone depth x_{01} (see Figure 6), where the denominator of Equation (11) becomes zero. That is

$$\sigma_{s1} - \frac{\rho_2}{\rho_1}\cdot\sigma_{s2} = 0 \rightarrow \overbrace{E_s\cdot\varepsilon_{s1}}^{\sigma_{s1}} - \frac{\rho_2}{\rho_1}\cdot\overbrace{E_s\cdot\varepsilon_{s2}}^{\sigma_{s2}} = 0 \rightarrow \varepsilon_{s1} = \frac{\rho_2}{\rho_1}\cdot\varepsilon_{s2}$$

Taking into account Equation (4), the above relation can be written as

$$\overbrace{\varepsilon_c\cdot(d-x_{01})/x_{01}}^{\varepsilon_{s1}} = \frac{\rho_2}{\rho_1}\cdot\overbrace{\varepsilon_c\cdot(x_{01}-d_2)/x_{01}}^{\varepsilon_{s2}} \rightarrow x_{01} = \left(d + \frac{\rho_2}{\rho_1}\cdot d_2\right)\bigg/\left(1 + \frac{\rho_2}{\rho_1}\right) \qquad (13)$$

Thus, at location 01, corresponding to compression zone depth x_{01} provided by Equation (13), both tensile reinforcement A_{s1} and bending moment M_d will be infinite (see Equations (11) and (12)). Let us name this location existing in the yield limit state "Asymptotic Location".

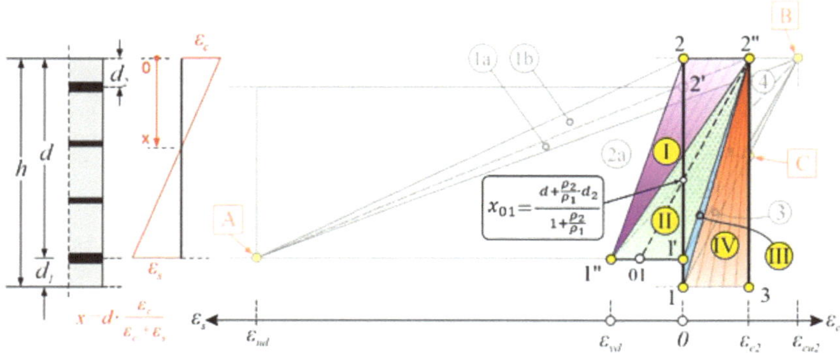

Figure 6. Permissible strain distributions in the yield limit state.

5. Solution Regions in the Failure Limit State

The regions comprising possible solutions in the failure limit state are presented in Figure 7. Region 1, where the steel reaches its failure limit [A], and region 2, where the concrete reaches its failure limit [B], are divided to subregions 1a, 1b and 2a, 2b respectively. Along the boundaries $2''$A and $1''$B, both M_d and A_{s1} tend to infinite values.

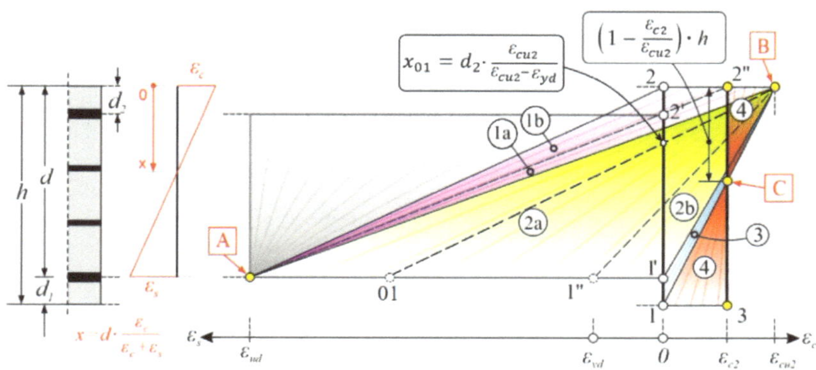

Figure 7. Permissible strain distributions in the failure limit state.

For $\rho_2/\rho_1 = 1$, the denominator of Equation (11) is written as $d\sigma = \sigma_{s1} - \sigma_{s2}$.

By observing the strain distributions in Figure 7, the following conclusions can be drawn:

(1) The boundary $1''B$ of the subregions 2a and 2b, has $\varepsilon_{s1} = \varepsilon_{yd} \rightarrow \sigma_{s1} = f_{yd}$ and $\varepsilon_c = \varepsilon_{cu2}$. Since it is usually $\varepsilon_{s2} \geq \varepsilon_{yd} \rightarrow \sigma_{s2} = f_{yd}$, we have $d\sigma = \sigma_{s1} - \sigma_{s2} = f_{yd} - f_{yd} = 0$.

(2) To the left of location $1''B$ will continue to be $d\sigma = 0$ until the specific location 01 with $\varepsilon_{s1} = \varepsilon_{s2}$.

From the last of Equations (4), for $\varepsilon_{s2} = \varepsilon_{yd}$ and $\varepsilon_c = \varepsilon_{cu2}$, the compression zone depth at location 01 is as follows:

$$\varepsilon_{yd} = \varepsilon_{cu2} \cdot (x_{01} - d_2)/x_{01} \rightarrow x_{01} = d_2 \cdot \varepsilon_{cu2}/(\varepsilon_{cu2} - \varepsilon_{yd}) \tag{14}$$

Thus, for the subregion extended between locations $1''B$ and 01, corresponding to compression zone depths x_{01} provided by Equation (14), both the bending moment M_d and the tensile reinforcement A_{s1} will be infinite (see Equations (11) and (12)). Let us name this region existing in the failure limit state for $\rho_2/\rho_1 = 1$ "Asymptotic Region". Notice that this asymptotic region is independent of the concrete class.

6. Application: The "Typical Rectangular Section" in Limit States

Consider a structural element made of C30/37 concrete and B500 steel with a 300 mm × 550 mm rectangular cross-section, as shown in Figure 8.

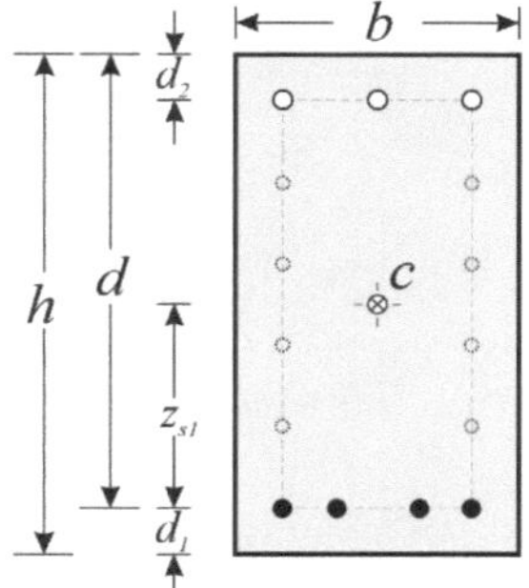

Figure 8. A typical rectangular section.

Provided:

$$b = 300 \text{ mm}, h = 550 \text{ mm}, d_2 = 50 \text{ mm}, d_1 = 50 \text{ mm},$$

$$f_{ck} = 30 \text{ MPa}, \gamma_c = 1.50, a_{cc} = 0.85, \varepsilon_{c2} = 2.0\text{‰}, \varepsilon_{cu2} = 3.5\text{‰},$$

$$f_{yk} = 500 \text{ MPa}, \gamma_s = 1.15, E_s = 200 \text{ GPa (for steel grades B500a,b,c)}$$

Derived:

$$d = h - d_1 = 500 \text{ mm}, z_{s1} = h/2 - d_1 = 0.225 \text{ m}$$

$$f_{cd} = f_{ck}/\gamma_c = 20.0 \text{ MPa}, k_F = a_{cc} \cdot b \cdot f_{cd} = 0.85 \cdot 0.30 \cdot 20.0 \cdot 10^3 = 5100 \text{ kN/m},$$

$$f_{yd} = f_{yk}/\gamma_s = 500/1.15 = 434.78 \text{ MPa}, \varepsilon_{yd} = f_{yd}/E_s = 434.78/(200 \times 10^3) = 2.174\text{‰}$$

For $\varepsilon_c = \varepsilon_{c2} = 2\text{‰}$, it is $\alpha = 0.6667$ and $\kappa = 0.375$, while for $\varepsilon_c = \varepsilon_{cu2} = 3.5\text{‰}$, it is $\alpha = 0.8095$ and $\kappa = 0.416$.

For B500c steel grade, $\varepsilon_{ud} = 20\text{‰}$ with $K = 1.0$ is used in the simplified stress–strain diagram, while $\varepsilon_{ud} = 67.5\text{‰}$ is used with $K = 1.15$ in the exact stress–strain diagram.

6.1. Yield Limit State

Using the regions for the yield limit state presented in Figure 6, we form Figure 9 for the "typical rectangular cross-section", where the strain-based region boundaries and the corresponding compression zones are clearly illustrated. We define the origin of the compression zone as the outermost upper fiber of the cross-section, while x_{ij} represents the compression zone depth corresponding to the location 'ij'. For practical representation reasons, we consider the cross-section to lie horizontally.

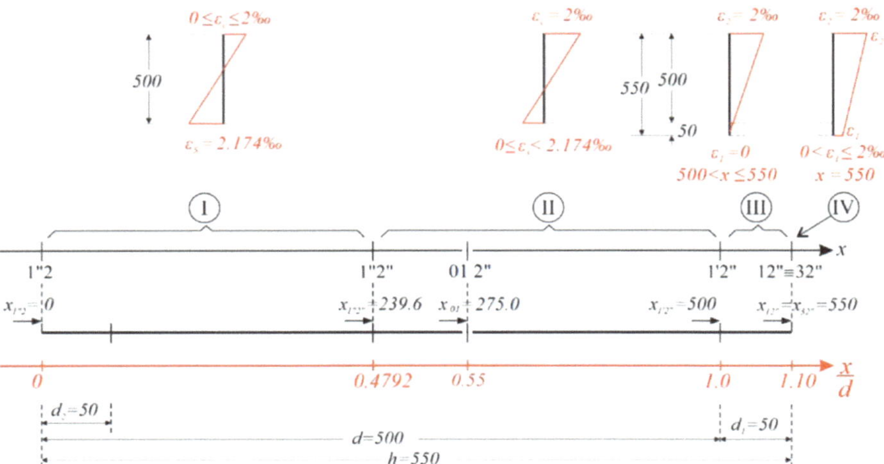

Figure 9. Strain-based region boundaries and corresponding compression zones in the yield limit state for the "typical rectangular section".

6.1.1. Strain and Curvature Diagrams in the Yield Limit State

The diagrams of strain ε_c, ε_s and the corresponding yield curvatures φ_y are shown in Figure 10. These values are independent of the axial force N_d and the reinforcement ratio ρ_2/ρ_1.

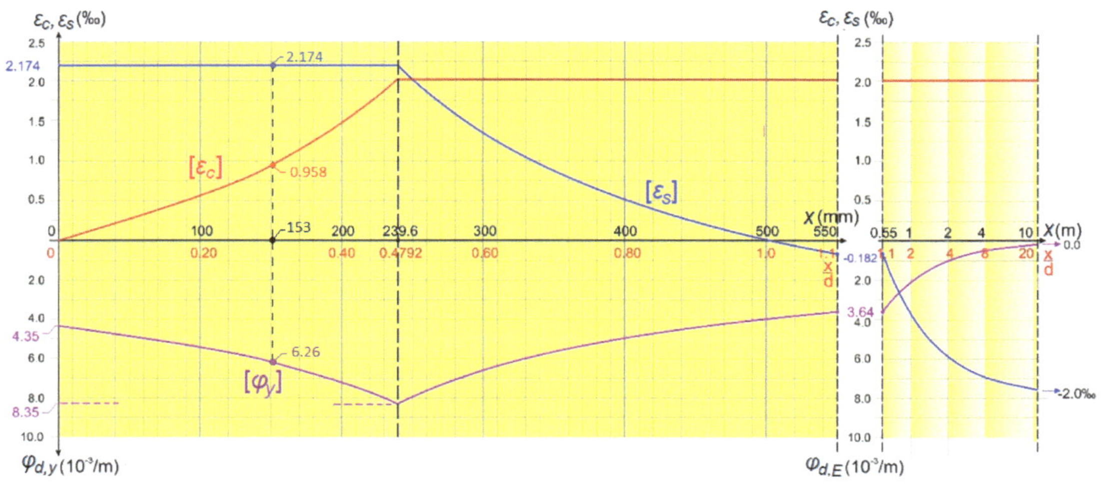

Figure 10. Strain diagrams and corresponding yield curvatures for the "typical rectangular section".

The compression zone depth is presented on the diagrams in millimeters up to a depth of $h = 550$ mm, and thereafter in meters on a logarithmic scale. In practice, the values of A_{s1} and M_d are required in each characteristic case of the cross-section with respect to x. These values, as obtained from Equations (11) and (12), depend on both the reinforcement ratio ρ_2/ρ_1 and the axial force N_d.

6.1.2. Asymptotic Location in Yield Limit State for $\rho_2/\rho_1 = 1$

For $\rho_2/\rho_1 = 1$, Equation (13) yields:

$$x_{01} = (500 + 1 \times 50)/(1 + 1) = 275 \text{ mm and } (d - x_{01})/x_{01} = 0.818.$$

Since the concrete reaches its critical value in this case (i.e., $\varepsilon_c = \varepsilon_{c2} = 2.0‰$), Equation (4) yield a tensile strain value for steel $\varepsilon_{s1} = 2.0‰ \times 0.818 = 1.636‰$. On the other hand, for $\varepsilon_c = \varepsilon_{c2} = 2‰$, it is $\alpha = 0.6667$ and $\kappa = 0.375$ (see Section 6). Consequently, from Equation (5) the compressive force F_c received by the concrete is $F_c = k_F \cdot x_{01} \cdot \alpha = 5100 \times 0.275 \times 0.6667 = 935.0$ kN.

6.1.3. Solution Nomogram in Yield Limit State for $\rho_2/\rho_1 = 1$

The method of reinforcement with $A_{s2} = A_{s1}$ (i.e., $\rho_2/\rho_1 = 1$) is used in cases where significant axial forces are exerted mainly on columns and/or in cases of beams with special anti-seismic requirements that entail high plasticity requirements [16].

Figure 11 presents a solution nomogram in the yield limit state for $\rho_2/\rho_1 = 1$, in the form of paired diagrams (M_d, A_{s1}) corresponding to different compression zone depths x and axial forces N_d. The asymptotic location here stands for $x_{01} = 275.0$ mm, and therefore, region II is divided into two subregions (see Figure 9). For this case, it is $\varepsilon_c = 2.0‰$ and $F_c = 935.0$ kN (see results in Section 6.1.2).

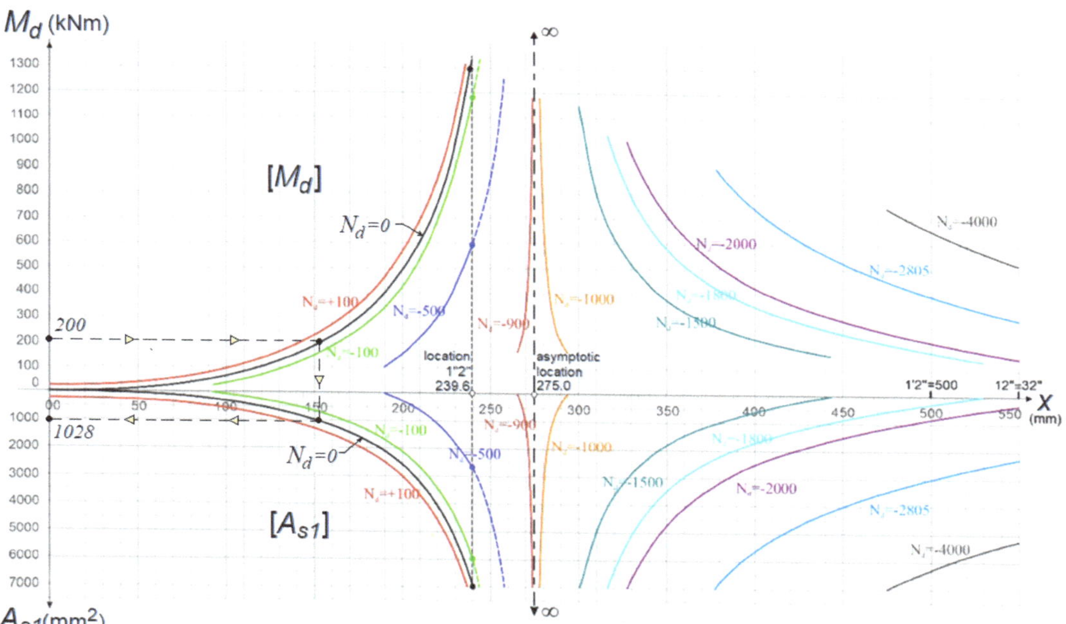

Figure 11. Paired diagrams (M_d, A_{s1}) corresponding to compression zone depths x and axial forces N_d in yield limit state for the "typical rectangular section" and $\rho_2/\rho_1 = 1$.

At each position x, there is a specific pair (A_{s1}, M_d) calculated from Equations (11) and (12). For example, for bending moment $M_d = 200$ kN·m and axial force $N_d = 0$ kN, the steel reaches the yield state first, so the required tensile reinforcement is $A_{s1} = 1028$ mm^2 (see Figure 11). The respective compression zone depth is found to be $x = 153.0$ mm, resulting in strains $\varepsilon_c = 0.958‰$ and $\varepsilon_s = 2.174‰$, clearly indicating that the steel has reached its yield point (see Figure 10). Then, Equation (2) provides the yield curvature (also presented in Figure 10).

$$\varphi_y = (0.958 + 2.174) \times 10^{-3}/(0.55 - 0.05) = 6.26 \times 10^{-3}/\text{m}.$$

6.2. Failure Limit State

Using the regions for the failure limit state presented in Figure 7, we form Figure 12 for the "typical rectangular cross-section" where the strain-based region boundaries and the corresponding compression zones are illustrated. In any algorithmic process adopted, the region boundaries should be first determined, because the upper and lower bounds of x and the corresponding values of the non-critical strain of the steel or concrete are needed. In the case of an accurate stress–strain diagram of the steel, for example, for $\varepsilon_{ud} = 67.5‰$ and $K = 1.15$, the region boundaries A2" and AB change significantly, but the calculation process remains the same. Furthermore, such differences in reinforcement design values are trivial in practice.

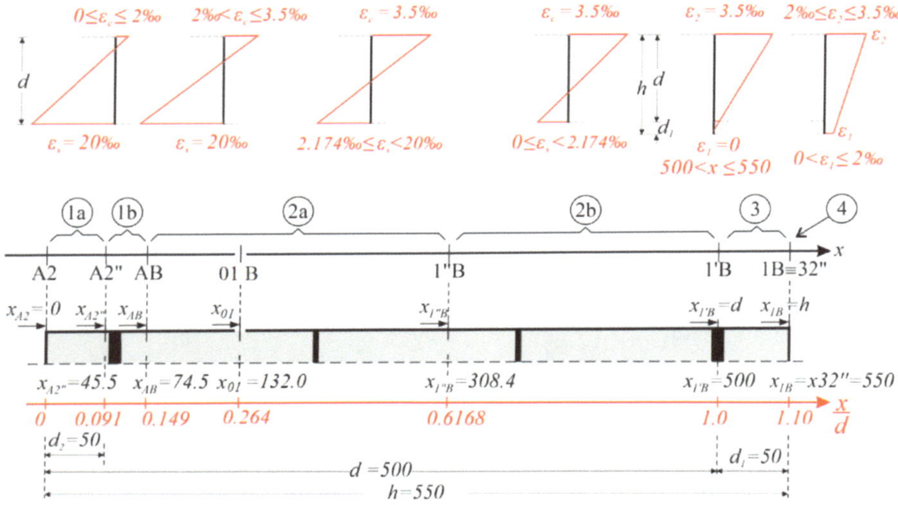

Figure 12. Strain-based region boundaries and corresponding compression zones in the failure limit state for the "typical rectangular section".

6.2.1. Strain and Curvature Diagrams in the Failure Limit State

The diagrams of strain ε_c, ε_s and the corresponding failure curvatures φ_u state are presented in Figure 13. For comparison reasons, the corresponding yield curvatures φ_y are also shown. Notice that all values are independent of the axial force N_d and the reinforcement ratio ρ_2/ρ_1. The compression zone depth x is given in millimeters up to the total depth of $h = 550$ mm, and from there on, in meters on a logarithmic scale. In practice, the values of A_{s1} and M_d are required at each characteristic location of the cross-section with respect to x. These values, obtained from Equations (11) and (12), depend on both the reinforcement ratio ρ_2/ρ_1 and the axial force N_d.

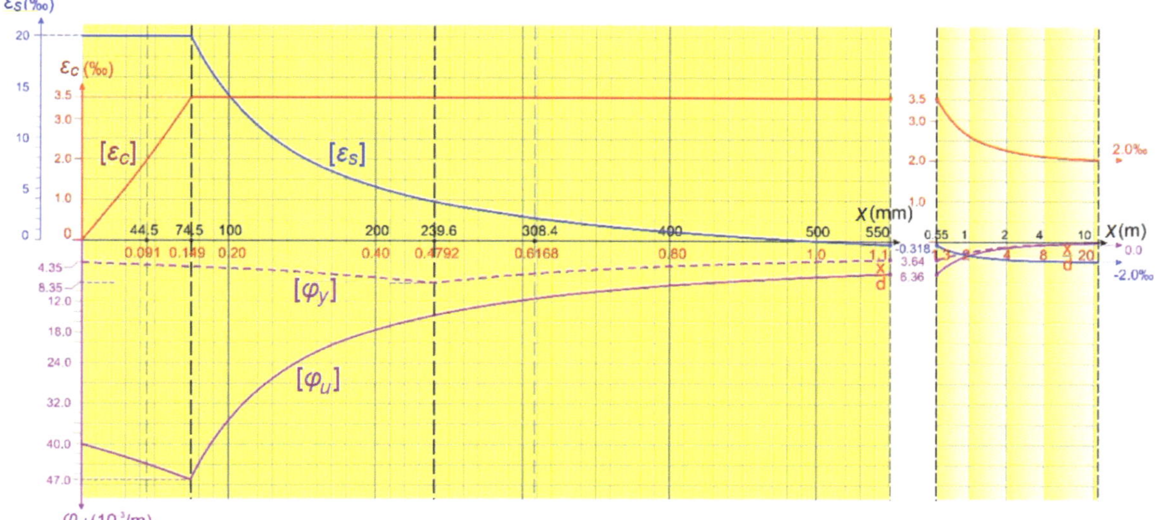

Figure 13. Strain diagrams and corresponding curvatures in the failure limit state for the "typical rectangular section".

6.2.2. Asymptotic Region in Failure Limit State for $\rho_2/\rho_1 = 1$

Equation (14) gives $x_{01} = 50 \times 3.5/(3.50 - 2.174) = 132.0$ mm, while Equation (5) give the force $F_c = 5100 \times 0.132 \times 0.8095 = 544.9$ kN received by the concrete at the location 01.

Equation (4) give $x_{1''B} = 500 \times 3.5/(3.5 + 2.174) = 308.4$ mm, while Equation (5) give the force $F_c = 5100 \times 0.3084 \times 0.8095 = 1273.2$ kN received by the concrete at the location 1″B.

Thus, Equation (11) maintains a zero denominator throughout the interval between $x_{01} = 132.0$ mm and $x_{1''B} = 308.4$ mm.

6.2.3. Solution Nomogram in Failure Limit State for $\rho_2/\rho_1 = 1$

Figure 14 presents a solution nomogram for $\rho_2/\rho_1 = 1$, in the form of paired diagrams (M_d, A_{s1}) corresponding to different compression zone depths x and axial forces N_d. At each position x, there is a specific pair (A_{s1}, M_{sd}) calculated from Equations (11) and (12). The diagram comprises the areas of dominant bending on the left and the areas of dominant compression on the far right. A multiple solution area is also apparent in the middle, theoretically extending to infinity. It should be pointed out that in our case, with $\rho_2/\rho_1 = 1$, there are two asymptotic boundary locations (in the sense of zeroing the denominator of Equation (11)), that is, 01 and 1″B, as determined in Section 6.2.2. Consequently, region 2a is divided into two subregions (AB, 01) and (01, 1″B) (see Figures 7 and 12).

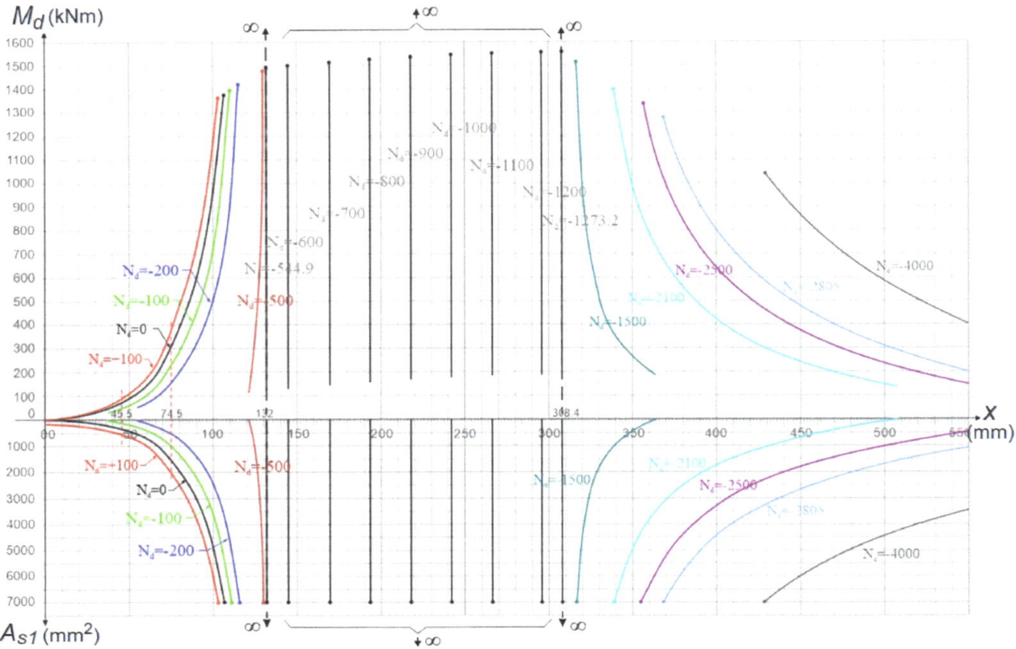

Figure 14. Paired diagrams (M_d, A_{s1}) corresponding to compression zones x and axial forces N_d in the failure limits state for the "typical rectangular section" and $\rho_2/\rho_1 = 1$.

6.2.4. Indeterminacy or Multiple Solution Region in Failure Limit State

This region extends between the two asymptotic locations 01 and 1″B, corresponding to compression zone depths $x_{01} = 132$ mm and $x_{1''B} = 308.4$ mm, respectively (see Figure 14).

By relating those to the respective forces, we can say that a cross-section is in the indeterminacy region when stressed by axial forces of 544.9 kN $\leq N_d \leq$ 1273.2 kN (see results in Section 6.2.2).

In this region, it is $\sigma_{s1} = \sigma_{s2}$. Thus, Equation (11) gives indeterminacy $A_{s1} = \infty$ for $F_c \neq -N_d$ and an infinite number of solutions for $F_c = -N_d$. For each N_d in the region, there is a certain x that gives $F_c = -N_d$. This location has $x = F_c/\alpha_{cc} \cdot f_{cd} \cdot b \cdot \alpha$, and the force F_c is exerted at the position $z_c = x \cdot \kappa$ (where $\alpha = 0.8095$ and $\kappa = 0.416$ because $\varepsilon_c = \varepsilon_{cu2} = 3.5‰$—see values in Section 6). Furthermore, the tensile strain is $\varepsilon_{s1} = \varepsilon_c \cdot (d-x)/x$, while the force is $F_{s2} = A_{s2} \cdot \sigma_{s2} = A_{s1} \cdot f_{yd}$.

Since $F_c = -N_d$, $\rho_2/\rho_1 = 1$, and $\sigma_{s2} = \sigma_{s1} = f_{yd}$, Equation (12) yields:

$$M_d = A_{s1} \cdot f_{yd} \cdot (d - d_2) - N_d \cdot (d - z_c - z_{s1}) \tag{15}$$

Notice that Equation (15) directly relates A_{s1} to M_d. So, when M_d is given in a problem, A_{s1} is uniquely calculated, while when A_{s1} is given, M_d is uniquely calculated.

Remark: In the multiple solution region, all pairs (M_d, A_{s1}) having the same N_d correspond to the same x, implying the same strains ε_{s1} and ε_c and, hence, constant failure curvatures φ_u.

6.2.5. Application in Failure Limit State for $\rho_2/\rho_1 = 1$ and $N_d = -1000$ kN

It is $x = 1000/(0.85 \times 20 \times 10^3 \times 0.30 \times 0.8095) = 0.2422$ m, $z_c = 0.2422 \times 0.416 = 0.101$ m, $\varepsilon_{s1} = 3.5 \times (0.50 - 0.2422)/0.2422 = 3.73‰$, $\varepsilon_c = 3.5‰$, and $\varphi_u = (3.5 + 3.73)/0.50 = 14.46‰/$m.

For $A_{s1} = A_{s2} = 0$ (when reinforcement is barely required), Equation (15) yields $M_d = 0 + 1000 \times (0.50 - 0.101 - 0.225) = 174$ kN·m. That is, for axial force $N_d = -1000$ kN and moment $M_d \leq 174$ kN·m, no reinforcement is required.

For $A_{s1} = A_{s2} = 7000$ mm^2 (selection of maximum reinforcement value A_{s1} in Figure 14), Equation (15) yields $M_d = 7000 \times 10^{-6} \times 434.78 \times 10^3 \times (0.50 - 0.05) + 1000 \times (0.50 - 0.101 - 0.225) = 1544$ kN·m.

For $M_d = 800$ kN·m, inverted Equation (15) yields $A_{s1} = [800{-}1000 \times (0.50 - 0.101 - 0.225)]/[434.78 \times 10^3 \times (0.50 - 0.05)] = 3.200 \times 10^{-3}$ m^2 = 3200 mm^2. Hence, the solution is determined as a pair (800, 3200) from the infinite number of pairs of specific solutions on the line $N_d = -1000$ kN.

7. Conclusions

- The influence of cracking on the curvature of RC elements is significant in relation to the corresponding elastic curvature, even in the yield state, so the corresponding yield stiffness is significantly smaller than the elastic stiffness.
- For specific flexural reinforcement, the yield curvature is much smaller than the failure curvature, while the yield moment of resistance is of the same order of magnitude as the failure moment of resistance.
- During the bending design of a cross-section in the failure limit state, there is an extended region of compressive axial forces where the curvature is practically constant regardless of the acting moment. Then, for any given axial force, the necessary flexural reinforcement is derived from the acting moment based on a first-order equation.
- It is proposed that the design process should take place in the following order:
    - (a)   Calculate reinforcement in the failure state.
    - (b)   Choose and apply reinforcement.
    - (c)   Calculate failure curvatures φ_u, yield curvatures φ_y, and elastic curvatures φ_e.
    - (d)   Estimate curvature ductilities from the ratios φ_u/φ_y and/or φ_u/φ_e.
    - (e)   Determine effective stiffness and resolve.

Author Contributions: Conceptualization, A.K.; Data curation, J.B.; Methodology, A.K.; Resources, J.B.; Software, J.B.; Formal analysis, A.K.; Validation, J.B.; Investigation, A.K.; Writing—original draft, A.K.; Writing—review & editing, J.B.; Visualization, A.K. All authors have read and agreed to the published version of the manuscript.

Funding: This research received no external funding.

Data Availability Statement: The raw data supporting the conclusions of this article will be made available by the authors on request.

Conflicts of Interest: Author Apostolos Konstantinidis is the director of the company BuildingHow PC. The remaining author declares that the research was conducted in the absence of any commercial or financial relationships that could be construed as a potential conflict of interest.

Nomenclature

The following symbols are employed in this paper:

A_s	total reinforcement
A_{s1}	tensile reinforcement
A_{s2}	compressive reinforcement
α_{cc}	long-term effects factor
b	beam width
c	ratio of compressive reinforcement to tensile reinforcement
d	beam effective depth
d_1	tensile reinforcement cover
d_2	compressive reinforcement cover
E	concrete modulus of elasticity
E_s	steel modulus of elasticity

F_c	concrete compressive force
F_{s1}	tensile reinforcement force
F_{s2}	compressive reinforcement force
f_{ck}	concrete compressive strength
f_{cm}	concrete mean compressive strength
f_{yk}	steel yield strength
f_{cd}	concrete design strength
f_{yd}	steel design strength
h	beam height
I	cross-section moment of inertia
K	strain hardening coefficient (ductility property)
k_F	concrete compressive force coefficient
M_d	bending moment acting in the cross-section center
M_{sd}	bending moment acting in the tensile reinforcement position
N_d	axial force acting in the cross-section center
R_e	radius of elastic curvature
R_y	radius of yield curvature
R_u	radius of failure curvature
x	compressive zone depth distance of the outermost upper fiber from the neutral axis
x_{ij}	compressive zone depth corresponding to the "ij" location
x_{01}	compressive zone depth corresponding to the asymptotic location
z_c	distance of the outermost upper fiber from the concrete compression center
z_{s1}	distance of the tensile reinforcement position from the cross-section center
α	distribution factor of concrete compressive force
γ_c	concrete safety factor
γ_s	steel safety factor
δ_1	yield state displacement due to shear
δ_2	yield state displacement due to bending
ε_c	concrete strain
ε_{c2}	concrete yield strain
ε_{cu2}	concrete ultimate strain
ε_s	steel strain
ε_{s1}	tensile reinforcement strain
ε_{s2}	compressive reinforcement strain
ε_{su}	steel ultimate strain
ε_{ud}	steel design ultimate strain
κ	position factor of concrete compressive force
ρ_1	tensile reinforcement percentage
ρ_2	compressive reinforcement percentage
σ_{s1}	tensile reinforcement stress
σ_{s2}	compressive reinforcement stress
φ_e	elastic curvature
φ_y	yield curvature
φ_u	failure curvature

References

1. Chen, C.C.; Hsu, S.M. Formulas for Curvature Ductility Design of Doubly Reinforced Concrete Beams. *J. Mech.* **2004**, *20*, 257–265. [CrossRef]
2. Hernández-Montes, E.; Aschheim, M.; Gil-Martín, L.M. Impact of Optimal Longitudinal Reinforcement on the Curvature Ductility Capacity of Reinforced Concrete Column Sections. *Mag. Concr. Res.* **2004**, *56*, 499–512. [CrossRef]
3. Chandrasekaran, S.; Nunziante, L.; Serino, G.; Carannante, F. Curvature Ductility of RC Sections Based on Eurocode: Analytical Procedure. *KSCE J. Civ. Eng.* **2010**, *15*, 131–144. [CrossRef]
4. Arslan, G.; Cihanli, E. Curvature Ductility Prediction of Reinforced High-strength Concrete Beam Sections. *J. Civ. Eng. Manag.* **2010**, *16*, 462–470. [CrossRef]
5. Lee, H.-J. Predictions of Curvature Ductility Factor of Doubly Reinforced Concrete Beams with High Strength Materials. *Comput. Concr.* **2013**, *12*, 831–850. [CrossRef]
6. Laterza, M.; D'Amato, M.; Thanthirige, A.P.; Braga, F.; Gigliotti, R. Comparisons of Codal Detailing Rules for Curvature Ductility and Numerical Investigations. *Open Constr. Build. Technol. J.* **2014**, *8*, 132–141. [CrossRef]

7.   Zhou, J.; He, F.; Liu, T. Curvature Ductility of Columns and Structural Displacement Ductility in RC Frame Structures Subjected to Ground Motions. *Soil. Dyn. Earthq. Eng.* **2014**, *63*, 174–183. [CrossRef]

8.   Baji, H.; Ronagh, H.R. Probabilistic Models for Curvature Ductility and Moment Redistribution of RC Beams. *Comput. Concr.* **2015**, *16*, 191–207. [CrossRef]

9.   Breccolotti, M.; Materazzi, A.L.; Regnicoli, B. Curvature Ductility of Biaxially Loaded Reinforced Concrete Short Columns. *Eng. Struct.* **2019**, *200*, 109669. [CrossRef]

10.   Kollerathu, J.A. Curvature Ductility of Reinforced Masonry Walls and Reinforced Concrete Walls. In *Sustainability Trends and Challenges in Civil Engineering: Select Proceedings of CTCS 2020*; Lecture Notes in Civil Engineering; Springer: Singapore, 2021; pp. 9–23. [CrossRef]

11.   Foroughi, S.; Yuksel, S.B. A New Approach for Determining the Curvature Ductility of Reinforced Concrete Beams. *Slovak. J. Civ. Eng.* **2022**, *30*, 8–20. [CrossRef]

12.   Karayannis, C.G. *Design and Behavior of Reinforced Concrete Structures for Seismic Actions: Chapter 8—Element Ductility*; Editions SOFIA: Sofia, Bulgaria, 2019; (In Greek). ISBN 978-960-633-005-6.

13.   Konstantinides, A. *Earthquake Resistant Buildings Made of Reinforced Concrete: The Art of Construction and the Detailing According to Eurocodes*; Alta Grafico SA: Ano Liossia, Greece, 2010; Volume A, ISBN 978-960-85506-3-6.

14.   *EN 1992-1-1: 2004*; Eurocode 2: Design of Concrete Structures. Part 1-1: General Rules and Rules for Buildings. British Standard Institution: London, UK, 2005.

15.   *EN 1998-1: 2004*; Eurocode 8: Design of Structures for Earthquake Resistance. Part 1: General Rules, Seismic Actions and Rules for Buildings. European Committee for Standardization: Brussels, Belgium, 2005.

16.   Konstantinides, A.; Bellos, J. *Earthquake Resistant Buildings Made of Reinforced Concrete: Static and Dynamic Analysis According to Eurocodes*; Alta Grafico SA: Ano Liossia, Greece, 2013; Volume B, ISBN 978-960-85506-4-3.

 *buildings*

Article

Bond-Damaged Prestressed AASHTO Type III Girder-Deck System with Retrofits: Parametric Study

Haoran Ni * and Riyad Aboutaha

Department of Civil and Environmental Engineering, Syracuse University, Syracuse, NY 13244, USA; rsabouta@syr.edu
* Correspondence: hni100@syr.edu

Abstract: This research describes an in-depth analysis of the flexural strength of a strengthened AASHTO Type III girder-deck system with debonding-damaged strands based on the finite element software ABAQUS 6.17. To investigate the stand-debonding impact and retrofit, two strengthening techniques by the separate use of carbon fiber-reinforced polymer (CFRP) and steel plate (SP) were proposed. A detailed finite element analysis (FEA) model considering strand debonding, material deterioration, and retrofitting systems was developed and verified against relevant experimental data obtained by other researchers. The proposed FEA model and the experimental data were in good agreement. The sensitivity of the numerical model to the mesh size, element type, dilation angle and coefficient of friction was also investigated. Based on the verified FEA model, 156 girder-deck systems were studied, considering the following variables: (1) debonding level, (2) span-to-depth ratio (L/d), (3) strengthening type, and (4) strengthening material amount. The results indicated that the debonding level and span-to-depth ratio had a major effect on both load–deflection behaviors and the ultimate strength. The relationships between the enhancement of the ultimate strength and the thickness of the strengthening material were obtained through regression equations with respect to the CFRP- and SP-strengthened specimens. The coefficient of determination (R^2) was 0.9928 for the CFRP group and 0.9968 for the SP group.

Keywords: prestressed concrete; bond damage; finite element analysis; CFRP; ABAQUS

Citation: Ni, H.; Aboutaha, R. Bond-Damaged Prestressed AASHTO Type III Girder-Deck System with Retrofits: Parametric Study. *Buildings* **2024**, *14*, 902. https://doi.org/10.3390/buildings14040902

Academic Editors: Atsushi Suzuki and Dinil Pushpalal

Received: 27 February 2024
Revised: 20 March 2024
Accepted: 25 March 2024
Published: 26 March 2024

1. Introduction

Prestressed concrete (PC) superstructures have been constructed in large numbers over the past several decades in the United States, particularly in the field of highway bridge girders with long spans. However, many existing highway bridges, particularly those near industrial facilities, experience large deflection issues caused by overloaded trucks or an increase in traffic density. Owing to the accumulative damage, large girder deflections and cracks result in bond reduction in the presence of an adequately large mechanical force. Over time, under frequent overloads, the bond between the strands and the surrounding concrete at critical bending sections deteriorates, which eventually results in debonding across the span section. Harries et al. [1] summarized the sources and types of observed damage for prestressed concrete girders. Almost all the cases were related to strand bonds. Strand bonds are unique interactions between prestressed steels and the surrounding concrete. Poor strand bonds cannot ensure that the strand and surrounding concrete function as a composite material with external loading [2]. FRP materials have many advantages over traditional repair materials in terms of their superior mechanical and chemical properties and easy constructability [3]. Additionally, it has been proven that bonded-steel plate strengthening is faster than other reinforcement methods and has a higher modulus of elasticity and ductility. Therefore, steel plates are considered as an effective strengthening technique.

For PC structures, many researchers [4–8] have developed bond-slip models based on the typical bond-slip relationship provided by the CEB-FIP Model Code [9]. Wang [8]

explained that the mechanism of bond failure is such that the bond strength and stiffness are the result of the adhesive force only under self-weight after strand release. When a PC member experiences an overload, the adhesive force gradually disappears. The surrounding concrete subsequently induces a friction force and mechanical interlocking force at the strand–concrete interaction. With an increase in the external overload, local crushing occurs, and the bond stress gradually decreases until strand bond slip occurs. Mohandoss et al. [10] described the bond failure mechanism by the force equilibrium at the strand–concrete interface. Researchers [11,12] demonstrated that the cracking number, width, and propagation in PC girders are affected by the debonding length and position. Girders with different bond conditions show a similar load–deflection behavior before the cracking load. Insufficient bonding results in ductility reduction. The strand debonding level also affects the cracking load and prestressing loss.

In recent decades, FEA has become a reliable tool for the nonlinear analysis of complex structures. Compared with conventional laboratory tests, FEA can save significant money on materials and labor. FEA modeling considers both geometrical and material nonlinearity, which ensures that FEA can handle complex geometries and materials and provide accurate and detailed solutions. Based on the available literature, many studies in the field of structural engineering have been conducted using FEA software. Garg et al. [13] numerically simulated the model of epoxy using cohesive elements at the concrete–CFRP interface. It was demonstrated that the proposed cohesive element was capable of capturing delamination failure. Qapo et al. [14] presented a 3D nonlinear FEM for PC girders strengthened with externally bonded CFRP. For the CFRP-to-concrete interface, the bond zone between the concrete and CFRP was modeled using eight-node plane quadrilateral interface elements, and the bond-slip model developed by Sato and Vecchio [15] was adopted for modeling the CFRP-to-concrete interface. Arab et al. [16] employed the extrusion technique to simulate the interface between strands and concrete and was modeled based on the friction surface-to-surface contact model in the software. Two friction coefficients (0.7 and 1.4) were used in the study, and the authors discussed their effects on modeling results. Kang et al. [17] numerically investigated the mechanical responses of fully bonded and unbonded post-tensioned concrete members considering the tension stiffening effect. The unbonded interface was described by frictionless tangential behavior; that is, node–surface contact. Kang and Huang [18] performed an FEA of post-tensioned concrete girders with both bonded and unbonded tendons. Three modeling methods were adopted to model the unbonded and bonded strand conditions: (1) a contact technique that reflects the actual physical strand conditions in the concrete; (2) a multiple-spring system; (3) a surface-to-surface contact formulation.

Based on the existing literature, both extrusion and embedment techniques are applicable to simulate the interface between the bonded strands and the surrounding concrete. The extrusion technique provides more details between the strand and the surrounding concrete, whereas the embedment technique is efficient in terms of the computational time. Three techniques can be employed for the contact between the unbonded strands and concrete. These are contact techniques that reflect actual physical interactions, spring systems that are more flexible in solving convergence problems, and surface-to-surface contacts with tangential and normal responses. Surface-to-surface contact is the most efficient method in terms of the computational cost.

In summary, numerical studies on strengthened PC members with debonding strands are very limited. Therefore, it is important to develop a reliable FEA model for debonding prestressed AASHTO Type III girder-deck systems using various strengthening techniques. This research contributes to structural engineering by providing an in-depth understanding of the partially debonding prestressed concrete girder-deck system with retrofitting, FEA models, sensitivity study, and parametric evaluation.

2. Cross-Section and Strengthening Systems

The prestressed girder-deck systems were designed according to the AASHTO LRFD Bridge Design Specifications [19]. The girder cross-sectional geometry followed the AASHTO Type III dimensions provided in the PCI Bridge Design Manual [20]. A 203 mm (8 in) × 1829 mm (72 in) rectangular concrete deck was casted on top of the girder. Figure 1a shows a CFRP-strengthened section. All the simply supported girder-deck systems were prestressed with a straight-strand profile. The maximum shear reinforcement was designed in this system to prevent shear failure before flexural failure.

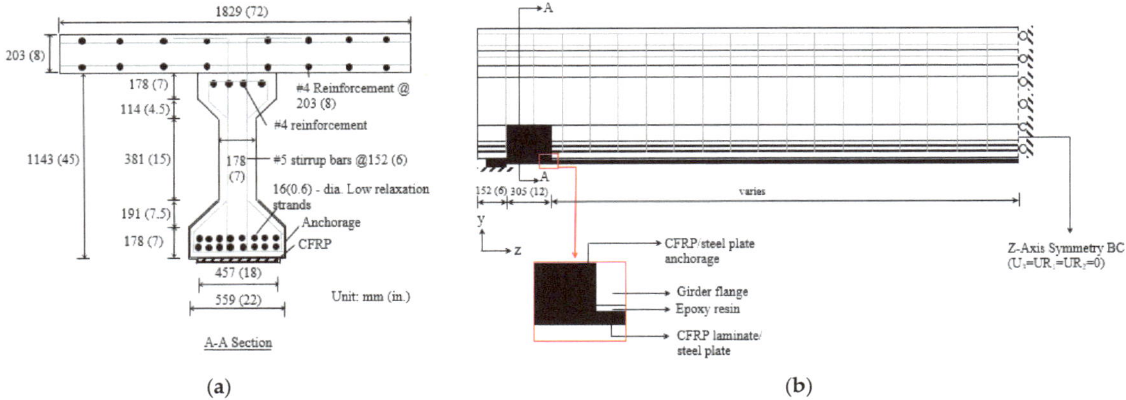

Figure 1. Details of specimens: (**a**) cross-sectional information; (**b**) strengthening systems.

Each of the girder-deck systems was reinforced with 16 Grade 270 seven-wire low-relaxation prestressing strands in the bottom flange. Each strand had a diameter of 15.2 mm (0.6 in). An effective prestressing force of 1206 Mpa (175 ksi) after prestress losses was applied to each strand. The ultimate strength of the strands adopted in the simulations was 1862 Mpa (270 ksi), the yield strength was 1675 Mpa (243 ksi), the modulus of elasticity was 198,569 Mpa (28,800 ksi), and Poisson's ratio was 0.3. In addition, US #4 Grade 60 steel was used for the non-prestressed reinforcement of both the girder and deck. Four rebars were placed on the top flange of the girder, and 16 rebars were located at both the top and bottom of the deck in the longitudinal direction. In the transverse direction of the deck, the US #4 rebars were spaced 203.2 mm (8 in) apart. To prevent shear failure before flexural failure, the maximum shear reinforcement in this system was designed with US #5 steel with a spacing of 152 mm (6 in).

Two strengthening techniques were designed to repair girder-deck systems using CFRP laminates and steel plates. For the CFRP-strengthened system, a CFRP laminate with U-shaped anchorages at both ends was externally bonded to the bottom of the concrete. The CFRP layout is shown in Figure 1b. The CFRP laminate was assumed to be perfectly bonded to the anchorage regions. A U-shaped CFRP was wrapped around the entire bottom flange with a width of 305 mm (12 in) located. Epoxy-reinforced composites with the carbon fiber volume content of 68%, a product of Sika® CarboDur® S, were employed in this study. The properties of the CFRP laminate, including the elastic behavior and the initiation and evolution criteria of damage, are described in Section 3.3. For the steel-plate-strengthened system, Grade 60 steel with the same width as that of the CFRP laminate was attached to the bottom of the concrete. Steel plates were anchored to both ends. To simplify the FEA model, the steel plate was considered to be perfectly bonded to the concrete in the anchored regions. The product Sikadur®-30, whose properties are presented in Section 3.4, was employed as epoxy resin which was used for attaching strengthening materials to the concrete. The adhesive layer thickness was 1 mm (0.04 in).

3. Material Modeling

3.1. Concrete

Concrete in compression is described as an elastic-plastic material with strain softening. The stress–strain model of concrete in compression proposed by Yang et al. [21] was employed to describe the compressive behavior of concrete, as shown in Figure 2a. Concrete stress is assumed to be elastically linear up to $0.4f_c'$, where f_c' is the ultimate concrete compressive strength. The modulus of elasticity E_c is a power law of f_c', as expressed in Equation (1).

$$E_c = A_1 \left(f_c'\right)^a \left(\frac{w_c}{w_0}\right)^b \tag{1}$$

where $A_1 = 8470$, $a = 1/3$, and $b = 1.17$. $w_c = 2400 \text{ kg/m}^3$ (150 lb/ft^3) is the concrete density and $w_0 = 2300 \text{ kg/m}^3$ (144 lb/ft^3) is the reference value proposed by Yang et al. [21].

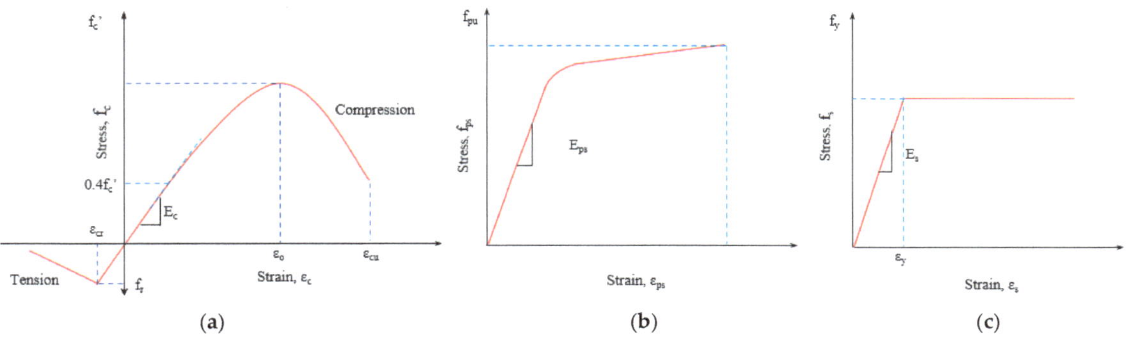

Figure 2. Constitutive relationships: (**a**) concrete in compression and tension [21]; (**b**) prestressing strand; (**c**) non-prestressing steel.

Above $0.4f_c'$, concrete is considered to be in a plastic stage. The calculation of the inelastic strain is shown in Equation (2). The plastic behavior of concrete consists of ascending and descending branches, given by Equation (3). In this equation, β_1 determines the slopes of the nonlinear branch. β_1 can be calculated using Equation (4) for both the ascending and descending segments.

$$\varepsilon_{in} = \varepsilon_c - \frac{f_c}{E_c} \tag{2}$$

$$f_c = \left[\frac{(\beta_1 + 1)\left(\frac{\varepsilon_c}{\varepsilon_0}\right)}{\left(\frac{\varepsilon_c}{\varepsilon_0}\right)^{\beta_1+1} + \beta_1} \right] f_c' \tag{3}$$

$$\beta_1 = \begin{cases} 0.2e^{0.73\xi} \ for \ \varepsilon_c \le \varepsilon_0 \\ 0.41e^{0.77\xi} \ for \ \varepsilon_c > \varepsilon_0 \end{cases} \tag{4}$$

$$\xi = \left(\frac{f_c'}{f_0}\right)^{0.67} \left(\frac{w_0}{w_c}\right)^{1.17} \tag{5}$$

$$\varepsilon_0 = 0.0016e^{240\left(\frac{f_c'}{E_c}\right)} \tag{6}$$

ε_{in} is inelastic strain; ε_c and f_c are the strain and stress variables; and ξ is given by Equation (5) and used to simplify the β_1 equations. Yang et al. (2014) set f_0 and w_0 to 10 MPa (1.5 ksi) and 2300 kg/m^3 (144 lb/ft^3). ε_0, expressed in Equation (6), is the strain at concrete maximum strength.

The tensile stress–strain relationship of concrete was identified using a bilinear model including the pre-cracking and post-cracking stages, as shown in Figure 2a. The first segment starts at the origin and ends at the maximum tensile strength, corresponding to the cracking strain. The second segment is characterized by a simplified linear softening branch. According to the AASHTO Bridge Design Specifications, for most regular concrete, the tensile strength can be estimated using Equation (7).

$$f_t = 0.23\sqrt{f_c'} \tag{7}$$

In this study, 10 and 20 times the cracking strain were adopted as the total strain increments in the post-cracking region for the debonding-damaged and fully bonded PC girders, respectively. The Poisson's ratio was taken as 0.2.

3.2. Prestressing Strand and Mild Steel

A 3D solid model was constructed for the strands in ABAQUS. The strands were meshed using a 6-node wedge element (C3D6). To be compatible with the points of the surrounding concrete, the cross-section of the strand was partitioned into 16 small triangles. An effective prestressing stress of 1206 MPa (175 ksi) was applied to the strands by defining a "predefined field" under a load module. The stress–strain behavior is shown in Figure 2b. Based on the PCI Design Handbook, this curve can be approximated using Equations (8) and (9). A yield strength of 1675 MPa (243 ksi) was measured at an elongation of 1%, and the approximate strain at rupture was 0.07 [20].

$$\varepsilon_{ps} \leq 0.0085 : f_{ps} = E_{ps}\varepsilon_{ps} \tag{8}$$

$$\varepsilon_{ps} > 0.0085 : f_{ps} = f_{pu} - \frac{0.04}{\varepsilon_{ps} - 0.007} \tag{9}$$

where ε_{ps} and f_{ps} are the strain and stress in strands, respectively; E_{ps} is the modulus of elasticity of strand, 198,569 MPa (28,800 ksi); and f_{pu} is the ultimate strength of the strand, 1862 MPa (270 ksi).

For mild reinforcement, the steel rebars were modeled using 2-node truss element (T3D2). A simplified bilinear elastoplastic model with a yield strength of 413 MPa (60 ksi) was employed to describe the stress–strain relationship, as shown in Figure 2c. The elastic behavior of mild steel is defined by its modulus of elasticity, E_s. Based on the ABAQUS User's Manual [22], ABAQUS approximates the smooth stress–strain behavior of a material with straight lines. Therefore, it is possible to use a straight line, which is a very close approximation of the actual material behavior, as the post-yield behavior of non-prestressed steel. The expressions for the bilinear model are given in Equations (10) and (11).

$$\varepsilon_s \leq 0.002 : f_s = E_s\varepsilon_s \tag{10}$$

$$\varepsilon_s \leq 0.002 : f_s = f_y \tag{11}$$

where ε_s and f_s are the strain and stress, respectively; E_s is the modulus of elasticity, 200,000 MPa (29,000 ksi); and f_y is yield strength, 413 MPa (60 ksi).

3.3. Constitutive Models for Undamaged and Damaged CFRP Laminate

Prior to damage, the CFRP laminate was modeled as a linear elastic orthotropic material with the constitutive behavior expressed in Equation (12).

$$\begin{bmatrix} \sigma_{11} \\ \sigma_{22} \\ \sigma_{33} \\ \sigma_{12} \\ \sigma_{23} \\ \sigma_{31} \end{bmatrix} = \begin{bmatrix} C_{11}^0 & C_{12}^0 & C_{13}^0 & 0 & 0 & 0 \\ C_{12}^0 & C_{22}^0 & C_{23}^0 & 0 & 0 & 0 \\ C_{13}^0 & C_{23}^0 & C_{33}^0 & 0 & 0 & 0 \\ 0 & 0 & 0 & C_{44}^0 & 0 & 0 \\ 0 & 0 & 0 & 0 & C_{55}^0 & 0 \\ 0 & 0 & 0 & 0 & 0 & C_{66}^0 \end{bmatrix} \begin{bmatrix} \varepsilon_{11} \\ \varepsilon_{22} \\ \varepsilon_{33} \\ \varepsilon_{12} \\ \varepsilon_{23} \\ \varepsilon_{31} \end{bmatrix} \tag{12}$$

where, the nine undamaged elastic constants are defined by Equations (12)–(22).

$$C_{11}^0 = E_1(1 - v_{23}v_{32})\Delta \tag{13}$$

$$C_{22}^0 = E_2(1 - v_{13}v_{31})\Delta \tag{14}$$

$$C_{33}^0 = E_3(1 - v_{12}v_{21})\Delta \tag{15}$$

$$C_{12}^0 = E_1(v_{21} + v_{31}v_{23})\Delta \tag{16}$$

$$C_{23}^0 = E_2(v_{32} + v_{12}v_{31})\Delta \tag{17}$$

$$C_{13}^0 = E_1(v_{31} + v_{21}v_{32})\Delta \tag{18}$$

$$C_{44}^0 = G_{12} \tag{19}$$

$$C_{55}^0 = G_{23} \tag{20}$$

$$C_{66}^0 = G_{13} \tag{21}$$

$$\Delta = 1/(1 - v_{12}v_{21} - v_{23}v_{32} - v_{13}v_{31} - 2v_{21}v_{32}v_{13}) \tag{22}$$

Epoxy-reinforced composites with a carbon fiber volume content of 68%, the product of Sika® CarboDur® S, were employed in this study. The main elastic parameters of CFRP followed the product data sheet of Sika® CarboDur® S, as listed in Table 1.

Table 1. Elastic properties of CFRP laminate.

E_1	E_2 (E_3)	G_{12} (G_{13})	G_{23}	v_{12} (v_{13})	v_{23}
165 GPa (23,900 ksi)	11 GPa (1595 ksi)	5.3 GPa (769 ksi)	3.9 GPa (566 ksi)	0.26	0.5

Damage initiation, which is defined as the onset of material degradation at a point, is modeled on Hashin's 3D failure criterion. The damage initiation criteria consider the following four damage initiation mechanisms: fiber tension, fiber compression, matrix tension, and matrix compression, given by Equations (23)–(26).

Tensile fiber mode $\sigma_{11} \geq 0$:

$$\left(\frac{\sigma_{11}}{X_T}\right)^2 + \frac{\sigma_{12}^2 + \sigma_{13}^2}{S_{12}^2} = \begin{cases} \geq 1 & failure \\ < 1 & no\ failure \end{cases} \tag{23}$$

Compressive fiber mode $\sigma_{11} < 0$:

$$\left(\frac{\sigma_{11}}{X_C}\right)^2 = \begin{cases} \geq 1 & failure \\ < 1 & no\ failure \end{cases} \tag{24}$$

Tensile matrix mode $\sigma_{22} + \sigma_{23} \geq 0$:

$$\frac{(\sigma_{22} + \sigma_{33})^2}{Y_T^2} + \frac{\sigma_{23}^2 - \sigma_{22}\sigma_{23}}{S_{23}^2} + \frac{\sigma_{12}^2 + \sigma_{13}^2}{S_{12}^2} = \begin{cases} \geq 1 & failure \\ < 1 & no\ failure \end{cases} \tag{25}$$

Compressive matrix mode $\sigma_{22} + \sigma_{23} < 0$:

$$\left[\left(\frac{Y_C}{2S_{23}}\right)^2 - 1\right]\left(\frac{\sigma_{22} + \sigma_{33}}{Y_C}\right) + \frac{(\sigma_{22} + \sigma_{33})^2}{4S_{23}^2} + \frac{\sigma_{23}^2 - \sigma_{22}\sigma_{23}}{S_{23}^2} + \frac{\sigma_{12}^2 + \sigma_{13}^2}{S_{12}^2} = \begin{cases} \geq 1 & failure \\ < 1 & no\ failure \end{cases} \tag{26}$$

where X_T and X_C are the tensile and compressive failure strengths in the fiber direction, respectively; Y_T and Y_C are the tensile and compressive failure strengths in the y-direction (transverse to the fiber direction); and S_{ij} is the shear failure strength in the i-j plane.

The input parameters required for the Hashin damage model in ABAQUS are presented in Table 2. The CFRP laminate was meshed using an 8-node continuum shell element (SC8R) with reduced integration, hourglass control, and a finite membrane.

Table 2. Parameters of the Hashin damage model.

X_T	X_C	Y_T	Y_C	S_{12}	S_{23}
2800 MPa	1654 MPa	110 MPa	240 MPa	115 MPa	40 MPa
(406 ksi)	(240 ksi)	(16 ksi)	(35 ksi)	(16.6 ksi)	(5.8 ksi)

3.4. Epoxy Resin

The cohesive element COH3D8, which is widely used in modeling adhesives or bonded interfaces [23,24], was adopted to model the CFRP–concrete interface. The mechanical constitutive response of the cohesive element is defined in terms of the traction–separation law (model). The traction–separation model in ABAQUS consists of a linear elastic response followed by the initiation and evolution of damage, as seen in Figure 3. The elastic behavior in the pre-damage stage can be expressed by Equation (27), in terms of nominal stress and nominal strain.

$$t = \begin{bmatrix} t_n \\ t_s \\ t_t \end{bmatrix} = \begin{bmatrix} K_{nn} & 0 & 0 \\ 0 & K_{ss} & 0 \\ 0 & 0 & K_{tt} \end{bmatrix} \begin{bmatrix} \varepsilon_n \\ \varepsilon_s \\ \varepsilon_t \end{bmatrix} = K\varepsilon \tag{27}$$

t and ε are the nominal traction stress vector and nominal strain vector, respectively; K is the stiffness matrix. If the cohesive layer has thickness T_c, and the stiffness and density of the adhesive material are E_c and ρ_c, respectively, the stiffness and density of the interface is given by $K_c = (E_c/T_c)$ and $\bar{\rho}_c = \rho_c T_c$, respectively. This is because the default constitutive thickness for modeling the response in terms of traction versus separation is 1.0, rather than the actual thickness of the cohesive layer. Therefore, in ABAQUS, K_c and $\bar{\rho}_c$ should be input as the material stiffness and density, respectively.

The damage to the cohesive layer is assumed to be initiated if a quadratic interaction function involving the nominal stress ratios, as expressed in Equation (28), reaches unity. Once the damage initiation criterion is fulfilled, the stiffness of the cohesive material degrades according to the damage evolution law. In this study, the damage evolution was defined based on the energy conjunction with linear softening, as shown in Figure 3. The Benzeggagh–Kenane fracture criterion (Equation (29)) was used.

$$\left\{\frac{\langle t_n \rangle}{t_n^o}\right\}^2 + \left\{\frac{\langle t_s \rangle}{t_s^o}\right\}^2 + \left\{\frac{\langle t_t \rangle}{t_t^o}\right\}^2 = 1 \tag{28}$$

$$G_n^C + \left(G_s^C - G_n^C\right)\left(\frac{G_S}{G_T}\right)^{\eta} = G_C \tag{29}$$

t_n, t_s, and t_t, respectively, are the normal and shear stresses of the cohesive material; t_n^0, t_s^0, and t_t^0 are the peak value of the corresponding normal and shear stresses; G_n^C and G_s^C are the critical fracture energies required to induce failure in normal and shear directions, respectively; $G_S = G_s + G_t$; $G_T = G_n + G_S$; and η is the material parameter set to 1.5.

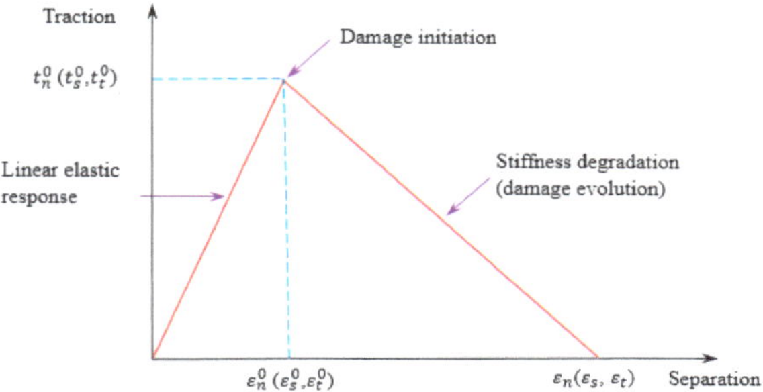

Figure 3. Typical traction–separation response [22].

The values of the main cohesive layer parameters, summarized in Table 3, were obtained from the product data sheet of Sikadur®-30, which is a high modulus, high strength, structural epoxy paste adhesive.

Table 3. Properties of cohesive material.

Elastic Properties					Damage Initiation			Damage Evolution		
E_n (GPa)	E_s (GPa)	E_t (GPa)	T_c (m)	ρ_c (kg/m³)	t_n^o (MPa)	t_s^o (MPa)	t_t^o (MPa)	G_n^C (J/m²)	G_s^C (J/m²)	G_t^C (J/m²)
4.5	11.7	11.7	10^{-3}	1650	24.8	16	16	355	280	280

4. Finite Element Modeling

4.1. Element Type, Interaction, Boundary Conditions

C3D8R was adopted considering the computational cost and shear locking. It has eight nodes and one integration point located at the center of the hexahedral brick element. A sensitivity study on the effects of different element types (C3D8, C3D8R, C3D20R) was performed, and discussed in the Sensitivity Study section. A 3D solid model was constructed for the strands in ABAQUS. The strands were meshed using a 6-node wedge element (C3D6). For mild reinforcement, the steel rebars were modeled using a 2-node truss element (T3D2). The CFRP laminate was meshed using an 8-node continuum shell element (SC8R) with reduced integration, hourglass control. The cohesive element COH3D8, which is widely used in modeling adhesives or bonded interfaces, was adopted to model the CFRP–concrete interface. The element types are summarized in Table 4.

Table 4. Summary in element type.

Part	Element Type	Description
Concrete, Steel Plate, rigid plate	C3D8R	• 8-node linear brick, reduced integration
Strands	C3D6	• 6-node linear triangular prism
Steel bars	T3D2	• 2-node linear displacement
CFRP	SC8R	• 8-node quadrilateral reduced integration
Epoxy	COH3D8	• 8-node cohesive element

The contact surface between the strands and concrete was simulated using a friction-governed model that included tangential and normal behaviors. As the transmission of shear and normal forces occurs across the contact interface, a frictional relationship exists between the contacting bodies. ABAQUS provides the classical isotropic Coulomb friction model, as shown in Figure 4, for modeling the tangential forces between the contact surfaces. The Coulomb friction model links the allowable frictional stress at the interface to the contact pressure between contacting bodies. According to the basic concept of the Coulomb friction model, the interface of two contacting bodies can carry shear stresses of a certain magnitude until they slide relative to each other, which is known as sticking. The critical transition point from sticking to sliding is determined by the coefficient of friction μ. Two methods are available in ABAQUS for defining the Coulomb friction model. The first method involves setting static and kinetic friction coefficients directly under the assumption that the friction coefficient decreases exponentially from the static status to the kinetic status. In the second method adopted in this study, the friction coefficient is defined as a function of the equivalent slip rate and contact pressure. Thus, the friction coefficient can be set to a non-negative value. Based on the research conducted by Arab et al. [16], a friction coefficient of 1.4 is suitable for simulating the interaction between fully bonded strands and surrounding concrete. For the debonding-damaged strands, the friction coefficient was set to zero, which meant that the strands could slide freely in the debonding region.

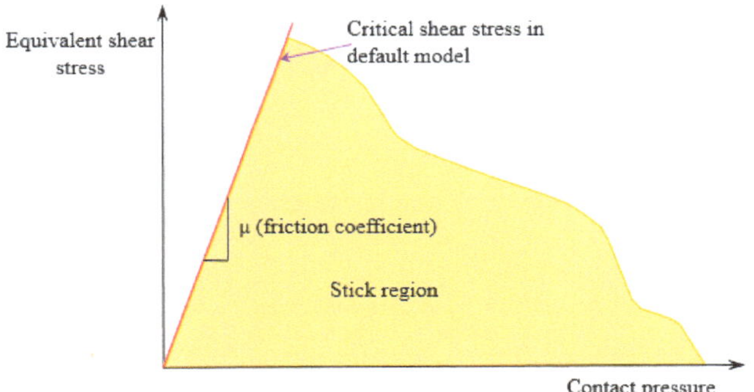

Figure 4. Basic Coulomb friction model [22].

Normal behavior was also defined to prevent strands from penetrating into surrounding concrete under overloads. The "hard" contact pressure–overclosure relationships, which can minimize the penetration of the slave surface into the master surface, were employed. In this case, the surfaces of the strands and concrete were identified as slave and master surfaces, respectively.

Owing to the doubly symmetric characteristics of the model, only a quarter of the girder-deck system was modeled considering the running time reduction, as shown in Figure 5. Using the z-direction symmetry, the girder-deck system was divided into two parts, with a plane parallel to the x-y plane. The displacement in the z-direction and rotation around the x- and y-directions were constrained by a roller support (U_3 = UR_1 = UR_2 = 0). Similarly, using x-direction symmetry, the girder-deck system was cut in a plane parallel to the y-z plane. The displacement in the x-direction and the rotation around the y- and z-directions were constrained by a roller support (U_1 = UR_2 = UR_3 = 0). In addition, the girder-deck system had a roller support that constrained the displacement in the y-direction (U_2 = 0) at the end.

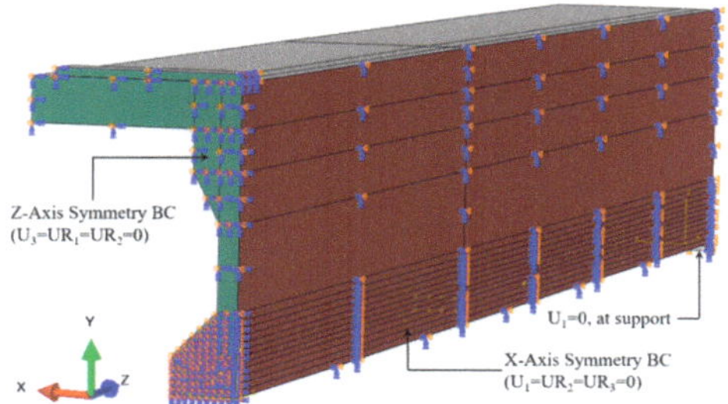

Figure 5. A quarter of a girder-deck system with BCs.

4.2. Verification of FEA Model

To validate the accuracy of the proposed FEA model, five experimental tests conducted by other researchers [25–27] were verified. These five tests included two fully bonded prestressed concrete beams, two prestressed concrete beams with debonding strands, and one CFRP-strengthened prestressed concrete beam.

ElSafty et al. [25] studied prestressed concrete girders that were laterally damaged by overheight vehicle collisions. The tested half-scale prestressed concrete girder was 20 ft long with a 4-inch-thick deck on top. Five low-relaxation-grade 270 seven-wire prestressing strands and three non-prestressed steel bars were used to reinforce the girder. The girder was loaded with four-point static loading. Ductile flexural failure was observed. The FEA model was verified using two T-section beams with debonding strands, as tested by Ozkul et al. [26]. Specimens 10 and 14 were selected for the ABAQUS model. Grade 270 (1862 MPa) seven-wire strands with a yield strength of 1689 MPa (245 ksi) were used as prestressing reinforcements. Both beams failed with concrete crushing. The two specimens tested by Meski and Harajli [27] were verified using the proposed FEA model. Beam UB2-H was used as a control beam. The other beam, UB2-H-F1 with fully debonding strands, was externally strengthened using a CFRP laminate. The thickness, modulus of elasticity, ultimate strength, and ultimate strain of the dry fibers were 0.37 mm, 230,000 MPa, 3800 MPa, and 1.7%, respectively. The corresponding values for the fiber–epoxy composites were 1.0 mm, 95,800 MPa, 986 MPa, and 1.0%. The dimensions of the CFRP laminates were 150 mm × 1.0 mm. The compressive strengths of UB2-H and UB2-H-F1 were 42 MPa (6.1 ksi) and 36 MPa (5.2 ksi), respectively. Both beams were prestressed with Grade 270 seven-wire strands. The failure mode of Specimen UB2-H was identified by concrete crushing, and the Specimen UB2-H-F1 failed with a combination of CFRP rupture and debonding.

The experimental data were compared with FEA results, and the results showed good agreement. The details of the experimental tests and the comparisons of load–deflection curves between experimental test and FEA can be found in the authors' other paper [28]. Table 5 and Figure 6 summarize the results of the comparison between the FEA and experimental tests. It was demonstrated that the FEA model could reasonably predict the flexural response of a CFRP/steel plate-strengthened prestressed concrete girder-deck system subjected to strand-debonding damage. Based on the correlation analysis in Figure 6, the root mean squared error (RMSE) is 2.46 kips.

Table 5. Comparison between exp. and FEA data.

Scholar	Test Label	$P_{cr\text{-}FEA}/P_{cr\text{-}Exp}$	$P_{u\text{-}FEA}/P_{u\text{-}Exp}$
ElSafty et al. (2012) [25]	---	1.14	1.05
Ozkul et al. (2008) [26]	No.10	1.04	1.1
	No.14	1.08	1.08
Meski and Harajli (2013) [27]	UB1-H	1	1.1
	UB1-H-F1	1.08	1.04

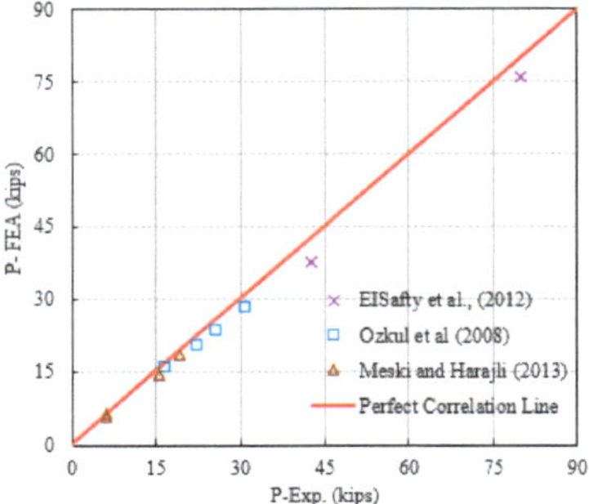

Figure 6. Comparison of the results obtained from exp. and FEA [25–27].

4.3. Sensitivity Study

The sensitivity of the numerical model was investigated regarding concrete parameters such as mesh size, element type, and dilation angle. The sensitivity of the model to the coefficient of friction between the strands and the surrounding concrete was also evaluated. The girder-deck system tested by EISafty et al. [25] was selected as the control specimen for the sensitivity analysis. The load–deflection curve, stiffness, and ultimate strength obtained from the FEA using different values of the above-mentioned parameters were compared with the data obtained from the experimental test. Figure 7 shows the results of the sensitivity study.

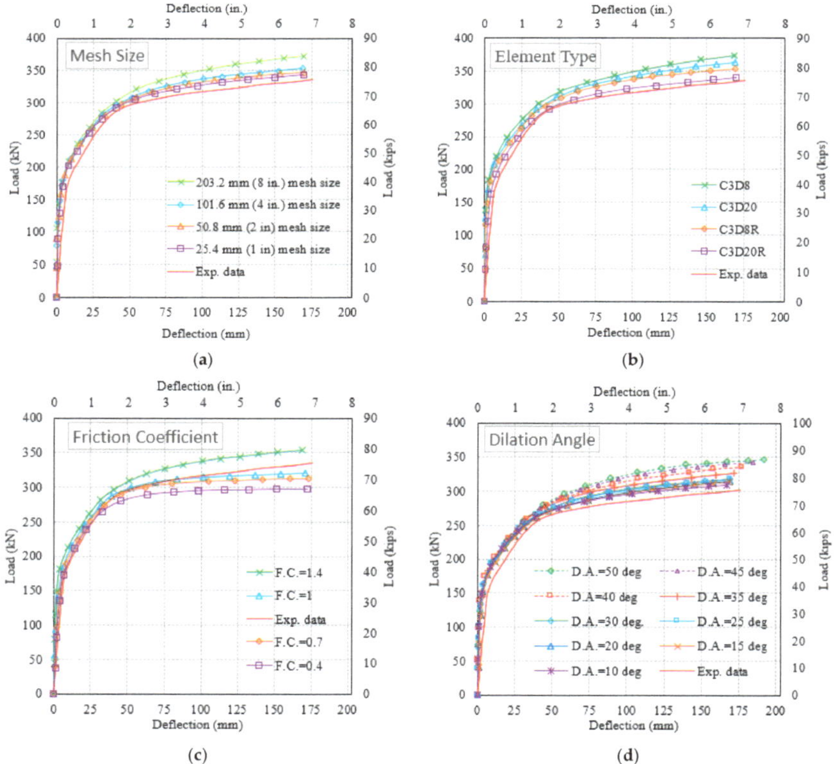

Figure 7. Sensitivity results: (**a**) mesh size; (**b**) element type; (**c**) friction coefficient; (**d**) dilation angle.

As shown in Figure 7a, a smaller mesh size produced a load–deflection curve closer to that of the experimental test. For various mesh sizes, slight differences were observed in the peak loads and deflections. The range of the differences was within the expected margin of error for numerical simulations. However, modeling with 25.4 mm (1 in) and 50.8 mm (2 in) mesh required more computational time, compared to modeling with 101.6 mm (4 in) and 203.2 mm (8 in) mesh. Therefore, considering the time consumption and numerical accuracy, a mesh size of 101.6 mm (4 in) for concrete was employed in this study.

To investigate the effect of the element type on the simulation results, the concrete was meshed individually using four element types: standard 8-node brick element with full integration (C3D8), 8-node brick element with reduced integration (C3D8R), 20-node brick element with full integration (C3D20), and 20-node brick element with reduced integration (C3D20R). As shown in Figure 7b, the fully integrated elements (C3D8 and C3D20) were stiffer to bend compared with elements with reduced integration (C3D8R and C3D20R). This phenomenon is produced by shear locking, which occurs when shear strain develops owing to the inability of the element edges to bend. Consequently, the elements are too stiff to be used in bending-dominant problems. However, elements C3D8R and C3D20R are not sensitive to shear locking because they have fewer Gaussian integration points, and these points are closer to the element's boundaries. The load–deflection curve of C3D20R was closer to the experimental curve than that of C3D8R. This is the result of the quadratic interpolation of C3D20R. Although a quadratic interpolation element such as C3D20R can prevent shear locking and provide a more accurate bending behavior, it induces a much higher computational cost. Therefore, C3D8R was the best choice for the concrete elements in this study.

The tangential behavior at the contact surface between the bonded strands and concrete was identified by the friction coefficient. In previous studies [29,30], the friction coefficient was set to 0.4. Arab et al. [16] revealed that the friction coefficient ranges from 0.7 to 1.4 for prestressed concrete members. To study the sensitivity to the friction coefficient, FEA modeling was performed with the friction coefficient values of 0.4, 0.7, 1.0, and 1.4. As shown in Figure 7c, the friction coefficient has a negligible impact on the elastic behavior of the PC girder. The results obtained from the models with friction coefficients of 1 and 1.4 show a good match with the experimental data. For the models with friction coefficients of 0.4 and 0.7, the numerical results indicate a lower ultimate strength than the experimental values. With more simulations and verifications against other experimental data, a value of 1.4 was adopted as the friction coefficient in this study.

The dilation angle, which refers to the deviation of a concrete element subjected to shear stress, was used to determine the failure surface. One of the input parameters in ABAQUS is to define the plastic flow potential and how it governs the volumetric strain in the plastic deformation stage. The dilation angle should be greater than zero and less than 56°. Many researchers [16,30,31] used dilation angles ranging from 25°–50°, whereas Chen and Graybeal [32] set the dilation angle to 15°. To investigate the sensitivity of the dilation angle to the flexural behavior of a concrete beam, a series of FEA were performed for ten values of the dilation angle: 10°, 15°, 20°, 25°, 30°, 35°, 40°, 45°, 50°, and 55°. As shown in Figure 7d, the dilation angle did not affect the flexural response during the elastic stage. However, the variation in the dilation angle significantly affected the post-cracking behavior. In the ultimate state, the effect of the dilation angle must be considered when it is greater than 35°. An increase in the dilation angle above 35° resulted in a higher ultimate load and corresponding deflection (Figure 8), while the computational cost decreased. No significant change in the flexural response was observed when the dilation angle was less than 35°. This variation in the effects occurs because the plastic flow potential is governed by the dilation angle and eccentricity [33]. Thus, considering the accuracy and computing time, the dilation angle was set to 30° in this study.

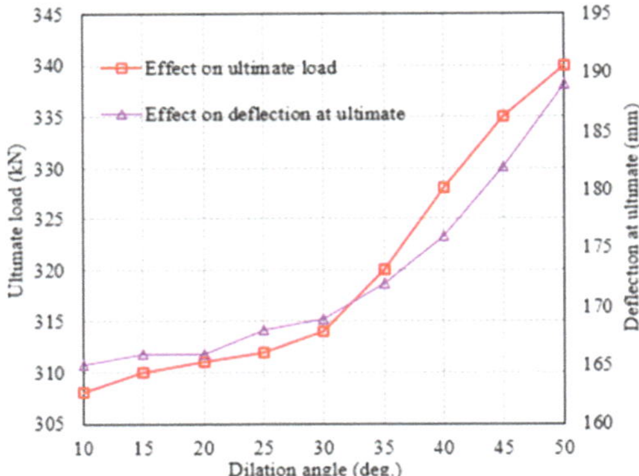

Figure 8. Effect of variation of dilation angle on ultimate load and deflection.

5. Parametric Study

5.1. Variables

Based on the FEA models presented in the previous sections, a parametric study was performed to investigate the effects of different parameters on the flexural strength of girder-deck systems. Four parameters were studied: span-to-depth ratio (10, 15, 20), debonding level (0, 20% L, 40% L, 60% L), type of strengthening material (CFRP laminate and steel plate), and amount of strengthening material (t_{CFRP} = 0, 1, 2, 3, 4, 5 mm for CFRP; t_{SP} = 0, 1, 2, 3, 6, 8, 12 mm for steel plate). The variables are summarized in Figure 9. The effects were divided into two categories: (1) un-strengthened specimens and (2) strengthened specimens. Additionally, the flexural capacity of each specimen (M_{db}) was compared with that of a fully bonded specimen (M_n) obtained from the AASHTO LRFD Bridge Design Specifications.

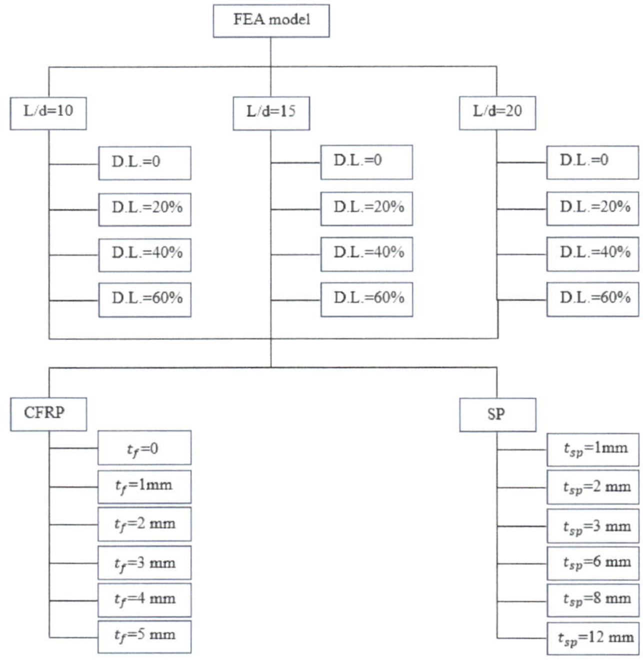

Figure 9. All cases studied by FEA model.

5.2. Modeling Process

The process to model the strengthened girder-deck system with bonded and/or debonding strands is as follows:

(1) Building the parts containing AASHTO Type III girder, concrete deck, strands, CFRP laminate, cohesive layer, steel plate, rigid support and rigid plate. (2) Applying the prestress to each strand in longitudinal direction before casting concrete by defining "PREDEFINED FIELD". In this step, the bond properties were invalid. (3) Casting the AASHTO Type III girder and releasing the strands by applying the bond properties between the fully bonded strands and concrete. The prestressing force was transferred from the strands to the concrete using the defined bond model, and the elastic shortening was considered using ABAQUS. In this stage, only the self-weight of the girder was applied. (4) Casting the concrete deck on top of girder by activating the elements of the concrete deck in the option of "MODEL CHANGE, REACTIVATE". (5) Applying the interface properties in the debonding region by switching the bond properties to debonding properties through the options of "CHANGE FRICTION" and "FIELD VARIABLE". (6) Attaching the cohesive layer and strengthening materials (CFRP or steel plate). (7) Applying the overloads until failure, which was identified by concrete crushing, strand rupture, and CFRP failure. The above steps are shown in the flowchart in Figure 10.

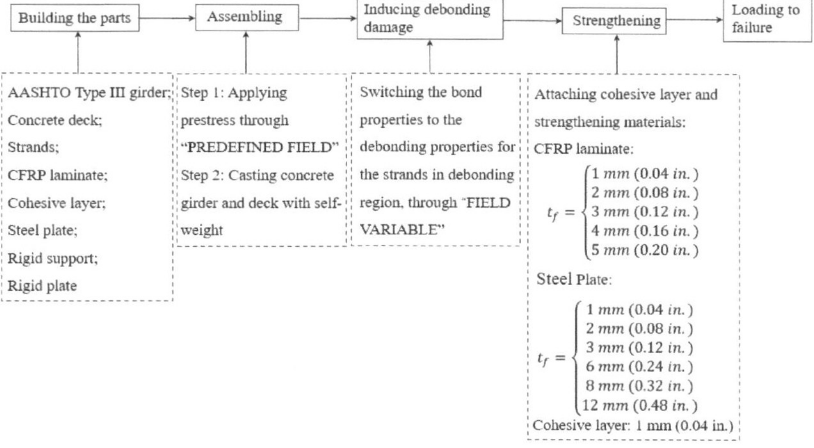

Figure 10. Modeling process.

6. Results and Discussion

6.1. Girder-Deck Systems without Strengthening

6.1.1. Effect of Debonding Level

The effects of the debonding level (λ) of un-strengthened girder-deck systems on flexural responses, including load–deflection behaviors and moment capacities, were investigated and evaluated. Figure 11 shows the load–deflection curves obtained by numerical simulations under different span-to-depth ratios (L/d = 10, 15, and 20). Under the same span-to-depth ratio, the specimen with a relatively longer debonding length exhibited a lower ultimate load and higher ductility. The degree of impact of the ultimate load decreased when the span-to-depth ratio increased from 10 to 20. For both span-to-depth ratios of 10 and 15, the debonding length resulted in higher ductility than that of the control specimen with fully bonded strands. However, when the span-to-depth ratio reached 20, the ductility achieved by debonding the specimens did not exceed the ductility obtained by the control specimen. Furthermore, the stiffness was not affected by the debonding level under the same span-to-depth ratio.

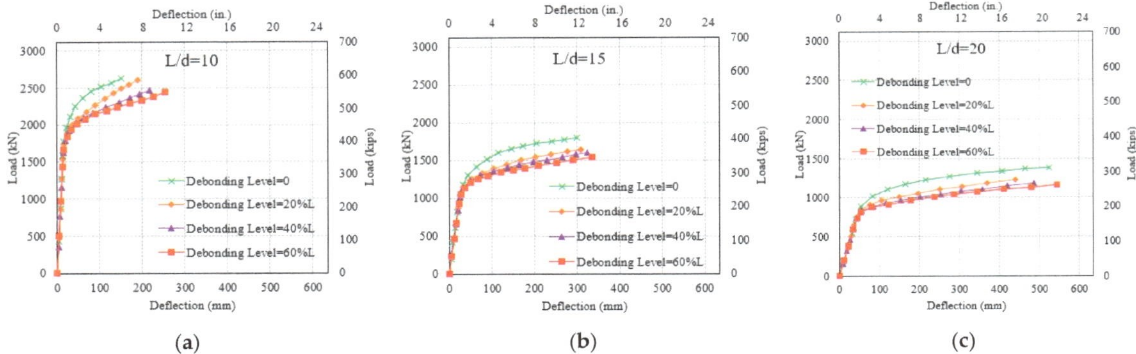

Figure 11. Load–deflection curves under the different span-to-depth ratios: (**a**) L/d = 10; (**b**) L/d = 15; (**c**) L/d = 20.

In addition to the load–deflection behaviors, the ultimate moment of the un-strengthened specimens was also evaluated. Figure 12 describes the effects of the debonding level on the ratio of M_{db} to M_n. A higher debonding level induced a lower value of Mdb/Mn. When the debonding level changed from 20% to 40%, the ultimate moment decreased by only 1%. However, when the debonding level increased from 40% to 60%, the ultimate moment decreased by 4.5%.

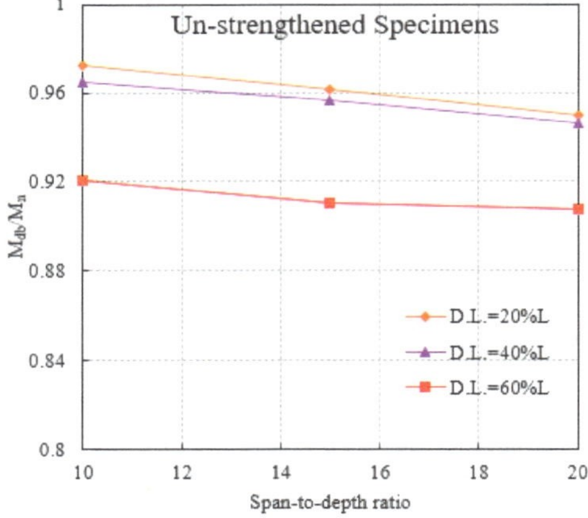

Figure 12. Effect of debonding level on M_{db}/M_n.

6.1.2. Effect of Span-to-Depth Ratio (L/d)

The effects of the span-to-depth ratio (L/d) on the load–deflection behaviors and the ultimate moment of the un-strengthened specimens with debonding strands were investigated. The debonding level of the studied specimens was 40% L and the prestressing reinforcement ratio (%) was 0.106. The load–deflection curves, shown in Figure 13a, indicate that a higher span-to-depth ratio induces a lower ultimate load and stiffness, whereas the ductility increases with an increase in the span-to-depth ratio. Figure 13b presents the influence on the ultimate moment of the un-strengthened girder deck systems with various debonding levels. The results revealed that the ultimate moment decreased with increasing span-to-depth ratio. At the same debonding level, the maximum reduction in M_{db}/M_n was 2.3% when the span-to-depth ratio increased from 10 to 20. The maximum reduction occurs at a debonding level of 20%. It is also noted that, with an increase in the debonding level, the effect of the span-to-depth ratio on the ultimate moment reduction slightly increases. At a debonding level of 60%, the reduction in the ultimate moment was as low as 1.2%. Therefore, the effect of the span-to-depth ratio on the ultimate moment of the un-strengthened girder-deck system was negligible and can be ignored.

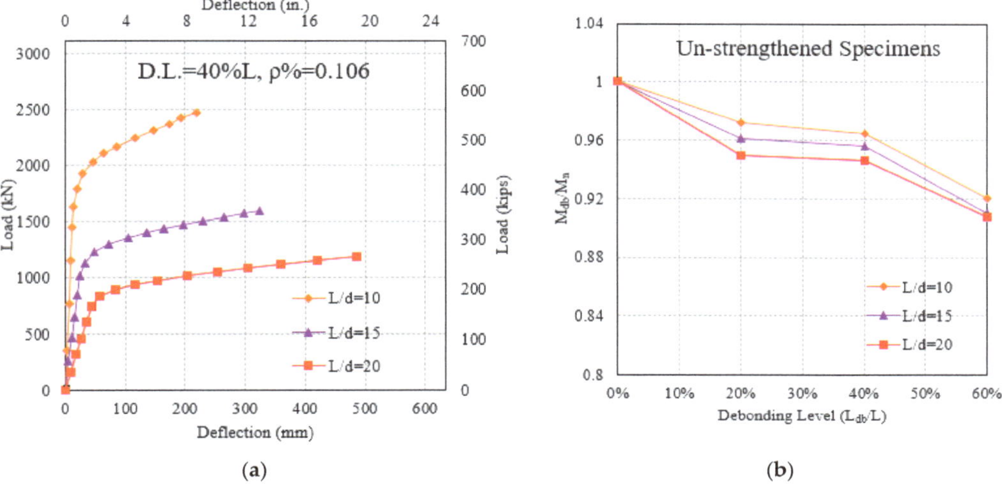

Figure 13. Effects of L/d for un-strengthened specimens: (**a**) on load–deflection; (**b**) on ultimate flexural capacity.

6.2. Girder-Deck Systems with Strengthening
6.2.1. Effect of Debonding Level

The effects of the debonding level on load–deflection behavior and the ultimate moment capacity of strengthened girder-deck systems with debonding strands were studied using two strengthening materials: CFRP laminate and SP.

Figure 14 presents the load-deflection curves of the debonding specimens strengthened using a 3 mm (0.12 in) thick CFRP laminate and SP, as an example. The specimens had a span-to-depth ratio of 15. Although the strengthening materials were different, the effect of the debonding level on the load-deflection behaviors of the strengthened specimens was similar. This indicates that a higher debonding level leads to a reduction in the ultimate load. The ultimate load decreased by approximately 5% for every 20% increment in the debonding level. The ductility increased slightly with the increase in the debonding level. The stiffness of the strengthened specimens was not affected by the debonding level, which was similar to the un-strengthened specimens.

Figure 15 shows the impact of the debonding level on the values of M_{db}/M_n for specimens with different strengthening materials: CFRP laminate and SP, respectively. The specimens had a span-to-depth ratio of 15, and were strengthened separately using CFRP laminate and SP. The thicknesses of the CFRP laminate and SP varied from 1 mm (0.04 in) to 5 mm (0.20 in), and from 1 mm (0.04 in) to 12 mm (0.48 in), respectively.

The CFRP-strengthened group, as shown in Figure 15a, indicates that the debonding level affected the ultimate moment capacity. As the debonding level increased, the ultimate moment capacity decreased considerably. However, the magnitude of this effect decreased when the debonding level increased. When the debonding level was less than 20% L, the value of M_{db}/M_n was reduced by up to 12%. When the debonding level exceeded 40% L, the maximum reduction was only 1.3%. For the specimens with thinner CFRP laminates, the debonding level significantly affected the ultimate moment capacity.

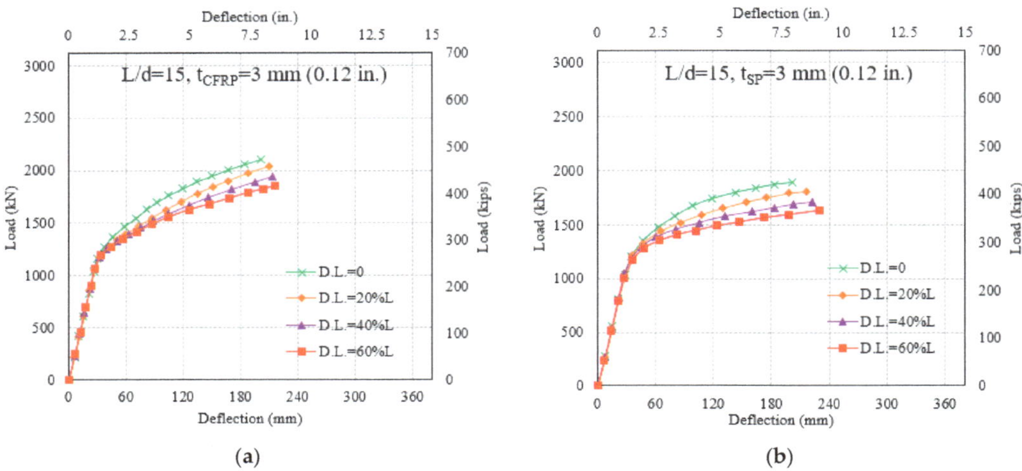

Figure 14. Load–deflection curves with various debonding levels: (**a**) obtained from CFRP-strengthened specimens; (**b**) obtained from SP-strengthened specimens.

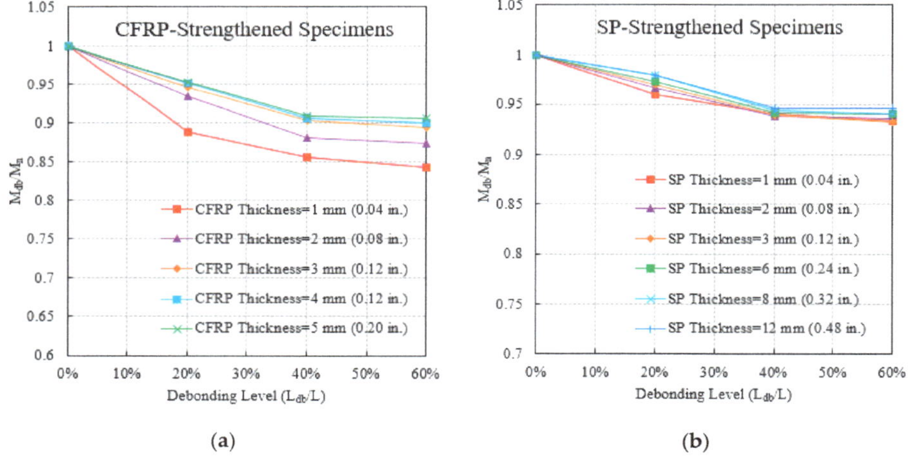

Figure 15. Effects of the debonding level on the ultimate flexural capacity: (**a**) for CFRP-strengthened specimens; (**b**) for SP-strengthened specimens.

In the SP-strengthened group, shown in Figure 15b, a higher debonding level resulted in a lower ultimate moment capacity. The percentage reduction ranged from 5.3% to 6.2% when the debonding level increased from 0 to 40%. When the debonding level exceeded 40%, the ultimate moment capacity was only slightly affected; the effect at this stage was less than 1% and can be neglected. SP thickness did not affect the degree of reduction when the debonding level increased.

6.2.2. Effect of Span-to-Depth Ratio (L/d)

The influence of the span-to-depth ratio of strengthened girder-deck systems with debonding strands on load-deflection behavior and ultimate moment capacity was studied. Specimens with a debonding level of 40% L were investigated for studying the load–deflection behavior. The specimens were separately strengthened using a 3 mm (0.12 in) CFRP laminate and SP. Figure 16a presents the load-deflection curves of the strengthened specimens with different span-to-depth ratios (L/d = 10, 15, and 20). A higher span-to-depth ratio can lower the stiffness and ultimate load for the same type and amount of strengthening material, whereas the ductility can be improved slightly.

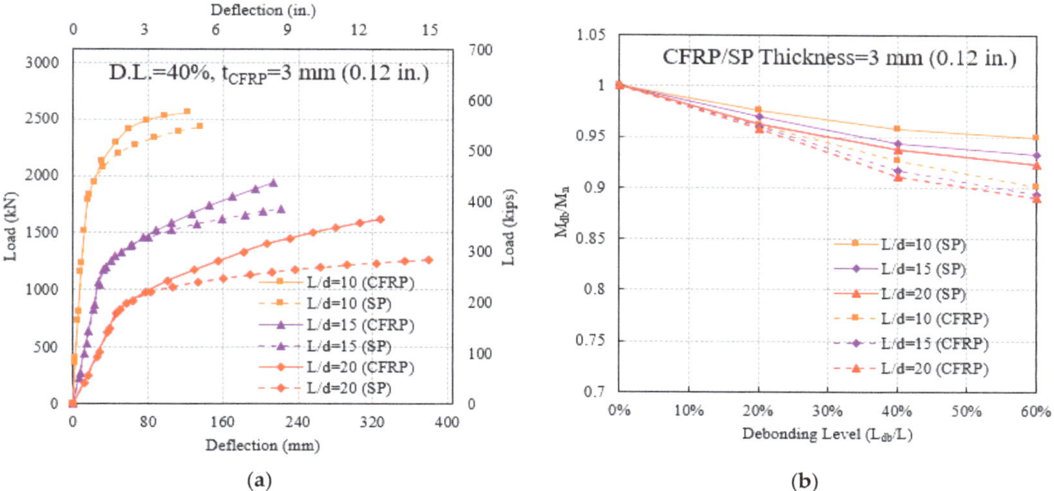

Figure 16. Effects of L/d for strengthened specimens: (**a**) on load–deflection; (**b**) on ultimate flexural capacity.

Figure 16b shows the effect of the span-to-depth ratio on the value of M_{db}/M_n, which is the ratio of the ultimate moment of the debonding specimen to that of the fully bonded (no debonding) specimen. Girder-deck systems with different debonding levels (0, 20% L, 40% L, and 60% L) and span-to-depth ratios (L/d = 10, 15, and 20) were strengthened using a 3 mm (0.12 in) CFRP laminate or SP. Compared to the CFRP-strengthened specimens, the span-to-depth ratio had a relatively larger effect on the ultimate moment of the SP-strengthened specimens when the span-to-depth ratio varied from 10 to 20. In particular, the maximum reduction in the ultimate moment at the same debonding level was 3% in the SP-strengthened group when the span-to-depth ratio increased from 10 to 20. In contrast, the ultimate moment of the CFRP-strengthened specimens decreased by only 1%. Although the span-to-depth ratio had a larger effect on the SP-strengthened specimens than on the CFRP-strengthened specimens, the 3% reduction was considered negligible. In addition, the effect of the span-to-depth ratio on the ultimate moment increased with a higher debonding level. For instance, the effect of the span-to-depth ratio on the ultimate moment of the specimens with a debonding level of less than 20% L was very small. With an increase in the debonding level, the span-to-depth ratio effect increased.

6.2.3. Effect of the Type of Strengthening Material

The effects of the type of strengthening material (CFRP laminate and SP) were investigated. Figure 17 shows the enhancements in the ultimate moment of the specimens with different debonding levels (0, 20% L, 40% L, and 60% L). Half of the specimens were strengthened using a CFRP laminate with a thickness of 1 mm (0.04 in), 2 mm (0.08 in), and 3 mm (0.12 in). For comparison, the other half of the specimens was strengthened using an equal amount of SP. The CFRP- and SP-strengthened systems had the same width of 457.2 mm (18 in). For the same amount of strengthening material, the specimens strengthened using CFRP laminate can achieve much higher enhancement percentages in the ultimate moment capacity than those strengthened using SP at a certain debonding level. For instance, the ultimate moment capacity of the specimens with a 60% L debonding level improved from 15.7% to 82.5% when the thickness of the CFRP laminate varied from

1 mm (0.04 in) to 3 mm (0.12 in). However, the specimens strengthened by SP achieved an enhancement of only up to 10%. SP has a relatively lower enhancement of the ultimate moment capacity with respect to the CFRP laminate. Therefore, in this study, SP with thicker dimensions, such as 6 mm (0.24 in), 8 mm (0.24 in), and 12 mm (0.24 in), was selected. Additionally, as shown in Figure 17, the enhancement percentage increased linearly with every 1 mm (0.04 in) increment in thickness for both the CFRP laminate and SP.

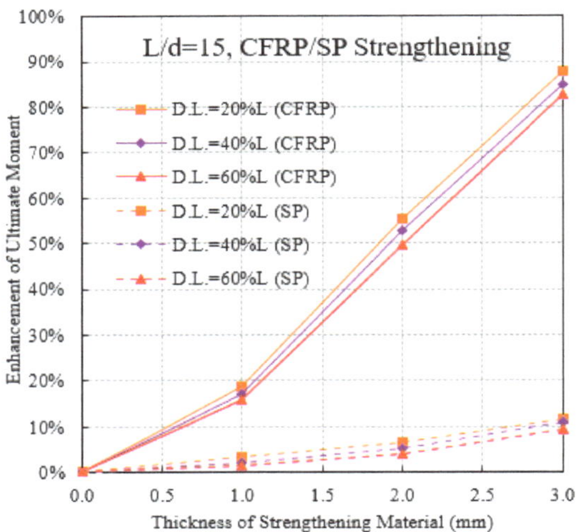

Figure 17. Enhancements on ultimate moment.

6.2.4. Effect of Amount of Strengthening Material

The effects of CFRP and/or SP amount on the ultimate load and moment capacity of the strengthened girder-deck systems with debonding strands was evaluated. Figure 18 shows the load–deflection curves of separately strengthened specimens with the span-to-depth ratio of 15 and debonding level of 60% L, and span-to-depth of 20 and deboning level of 40%, as an example in our case study. These specimens were strengthened using CFRP with thicknesses of 1 mm (0.04 in), 2 mm (0.08 in), 3 mm (0.12 in), 4 mm (0.16 in), and 5 mm (0.20 in) and using SP with thicknesses of 1 mm (0.04 in), 2 mm (0.08 in), 3 mm (0.12 in), 6 mm (0.16 in), 8 mm (0.20 in), and 12 mm (0.48 in). The specimens with higher amounts of CFRP and SP achieved a higher ultimate load than the reference specimen without any strengthening. The ultimate loads of the CFRP- and SP-strengthened specimens increased by approximately 15% and 10%, respectively. Ductility was affected by the strengthening amount. The higher the strengthening amount, the lower the ductility. The stiffness was not affected by the amount of strengthening.

The relationship between the enhancement of the ultimate moment capacity and the thickness of the strengthening material, shown in Figure 19, can be expressed using linear regression equations with respect to CFRP- and SP-strengthened specimens. The coefficients of determination (R^2) were 0.9928 and 0.9893 for the CFRP group, 0.9968 and 0.9941 for the SP group.

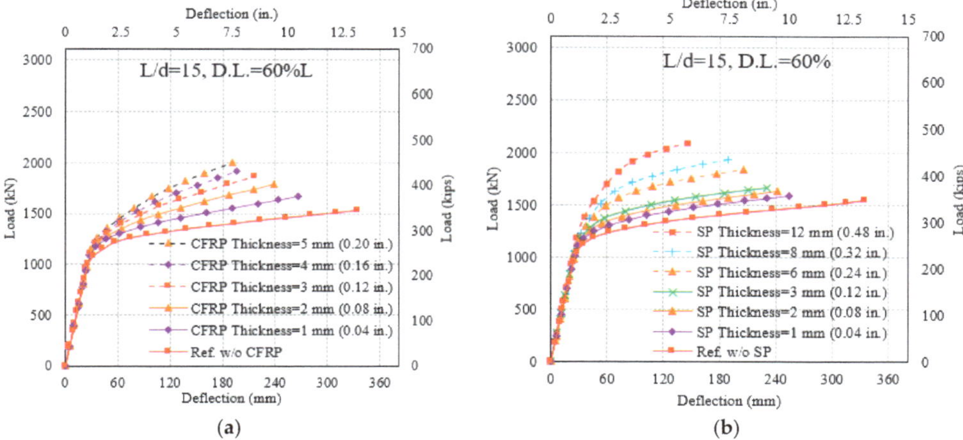

Figure 18. Load–deflection curves with various strengthening amounts: (**a**) obtained from CFRP-strengthened specimens; (**b**) obtained from SP-strengthened specimens.

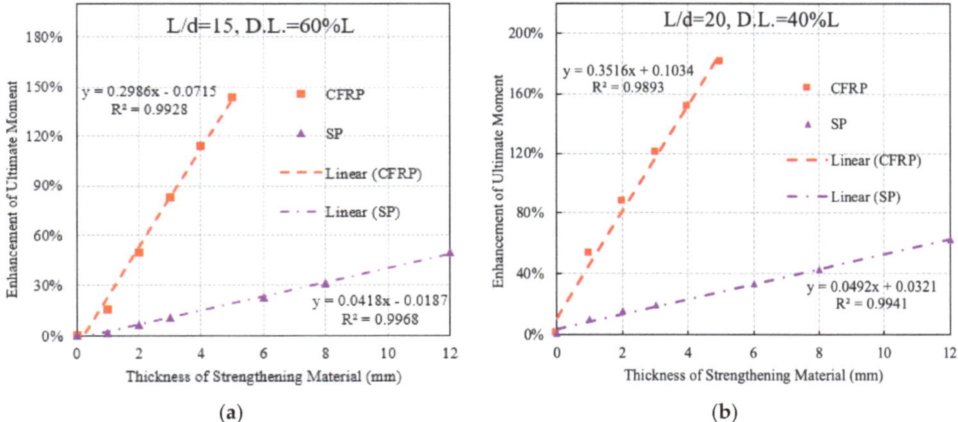

Figure 19. Relationships between ultimate moment enhancement and strengthening material thickness: (**a**) L/d = 15, D.L. = 60%; (**b**) L/d = 20, D.L. = 40%.

7. Summary and Conclusions

This paper presented an in-depth investigation of the flexural behaviors of CFRP- or SP-strengthened AASHTO Type III girder-deck systems with partially debonding strands based on the numerical method. A detailed FE model was proposed to predict the flexural responses of a PC member. The FE model was validated against the data from available experimental tests. The FE results were in good agreement with the experimental data by others. The FEA model was utilized to evaluate 156 strengthened and/or un-strengthened girder-deck systems with partially debonding strands. The 156 specimens included the following parameters: span-to-depth ratio, debonding level, type of strengthening material, and amount of strengthening material. A parametric study was conducted to investigate the effects of these parameters on the flexural response of girder-deck systems.

Based on the parametric study, the following conclusions could be drawn:

- Girder-deck systems without strengthening

  (1) For the same span-to-depth ratio, the specimen with a longer debonding length achieved a lower ultimate load and higher ductility.

(2) A higher debonding level induced a lower value of M_{db}/M_n. When the debonding level increased from 40% L to 60% L, the ultimate moment decreased by 4.5%.

(3) A higher span-to-depth ratio lowered the ultimate load and stiffness, whereas the ductility increased.

(4) The ultimate moment decreased with an increasing span-to-depth ratio. With an increase in the debonding level, the effect of the span-to-depth ratio on the ultimate moment reduction increased slightly.

- Girder-deck systems with strengthening

(1) A higher debonding level led to a reduction in the ultimate load of the strengthened specimens. The ductility increased slightly with an increase in the debonding level. The stiffness of the strengthened specimens was not affected by the debonding level.

(2) With an increase in the debonding level, the ultimate moment capacity of the CFRP-strengthened specimens decreased considerably. The degree of this effect reduces when the debonding level increases. For specimens with thinner CFRP laminates, the debonding level significantly affected the ultimate moment capacity.

(3) A higher debonding level resulted in a lower ultimate moment capacity of the SP-strengthened specimens. The percentage reduction ranged from 5.3% to 6.2% when the debonding level increased from 0 to 40%. Above 40%, the ultimate moment capacity was affected at a negligible 1% level.

(4) A higher span-to-depth ratio lowered the stiffness and ultimate load under the same type and amount of strengthening material, whereas the ductility improved slightly.

(5) Compared with the CFRP-strengthened specimens, the span-to-depth ratio had a larger effect on the ultimate moment of the SP-strengthened specimens when the span-to-depth ratio varied from 10 to 20. In addition, the effect of the span-to-depth ratio on the ultimate moment increased with a higher debonding level.

(6) For the same amount of strengthening material, the specimens strengthened by the CFRP laminate achieved a much higher increase in the ultimate moment capacity than those strengthened by SP at a certain debonding level.

(7) The specimens with higher amounts of CFRP or SP achieved a higher ultimate load than the reference specimen without any strengthening. The ultimate loads of the CFRP- and SP-strengthened specimens increased by approximately 15% and 10%, respectively. The higher the strengthening amount, the lower the ductility.

(8) The relationship between the enhancement of the ultimate moment capacity and the thickness of the strengthening material can be expressed using linear regression equations with respect to the CFRP- and SP-strengthened specimens.

Author Contributions: Conceptualization, R.A.; Methodology, H.N.; Software, H.N.; Validation, H.N.; Formal analysis, H.N.; Investigation, H.N.; Resources, R.A.; Data curation, H.N.; Writing—original draft preparation, H.N.; Writing—review and editing, H.N. and R.A.; Visualization, H.N.; Supervision, R.A.; Project administration, R.A. All authors have read and agreed to the published version of the manuscript.

Funding: This research received the support of funding and computational facilities provided by the Department of Civil and Environmental Engineering at Syracuse University.

Data Availability Statement: The data used to support the findings of this study can be accessed by emailing to the corresponding author.

Acknowledgments: The authors are truly grateful to the support provided by the Department of Civil and Environmental Engineering at Syracuse University.

Conflicts of Interest: The authors declare no conflicts of interest.

Nomenclature

The symbols used in this paper are as follows:

A_1	constant parameter
C	undamaged elastic constant
d	depth from extreme top fiber of concrete to the centroid of strands
E_c	modulus of elasticity of concrete
E_{ps}	modulus of elasticity of strand
E_s	modulus of elasticity of mild steel
f_c	concrete stress
f_c'	release concrete strength
f_0	reference value
f_{ps}	stress in strand
f_{pu}	ultimate stress of strand
f_s	stress in mild steel
f_t'	tensile strength of concrete
f_y	yield strength of mild steel
G_n^C	critical fracture energy required to induce failure in normal
G_s^C	critical fracture energies required to induce failure in shear directions
K	stiffness matrix
K_c	invariant stress ratio
L	Total span of specimen
M_n	flexural capacity of specimens
M_{db}	flexural capacity of specimens with debonding strands
P_{cr-Exp}	cracking load from experimental test
P_{cr-FEA}	cracking load from FEA model
P_{u-Exp}	ultimate load from experimental test
P_{u-FEA}	ultimate load from FEA model
t_{CFRP}	the thickness of CFRP
t_n	normal stress of the cohesive material
t_n^o	peak value of the normal stresses
t_{sp}	the thickness of steel plate
t_t	shear stresses of the cohesive material
t_t^o	peak value of the shear stresses
w_c	the density of concrete
w_0	reference density of concrete
X_C	compressive failure strength in fiber direction
X_T	tensile failure strength in fiber direction
Y_T	tensile failure strength in direction Y (transverse to fiber direction)
Y_C	compressive failure strength in direction Y (transverse to fiber direction)
η	viscosity parameter representing the relaxation time of the viscoplastic system
μ	friction coefficient
σ	reinforcement stress
σ_c	stress of concrete
λ	debonding level
ε_c	concrete strain
ε_{in}	inelastic strain
ε_{ps}	the strain in strand
ε_s	strain in mild steel
β_1	a key parameter that determines slopes of nonlinear branches of concrete constitutive model
ζ	a parameter which is used to simplify Equation (4)
ψ	dilation angle
S_{ij}	shear failure strength in i-j plane
ρ_c	density of adhesive material
Δ	a parameter which used to simplify the equation of C

References

1. Harries, K.A.; Kasan, J.; Aktas, J. *No. FHWA-PA-2009-008-PIT 006*; Repair Methods for Prestressed Girder Bridges; Department of Transportation: Harrisburg, PA, USA, 2009.
2. Dang, C.N.; Murray, C.D.; Floyd, R.W.; Hale, W.M.; Martí-Vargas, J.R. Analysis of bond stress distribution for prestressing strand by Standard Test for Strand Bond. *Eng. Struct.* **2014**, *72*, 152–159. [CrossRef]
3. Sim, J.; Park, C. Characteristics of basalt fiber as a strengthening material for concrete structures. *Compos. Part B Eng.* **2005**, *36*, 504–512. [CrossRef]
4. Haskett, M.; Oehlers, D.J.; Ali, M.M. Local and global bond characteristics of steel reinforcing bars. *Eng. Struct.* **2008**, *30*, 376–383. [CrossRef]
5. Yu, Q.; Bazant, Z.P.; Wendner, R. Improved algorithm for efficient and realistic creep analysis of large creep-sensitive concrete structures. *ACI Struct. J.* **2012**, *109*, 665. [CrossRef]
6. Xu, L.; Hai, T.K.; King, L.C. Bond stress-slip prediction under pullout and dowel action in reinforced concrete joints. *ACI Struct. J.* **2014**, *111*, 977. [CrossRef]
7. Zhang, X.; Wang, L.; Zhang, J.; Liu, Y. Model for flexural strength calculation of corroded RC beams considering bond-slip behavior. *J. Eng. Mech.* **2016**, *142*, 04016038. [CrossRef]
8. Wang, L. *Strand Corrosion in Prestressed Concrete Structures*, 1st ed.; Springer Nature: Singapore, 2023; pp. 105–135. [CrossRef]
9. Code, M. *FIB Model Code for Concrete Structures*, 1st ed.; Ernst & Sohn: Iselin, NJ, USA, 2013.
10. Mohandoss, P.; Pillai, R.G.; Gettu, R. Determining bond strength of seven-wire strands in prestressed concrete. *Structures* **2021**, *33*, 2413–2423. [CrossRef]
11. Wang, L.; Zhang, X.; Zhang, J.; Ma, Y.; Xiang, Y.; Liu, Y. Effect of insufficient grouting and strand corrosion on flexural behavior of PC beams. *Constr. Build. Mater.* **2014**, *53*, 213–224. [CrossRef]
12. Mousa, M.I.; Mahdy, M.G.; Abdel-Reheem, A.H.; Yehia, A.Z. Mechanical properties of self-curing concrete (SCUC). *HBRC J.* **2015**, *11*, 311–320. [CrossRef]
13. Garg, S.; Matsagar, V.; Marburg, S. Nonlinear finite element analyses of prestressed concrete beams strengthened with CFRP laminate. *Struct. Eng. Int.* **2023**, *33*, 258–264. [CrossRef]
14. Qapo, M.; Dirar, S.; Yang, J.; Elshafie, M.Z. Nonlinear finite element modelling and parametric study of CFRP shear-strengthened prestressed concrete girders. *Constr. Build. Mater.* **2015**, *76*, 245–255. [CrossRef]
15. Sato, Y.; Vecchio, F.J. Tension stiffening and crack formation in reinforced concrete members with fiber-reinforced polymer sheets. *J. Struct. Eng.* **2003**, *129*, 717–724. [CrossRef]
16. Arab, A.A.; Badie, S.S.; Manzari, M.T. A methodological approach for finite element modeling of pretensioned concrete members at the release of pretensioning. *Eng. Struct.* **2011**, *33*, 1918–1929. [CrossRef]
17. Kang, T.H.K.; Huang, Y.; Shin, M.; Lee, J.D.; Cho, A.S. Experimental and numerical assessment of bonded and unbonded post-tensioned concrete members. *ACI Struct. J.* **2015**, *112*, 735–748. [CrossRef]
18. Kang, T.H.K.; Huang, Y. Nonlinear finite element analyses of unbonded post-tensioned slab-column connections. *PTI J.* **2012**, *8*, 4–19.
19. American Association of State Highway and Transportation Officials (AASHTO LRFD). *AASHTO LRFD Bridge Design Specifications*, 9th ed.; AASHTO LRFD: Washington, DC, USA, 2020.
20. Precast/Prestressed Concrete Institute (PCI). *Precast and Prestressed Concrete Handbook*, 8th ed.; PCI: Chicago, IL, USA, 2020.
21. Yang, K.H.; Mun, J.H.; Cho, M.S.; Kang, T.H. Stress-strain model for various unconfined concretes in compression. *ACI Struct. J.* **2014**, *111*, 819. [CrossRef]
22. *ABAQUS User's Manual, ABAQUS 6.17*; Dassault Systèmes Simulia Corp: Providence, RI, USA, 2016.
23. Jiang, J.; Zou, Y.; Yang, J.; Zhou, J.; Zhang, Z.; Huang, Z. Study on bending performance of epoxy adhesive prefabricated UHPC-steel composite bridge deck. *Adv. Civ. Eng.* **2021**, *2021*, 6658451. [CrossRef]
24. Hu, B.; Li, Y.; Jiang, Y.T.; Tang, H.Z. Bond behavior of hybrid FRP-to-steel joints. *Compos. Struct.* **2020**, *237*, 111936. [CrossRef]
25. ElSafty, A.; Graeff, M.K.; Fallaha, S. Behavior of laterally damaged prestressed concrete bridge girders repaired with CFRP laminates under static and fatigue loading. *Int. J. Concr. Struct. Mater.* **2014**, *8*, 43–59. [CrossRef]
26. Ozkul, O.; Nassif, H.; Tanchan, P.; Harajli, M. Rational approach for predicting stress in beams with unbonded tendons. *ACI Struct. J.* **2008**, *105*, 338. [CrossRef]
27. El Meski, F.; Harajli, M. Flexural behavior of unbonded posttensioned concrete members strengthened using external FRP composites. *J. Compos. Constr.* **2013**, *17*, 197–207. [CrossRef]
28. Ni, H.; Li, R.; Aboutaha, R.S. Strengthening of prestressed girder-deck system with partially debonding strand by the use of CFRP or steel plates: Analytical investigation. *Comput. Concr.* **2023**, *31*, 34–358. [CrossRef]
29. Abdelatif, A.O.; Owen, J.S.; Hussein, M.F. Modelling the prestress transfer in pre-tensioned concrete elements. *Finite Elem. Anal. Des.* **2015**, *94*, 47–63. [CrossRef]
30. Yapar, O.; Basu, P.K.; Nordendale, N. Accurate finite element modeling of pretensioned prestressed concrete beams. *Eng. Struct.* **2015**, *101*, 163–178. [CrossRef]
31. Mercan, B. Modeling and Behavior of Prestressed Concrete Spandrel Beams. Ph.D. Thesis, University of Minnesota, Minnesota, MN, USA, 2011.

32. Chen, L.; Graybeal, B.A. Modeling structural performance of ultrahigh performance concrete I-Girders. *J. Bridge Eng.* **2012**, *17*, 754–764. [CrossRef]
33. Wosatko, A.; Winnicki, A.; Polak, M.A.; Pamin, J. Role of dilatancy angle in plasticity-based models of concrete. *Arch. Civ. Mech. Eng.* **2019**, *19*, 1268–1283. [CrossRef]

Article

Study on the Mechanical Properties and Design Method of Frame-Unit Bamboo Culm Members Based on Semi-Rigid Joints

Guojin Wang [1,2], Xin Zhuo [2,3,*], Shenbin Zhang [3] and Jie Wu [3]

[1] Center for Balance Architecture, Zhejiang University, Hangzhou 310028, China; 22112246@zju.edu.cn
[2] Department of Civil Engineering, Zhejiang University, Hangzhou 310058, China
[3] Architectural Design & Research Institute of Zhejiang University Co., Ltd., Hangzhou 310028, China; zsb1@zuadr.com (S.Z.); wuj@zuadr.com (J.W.)
[*] Correspondence: zhuoxin@zju.edu.cn

Abstract: The frame-unit bamboo culm structure system offers a novel approach to bamboo structure, combining advantages like reduced construction times and simplified joint designs. Despite its benefits, there is limited research on its mechanical properties and computational methodologies. This study conducted bending performance tests on simply supported frame-unit bamboo culm structures, revealing that the bending stiffness of the structure increases with the number of bolts in the edge joints, though with diminishing efficiency. Based on the experimental observations, a calculation model for this type of structure was established, proposing formulas to describe the stiffness relationships between the corner joints, edge joint, and the overall structure. Numerical simulations calculated the stiffness of the edge joint as a function of the number and placement of bolts, indicating that positioning bolts closer to the outer side enhances edge joint stiffness. By inputting the various rotational stiffness values of corner joints into the simulations and stiffness formulas, consistent total stiffness values were obtained, validating the proposed stiffness relationship formulas. The average stiffness values of the corner joints were derived from these formulas and experimental data, and the rotational stiffness of other types of corner points can also be obtained using this method. Furthermore, a finite element computational method tailored for this structural system was introduced, converting the actual structure into a beam element model for calculation. The equivalent joint forces can be distributed to various components of the actual structure, resulting in the internal force distribution of bamboo culms and bolts in the actual structure, thus achieving the design of the components. The calculated displacement values obtained from this method are close to the displacement values in the experiment, proving the feasibility of this method.

Keywords: frame-unit prefabricated bamboo culm structure; semi-rigid joint; mechanical property; numerical simulation methods; design method

Citation: Wang, G.; Zhuo, X.; Zhang, S.; Wu, J. Study on the Mechanical Properties and Design Method of Frame-Unit Bamboo Culm Members Based on Semi-Rigid Joints. *Buildings* **2024**, *14*, 991. https://doi.org/10.3390/buildings14040991

Academic Editors: Atsushi Suzuki and Dinil Pushpalal

Received: 5 March 2024
Revised: 27 March 2024
Accepted: 29 March 2024
Published: 3 April 2024

1. Introduction

The production process of building materials such as steel bars and cement generates a large amount of carbon and harmful gas emissions [1], exacerbating global climate issues such as climate change and environmental pollution. Promoting natural building materials can help alleviate these problems. Bamboo and wood are two common natural building materials that have been widely used in construction projects, where bamboo has higher carbon sequestration efficiency [1] and better mechanical properties than wood [2,3]. A series of studies and engineering examples have proved the feasibility of bamboo as a building material [4,5]. Compared with engineered bamboo, raw bamboo culms do not require the use of adhesives and are more environmentally friendly as a building material.

The different cross-sectional geometries of natural bamboo culms, the anisotropy of bamboo materials, the different mechanical properties of different types of bamboo [6,7], the different mechanical properties of different parts of the same bamboo culm [8], and the different orientations of different joints connecting different bamboo tubes in the bamboo tube structure constrain the standardization of the design of the bamboo tube structures and bring difficulties to the promotion of the bamboo tube structure. Amede et al. [9] found that current standards for bamboo structures are inadequate for its full use as a structural material, indicating a need for a universally applicable design and construction specification to overcome design and construction challenges of bamboo structures. Ma et al. [10] proposed a method for grading based on the minimum outer diameter of bamboo. They measured and analyzed the geometric, physical, and mechanical properties of multiple bamboo culms, and graded the bamboo culms with different processing techniques based on the measurement data. This study contributes to the standardization design of bamboo tube structures.

Joint design is an important part of bamboo tube structure design. Hong et al. [11] divided the common bamboo tube connection joints into two categories of traditional joints and modern joints, where the traditional joints include tied connections and mortise and tenon connections, and the modern joints include types such as bolts, steel components, and fillers. In the German–Chinese House at the Shanghai World Expo 2010 [12], the joints are made of prefabricated steel parts with grouting, and the prefabricated steel components are capable of connecting bamboo tubes in different directions. Huang et al. [13] proposed a bamboo tube joint using grouting and built-in steel sheet connection, and the tests showed that this type of joint has high load-bearing capacity. These improved grouted joints showed better mechanical properties, but the grouting significantly increased the self-weight of the structure and could not take advantage of the light weight of the bamboo tube structure. Richard et al. [14] proposed a new type of joint, connecting bamboo tubes with steel rings and pieces, and the stiffness and strength of this type of joint were proved to be more than that of the traditional grouted bamboo tube joint through tests. Benoit et al. [15] proposed a bamboo tube joint using a combination of wood plugs and metal clamps, and demonstrated through experiments and numerical simulations that the joint could be used for the longitudinal extension of bamboo tubes with good strength, providing a new idea for lightweight joints of bamboo tube structures.

Bolt connection is one of the most practical metal connectors, which has the advantages of simple construction and low cost. It is widely used for connecting round bamboo joints and specimens. Bolts can be used to assemble scattered bamboo tubes into a bamboo tube bundle column with a strong axial compression ability. For instance, in the studies by Nie [16] and Yang [17], long bolts were used to laterally connect multiple parallel bamboo tubes into a bamboo tube column specimen, which demonstrated a strong axial compression performance in tests. Bolts can also be used as embedded parts to form bolted-mortar infill connections (BMIs). Correal et al. [18] proposed a new model to calculate this type of joint, and this model has been validated through experimentation. This research has contributed to the improvement of design procedures and national regulations for BMI connections in Guadua Angustifolia Kunth and other bamboo species.

In this paper, a lightweight joint for the cross-connection of bamboo tubes is proposed, as shown in Figure 1a, where wood plugs are embedded in each of the two sides of the bamboo tubes, and each screw is screwed from the outside of one side of the bamboo tube into the wood plug inside the other side of the bamboo tube, and the bamboo tube and the wood plugs are cut at the same angle in the joint part. The joint has the advantages of being of light weight, easy production, and low cost.

The existing bamboo tube space structures can be divided into two types: [19,20] bamboo tube arch structure and bamboo tube lattice structure. Bending and connecting one or more long bamboo tubes forms a large-span bamboo tube arch structure, the disadvantage being that the bending process uses manual labor which makes it difficult to control processing accuracy. The bamboo tube lattice structure uses straight bamboo tubes, which are easy to process and low in cost, but the processing difficulty of multi-directional joints and the customization cost is high. Zhuo et al. [21,22] proposed a framed bamboo culm grid structure system (a frame-unit prefabricated bamboo culm structure), the formation principle of which is shown in Figure 1c,d. Straight bamboo culms are assembled into bamboo culm frame units, which can be in polygonal geometries such as triangles, rectangles, trapezoids, hexagons, and other polygonal geometries; the joints of neighboring culms are called corner joints, which can be connected by a combination of grouting, screws, and wood plugs (Figure 1a), and the joints between neighboring bamboo culm frames are called edge joints, which use bolts as connectors (Figure 1b).

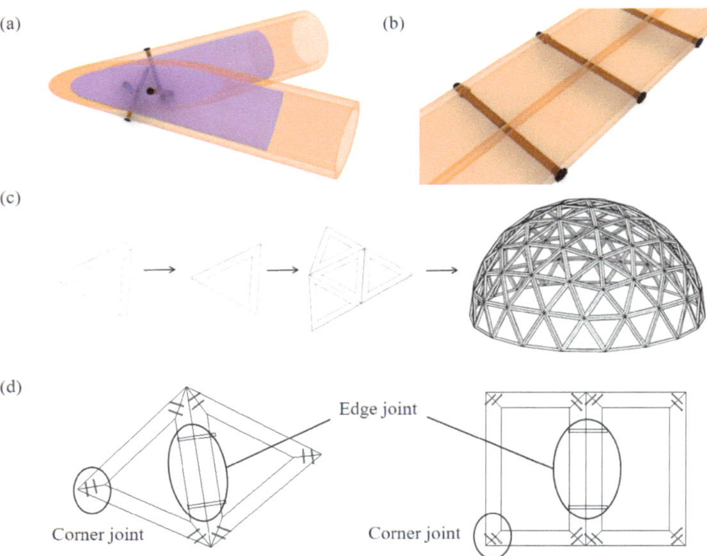

Figure 1. Structural forming principles and two types of joints. (**a**) corner joint. (**b**) edge joint. (**c**) formation of the structure. (**d**) the positions of the corner joint and edge joint.

After all the bamboo culm frame units are connected, the overall structure can be formed. This paper employs experimental, numerical simulation and theoretical analysis to investigate the flexural performance of frame-unit bamboo culm joints. The focus is on studying the semi-rigid bearing mechanism and calculation method for joints when materials are in the elastic phase at edge and corner joints. The research process of this paper is illustrated in Figure 2.

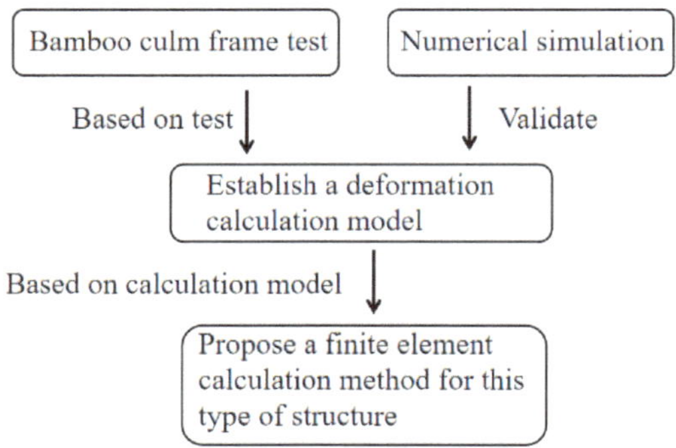

Figure 2. Research process.

2. Bamboo Culm Frame Specimen Bending Performance Test

2.1. Overview of the Test

Moso bamboo is one of the most widely distributed bamboo species in China, mainly found in the southern regions of China. It is extensively used in bamboo architectural structures [23].

The bamboo used for the experiment is Moso bamboo grown in the southern mountainous areas of China. It has a growth age of about 4 years, with internode spacing between 20–40 cm and a wall thickness between 7.5–8.5 mm. The outer diameter of the bamboo culm was measured using a vernier caliper and ranged from 75 mm to 95 mm.

The specimen in this study is composed of two adjacent rectangular bamboo culm frames (Figure 3a). Each bamboo culm frame is handmade, and wooden plugs are inserted into the ends of the bamboo culms. Screws with a diameter of 4 mm are driven into the exterior wall of one side of the bamboo culm, penetrating through to the wooden plug located within the bamboo culm on the opposite side; a single corner joint is formed using four screws (Figure 3b). The experiment was conducted in an environment with a temperature of 25 °C and an air humidity of about 45%. Place two bamboo culm frames side by side. After drilling, use long bolts to connect the adjacent bamboo culms placed in parallel to form a test specimen. (The diameter of the holes in the bamboo culms and the diameter of the long bolts are the same, at 10 mm. The long bolts are made of Q235 steel, with a length of approximately 200 mm.)

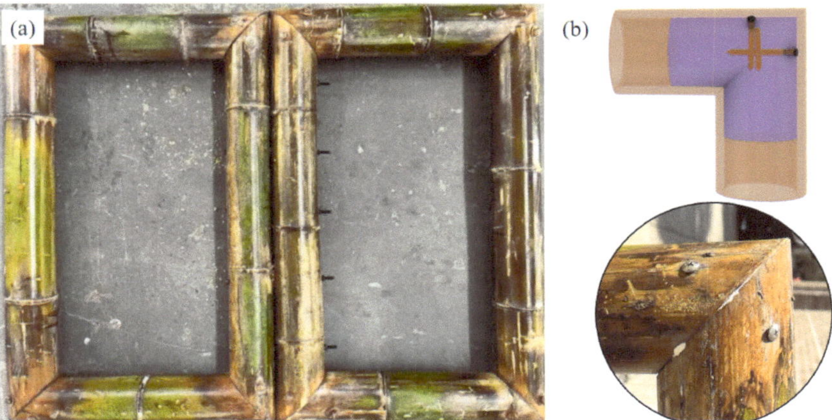

Figure 3. Corner joint and edge joint. (**a**) edge joint. (**b**) corner joint.

The specimens were rested in a simply supported manner on the bull legs of the steel frame test rig (shown in Figure 4).

The positions of bolt holes are marked as A, B, C, B′, and A′, respectively. The dimensions of the assembled specimens are shown in Figure 5. The loading position is in the middle of the specimen (Figure 4). Loading is performed by stacking sandbags, with each sandbag weighing 10 kg (±0.5 kg). When the sixth sandbag was placed, the structure experienced significant displacement and could no longer continue to accommodate more sandbags. For safety reasons, the load at this point is considered the maximum load. Loading tests were conducted on specimens with 2, 3, and 5 bolts connected to the edge joints; the experimental parameters for each model are shown in Table 1. The specimen number "Sn" indicates that the number of bolts in the edge joint is n.

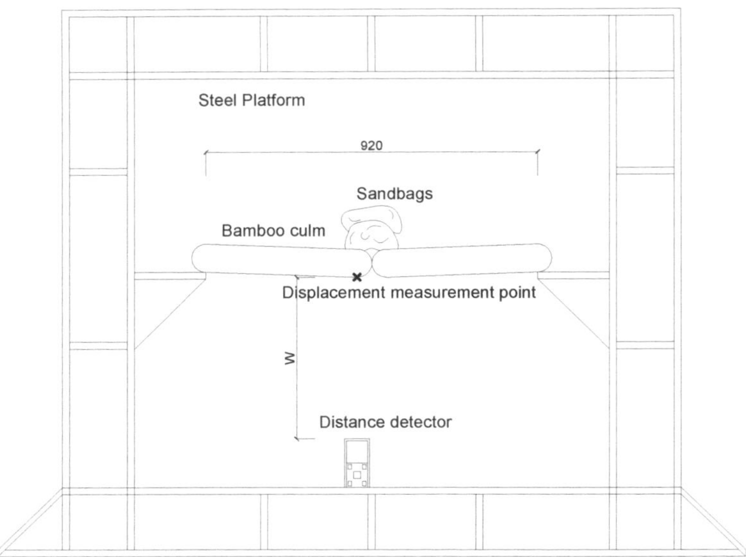

Figure 4. Loading device (mm).

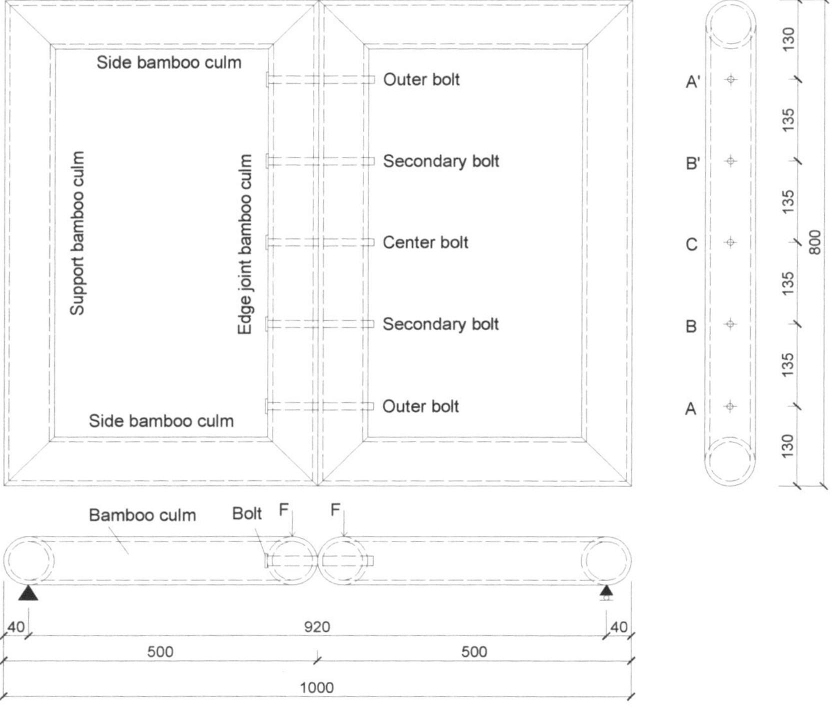

Figure 5. Dimensions of specimens (mm).

Table 1. Test variables.

Specimen Numbers	$S2_{test}$	$S3_{test}$	$S5_{test}$
Bolt positions	A, A′	A, C, A′	A, B, C, B′, A′
Maximum load applied (N)	620.3	620.3	620.3

2.2. Experimental Process and Result Analysis

The two bamboo culm frames of the specimen are in the same plane when unloaded. Vertical displacement occurs at the edge joint positions of the specimen, gradually increasing with the number of sandbags (Figure 6a), and there is a gap between the two bamboo culms at the corner joint (Figure 6b). The measuring point for vertical displacement is at the midpoint of the lower edge of the edge joint bamboo culm (Figure 4), measured using a laser distance detector with an accuracy of 0.1 mm. During loading, sandbags are placed one by one, and data from the displacement detector are read only after the vertical displacement stabilizes after placing a sandbag (and the unloading stage is similar). The deflection of the measuring point under each stage of loading is:

$$u_i = w_i - w_0 \tag{1}$$

where w_i is the reading at stage i, and w_0 is the reading without load. The data recorded at each loading step are used to create the load–displacement relationship shown in Figure 7.

Before loading, due to the presence of bamboo nodes, it is impossible to achieve a complete fit between the outer walls of adjacent bamboo culms. In addition, it is also impossible to achieve a complete fit between the bamboo culm bolt hole wall and the bolt thread, so the deformation during the loading process includes measurement errors caused by these factors. In contrast, the unloading process has achieved close contact between different materials, and the corresponding load–displacement relationship can better reflect the true mechanical properties of the specimens.

Figure 6. Deformation of test specimen joints after loading. (**a**) deformation of the edge joint. (**b**) a gap has occurred in the corner joint.

The total bending linear stiffness K_l (hereafter referred to as total linear stiffness) of the bamboo culm frame in the elastic phase is defined as the ratio of the out-of-plane load at the edge joints to the vertical displacement occurring in the direction of action:

$$K_l = \frac{\Delta F}{\Delta d} \tag{2}$$

where ΔF and Δd are the incremental load and the displacement of the measuring point.

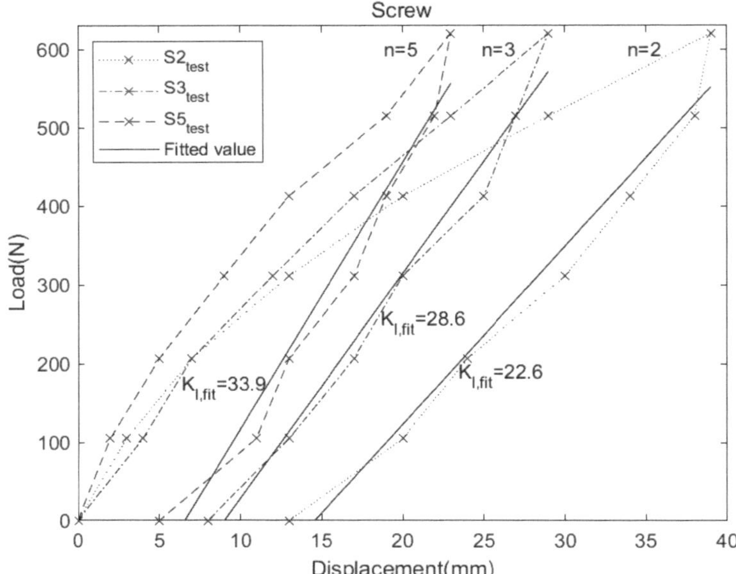

Figure 7. Experimental data and fitted values.

The test data from the unloading stage were selected for linear fitting (using the polyfit function in MATLAB.R2020b) to obtain the fitted values of total linear stiffness $K_{l,fit}$ (Figure 7). This shows that the total bending linear stiffness is positively correlated with the number of bolts. However, as the number of bolts increases, the efficiency of stiffness increases decreases gradually.

2.3. Synergistic Relationship between Corner Joints and Edge Joints

During the loading process, no visible bending occurred in each bamboo culm, while vertical displacement occurred at the edge joints. At the same time, the gap at the corner joints gradually widened with the increase of the load (Figure 6b), indicating that the corner joints and edge joints are not completely rigid (semi-rigid). Define k_{sa} and k_{ca} as the rotational stiffness of the edge joints and corner joints, respectively:

$$k_{sa} \overset{\text{def}}{=} \frac{\Delta M_s}{\Delta \theta_s} \tag{3}$$

$$k_{ca} \overset{\text{def}}{=} \frac{\Delta M_c}{\Delta \theta_c} \tag{4}$$

where M_s, θ_s are the bending moments and turning angles of the edge joints, M_c, θ_c are the bending moments and turning angles of the corner joints, and the units of bending moments and turning angles are N·mm, Rad, respectively.

To convert the relationship between force and displacement into the relationship between bending moment and rotation, the various bamboo culms are simplified into rigid members, and a simplified diagram for the internal force calculation of the frame-unit bamboo culm members shown in Figure 8 is established.

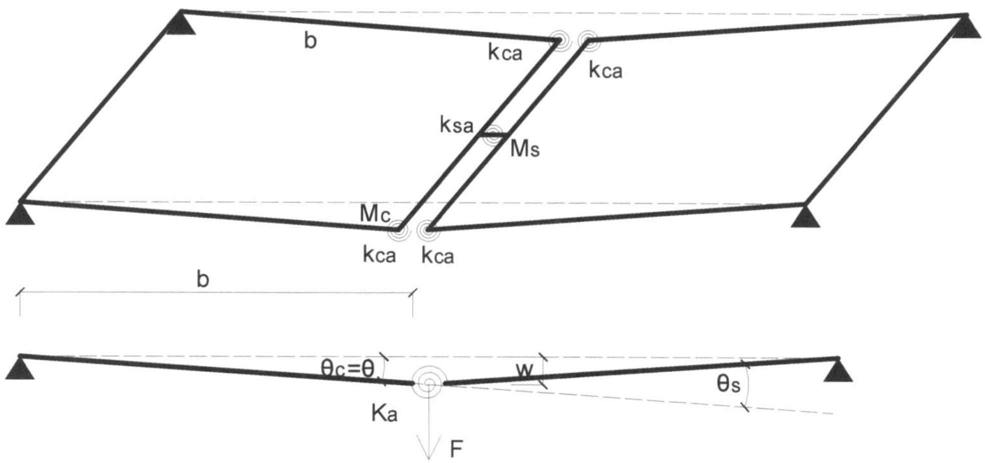

Figure 8. Schematic diagram for structural deformation calculation.

In Figure 8, F is the vertical force acting on the upper surface of the edge joint bamboo culms, b is the distance between the axis of the edge joint bamboo culm and the support bamboo culm (hereinafter referred to as the shear span). Under the action of F, both the edge joint and the corner joints produce vertical displacement.

When the corner joints are completely rigid, define the edge joint linear stiffness as follows:

$$k_{sl} \stackrel{\text{def}}{=} \frac{\Delta F}{\Delta w} \tag{5}$$

When the edge joint is completely rigid, define the corner joint linear stiffness as follows:

$$k_{cl} \stackrel{\text{def}}{=} \frac{\Delta F}{\Delta w} \tag{6}$$

According to Figure 8:

$$\begin{cases} M_s = 0.5Fb \\ \theta_s = 2\theta \end{cases} \tag{7}$$

$$\begin{cases} M_c = 0.25Fb \\ \theta_c = \theta \end{cases} \tag{8}$$

When the structural deformation is small:

$$\theta \sim tan\theta = \frac{w}{b} \tag{9}$$

The numerator and denominator in Equation (3) are represented by ΔF and Δw, respectively, and the relationship between the edge joint linear stiffness and its rotational stiffness can be obtained:

$$k_{sa} = \frac{0.5\Delta Fb}{2\Delta\theta} = \frac{0.5\Delta Fb}{2\frac{\Delta w}{b}} = \frac{b^2}{4}k_{sl} \tag{10}$$

Similarly, the relationship between the corner joint linear stiffness and its rotational stiffness can be obtained from Equation (4).

$$k_{ca} = \frac{b^2}{4}k_{cl} \tag{11}$$

The unit of edge and corner joint linear stiffness is N/mm. From Equations (5) and (6), the linear stiffness is not a property of the joint itself, and its value is related to structural parameters such as rotational stiffness and span.

In structural mechanics, the shear distribution method is solved by analyzing the series–parallel relationship and stiffness relationship of the members [24]. In the structure shown in Figure 8, the two corner joints in a bamboo culm single-sided frame structure are in a parallel relationship, and the relationship between the corner joints and the edge joint constitutes a series relationship; the series and parallel relationships of the assembled joints are shown in Figure 9.

Figure 9. Series and parallel relationships between the joints.

According to Figure 9, establish the relationship between the overall rotational stiffness and that of the edge and corner joints.

$$\frac{1}{K_a} = \frac{1}{2k_{ca}} + \frac{1}{k_{sa}} + \frac{1}{2k_{ca}} = \frac{1}{k_{ca}} + \frac{1}{k_{sa}} \tag{12}$$

Similarly, the relationship between total linear stiffness and the linear stiffness of the edge and corner joints can be established.

$$\frac{1}{K_l} = \frac{1}{k_{cl}} + \frac{1}{k_{sl}} \tag{13}$$

3. Numerical Simulation

3.1. Overview

3.1.1. Setting of Numerical Simulation

1. The finite element software Abaqus2021 is used for modelling and the calculation of numerical simulation. Since the two neighboring bamboo culm frames have symmetry, in order to improve the computational efficiency, only a single bamboo culm frame is established, and symmetrical constraints are applied to the mid-span cross-sections of truncated bolts;
2. The influence of bamboo nodes is not taken into account;
3. The actual bamboo culm wall has a different elastic modulus along the axial, lateral, and radial directions, which is simplified by setting up bamboo culms as orthotropy material in both axial and lateral directions and ignoring the difference in the mechanical parameters of the bamboo culms along the radial direction;
4. The diameter of the opening of the bamboo culm is the same as the diameter of the bolt, which is taken as 10 mm, ignoring the effect of the thread on the bamboo culm;
5. The loading plate for applying load is set to simulate the sandbag load in the test.

3.1.2. Material Properties

According to the test data of García et al. [25], the mechanical properties of bamboo are shown in Table 2, where direction two is the bamboo culm axis (fiber) direction, and directions one and three are perpendicular to the axis direction (the bamboo culm material property setting direction is the local coordinate system of the bamboo culm). E_{mn}, v_{mn}, and G_{mn} are the elastic modulus, Poisson's ratio, and shear modulus. The yield strength of the bolt is 235 MPa, the elastic modulus is 206 GPa, and the Poisson's ratio v is 0.3. According to the axial compression test conducted by Zhang et al. [26,27], the axial compressive yield strength is 50.3 Mpa, and the shear yield strength is 25 Mpa. The compressive strength σ_c

of the bamboo culm groove is calculated according to the formula proposed by Peng Hui et al. [28]:

$$F_s = 0.8\sigma_1 dt \tag{14}$$

where F_s is the yield load of the bearing compression of the bolt hole, σ_1 is the compressive strength of bamboo culm with grain, d is the diameter of the hole, and t is the thickness of the bamboo culm. Rewriting Equation (14) can obtain the compressive strength of bamboo culm at the location of the bolt hole wall:

$$\sigma_c = \frac{F_s}{dt} = 0.8\sigma_1 = 40.2 \text{ MPa} \tag{15}$$

Table 2. Properties of bamboo materials.

E_1	E_2	E_3	v_{23}	v_{13}	v_{12}	G_{23}	G_{13}	G_{12}
398	14700	398	0.3	0.14	0.008	581	175	581

3.1.3. Calculation Model

The friction coefficient between the bolt and the bamboo culm wall is set to 0.3. The loading plate is used solely for loading and is not a component of the structure, so the coefficient of friction between it and the bamboo culm surface is 0. Only compressive forces are transmitted in the normal direction between the loading plate, bolt, and bamboo culm. The locations of the bolts in each model are the same as those in the experimental specimens (Table 1). The components established include bamboo culms and bolts, and the outer diameter and wall thickness of the bamboo culms are 80 mm and 8 mm, respectively. The dimensions of the bamboo culm frames are shown in Figure 5, and the diameter of the bolts is 10 mm. The displacement in the y-direction and z-direction of the lower edge of the support bamboo culm, and the x-direction displacement at various points on the cross-section of each bolt is restrained, forming a semi-span simple supported structure. The mesh consisted of the eight-node hexahedral elements. The loading position on the edge joint bamboo culm does not include the angled region at the ends of the bamboo culm. A linearly increasing concentrated force is applied at the center position of the loading plate (the load remains consistent with the experiment direction: -y, maximum value: 620.3 N).

The assembled structure is shown in Figure 10.

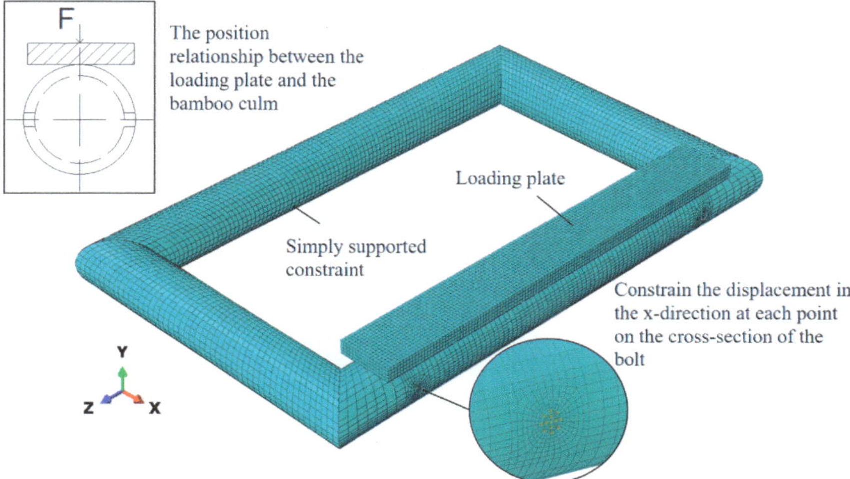

Figure 10. Loading surface and boundary conditions.

3.2. Semi-Rigid Corner Joint

The spatial relationship between the various materials at corner joint positions is complex, making it difficult to accurately model according to the actual situation. To obtain the average rotational stiffness value of the corner joints of the test specimens, the calculations are first performed assuming the corner joints are completely rigid, to obtain the stiffness values at the edge joint. Then, the corner joints are considered semi-rigid to determine the overall stiffness of the structure. The obtained values are compared with the calculated values obtained from the stiffness relationship formula to validate the accuracy of the formula. Finally, based on the experimental data and the stiffness relationship formula, the average rotational stiffness value of the corner joints of the test specimens is determined.

3.2.1. Set the Corner Joints as Rigid

The points at the ends of both bamboo culms are set as coupling constraints, with the highest point of the side bamboo culm serving as the control point. This ensures that there is no relative displacement between the points at the ends of the bamboo culms, thereby forming a rigid connection (Figure 11).

In the Visualization module, view the step time and corresponding load when the bolt yields (yield load). The three models with two, three, and five bolts, respectively, have yield load values of 245 N, 330 N, and 460 N. When the bolt yields, the Mises stress distribution and the shape of the bending moment for the bolt are shown in Figure 12; the shape of the bending moment diagram is the same for each bolt, but the values differ. The distribution ratios of bolt moments for each model are presented in Table 3.

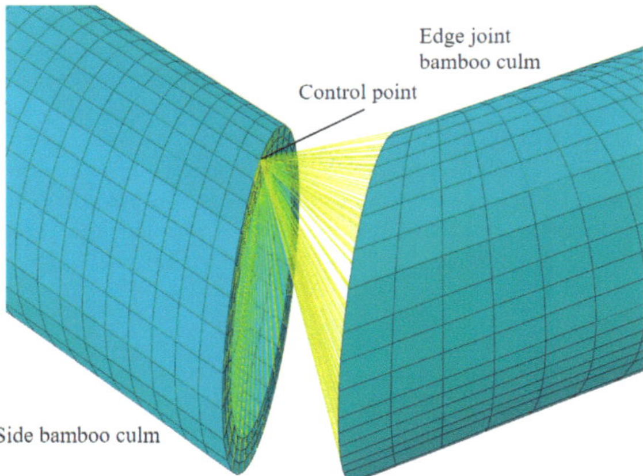

Figure 11. Fully rigid corner joints.

The displacement measurement point is located at the midpoint of the lower edge of the edge joint bamboo culm, consistent with the experimental setup. The total linear stiffness value $K_{l,sim}$, shown in Table 4, is calculated using Equation (16). Here, w_{50} and w_{100} represent the vertical displacements at the measurement point under loads of 50 N and 100 N, respectively. Comparing $K_{l,sim}$ with $K_{l,fit}$ (Figure 7) reveals that the total stiffness values calculated, assuming fully rigid corner joints, significantly deviate from the experimental results, underscoring the importance of accounting for the semi-rigidity characteristics of the corner joints.

$$K_{l,sim} = 2\frac{\Delta P}{\Delta w} = \frac{100}{w_{100} - w_{50}} \tag{16}$$

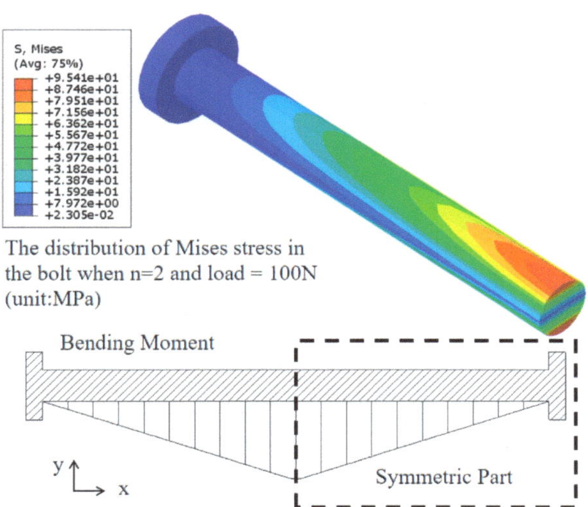

The distribution of Mises stress in the bolt when n=2 and load = 100N (unit:MPa)

Figure 12. Stress distribution and bending moment diagram of a bolt.

Table 3. Bending moment ratio of each bolt.

Bolt Position	$n = 2$	$n = 3$	$n = 5$
A	0.5	0.366	0.258
B	/	/	0.169
C	/	0.268	0.146

Table 4. Linear stiffness values of edge joints obtained by numerical simulation. (N/mm).

n	2	3	5
$K_{l,sim}$	51.8	66.0	84.3
$K_{l,fit}$	22.6	28.6	33.9

According to Equation (13), the total linear stiffness is the edge joint linear stiffness when the corner joint stiffness is infinite:

$$\lim_{k_{cl} \to \infty} \frac{1}{K_l} = \frac{1}{k_{sl}} \tag{17}$$

3.2.2. Set the Corner Joints as Semi-Rigid

The steps to modify the corner joints to be semi-rigid are as follows (Figure 13):

1. Delete the rigidity constraint from the corner joints.
2. Set coupling constraints between various points at the end of each bamboo culm and its respective highest point (control point). Wire feature was established between control points of adjacent bamboo culms.
3. Create a hinge-type connector and assign different rotational stiffness values.
4. This connector was applied to the wire features, forming a semi-rigid joint.
5. Move the bamboo culm to the starting position.

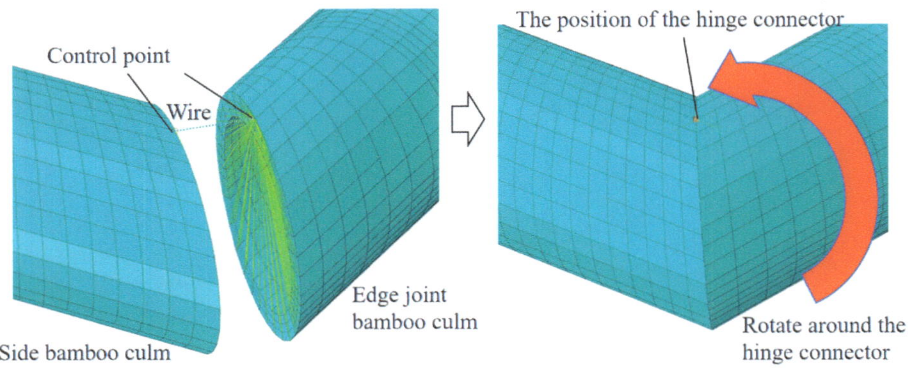

Figure 13. Set the corner joints to be semi-rigid.

By incorporating the rotational stiffness values of different corner joints into numerical simulations, combined with Equation (16), the total stiffness value of the structure can be obtained (simulated values $K_{l,sim}$).

These varying rotational stiffness values of corner joints were then converted into linear stiffness values, which were substituted into Equation (13) (the stiffness values of the edge joints are $K_{l,sim}$, shown in Table 4) to derive a series of total linear stiffness values (theoretical values $K_{l,equ}$).

By comparing the values of $K_{l,sim}$ and $K_{l,equ}$, it was found that the total linear stiffness values calculated by the two methods are close (Table 5), proving the accuracy of the proposed stiffness relationship formula.

Table 5. Comparison between the simulated and theoretical values of total stiffness obtained by substituting different corner joint stiffness (the units of k_{ca} and K_l are 10^4 N·mm/Rad and N/mm, respectively).

	$n = 2$			$n = 3$			$n = 5$	
k_{ca}	$K_{l,sim}$	$K_{l,equ}$	k_{ca}	$K_{l,sim}$	$K_{l,equ}$	k_{ca}	$K_{l,sim}$	$K_{l,equ}$
100	15.6	15.8	100	16.7	16.9	100	17.7	17.9
200	24.0	24.2	200	26.7	26.9	200	29.3	29.5
300	29.3	29.4	300	33.4	33.5	300	37.5	37.6
400	32.9	33.0	400	38.1	38.2	400	43.5	43.7

The values of linear stiffness k_{cl} for the considered semi-rigid corner joints can be obtained by bringing the first and second rows of Table 4 into Equation (13) as k_{sl} and K_l, respectively. The value of k_{cl}, after substituting into Equation (11), is transformed into the value of k_{ca}, as shown in Table 6. This value is the average rotational stiffness of the four corner joints connecting the edge joints in Figure 7, not the rotational stiffness of the corner joints in a specific certain position.

Table 6. The average rotational stiffness values of the corner joints obtained by substituting the total stiffness from the experiment into the stiffness relationship formula (10^4 N·mm/Rad).

n	2	3	5
k_{ca}	176.8	222.6	250.1

3.3. Semi-Rigid Edge Joint

To analyze the effect of positions on the stiffness of the edge joint when the number of bolts remains the same, s is defined as the distance from the outermost bolt to the axis of the side bamboo culm (defined as "side distance", Figure 14), and $k_{sl}(s)$ is the linear stiffness of the edge joint as a function of s, the bolts arranged at equal distances. Through the numerical simulation method established in this paper, the k_{sl} corresponding to different values of s can be obtained, as shown in Table 7. The linear regression equations of k_{sl} and s are established as follows:

$$k_{sl}(s) = As + B \tag{18}$$

where A and B are coefficients to be determined.

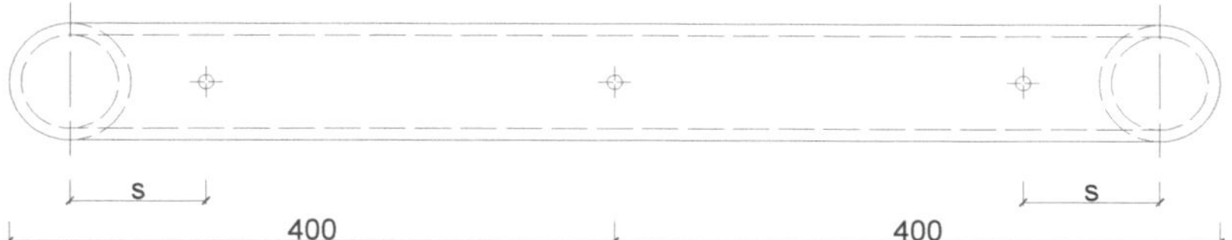

Figure 14. Side distance s.

Table 7. Linear stiffness values of edge joints obtained by numerical simulation.

n	2			3			5		
s(mm)	90	180	270	90	180	270	90	180	270
k_{sl}(N/mm)	51.8	42.4	36.3	66.0	54.0	45.3	84.3	67.8	56.0

Matlab.R2020b was used to linearly fit the (s,k_{sl}) points of Table 7, according to Equation (18); the $k_{sl} - s$ relationship can be obtained for different numbers of bolts (Figure 15). The stiffness values calculated from the fitted relationship equation have an error of less than 3% compared to the stiffness values obtained from the numerical simulation, which indicates that the regression effect is good. It was also found that, under the condition of the same number of bolts, k_{sl} increases linearly with the decrease of s, which indicates that the outer bolt closer to the corner joint is more beneficial for stiffness improvement.

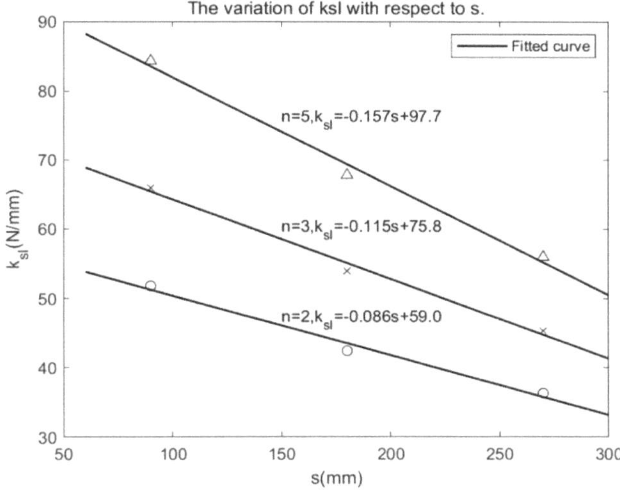

Figure 15. *K-s* relationship for different number of bolts.

By substituting the fitting equation in Figure 15 and $b = 420$ mm into Equation (10), the calculation formula for the rotational stiffness of the edge joint can be obtained when the diameter of the bolts in the edge joint is set to 10 mm.

$$k_{sa}(s) = \begin{cases} -3797.5s + 260.2 \times 10^4, & n = 2 \\ -5071.5s + 334.3 \times 10^4, & n = 3 \\ -6933.5s + 430.7 \times 10^4, & n = 5 \end{cases} \tag{19}$$

From Equations (3) and (4), the rotational stiffness of the joint is the relative angle of rotation produced under the action of unit bending moment, which is a property of the joint itself. Therefore, Equation (19) is a general formula for the rotational stiffness of the corner joint when the diameters of the bolts of the edge joint are 10 mm. Similarly, the rotational stiffness values k_{ca}, in Table 6, are generic values for corner joints in a specific configuration (using four screws combined with wooden plugs). These values and formulas derived from experimental data, numerical simulations, and theoretical derivations provide a basis for the setting of joint semi-stiffness in structural finite element analyses.

4. Finite Element Analysis of Structures and Design Methods for Components

The finite element analysis method of the structure models the components according to the axes and, since the calculation formulas and parameters obtained in Section 3 of this paper are based on solid modelling, a practical method for the finite element analysis and component design of the bamboo culm and frame-unit bamboo culm structure will be presented in this chapter. The equivalent joint forces and joint deformations in the structure are first analyzed according to the computational model in Figure 8, and then the internal forces of the bamboo culm and bolts are calculated in combination with the computational model. Then the allowable stresses and deformations of each component are verified.

Taking the model shown in Figure 5 as an example, the material, size, and number of bolts of the two bamboo culm frame units are the same as those of the S5test in the test, i.e., the outer diameter D of the bamboo culm is 80 mm, the thickness of the bamboo wall t is 8 mm, and the size of the outer frame of the bamboo culm frame: $L \times H = 800$ mm $\times$ 500 mm (axial figure: $a = 720$ mm, $b = 420$ mm). The edge joint is connected by five bolts with $d = 10$ mm, the side distance s is 90 mm, and the rest of the bolts are arranged at an equal distance from each other. The corner joints are connected by four screws and wooden plugs. The support bamboo culm is used as the simply supported side. The top of the edge joint bamboo culm is subjected to a uniform load, and the total force F is consistent with the experimental value in Figure 7, which is taken as 413 N.

4.1. Structural Finite Element Analysis Methods

4.1.1. Structural Modelling

The equivalent computational model of each member is shown in Figure 16:

1. Select the axis of each bamboo culm as the position of the equivalent members;
2. An equivalent bolt is set at the midpoint of the edge joint bamboo culm, and the connection between the equivalent bolt and the edge joint bamboo culm is a rigid connection. The middle position of the equivalent bolt is disconnected, and the disconnection is set as a semi-rigid connection;
3. All corner joints are set to semi-rigid connections;
4. The two ends of the support bamboo culms are set as hinged.

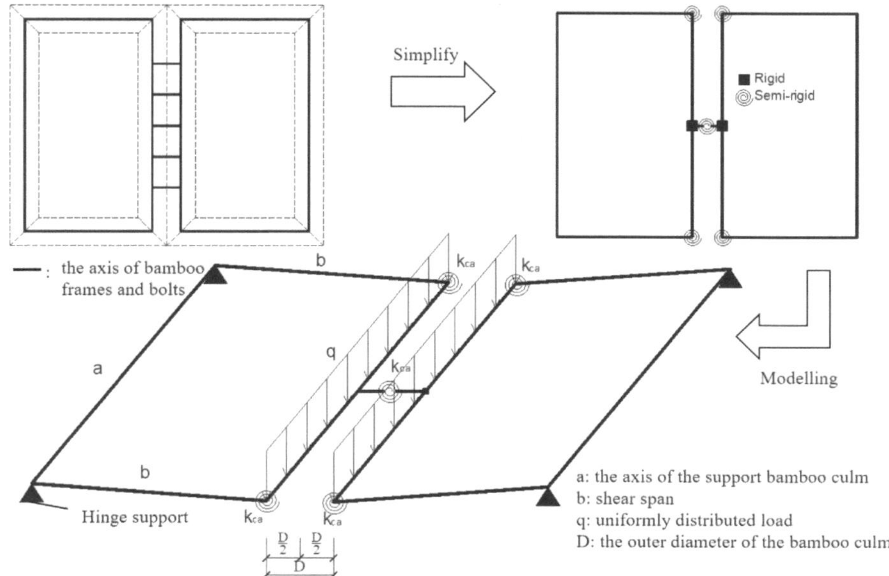

Figure 16. Equivalent beam element model.

4.1.2. Parameter Settings

Since the corner and edge joint stiffnesses studied in this paper already include the deformation of the material itself, to avoid repeated calculations, the members should be set as rigid bodies during the finite element analysis of the structure, i.e., the stiffness and elastic modulus of various materials are set to infinity. The semi-rigid joints are set as follows.

From Table 6, the rotational stiffness of the corner joints is:

$$k_{\mathrm{ca}} = 250.1 \times 10^4 \text{ N·mm/Rad}$$

From Equation (19), the edge joint rotational stiffness is:

$$k_{\mathrm{sa}} = -6933.5 \times 90 + 430.7 \times 10^4 = 368.3 \times 10^4 \text{ N·mm/Rad}$$

Taking the test data at $n = 5$ in Figure 7 when the load is 413 N, the uniform load on the edge joint bamboo culms is:

$$q = \frac{F}{2(L - 2D)} = \frac{413}{2 \times (800 - 2 \times 80)} = 0.32 \text{ N/mm}$$

4.1.3. Calculation Result

The maximum deflection obtained by using Abaqus2021 is 12.2 mm, while the measured value of the test in Figure 7 is 13.0 mm, which indicates that the calculation method in this paper has a good calculation accuracy.

4.2. Design Methods for Components

4.2.1. Internal Force Distribution

The software calculates the maximum bending moment on the equivalent edge joint bolt to be 8.68×10^4 N·mm, which is the value of the combined cross-sectional internal force in the mid-span of each bolt in the actual structure. So, this equivalent joint force needs to be assigned to each bolt.

The bending moment at the equivalent edge joint was distributed to each bolt according to the distribution ratio coefficients in Table 3, and the values of bending moments on the five bolts were obtained as 2.24, 1.47, 1.27, 1.47, and 2.24 $(10^4$ N·mm), respectively. Substituting these values into the torque calculation sketch in Figure 17a gives the torque on each section of the bamboo culm at the edge joints (Figure 17b).

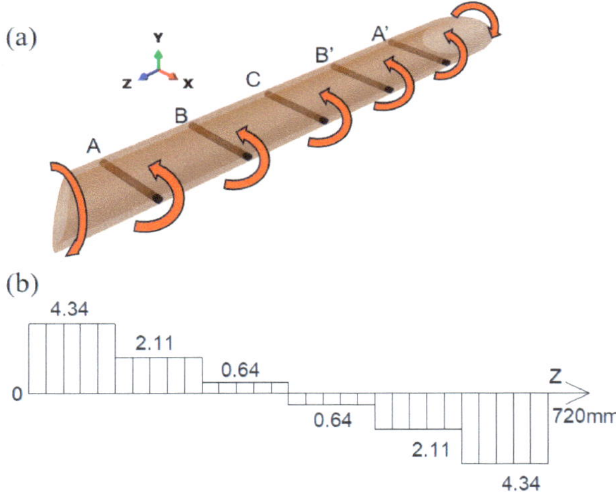

Figure 17. Torque calculation of edge joint bamboo culm. (**a**) simplified diagram of torque calculation of bamboo culm with edge joints. (**b**) torque diagram for edge joint bamboo culm.

4.2.2. Strength Calculation of Components

Figure 18 shows the force sketch of the bolt in bending. Due to the thin wall of the bamboo culm, the constraint of the bamboo culm wall on the bolt can be considered as simple support, and the force between the outermost bolt and the walls of the bamboo culm are as follows:

$$0.5P = \frac{M}{D} = \frac{2.24 \times 10^4}{80} = 280 \text{ N}$$

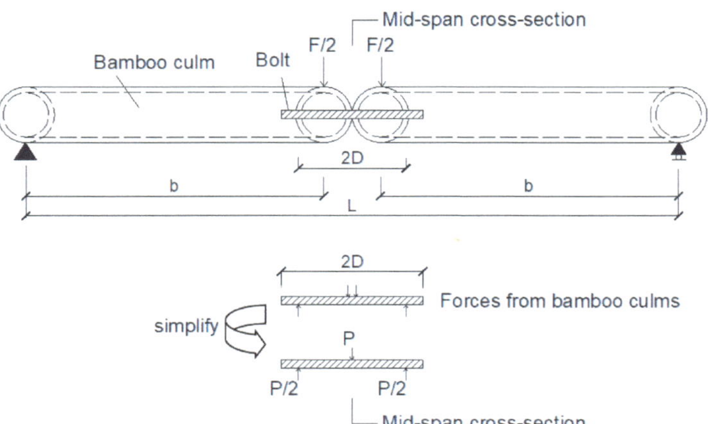

Figure 18. Bolt calculation diagram.

Calculate the bearing compression stress at the location of the bamboo culm hole wall:

$$\sigma_{xc} = \frac{0.5P}{dt} = \frac{280}{10 \times 8} = 3.5 \text{ MPa} < 40.2 \text{ MPa}$$

Calculate the bending strength of the outermost bolt:

$$\sigma_{max} = \frac{M}{W} = \frac{2.24 \times 10^4}{\frac{\pi \times 10^3}{32}} = 228.2 \text{ MPa} < 235 \text{ MPa}$$

Calculate the torsional strength of the bamboo culm by taking the maximum torque of the bamboo culm:

$$\tau_{max} = \frac{T}{W_P} = \frac{4.34 \times 10^4}{\frac{\pi}{16 \times 80}(80^4 - 72^4)} = 1.3 \text{ MPa} < 25 \text{ MPa}$$

5. Conclusions

This study combined experimental and numerical simulation methods to conduct a comprehensive analysis of the two main types of joints in this structural system, and has successfully established corresponding calculation models. Furthermore, a new finite element calculation method for this structure has been proposed.

Experiments on specimens with bolt diameters of 10 mm and bolt counts of two, three, and five have been conducted, recording the vertical deformation and load relationship curves. Linear fitting of the unloading phase data was performed to obtain the total stiffness. The results indicate that an increase in the number of bolts leads to an increase in total stiffness, but with diminishing efficiency. Observations from the experiments reveal that, despite no visible deformation in the bamboo culms during the loading process, both the corner and edge joints connecting the bamboo culms underwent varying degrees of deformation, indicating that these joints are not completely rigid. Based on these observations, a calculation model for the frame-unit bamboo culm considering the semi-rigidity of the joints has been further proposed. This study has derived formulas for converting linear stiffness to rotational stiffness for both corner and edge joints, as well as a universal stiffness relationship formula between joints.

This study also explores a numerical simulation method for frame-unit bamboo culm structure. By setting the corner joints as fully rigid for numerical simulation, the calculation formulas for edge joints under different bolt configurations were obtained, showing that the placement of outer bolts closer to the corner joints significantly increases the stiffness of the edge joints. By treating corner joints as semi-rigid and incorporating different values of rotational stiffness for corner joints into both numerical simulations and the stiffness relationship formula, the consistency in the total stiffness values obtained from both methods confirms the accuracy of the proposed stiffness relationship formula. By applying the proposed stiffness relationship formula, this study has calculated the average rotational stiffness values for the corner joints of the experimental specimens. This provides a new method for determining the rotational stiffness values of corner joints in various construction forms.

Lastly, based on the calculation model of the frame-unit bamboo culm structure established in this paper, a finite element analysis method for this type of structure based on semi-rigid joints is proposed. This method converts all members into rigid beam elements, and remodels accordingly. Following the principle of equal stiffness, establish equivalent semi-rigid corner and edge joints to connect beam elements. After calculation, the equivalent joint internal force and joint deformation can be obtained. The equivalent joint internal force can be distributed to obtain the distribution of the internal force of the bamboo culms and bolts in the actual structure. The equivalence of the joint deformation values to the experimental values demonstrates the feasibility of this method.

Author Contributions: Conceptualization, G.W. and X.Z.; methodology, G.W. and X.Z.; software, S.Z.; validation, J.W. and S.Z.; experiment, J.W. and S.Z.; formal analysis, G.W.; resources, J.W. and S.Z.; data curation, J.W.; writing—original draft preparation, G.W.; writing—review and editing, G.W. and X.Z.; supervision, X.Z.; project administration, X.Z.; funding acquisition, X.Z. All authors have read and agreed to the published version of the manuscript.

Funding: This research was funded by the National Key Research and Development Program of China, Grant No. 2017YFC0703500.

Data Availability Statement: Data will be made available on request.

Conflicts of Interest: Authors Xin Zhuo, Shenbin Zhang and Jie Wu was employed by the company Architectural Design & Research Institute of Zhejiang University Co., Ltd. The remaining authors declare that the research was conducted in the absence of any commercial or financial relationships that could be construed as a potential conflict of interest.

References

1. Lorenzo, R.; Mimendi, L. Digitisation of bamboo culms for structural applications. *Build. Eng.* **2020**, *29*, 101193. [CrossRef]
2. Escamilla, E.Z.; Habert, G. Environmental impacts of bamboo-based construction materials representing global production diversity. *J. Clean. Prod.* **2014**, *69*, 117–127. [CrossRef]
3. Anokye, R.; Bakar, S.E.; Ratnansingam, J.; Awang, B.K. Bamboo properties and suitability as a replacement for wood. *Pertanika J. Sch. Res. Rev.* **2016**, *2*, 63–79. [CrossRef]
4. Chung, K.F.; Yu, W.K. Mechanical properties of structural bamboo for bamboo scaffoldings. *Eng. Struct.* **2002**, *24*, 429–442. [CrossRef]
5. Adier, M.F.V.; Sevilla, M.E.P.; Valerio, D.N.R.; Ongpeng, J.M.C. Bamboo as Sustainable Building Materials: A Systematic Review of Properties, Treatment Methods, and Standards. *Buildings* **2023**, *13*, 2449. [CrossRef]
6. Liu, P.; Zhou, Q.; Fu, F.; Li, W. Effect of bamboo nodes on the mechanical properties of p. Edulis (Phyllostachys edulis) bamboo. *Forests* **2021**, *12*, 1309. [CrossRef]
7. Chen, M.; Ye, L.; Li, H.; Wang, G.; Chen, Q.; Fang, C.; Dai, C.; Fei, B. Flexural strength and ductility of moso bamboo. *Constr. Build. Mater.* **2020**, *246*, 118418. [CrossRef]
8. Li, R.R.; He, C.J.; Peng, B.; Wang, C.G. Differences in fiber morphology and partial physical properties in different parts of Phyllostachys edulis. *J. Zhejiang AF Univ.* **2021**, *38*, 854–860. (In Chinese) [CrossRef]
9. Amede, E.A.; Hailemariama, E.K.; Hailemariam, L.M.; Nuramo, D.A. A Review of Codes and Standards for Bamboo Structural Design. *Adv. Mater. Sci. Eng.* **2021**, *2021*, 4788381. [CrossRef]
10. Ma, R.; Chen, Z.; Du, Y.; Jiao, L. Structural Grading and Characteristic Value of the Moso Bamboo Culm Based on Its Minimum External Diameter. *Sustainability* **2023**, *15*, 11647. [CrossRef]
11. Hong, C.K.; Li, H.T.; Lorenzo, R.; Wu, G.; Corbi, I.; Corbi, O.; Xiong, Z.H.; Yang, D.; Zhang, H.Z. Review on connections for original bamboo structures. *J. Renew. Mater.* **2019**, *7*, 713–730. [CrossRef]
12. Dai, P.Q.; Luo, Z.Y.; He, M.J. Structural design and analysis for shanghai expo special project DuC. *Struct. Eng.* **2011**, *27*, 6–11. (In Chinese) [CrossRef]
13. Huang, T.; Zhuo, X. Experimental Study on the Bending Properties of Grouting Butt Joints Reinforced by Steel Plate Embedded in Bamboo Tube. *J. Renew. Mater.* **2022**, *10*, 993–1005. [CrossRef]
14. Moran, R.; García, J.J. Bamboo joints with steel clamps capable of transmitting moment. *Constr. Build. Mater.* **2019**, *216*, 249–260. [CrossRef]
15. Lefevre, B.; West, R.; O'Reilly, P.; Taylor, D. A new method for joining bamboo culms. *Eng. Struct.* **2019**, *190*, 1–8. [CrossRef]
16. Nie, S.D.; Yu, P.; Huang, Y.Z.; Luo, Y.; Wang, J.L.; Liu, M.; Elchalakani, M. Experimental study on compressive performance of the multiple-culm bamboo columns connected by bolts. *Eng. Struct.* **2024**, *303*, 117525. [CrossRef]
17. Yang, C.C.; Zhuo, X. Research on the Axial Compressive Experiment of Single Tube and Four-tube Bundle of Moso Bamboo. *Chin. Overseas Archit.* **2020**, *10*, 200–203. (In Chinese) [CrossRef]
18. Correal, J.F.; Prada, E.; Suárez, A.; Moreno, D. Bearing capacity of bolted-mortar infill connections in bamboo and yield model formulation. *Constr. Build. Mater.* **2021**, *305*, 124597. [CrossRef]
19. Minke, G. *Building with Bamboo*; Birkhäuser: Basel, Switzerland, 2012; pp. 47–66.
20. Liu, K.W.; Xu, Q.F.; Wang, G.; Chen, F.M.; Leng, Y.B.; Yang, J.; Harries, K.A. *Contemporary Bamboo Architecture in China*; TsingHua University Press: Beijing, China, 2022; pp. 31–55.
21. Zhuo, X.; Dong, S.L. Bamboo tube bundle spatial lattice structure system and construction technology. *Spat. Struct.* **2021**, *27*, 3–8. (In Chinese) [CrossRef]
22. Zhuo, X.; Dong, S.L. Frame-unit prefabricated bamboo culm lattice structure system and engineering practices. *J. Build. Struct.* **2024**, *45*, 43–51. (In Chinese) [CrossRef]
23. Yu, L.; Wei, J.; Li, D.; Zhong, Y.; Zhang, Z. Explaining Landscape Levels and Drivers of Chinese Moso Bamboo Forests Based on the Plus Model. *Forests* **2023**, *14*, 397. [CrossRef]

24. Zhu, C.M.; Zhang, W.P. *Structural Mechanics*; Higher Education Press: Beijing, China, 2016; Volume 2, pp. 51–54. (In Chinese)
25. García, J.; Rangel, C.; Ghavami, K. Experiments with rings to determine the anisotropic elastic constants of bamboo. *Constr. Build. Mater.* **2010**, *31*, 52–57. [CrossRef]
26. Zhang, X.X.; Yu, Z.X.; Yu, Y.; Wang, H.K.; Li, J.H. Axial compressive behavior of Moso Bamboo and its components with respect to fiber-reinforced composite structure. *J. For. Res.* **2019**, *30*, 2371–2377. [CrossRef]
27. Walter Liese; Michael Köhl. *Bamboo the Plant and Its Use*; Springer International Publishing: Cham, Switzerland, 2015; pp. 251–253.
28. Peng, H.; Zhuo, X. *Research on Mechanical Behavior of the Screwed Connection at the End of Bamboo Strip*; Zhejiang University: Hangzhou, China, 2022. (In Chinese) [CrossRef]

 *buildings*

Article

Fundamental Properties of Sub-THz Reflected Waves for Water Content Estimation of Reinforced Concrete Structures

Akio Tanaka [1,*], Koji Arita [2], Chihiro Kobayashi [3], Tomoya Nishiwaki [3], Tadao Tanabe [4] and Sho Fujii [5]

[1]   Department of Architecture, Nippon Institute of Technology, Saitama 345-8501, Japan
[2]   Constec Engi,Co., Tokyo 143-0006, Japan; arita-koji@cons-hd.co.jp
[3]   Graduate School of Engineering, Tohoku University, Sendai 980-8577, Japan;
      chihiro.kobayashi.t7@dc.tohoku.ac.jp (C.K.); tomoya.nishiwaki.e8@tohoku.ac.jp (T.N.)
[4]   College of Engineering and Design, Shibaura Institute of Technology, Tokyo 135-8548, Japan;
      tanabet@shibaura-it.ac.jp
[5]   Faculty of Science, Yamagata University, Yamagata 990-0021, Japan; sfujii@sci.kj.yamagata-u.ac.jp
*   Correspondence: tanaka.akio@nit.ac.jp

Abstract: Water plays a significant role in the deterioration of reinforced concrete buildings; therefore, it is essential to evaluate the water content of the cover concrete. This study explores a novel non-destructive method for assessing the water content using sub-terahertz (sub-THz) waves. Among the four frequencies selected to evaluate the water content, an increase in reflectance was observed as the unit volume water content increased, and smaller data scatter was confirmed as the frequency increased. The derived empirical equation can classify the corrosion risk of the rebar environment based on the water content obtained using reflectance measurements. In other words, it can contribute to the diagnosis of the building integrity associated with rebar corrosion.

Keywords: sub-THz waves; reflection resistivity; measurement angle; volume water content; rebar corrosion

Citation: Tanaka, A.; Arita, K.; Kobayashi, C.; Nishiwaki, T.; Tanabe, T.; Fujii, S. Fundamental Properties of Sub-THz Reflected Waves for Water Content Estimation of Reinforced Concrete Structures. *Buildings* **2024**, *14*, 1076. https://doi.org/10.3390/buildings14041076

Academic Editors: Eva O.L. Lantsoght and Mizan Ahmed

Received: 7 March 2024
Revised: 3 April 2024
Accepted: 10 April 2024
Published: 12 April 2024

1. Introduction

Reinforced concrete structures, recognized for their exceptional durability and economic efficiency, began to be put to practical use in the world in the 20th century. In Japan, the extensive adoption of reinforced concrete structures was catalyzed by the Great Kanto Earthquake in 1923, the period of rapid economic growth beginning in 1955, and the Tokyo Olympics in 1964. Nowadays, many buildings constructed during these backgrounds have reached the end of their service life due to deterioration.

The critical condition of reinforced concrete structures is defined by losing their alkaline atmosphere due to the carbonation of the cover concrete, and this carbonation reaches the location of the steel bars [1]. Rebar corrosion depends on the environmental conditions of the building and is strongly observed in areas subject to repetitive dry and wet conditions [2]. In addition, many buildings have been damaged or deteriorated due to natural disasters that have occurred frequently in recent years. Damaged buildings may not be structurally safe enough to be surveyed from a close distance. Therefore, an inspection method is needed to capture the damaged state of the interior of building materials by non-destructive inspections from a long distance.

Traditional measurement methods for detecting the internal deterioration of reinforced concrete structures, such as hammering methods [3], percussion acoustic methods represented by ASTM C1383-15 (2022) [4–6], ultrasonic methods [7,8], and ground penetrating radar (GPR) methods [9–13], require direct contact or close proximity to the target object and are constrained by the nature of the target. Infrared thermography is widely used in non-contact diagnoses of the external surfaces of buildings [14–16]. However, solar radiation is a necessary condition for the survey. For this method, it is imperative to consider the variations in height, time, and measurement angle relative to the temperature fluctuations of the target building, and the survey timing is limited due to the influence of climatic conditions and the surrounding environment. In addition, the thermography technique makes it difficult to detect delamination as the thickness of the concrete increases. It is impossible to detect cracks or evaluate the environment for accelerated deterioration based on the water content. X-ray methods are promising alternatives with non-contact procedures to inspect the interior of concrete with high resolution [17–19]. On the other hand, this method uses high energy that is dangerous to the human body and is not easy to handle in the field.

In this study, measurements were made using low-frequency sub-terahertz (hereafter, sub-THz) waves (8.5 to 12.5 GHz) with a millimeter wave spectrum, which are used in electromagnetic radar techniques for rebar detection and meteorological radar. The sub-THz wave is minimal and safe energy, yet it can penetrate the concrete interior. The high absorption characteristics of the sub-THz wave by water may offer the potential to measure the interior water content of concrete, which has a significant impact on rebar corrosion. To evaluate the water content of concrete, the sub-THz measurement system was developed and applied. This study scrutinized the influence of the measurement conditions of the optical system equipment through a three-step experimental process: first, the measurement distance, area, and angle are determined (STEP1); then, the water content of concrete is determined (SETP2); finally, the condition of the steel bars in reinforced concrete is determined (STEP3). Through this process, we aim to examine the evaluation of water content using sub-THz waves.

2. Outline of Sub-THz Waves

2.1. Definition of Sub-THz Wave

Terahertz (THz) waves are electromagnetic waves characterized by frequencies and wavelengths in the range of 0.3 to 10 THz and 30 μm to 1 mm in a vacuum, respectively. They exhibit high penetration capabilities through non-polar materials, while demonstrating significant absorption characteristics when interacting with polar materials like water. Sub-terahertz waves, with frequencies and wavelengths spanning from 0.03 to 0.3 THz and 1 to 10 mm in a vacuum, respectively, share similar attributes to THz waves but offer superior transmission properties. The frequencies employed in our study, 8.5 to 12.5 GHz, fall slightly below the conventional sub-THz range of 0.1 to 1 THz. However, our research group has been investigating the application potential across a broad frequency spectrum from approximately 0.01 to 10 THz, encompassing frequencies higher than those discussed in this paper [20–25]. Based on this foundation, we have classified this spectrum as the "sub-THz" range.

2.2. Sub-THz Wave Optical System

Figure 1 shows schematic diagrams of the (a) transmission and (b) reflection methods in the sub-THz wave optical system. There are two measurement methods for sub-THz wave optics: the transmission method [20] and the reflection method [21]. The transmission method requires the oscillator and detector to be placed in a straight line, which is difficult to align in measurements on real structures. Therefore, in this experiment, the reflection method with a half-mirror was mainly employed.

In the reflection method, as shown in Figure 1b, sub-THz waves emitted from the oscillator with the horn antenna pass through a Teflon lens and a half-mirror. The target object reflects the wave with a specific intensity of reflectance ratio corresponding to the conditions. The detector receives the propagated wave through the half-mirror and Teflon lens. Teflon lenses convert sub-THz waves into collimated beams. The half-mirror plays the role of straight transmitting the incident wave and reflecting to change the angle of the reflected wave from the target specimen. Table 1 shows the specifications of the sub-THz wave optical system used in this study.

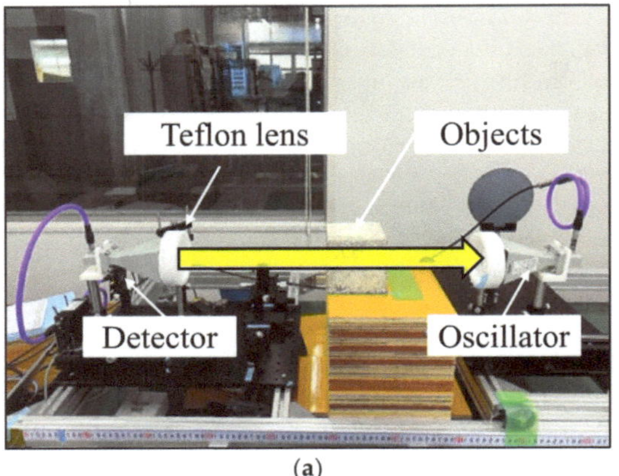

(a)

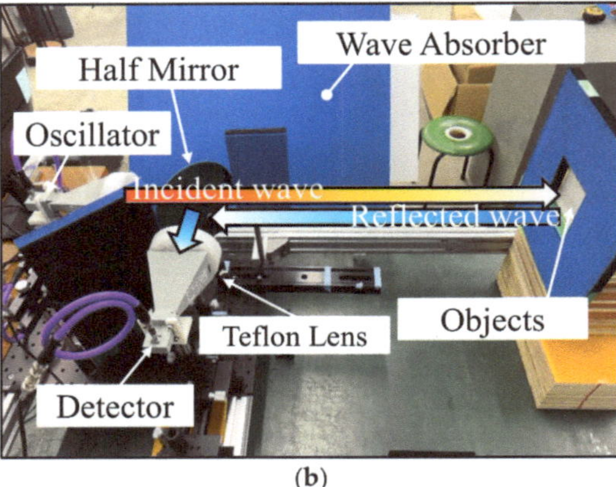

(b)

Figure 1. Sub-THz wave optical system. (**a**) Transmission method. (**b**) Reflection method.

Table 1. Specification of the sub-THz wave optical system.

Device	Specification
Oscillator	Frequency range: 7.5 to 15.0 GHz
AMP	0 to 31.5 dBm at 0.5 dBm, Max 1 W, Frequency range 5 to 20 GHz
Hone Antenna	Frequency range: 8.2 to 12.4 GHz VSWR: 1:15:1 Typ
Teflon Lens	φ76.2 mm
Half Mirror	Silicon wafer, Type: Non-dope, Resistivity: over 1000 ohm.cm
Detector	Frequency range:10 to 18.5 GHz Voltage Sensitivity: 450 mV/mW
Wave Absorber	Supported frequency bands: C, A, B, X, Ku, K

Here, reflectivity is defined in terms of Equation (1) below. The numerator is the absolute value of the reflection intensity as measured with an oscilloscope. The denominator is the absolute value of the reflection intensity as measured with an aluminum plate, which is considered to have 100% reflection or low electromagnetic wave loss. In general, reflectivity is the ratio of reflected energy to incident energy when the incident is perpendicular to an object. However, the sub-THz waves used in this study include several unclear aspects such as the amount of the incidence, reflection, and scattering. Therefore, the reflectivity was calculated using Equation (1).

$$\text{Reflectivity}(\%) = \frac{\text{Detection intensity of specimens (mV)}}{\text{Detection intensity of aluminum plate (mV)}} \tag{1}$$

3. Experimental Outlines

3.1. Influence of Measurement Conditions of Optical System Equipment (STEP1)

3.1.1. Attenuation Effects by Measurement Distance

In general, electromagnetic waves are attenuated by atmospheric moisture and air density. Therefore, the farther the distance to the target object, the lower the measured intensity. In addition, the energy required depends on the depth of the defect to be detected. Thus, it is necessary to determine an appropriate emitting intensity according to the distance. In order to understand the electromagnetic characteristics of sub-THz waves, the attenuation effect of sub-THz waves with distance was measured with the transmission and reflection methods. The employed measurement frequencies ranged from 8.5 to 12.5 GHz at 0.5 GHz intervals. For the transmission method, the measurement distance was 610 to 3000 mm at 10 mm intervals. For the reflection method, the measurement distance was 420 to 650 mm at 5 mm spacing, 650 to 1000 mm at 10 mm spacing, and 1000 to 3000 mm at 500 mm spacing.

The measurable strength limit of the sensor used in the detector is approximately 2000 mV. In addition, there is a risk that a higher intensity can exceed the intensity limit due to interference and other effects. Therefore, sub-THz waves were oscillated with an output of 18.5 dBm to obtain an intensity of 1000 mV at a measurement distance of 420 mm for both the transmission and reflection methods. Since electromagnetic waves generally diffuse with distance, evaluation on a logarithmic axis is desirable. Therefore, we substituted the obtained intensity (mV) into Equation (2) to obtain and evaluate the radio wave intensity.

$$C_{ss} = 20\log(rf) - 20\log(230) \tag{2}$$

where, C_{ss} is the converted intensity (dBm), and rf is the measured intensity (mV).

3.1.2. Irradiation Range

In general, electromagnetic waves diffuse with distance even when polarized into a collimated beam using the Teflon lens. Therefore, the irradiation range according to distance was experimentally investigated. Figure 2 shows a schematic diagram of the irradiation range measurement. The measurement distance is from 600 to 2400 mm, with the interval of 600 mm. The frequency range was 8.5 to 12.5 GHz. The measurement target to be measured was a 300×300 mm plate made of an aluminum alloy (A5052 according to JIS H 4000:2022 "Aluminium and aluminium alloy sheets, strips and plates" of Japan Industrial Standards [26], corresponding to AlMg2.5 of ISO 209 [27]) with a high reflectance and thickness accuracy of ± 0.05 mm or less. The blue and green areas shown in Figure 2 represent electromagnetic wave absorbers based on urethane foam, possessing absorption performance for frequencies ranging from 0.6 to 50 GHz. The purpose of the blue absorber was to eliminate the influence on the measurement results caused by the divergence of electromagnetic waves reflected by the surrounding environment. The green absorber was placed in front of the 300×300 mm aluminum plate to examine the irradiation range. The green absorber was enlarged by 20 mm each time to investigate the irradiation range based on the attenuation of intensity. The minimum visible size of the target aluminum plate was set to 100×100 mm. The reflectivity R_{ir} was calculated using Equation (3).

$$R_{ir}(\%) = \frac{\text{Intensity with the absorbers (mV)}}{\text{Intensity without any absorbers (mV)}} \tag{3}$$

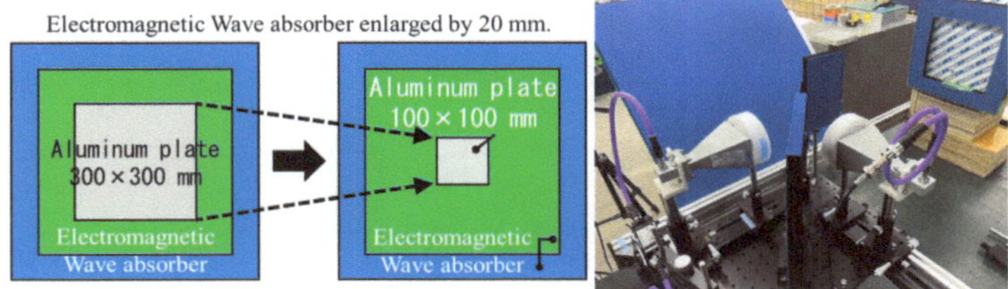

Figure 2. Schematic diagram of the irradiation range measurement.

3.1.3. Measurement Angle

When measuring actual building structures, the measurement environment, including the location conditions of the target structures, often precludes the possibility of conducting measurements directly from the front. Therefore, this section aims to experimentally verify the influence of the measurement angle. Measurements were conducted using an aluminum plate to simulate the target object. Figure 3 shows a schematic diagram of the measurement to assess the effect of angles. The aluminum plate was rotated by 1 degrees in each direction, with clockwise rotation as positive and counterclockwise rotation as negative. The measurement distance ranged from 1200 to 2400 mm, with measurements taken at 600 mm intervals. The reflectivity R_{an} was calculated using Equation (4).

$$R_{an}(\%) = \frac{\text{Intensity with the inclined angle (mV)}}{\text{Intensity with the angle of 0 degrees (mV)}} \tag{4}$$

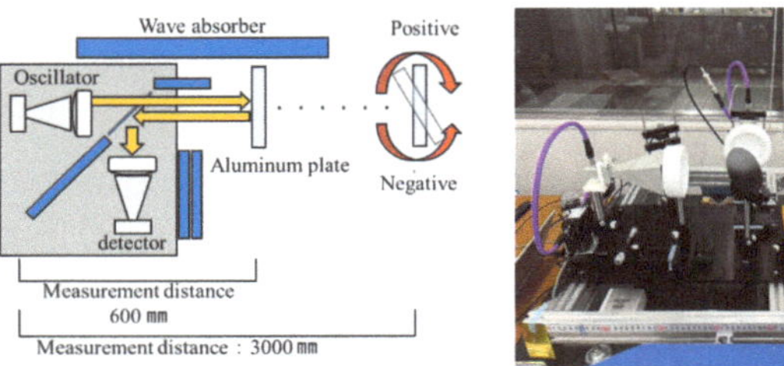

Figure 3. Schematic diagram of the effect of measurement angle.

3.2. Determine Water Content of Concrete (STEP2)

Outlines of Concrete Specimen

In this step of the process, concrete specimens were prepared to measure their water content. Table 2 shows the mix proportion of the concrete specimens. Tap water was used as mixing water (W). Ordinary Portland cement (density: 3.16 g/cm^3) was used for cement (C). The fine aggregate (S) was land sand (density: 2.58 g/cm^3), and the coarse aggregate (G) was crushed hard sandstone with a size from 5 to 20 mm (density: 2.65 g/cm^3). In Japan, concrete buildings typically employ concrete with a compressive strength ranging from 18 to 45 N/mm^2. The average water–cement ratio corresponding to these compressive strengths is approximately 50 to 58%; our study utilized concrete with a W/C ratio of 50 to 58%.

Table 2. Mix proportion of concrete.

W/C (%)	s/a (%)	Unit Amount (kg/m^3)				Geometry (mm)
		W	C	S	G	
50	49	172	344	851	912	$100 \times 100 \times 100$
53	49	172	287	875	936	$200 \times 200 \times 200$
55	47.4	162	295	870	965	$780 \times 780 \times 150$
58	44	176	303	780	1012	$780 \times 780 \times 150$

The geometries of the specimens were cube types of $100 \times 100 \times 100$ and $200 \times 200 \times 200$ mm, and wall types of $780 \times 780 \times 150$ mm. The specimens were demolded at 7 days from casting and cured in water until 28 days of age. The 200 mm cube specimens were then cut into plate specimens with a thickness of 30 and 50 mm using a concrete saw. All specimens were dried at 105 °C for 2 weeks to define the absolute dried condition. For the air-dried condition, the specimens after drying were maintained in a thermostatic chamber at 20 °C and 60% humidity. Moreover, for the wet condition, the specimens after drying were allowed to absorb moisture in a desiccator at 20 °C and 100% humidity.

The 100 mm cube specimens were subjected to three levels of water content: dry (0.4% water content), air-dry (2.0% water content), and wet (4.3% water content). The water content of the actual structure tends to increase from the surface towards the interior.

Therefore, the $200 \times 200 \times 30$ and $200 \times 200 \times 50$ mm specimens were allowed to absorb water from the cut surfaces to simulate the water content trend of the actual structure.

Here, measurements were conducted using frequencies of 9.4, 10.9, 11.5, and 12.3 GHz, where a preliminary investigation confirmed a trend of increasing the reflectance with the higher water content [28]. Measurements were carried out at a 900 mm distance.

In case of the wall specimens, the risk of corrosion in the reinforcing rebars inside concrete increases with the high water content surrounding them. To simulate the corrosive environments at any given point within the concrete specimens, we adopted the water injection method to control the water content. The water content can be adjusted by holes which are 20 mm in diameter and 100 mm deep; these holes were drilled at 100 mm intervals from the center of the specimen. Water was injected to an area of approximately 300×300 mm from the side opposite to the measurement surface of the specimen to partially change the water content. Aluminum tape was attached to the left edge of the specimen to serve as the reflectivity reference.

3.3. Effect of Rebar in Reinforced Concrete Members (STEP3)

3.3.1. Outlines of Reinforce Concrete Specimen

In this step of the process, reinforced concrete specimens were used to investigate the effect of rebar within concrete on the measurement results. The specimens were made of concrete with a water–cement ratio of 58% (shown in Table 2). Figure 4 shows an overview of the reinforced concrete member. In this study, the influence of the reinforcement condition on the water content estimation has been considered as an experimental parameter. Therefore, the over-dense to sparse arrangement was employed. The geometry of the wall specimen was $780 \times 780 \times 150$ mm with D13 deformed bars embedded at a cover depth of 30 mm and spacing of 40, 125, 200, and 250 mm to ensure non-uniform reinforcement. The longitudinal rebars were placed at a cover thickness of 30 mm.

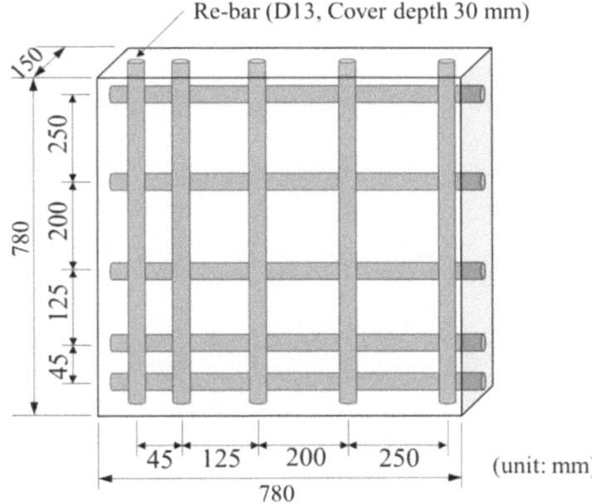

Figure 4. Overview of the reinforced concrete member.

3.3.2. Experimental Procedures

The measurement distance was 900 mm, and the frequency was 12.3 GHz, which is considered optimal for determining the water content of concrete surfaces as described below. The measurement angle was set at 90 degrees from the front of the specimen, which is directly in front of the specimen. The angle of the optical system was changed by 5 degrees to 15 degrees (the measurement angles: 90, 85, 80, and 75 degrees, as shown in Figure 5). The optical system was moved at 100 mm intervals in the vertical and horizontal directions to measure each area of the wall specimen, and a total of 42 points were measured (6 rows (A to F) in the vertical direction and 7 (1 to 7) columns in the horizontal direction) to evaluate the water content of the entire concrete wall.

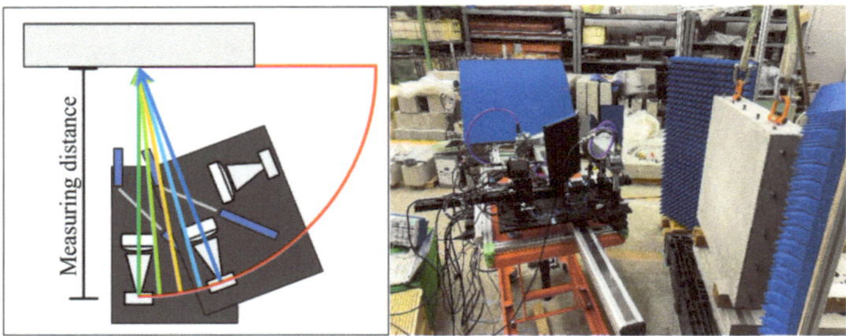

Figure 5. Measuring situation of wall specimens.

The water content of the specimens was measured using a contact-type electrical resistance water content meter, which can be easily used for measurements by pressing it against the concrete surface. The objective of the employed device was to qualitatively measure the water content. The specimen was dried up to the same measured values of the controlled specimen with the water content of 3.5%.

4. Experimental Results and Discussion

4.1. Influence of Measurement Conditions of Optical System Equipment (STEP1)

4.1.1. Attenuation Effects by Measurement Distance

Figures 6 and 7 show the results of distance attenuation by using the transmission and reflection methods, respectively. For both the transmission and reflection methods, it was confirmed that the intensity attenuates as the measurement distance increases. In the reflection method, the detected intensity attenuates with significant amplitude fluctuations compared to the transmission method. This is due to interference between the incident and reflected waves. In the reflection method, detection intensities higher than the output of 18.5 dBm were obtained at 10, 0.5, and 11.5 GHz up to a measurement distance of approximately 600 mm. This is presumably due to the interference between the incident and reflected waves, and also because the detection sensitivity of the detector increases with the high frequency.

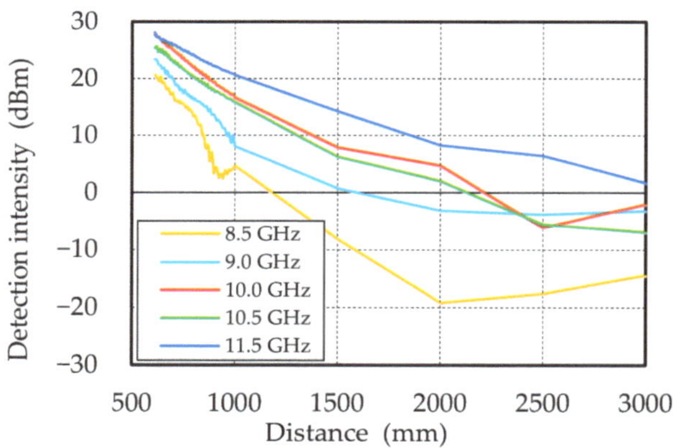

Figure 6. Distance attenuation (transmission method).

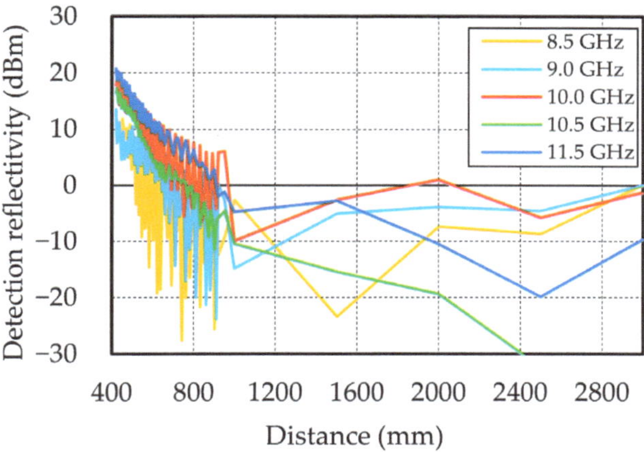

Figure 7. Distance attenuation (reflection method).

4.1.2. Relationship between the Measurement Distance and Irradiation Area

Figure 8 shows the irradiation range according to the measurement distance. The scale bar of reflectivity R_{ir} was calculated using Equation (3). A high R_{ir} indicates a red color, and a large red area indicates the diffusion of the sub-THz waves. In contrast, green indicates a low R_{ir}, and a large green area means high directivity. In the case where the measurement distance was 600 mm, a high reflectivity was obtained in the 100×100 mm range, confirming that the Teflon lenses were able to provide generally parallel beams. When the measurement distance is more than 1200 mm, the electromagnetic wave diffuses down to about 200 mm in diameter, and the irradiation area is expanded.

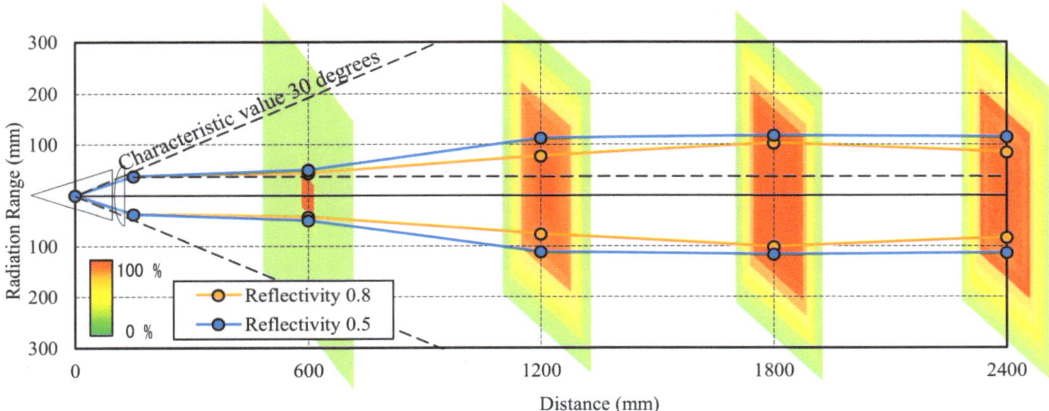

Figure 8. Relationship distance and radiation range.

When this method is practically used on-site, a greater distance to the target structure building might be needed. Therefore, it is essential to examine the dispersion of the sub-THz wave at further measurement distances.

4.1.3. Relationship between the Measurement Distance and Measurement Angle

Figure 9 shows the effect of the measurement angle. The positive and negative rotation angles produced a reflectivity R_{an} that was symmetrical with respect to 0 degrees. Within an angle range of ± 2 degrees, R_{an} shows approximately more than 80%. Furthermore, within an angle range of ±5 degrees, a linear decreasing trend of R_{an} is observed. Beyond these angles, fluctuations in R_{an} due to the interference of sub-THz waves can be confirmed.

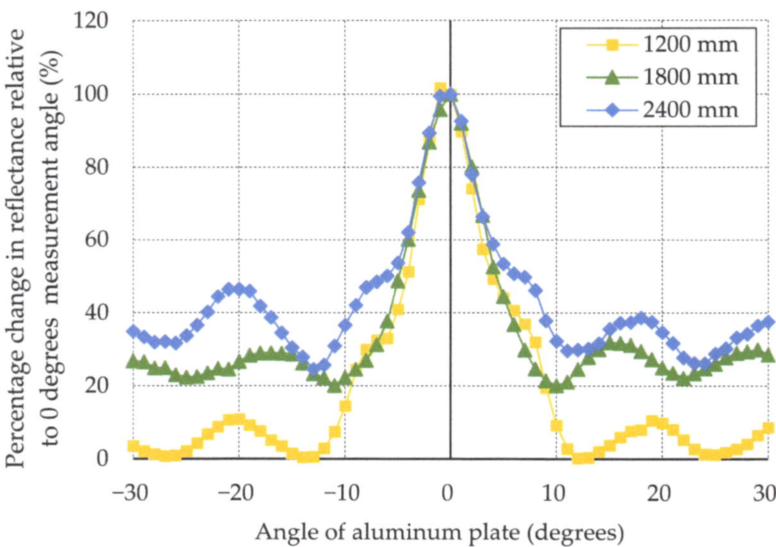

Figure 9. Effect of the measurement angle.

It is therefore predicted that the oblique incidence of sub-THz waves at different measurement angles will reduce the accuracy of detection. Therefore, the decrease in reflectance due to a 1 degrees deviation in measurement angle must be fully considered, and further studies on the allowable range of measurement angles and correction methods are needed.

4.2. Study to Determine Water Content of Concrete (STEP2)

4.2.1. Relationship between Unit Volume Water Content and Reflectance

The results of the measurements of concrete with different water contents suggest that in the transmission method, the strength decreases as the water content increases, while in the reflection method, as the water content increases, the reflected strength also tends to increase [28].

Figure 10 shows the relationship between the unit volume water content and reflectivity calculated using Equation (1). The result of 9.4 GHz shows a linear trend between the unit volume water content and reflectivity when the three specimens are evaluated individually. However, the relationship includes a large variation. The reflectivity tended to increase as the unit volume water content increased, regardless of the test specimen. Within the range of this experiment, the variation in the reflectivity tended to decrease as the frequency increased, with the smallest variation at 12.3 GHz. It can be inferred from this that the measurement at 12.3 GHz is the most suitable among the employed frequencies for determining the water content at the cover concrete.

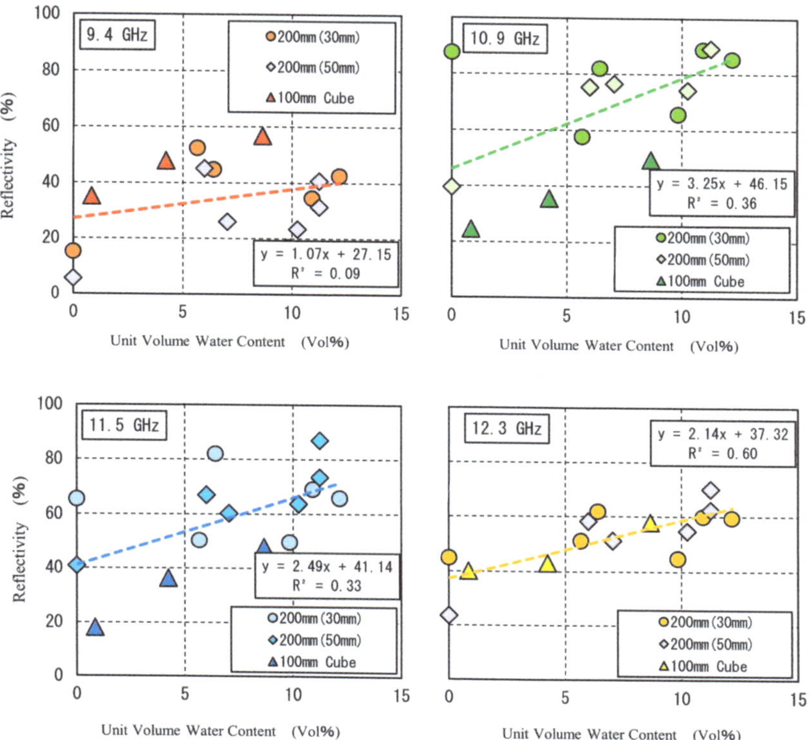

Figure 10. Relationship between the unit volume water content and reflectance.

4.2.2. Estimation of Water Content Distribution at the Water Inlet Location

The results suggest that it may be possible to identify the location of the water injection and the location of the dry state in the transmission method [20]. The same results are expected to be obtained in the present study; however, they do not necessarily coincide with those obtained for a specimen simulating a real structure. Therefore, experiments were conducted under conditions in which the moisture content and reflected intensity could be determined, and the intensity obtained from the reflection method was used to stockpile basic data for the water content evaluation as follows.

Figure 11 shows the estimated distribution of water content at the locations of the water injections. The yellow framed areas indicate the locations where water was injected from the backside of the specimen. The color scale of this figure shows that the reflectivity increases in the order of red, white, and blue.

At the measurement results with 9.4 GHz, high reflectivity was observed at the right end of the specimens and low reflectivity was observed at the left side of the specimens; at 10.9 GHz, no high reflectivity was observed at the water injection point and high reflectivity was observed at the center and bottom of the specimens; at 11.5 and 12.3 GHz, high reflectivity was observed near the water injection point, suggesting that water was detected. These results suggest that differences in the understanding of water distribution were obtained due to the relationship between the penetration depth of the injected water and the wavelength of the frequency, as well as differences in the transmittance of the frequency.

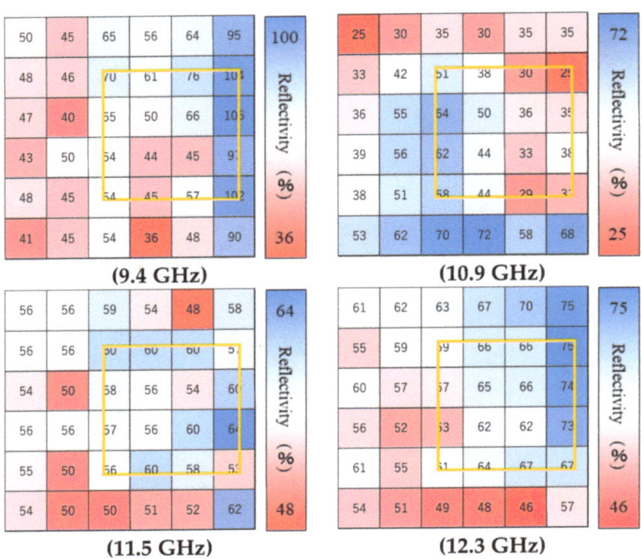

Figure 11. Estimated distribution of water content at the water inlet locations.

4.3. Effect of Measurement Angle on Reinforced Concrete Members (STEP3)

Figure 12 shows the intensity distribution of the measured angles of the non-uniformly reinforced concrete specimen measured using the frequency of 12.3 GHz. The color scale in the figure indicates the ascending order of the detection intensity from green to white to red.

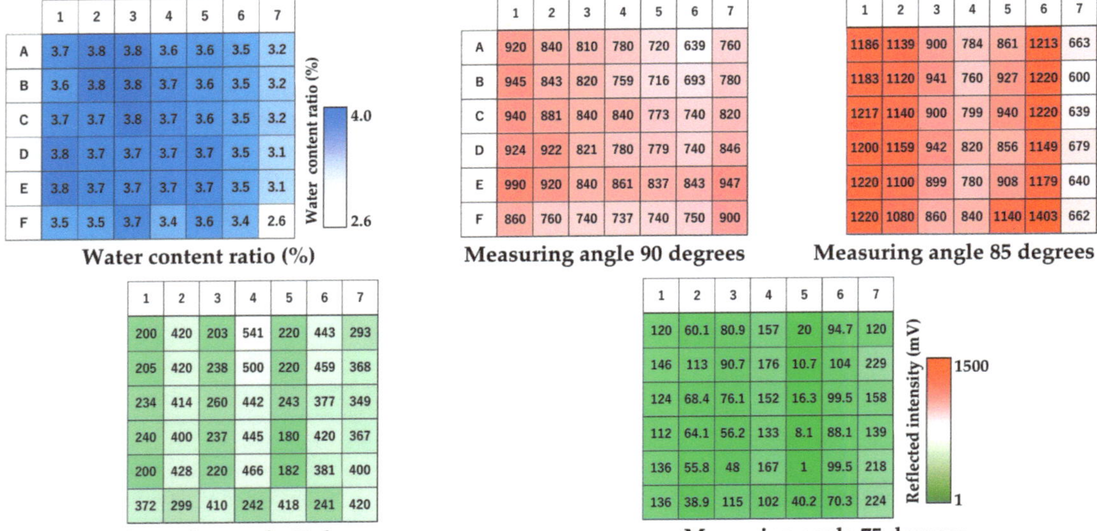

Figure 12. Reflected intensity as a function of measurement angle for reinforced concrete members.

The highest detection intensity was obtained at a measurement angle of 85 degrees, beyond which the detection intensity diminished with a decreasing measurement angle. The distribution of detection intensity at a 90 degrees angle was similar to the distribution of water content. However, no correlation between the detection intensity and water content distribution was noted at the other measurement angles.

At measurement angles of 85, 80, and 75 degrees, detection intensities resembling vertical stripes were observed, predominantly at the locations corresponding to the rebar, indicating its detection. Although 12.3 GHz has no significant penetrating performance through concrete, Figure 13 suggests that the reflected sub-THz waves from the concrete surface follow a different propagation path compared to the incident waves, leading to the reduced detected intensity. However, the sub-THz waves that penetrated the concrete were reflected by the rebar and traced the same path as the incident wave, which is assumed to have resulted in a clear difference in detection intensity.

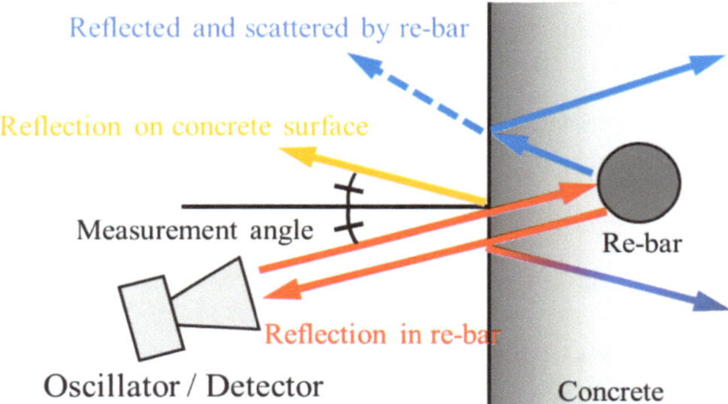

Figure 13. Propagation path conceptual diagram.

5. Conclusions

The following findings were obtained regarding the selection of frequencies and the applicability of the selected frequencies for the purpose of determining the water content of concrete from a distance using sub-THz waves:

(1) The effects of the measurement angle were systematically investigated: when the angle between the object and the measurement device was established at 90 degrees as the reference, variations of ±2 degrees resulted in a reduction of up to 80% in reflection intensity. This phenomenon exhibited a linear trend up to a deviation of ±5 degrees; beyond a deviation of ±10 degrees, an accurate assessment of the object's condition might be compromised due to the effects of secondary reflections and electromagnetic wave scattering. These findings suggest that, for the purpose of evaluating the condition of the concrete based on the reflectivity data, the measurement angle should ideally be maintained within a range of ±5 degrees from the 90-degree reference.

(2) The irradiation range of the sub-THz wave differs depending on the measurement distance. The farther the measurement distance, the larger the irradiation range, and a more diffuse electromagnetic wave was observed. However, it was confirmed that the light was generally focused at about 200×200 mm.

(3) In the estimation of the water content of concrete using the reflection method, the trend of increasing reflectance with increasing water content was confirmed by selecting frequencies (9.4, 10.3, 11.5, and 12.3 GHz). The results indicate that when moisture content distribution is generated in concrete wall members, the selected frequencies may be used to identify locations with high water content.

(4) Upon conducting measurements of the reinforced concrete wall specimens from varying angles, it was discerned that the relative impact of the concrete surface layer as opposed to that of the reinforcing bars exhibited variance at a measurement angle of approximately 90 ± 10 degrees, contingent upon the specific angle of measurement employed. This variance is ostensibly attributed to the propagation pathways of the reflected waves. Consequently, the findings indicate that, for the purpose of assessing

the moisture content within the concrete surface layer, the measurement angle ought to be meticulously maintained within a range of 90 ± 5 degrees.

In this paper, the potential of using sub-THz waves for estimating the water content of concrete surface layers from a distance was demonstrated. The use of sub-THz waves is not limited to estimating water content but is also applicable for identifying cracks inside concrete caused by reinforcing bar corrosion, fire loading, and so on. However, since studying cracks requires measuring defects smaller than the wavelength of the frequency used, it is crucial to understand electromagnetic wave characteristics, including interference, scattering, and absorption by the rebar and propagation paths.

Therefore, as a direction for future research, to precisely obtain information on the interior of concrete, especially targeting cracks, it is essential to comprehend the electromagnetic wave characteristics, construct a measurement system that can disregard the effects of interference, and process information using AI.

Author Contributions: Conceptualization, A.T. and T.N.; methodology, S.F. and T.T.; validation, K.A. and C.K.; formal analysis, K.A.; investigation, K.A.; data curation, K.A.; writing—original draft preparation, A.T.; writing—review and editing, T.N.; visualization, K.A.; supervision, T.N.; project administration, T.N.; funding acquisition, A.T. All authors have read and agreed to the published version of the manuscript.

Funding: The APC was funded by JAEA Nuclear Energy S&T and Human Resource Development Project Grant Number JPJA21P21458909.

Data Availability Statement: Data are contained within the article.

Conflicts of Interest: Koji Arita was employed by Constec Engi,Co. The remaining authors declare that the research was conducted in the absence of any commercial or financial relationships that could be construed as a potential conflict of interest.

References

1. Architectural Institute of Japan (AIJ). *Recommendations of Durability Design and construction Practice of Reinforced Concrete Buildings*, 2nd ed.; AIJ: Tokyo, Japan, 2016. (In Japanese)
2. Beushausen, H.; Luco, L.F. Performance-Based Specifications and Control of Concrete Durability: State-of-the-Art Report RILEM TC 230-PSC. In *Performance-Based Specifications and Control of Concrete Durability: State-of-the-Art Report RILEM TC 230-PSC*; Springer: Berlin/Heidelberg, Germany, 2015; Volume 18, pp. 1–373. [CrossRef]
3. Kazemi, M.; Madandoust, R.; de Brito, J. Compressive strength assessment of recycled aggregate concrete using Schmidt rebound hammer and core testing. *Constr. Build. Mater.* **2019**, *224*, 630–638. [CrossRef]
4. *ASTM C1383-15:2022*; Standard Test Method for Measuring the P-Wave Speed and the Thickness of Concrete Plates Using the Impact-Echo Method. American Society for Testing and Materials (ASTM): West Conshohocken, PA, USA, 2022.
5. Kumar, A.; Raj, B.; Kalyanasundaram, P.; Jayakumar, T.; Thavasimuthu, M. Structural integrity assessment of the containment structure of a pressurised heavy water nuclear reactor using impact echo technique. *NDT E Int.* **2002**, *35*, 213–220. [CrossRef]
6. Zheng, L.; Cheng, H.; Huo, L.; Song, G. Monitor concrete moisture level using percussion and machine learning. *Constr. Build. Mater.* **2019**, *229*, 117077. [CrossRef]
7. Lawson, I.; Lawson, I.; Danso, K.A.; Odoi, H.C.; Adjei, C.A.; Quashie, F.K.; Mumuni, I.I.; Ibrahim, I.S. Non-Destructive Evaluation of Concrete using Ultrasonic Pulse Velocity. *Res. J. Appl. Sci. Eng. Technol.* **2011**, *3*, 499–504. Available online: https://www.researchgate.net/publication/265144573 (accessed on 10 February 2024).
8. Choi, P.; Kim, D.-H.; Lee, B.-H.; Won, M.C. Application of ultrasonic shear-wave tomography to identify horizontal crack or delamination in concrete pavement and bridge. *Constr. Build. Mater.* **2016**, *121*, 81–91. [CrossRef]
9. Hugenschmidt, J.; Mastrangelo, R. GPR inspection of concrete bridges. *Cem. Concr. Compos.* **2006**, *28*, 384–392. [CrossRef]
10. Chang, C.W.; Lin, C.H.; Lien, H.S. Measurement radius of reinforcing steel bar in concrete using digital image GPR. *Constr. Build. Mater.* **2009**, *23*, 1057–1063. [CrossRef]
11. Laurens, S.; Balayssac, J.P.; Rhazi, J.; Klysz, G.; Arliguie, G. Non-destructive evaluation of concrete moisture by GPR: Experimental study and direct modeling. *Mater. Struct.* **2005**, *38*, 827–832. [CrossRef]
12. Wiwatrojanagul, P.; Sahamitmongkol, R.; Tangtermsirikul, S.; Khamsemanan, N. A new method to determine locations of rebars and estimate cover thickness of RC structures using GPR data. *Constr. Build. Mater.* **2017**, *140*, 257–273. [CrossRef]
13. Liu, H.; Lin, C.; Cui, J.; Fan, L.; Xie, X.; Spencer, B.F. Detection and localization of rebar in concrete by deep learning using ground penetrating radar. *Autom. Constr.* **2020**, *118*, 103279. [CrossRef]
14. Janků, M.; Březina, I.; Grošek, J. Use of Infrared Thermography to Detect Defects on Concrete Bridges. *Procedia Eng.* **2017**, *190*, 62–69. [CrossRef]

15. Cheng, C.-C.; Cheng, T.-M.; Chiang, C.-H. Defect detection of concrete structures using both infrared thermography and elastic waves. *Autom. Constr.* **2008**, *18*, 87–92. [CrossRef]
16. Omar, T.; Nehdi, M.L. Remote sensing of concrete bridge decks using unmanned aerial vehicle infrared thermography. *Autom. Constr.* **2017**, *83*, 360–371. [CrossRef]
17. Michel, A.; Pease, B.J.; Geiker, M.R.; Stang, H.; Olesen, J.F. Monitoring reinforcement corrosion and corrosion-induced cracking using non-destructive x-ray attenuation measurements. *Cem. Concr. Res.* **2011**, *41*, 1085–1094. [CrossRef]
18. Skarżyński, Ł; Tejchman, J. Experimental Investigations of Fracture Process in Concrete by Means of X-ray Micro-computed Tomography. *Strain* **2016**, *52*, 26–45. [CrossRef]
19. Lukovic, M.; Schlangen, E.; Ye, G.; van Breugel, K. Moisture exchange in concrete repair system captured by X-ray absorption. In *Concrete Repair, Rehabilitation and Retrofitting IV*; CRC Press: Boca Raton, FL, USA, 2016.
20. Nishiwaki, T.; Shimizu, K.; Tanabe, T.; Gardner, D.; Maddalena, R. Terahertz (THz) Wave Imaging in Civil Engineering to Assess Self-Healing of Fiber-Reinforced Cementitious Composites (FRCC). *J. Adv. Concr. Technol.* **2023**, *21*, 58–75. [CrossRef]
21. Kobayashi, C.; Nishiwaki, T.; Hara, S.; Tanabe, T.; Oohashi, T.; Hamasaki, H.; Hikishima, S.; Tanaka, A.; Arita, K. Fundamental research on non-destructive testing of reinforced concrete structures using sub-terahertz reflected waves. *MATEC Web Conf.* **2023**, *378*, 04007. [CrossRef]
22. Oyama, Y.; Yamagata, T.; Kariya, H.; Tanabe, T.; Saito, K. Non-destructive Inspection of Copper Corrosion via Coherent Terahertz Light Source. *ECS Trans.* **2013**, *50*, 89–98. [CrossRef]
23. Kariya, H.; Sato, A.; Tanabe, T.; Saito, K.; Nishihara, K.; Taniyama, A.; Oyama, Y. Non-destructive Evaluation of a Corroded Metal Surface Using Terahertz Wave. *ECS Trans.* **2013**, *50*, 81–88. [CrossRef]
24. Nakamura, Y.; Kariya, H.; Sato, A.; Tanabe, T.; Nishihara, K.; Taniyama, A.; Nakajima, K.; Maeda, K.; Oyama, Y. Nondestructive Corrosion Diagnosis of Painted Hot-Dip Galvanizing Steel Sheets by Using THz Spectral Imaging. *Corros. Eng.* **2014**, *63*, 411–416. [CrossRef]
25. Tanabe, T.; Watanabe, K.; Oyama, Y.; Seo, K. Polarization sensitive THz absorption spectroscopy for the evaluation of uniaxially deformed ultra-high molecular weight polyethylene. *NDT E Int.* **2010**, *43*, 329–333. [CrossRef]
26. *JIS H 4000:2022*; Aluminium and Aluminium Alloy Sheets, Strips and Plates. Japanese Industrial Standards (JIS): Tokyo, Japan, 2022.
27. *ISO 209:2007*; Aluminium and Aluminium Alloys—Chemical Composition. International Organization for Standardization (ISO): Geneva, Switzerland, 2007.
28. Arita, K.; Tanaka, A.; Nishiwaki, T. Fundamental Study of Internal Defect Detection by Sub-THz Waves for Long-Range Measurement. *Proc. Jpn. Concr. Inst.* **2023**, *45*, 1450–1455. (In Japanese)

Article

Rotational Stiffening Performance of Roof Folded Plates in Torsion Tests and the Stiffening Effect of Roof Folded Plates on the Lateral Buckling of H Beams in Steel Structures

Yuki Yoshino [1] and Yoshihiro Kimura [2,*]

[1] National Institute of Technology, Sendai College, Sendai 981-1239, Japan; yoshinoy@sendai-nct.ac.jp
[2] Graduate School of Engineering, Tohoku University, Sendai 980-8577, Japan
* Correspondence: kimura@tohoku.ac.jp; Tel.: +81-22-795-7865

Abstract: Non-structural members, such as roofs and ceilings, become affixed to main beams that are known as structural members. When such main beams experience bending or compressive forces that lead to lateral buckling, non-structural members may act to restrain the resulting lateral buckling deformation. Nevertheless, neither Japanese nor European guidelines advocate for the utilization of non-structural members as lateral buckling stiffeners for beams. Additionally, local buckling ensues near the bolt apertures in the beam–roof folded plate connection due to the torsional deformation induced by the lateral buckling of the H beam, thereby reducing the rotational stiffness of the roof folded plate to a percentage of its ideal stiffness. This paper conducts torsional experiments on roof folded plates, and with various connection methods between these plates and the beams, to comprehend the deformation mechanism of roof folded plates and the relationship between their rotational stiffness and the torsional moment. Then, the relationship between the demand values against restraining the lateral buckling of the main beam and the experimentally determined bearing capacity of the roof folded plate is elucidated. Results indicate the efficacy of utilizing the roof folded plate as a continuous brace. The lateral buckling design capacity of H beams that are continuously stiffened by roof folded plates is elucidated via application of a connection method that ensures joint stiffness between the roof folded plate and the beam while using Japanese and European design codes.

Keywords: non-structural members; lateral buckling strength; roof folded plate; rotational stiffness; torsional moment

Citation: Yoshino, Y.; Kimura, Y. Rotational Stiffening Performance of Roof Folded Plates in Torsion Tests and the Stiffening Effect of Roof Folded Plates on the Lateral Buckling of H Beams in Steel Structures. *Buildings* **2024**, *14*, 1158. https://doi.org/10.3390/buildings14041158

Academic Editor: Francisco López-Almansa

Received: 14 March 2024
Revised: 11 April 2024
Accepted: 14 April 2024
Published: 19 April 2024

1. Introduction

The collapse or detachment of non-structural members, such as roofs and ceilings, within school gymnasiums during seismic occurrences [1,2] or typhoons [3,4] poses impediments to their utilization as evacuation shelters [1–4]. The collapse or detachment of non-structural members is instigated by the imposition of bending moments or compressive forces upon main beams, which are structural members, amidst seismic forces or typhoons. These forces may incite lateral buckling within the beams. In their investigation of the lateral buckling of beams, Timoshenko and Bleich derived elastic lateral buckling load equations [5,6] based on elastic theory and analyzed them using elastic eigenvalue analysis. Notably, Nethercott studied ideal boundary conditions [7] for beam ends in moment resisting frames, while Suzuki et al. examined the effect of web deformation on the lateral buckling of beams [8]. Previous research [9–14] has demonstrated the influence of moment distribution on a range of bending moments, from uniform bending to inversely symmetric, on the lateral buckling load of beams. Research into the effect of stiffeners [15] on the restraining of lateral buckling deformations in beams has focused on stiffeners positioned at the center span [16–19] as well as discrete stiffeners, like small beams, and has been undertaken through numerical analyses [20–26] and experimental studies [27–30].

These investigations into discrete stiffening have yielded elastic lateral buckling load equations, illuminating the effects of stiffening spacing [21,22], encompassing equally spaced or end-only stiffening [18], and the precise positioning of stiffeners [23] within the beam section. Furthermore, these inquiries have elucidated the lateral buckling capacity [25] and post-buckling deformation behavior [26] of beams under varied loading conditions, with parameters including the rigidity [24] of the stiffeners. Additionally, experimental studies have delved into the spacing of stiffeners [28] and the deformation performance [29] of beams subjected to lateral buckling under diverse loading conditions [27], exploring the effect of beam end restraints [30] by columns on lateral buckling. One member that is continuously affixed to a beam in an actual structure is a floor slab. The lateral buckling of beams, with floor slabs serving as composite members of steel and concrete, has been examined [31–35]. Here, the floor slab is replaced by a continuous spring, facilitating the numerical analysis of the effects of horizontal deformation of the compression flange [36] and the bending and torsional deformation of the beam [37] on the lateral buckling load.

When the main beam in an actual structure experiences lateral buckling, stress is transferred to non-structural members attached to the beams. In essence, the nonstructural members resist the stress transmitted from the main beams, thereby contributing to the restraint of lateral buckling deformation [38–43]. However, neither Japanese design codes [44–46] nor the European Eurocode [47] advocate for the utilization of non-structural members as lateral buckling stiffeners for beams, and neither do they specify the requisite stiffener values.

Non-structural elements, like the roof folded plates [48,49] (referred to as profile sheets in the Eurocode [47]) depicted in Figures 1 and 2, are affixed to the upper flange of the beam and are installed continuously along its length. During lateral buckling deformation of the beam in particular, and as illustrated in Figure 3a, stresses (horizontal forces: F or rotational moments: M) occur at the junction between the beam and the roof folded plate. At this junction, horizontal and rotational deformations of the beam are constrained. However, the stiffness of the roof folded plate, which restrain the horizontal and rotational deformations of the beam, is lower than the stiffness of the buckling reinforcement, such as that provided by a small beam affixed to the beam. Furthermore, concrete slabs exhibit restraining properties in both the in-plane and out-of-plane directions of the beam [50], while roof folded plates exert minimal in-plane forces on the beam and provide continuous stiffening in the out-of-plane direction of the beam.

Figure 1. Beam and roof folded plate for actual structures.

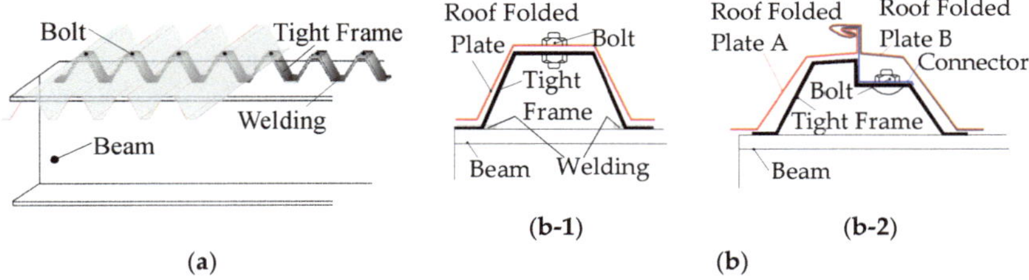

Figure 2. Detail of beam–roof plate joint. (**a**) Beam length direction and (**b**) cross-sectional direction: (**b-1**) cross section A and (**b-2**) cross section B.

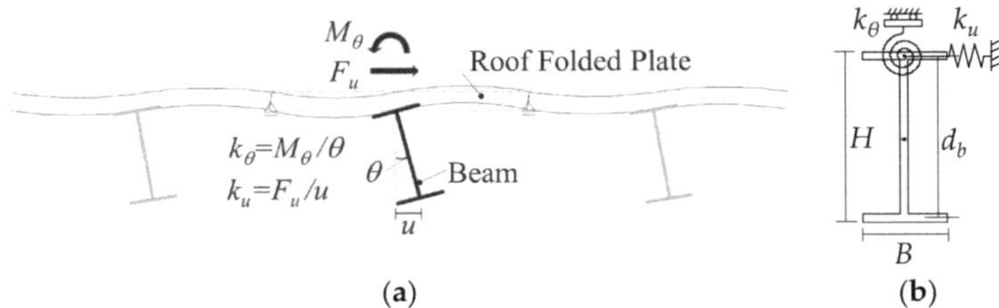

(**a**)   (**b**)

Figure 3. Continuous stiffening for lateral buckling deformation of H beams. (**a**) Stresses in roof folded plates and (**b**) stiffness of the spring.

Kimura et al. have employed numerical analysis to postulate that a beam and a roof folded plate should be rigidly interconnected, subsequently reconfiguring the roof folded plate with evenly distributed horizontal and rotational springs, as depicted in Figure 3b, in order to scrutinize the effects of the loading conditions [38–40] and the beam edge restraint [41–43] imposed by the column on the lateral buckling of the H beam with the roof folded plate. The horizontal and rotational spring stiffnesses, denoted as k_u and k_θ in Figure 3b, respectively, denote the elastic stiffness of the roof folded plate.

Yoshino et al. [51] delved into the stiffening effect of a roof folded plate on the reverse buckling of a beam through partial frame loading tests. In the depicted actual structure in Figure 2a, tight frames are meticulously welded to the H beams in a linear configuration along the beam length at the center of the top flange and are fastened to each top flange of the roof folded plate. As per the findings from the partial frame loading tests [51], the out-of-plane torsional deformation resulting from the lateral buckling of the H beams induced localized buckling near the bolt holes in the beam–roof folded plate junctions. Consequently, the rotational stiffness of the roof folded plate diminished to a fractional percentage of its theoretical stiffness due to this localized buckling. the rotational stiffness of the roof folded plate could be enhanced by refining the connector or connection method between the beam and the roof folded plate.

In the exploration of roof folded plates, the shear stiffness and shear seating behavior of folded plates under various edge support conditions have been elucidated through numerical analysis [52,53] and experiments [54]. Furthermore, experiments assessing the compression and shear strength [55,56] of thin steel folded plates as wall members have been conducted, while investigations into buckling tests [57–59] and numerical analyses [60,61] have centered on corrugated steel plate web beams.

Studies focusing on the bending performance of roof folded plates as roof members have unveiled their bending strength, considering variants with and without holes for equipment [62–65] and their reference manuals [48,49,66–68].

Moreover, studies leveraging the joints between roof members and beams as variables have revealed shear tests of roof folded plates when these joints are welded [69,70], as well as the bearing capacity of the joints of thin steel folded plates under distributed loads, such as wind loads [71].

These studies have exposed the holding performance of roof folded plates [52–68] and the mechanical behavior of roof folded plate–beam joints [69–71] within actual structures. However, they did not endeavor to employ the roof folded plate, a non-structural member, as a lateral buckling stiffener for beams. Furthermore, the mechanism of stress transfer from the beam to the folded plate due to variations in joint geometry remains undisclosed.

This paper conducts torsional experiments on roof folded plates, with various connection methods between them, in order to comprehend the deformation mechanism of roof folded plates and the relationship between their torsional moment and rotational stiffness. Next, the structures of roof members, assumed as continuous braces, and structural elements, such as large and small beams, will be investigated. Then, the relationship between the demand values against restraining the lateral buckling of the large beam and the experimentally determined bearing capacity of the roof folded plate will be elucidated. This aims to verify the feasibility of utilizing the roof folded plate as a continuous brace.

Additionally, Yoshino et al. [38–41,51] have investigated the design capacity associated with the use of the Japanese design code [46]. In this paper, the lateral buckling design capacity of H beams continuously stiffened by roof folded plates is elucidated by applying a connection method that ensures the joint stiffness between the roof folded plate and the beam using Japanese and European design codes [46,47].

2. Outline of Torsional Experiment on Roof Folded Plates

2.1. Outline of Torsional Experiment Apparatus

Figure 4 illustrates the experimental setup for conducting a torsion test on a roof folded plate. The specimen depicted in Figure 5 exemplifies a roof folded plate utilized in practical applications. With the objective of this paper being to determine the rotational stiffness of the connection between the roof folded plate and the beam, a cross-sectional profile [48,49], conducive to bolting onto a tight frame, as portrayed in Figure 2(b-1), and commonly encountered in actual structures in Japan, is employed. The length of the roof folded plates is L = 1800 mm. The width of the cross-section is 600 mm, equivalent to the combined width of three layers of a single roof folded plate. The boundary condition at the edge of the specimen in the z-direction is a pin on both the left and right sides (pin A, as shown in Figure 4). The loading beam possesses an H-shaped cross-section, H-300 × 150 × 6.5 × 9. The left side or the right side is identified as either the L side or R side from the center of pin B, which corresponds with the position of the loading beam web. Pin B, as delineated in Figure 4, is situated at the base of the loading beam. Pin B can pivot in two directions: vertically and horizontally. The pin plate of pin B is linked to the bottom flange of the loading beam and to the jack and slide bearing. Pin B is located atop the slide bearing, thus assuming the role of a pin roller support. The extremities of the test specimen in Figure 6, a roof folded plate, are vertically clamped between two plates, top and bottom, as exemplified at the a–a′ line of Figure 4. This is enacted to forestall local buckling at the peripheries of the roof folded plate. The roof folded plate is fastened 30 mm from the edge. The distance between the pins at both ends is $L_r = L - 2 \times 30$ mm = 1740 mm.

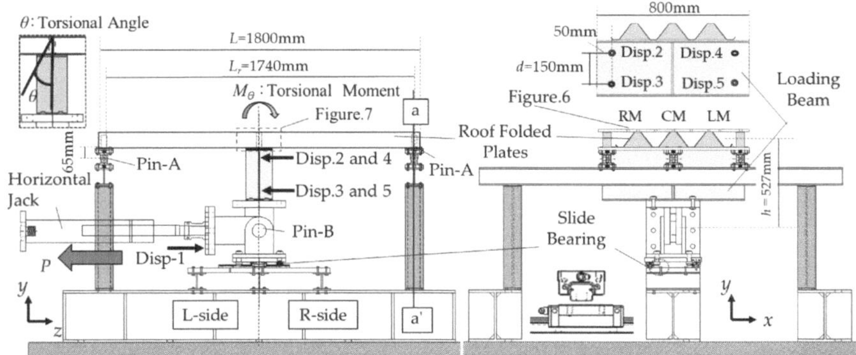

Figure 4. Specimen and loading instrument.

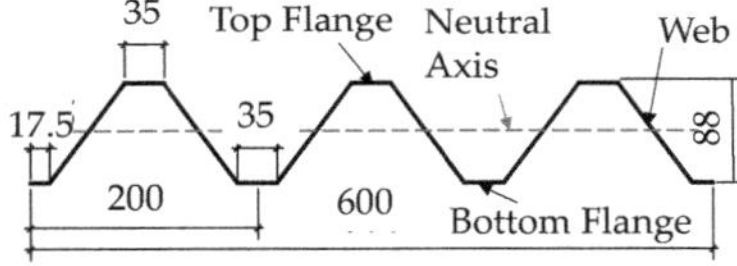

Figure 5. Detail of the specimen.

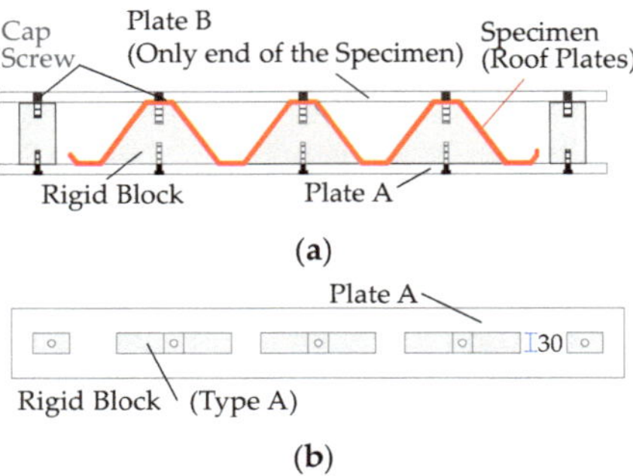

Figure 6. Connector at the edge of the specimen (a–a′ line). (**a**) Elevation and (**b**) ground plan.

2.2. Loading Protocols

The jack draws the loading beam in the *z*-axis direction, displacing pin B horizontally and inducing rotation. The specimen undergoes a torsional moment due to the rotation of the loading beam. The loading is monotonous.

2.3. Specimen Configuration

Figure 7 illustrates the method of connection for the joint between the loading beam and the specimen. The H beam and the roof folded plate are connected via the plate (PL-800 × 150 × 15) and connector depicted in Figure 7. Bolts are used to join the plates and loading beam. Typically, a tight frame acts as a connector between beams and folded plates, as observed in actual structures. Connectors are bolted to the top flange of the folded plate.

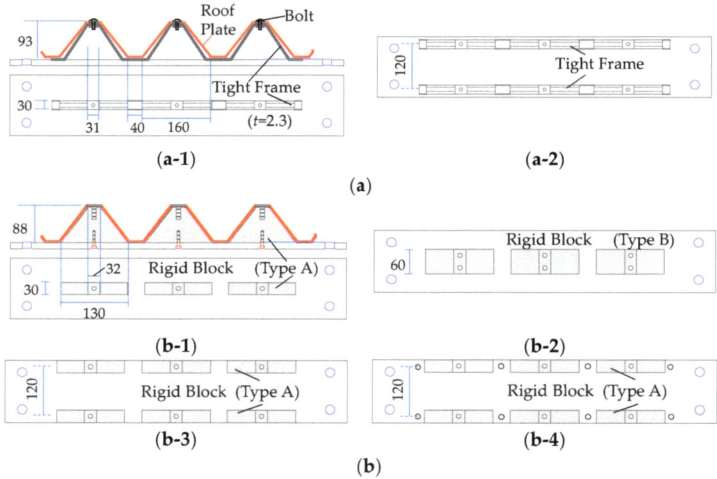

Figure 7. Detail of connector (unit: mm). (**a**) Tight frame: (**a-1**) TF11 and (**a-2**) TF21. (**b**) Rigid block: (**b-1**) RA11, (**b-2**) RB12, (**b-3**) RA21, and (**b-4**) RA22.

The connector depicted in Figure 7a is a tight frame (2.3 mm thick) welded to the plate. The connection method TF11 illustrated in Figure 7(a-1) resembles the actual structure depicted in Figure 2a, where the tight frame is linearly welded in a single row along the beam's length at the midpoint of the top flange. However, the connectors in TF11 may concentrate stress only at the bolted joints during torsional deformation of the beam. Therefore, TF21 in Figure 7(a-2) is a connector in which the tight frame is welded in two rows along the length of the beam on the top flange, increasing the number of bolted joint locations. Here, previous experiments [51] have confirmed the deformation of the tight frame. Additionally, the shape of the roof folded plate differs from that of the tight frame; the height of the roof folded plate shown in Figure 5 is 88 mm, while the height of the tight frame is 93 mm. Consequently, even when the roof folded plate and tight frame are bolted together, a 5 mm gap exists between the bottom flange of the roof folded plate and the tight frame. Furthermore, the roof folded plate is positioned 7.3 mm higher than the top flange of the loading beam.

This paper also introduces connectors that are designed to enhance the contact surface between the roof folded plate and the connector, when compared with TF11, and to increase rotational stiffness. The connector depicted in Figure 7b is a rigid block machined to match the cross-section and the height of the roof folded plate. When the roof folded plate and the connector are joined, the bottom flange of the roof folded plate contacts the plate. Type A employs rigid steel blocks 30 mm wide, while type B uses blocks 60 mm wide. Rigid blocks offer greater stiffness compared with tight frames. Roof folded plates are affixed to rigid blocks using high tension bolts. Connectors RA11 and RB12, shown in Figure 7(b-1,b-2), are positioned at the center of the top flange of the loading beam. Connector RA21, depicted in Figure 7(b-3), is positioned in two rows on the top flange of the loading beam. Furthermore, with the goal of achieving a closer approximation of rigid contact when compared with connector RA21 and to elucidate the effect of the presence or absence of a joint on the bottom flange of the roof folded plate, connector RA22, shown in Figure 7(b-4), joins the lower flange in addition to the joints of connector RA21.

Table 1 presents the list of specimens. Ten specimens were evaluated in the experiment for the following three variables: (1) plate thickness (t = 0.8, 1.0 mm), (2) connector (tight frame, rigid block-type A, B), and (3) bolt joint positions (top flange, top and bottom flanges).

Table 1. Details of specimens.

Specimen	Roof Folded Plate's Thickness	Roof Folded Plate—Joint		
		Material	Position	Bolt
	(mm)	—	—	—
TF11-0.8	0.8	Tight frame	One line	One bolt/top flange
TF11-1.0	1.0			
TF21-0.8	0.8		Two line	
TF21-1.0	1.0			
RA11-0.8	0.8	Rigid block (Type A)	One line	Two bolt/top flange
RB12-0.8	0.8	Rigid block (Type B)	One line	
RB12-1.0	1.0			
RA21-0.8	0.8	Rigid block (Type A)	Two line	One bolt/top flange
RA21-1.0	1.0			
RA22-1.0	1.0			Two bolt/top and bottom flange

Example of specimen name

$$\underset{①\ \ ②\ \ \ ③}{\underline{\mathrm{TF11\text{-}0.8}}}$$

①: Connector ⎰ TF : Tight Frame / RA : Rigid Block(Type A) / RB : Rigid Block(Type B)

②: Placement of restraints and Number of Bolts ⎰ 11 : One Line / One Bolt / 12 : One Line / Two Bolt / 21 : Two Line / One Bolt / 22 : Two Line / Two Bolt

③: Thickness of Folded Roof Plate ⎰ 0.8 : t=0.8 / 1.0 : t=1.0

2.4. Material Properties

The material of the roof folded plate is galvanized steel sheet [72], selected for its corrosion resistance which makes it suitable for roofing applications. The material test results for the roof folded plate are presented in Table 2. The tensile strength test of steel members was conducted according to JIS Z 2241 [73]. The yield and ultimate strength of the roof folded plate are, respectively, 309 to 322 N/mm² and 380 to 483 N/mm².

Table 2. Material properties of the roof folded plates.

Thickness (mm)	Young's Modulus ($\times 10^3$ N/mm²)	Yield Strength (N/mm²)	Ultimate Strength (N/mm²)
0.8	185	322	380
1.0	178	309	383

2.5. Measurement Methods

Figure 8 elucidates the locations for strain measurement. Strain gauges are affixed to both facets of the roof folded plate to quantify plate bending, as depicted in Figure 8b.

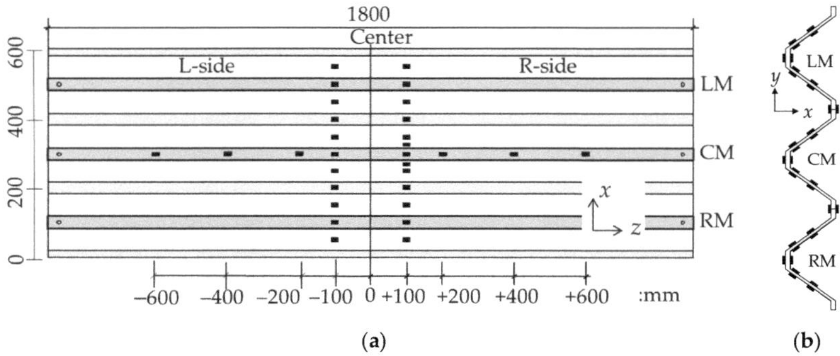

Figure 8. Strain measurement position. (**a**) Ground plan and (**b**) elevation.

3. Results of Torsional Experiment on Roof Folded Plate

3.1. Rotational Stiffness and Torsional Moment of Roof Folded Plate

This section delineates the load–displacement correlation for each specimen to elucidate the impacts of the experimental parameters.

Figure 9 illustrates the correlation between torsional moment $M_{\theta,i}$ and rotation angle for the torsion experiment. The vertical axis depicts the ratio of the torsional moment experienced by the roof folded plate during the experiment to the yield bending moment of the roof folded plate $M_{y,r}$. $M_{\theta,i}$ and $M_{y,r}$ can be calculated using the following equation.

$$M_{\theta,i} = P \cdot h \tag{1}$$

$$M_{y,r} = \sigma_{y,r} \cdot Z_r \tag{2}$$

where h signifies the distance between the pin and the roof folded plate, $h = 527$ mm (refer to Figure 4); $\sigma_{y,r}$ represents the yield stress of the roof folded plate in Table 2; and Z_r denotes the section modulus. The horizontal axis represents the ratio of the rotation angle θ of the loading beam to the rotation angle $\theta_{y,r}$ at yield bending moment $M_{y,r}$. The rotation angle θ of the loading beam is calculated as the average of the difference in horizontal displacements u measured at two points ("disp.3–disp.2" and "disp.5–Disp.4") on either side of the beam, as shown in Figure 4, divided by the distance d between measurements. This is obtained from the following equation:

$$\theta = \left\{ \left(u_{disp.3} - u_{disp.2}/d \right) + \left(u_{disp.5} - u_{disp.4}/d \right) \right\}/2 \tag{3}$$

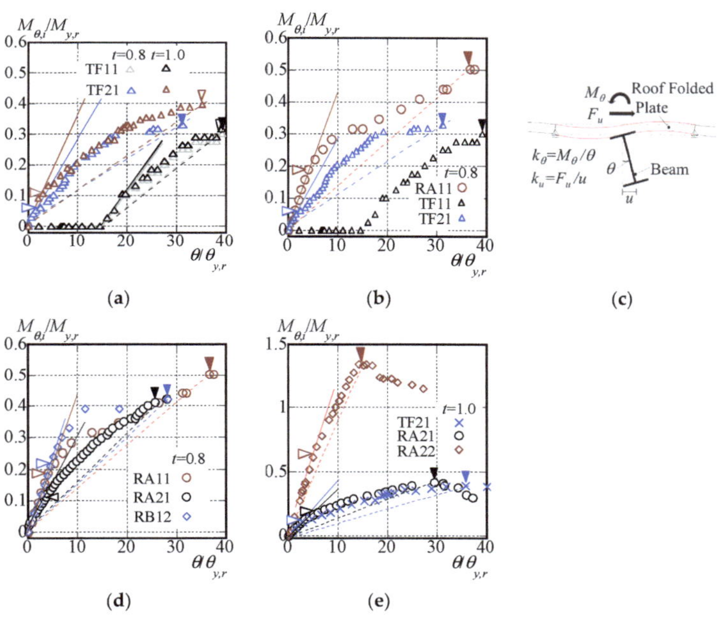

Figure 9. Hysteresis curves (torsional moment–angle). (**a**) Thickness, (**b**) shape of connector, (**c**) stresses in roof folded plates, (**d**) number of bolts, and (**e**) number of joints.

This experiment considered a situation in which beams linked to roof folded plates laterally buckled as illustrated in Figure 9c The rotational stiffness $k_{\theta,i}$ of the roof folded plate is defined as the ratio of the torsional moment $M_{\theta,i}$ to the torsional deformation θ occurring in the roof folded plate. $k_{\theta,i}$ corresponds with the slope of the $M_{\theta,i}-\theta$ relationship in Figure 9a,b,d,e. The symbol "i" in $k_{\theta,i}$ represents the three rotational stiffness values (i, =; 0; 1; 2), and is derived from the equation below.

$$k_{\theta,i} = \frac{M_{\theta,i}}{\theta} \tag{4}$$

The solid line in Figure 9 represents the tangent line of the measured data. The inclination of the tangent line is defined as the initial rotational stiffness $k_{\theta,1}$. The ▷ horizontal triangle plot indicates the juncture at which the slope is more than 5% lower than the tangent slope. The ▼ vertical triangle plot signifies the point of maximum moment. The dashed line delineates the slope connecting the origin and the ▼ plot. The slope of the dashed line is defined as the secant rotational stiffness $k_{\theta,2}$. The theoretical rotational stiffness $k_{\theta,0}$ can be calculated from Equation (5).

$$k_{\theta,0} = \frac{12E_r I_r}{L_r} \tag{5}$$

where $E_r I_r$ represents the bending stiffness of the roof folded plate.

Figure 9a illustrates the experimental results for specimens with varying plate thicknesses. In the TF11 specimen, where a single row of tight frames is welded along the length of the beam at the center of the top flange, the difference in rotational stiffness, indicated by the slope of the tangent line, shows minimal variation with plate thickness. This is attributed to the small stiffness of the roof folded plate–beam joint, making it difficult to transmit torsional moments through the joint effectively. Conversely, in the TF21 specimen with two rows of tight frames welded to the top flange of the loading beam, torsional moments are induced in the roof folded plate via the two bolted joints, thereby demonstrating the bending stiffnesses of the roof folded plates at each plate thickness according to the stiffness of the roof folded plate–beam joint. Consequently, there exists a difference in rotational stiffness among different plate thicknesses. Specifically, the increase in rotational stiffness of the specimen with $t = 1.0$ relative to $t = 0.8$ is 1.2 times greater than that of the specimen with $t = 0.8$.

However, the theoretical rotational stiffness due to plate thickness, as indicated in Equation (5), is expected to increase by about 1.5 times. This discrepancy arises because the joints assumed for the theoretical rotational stiffness are rigid, whereas the joints between the roof folded plate and the beam in the experiment are not rigid. Hence, an increase in the rate of rotational stiffness in the experiment is deemed to be smaller than the equivalent theoretical value.

Similarly, while the yield moment calculated from Equation (3) escalates by a factor of 1.4 with plate thickness, the rate of maximum load increase with plate thickness for the TF21 specimen is 1.2. This discrepancy is attributed to localized fracture at the bolted joints of the TF21 specimen, as depicted in Figure 10, which is discussed subsequently.

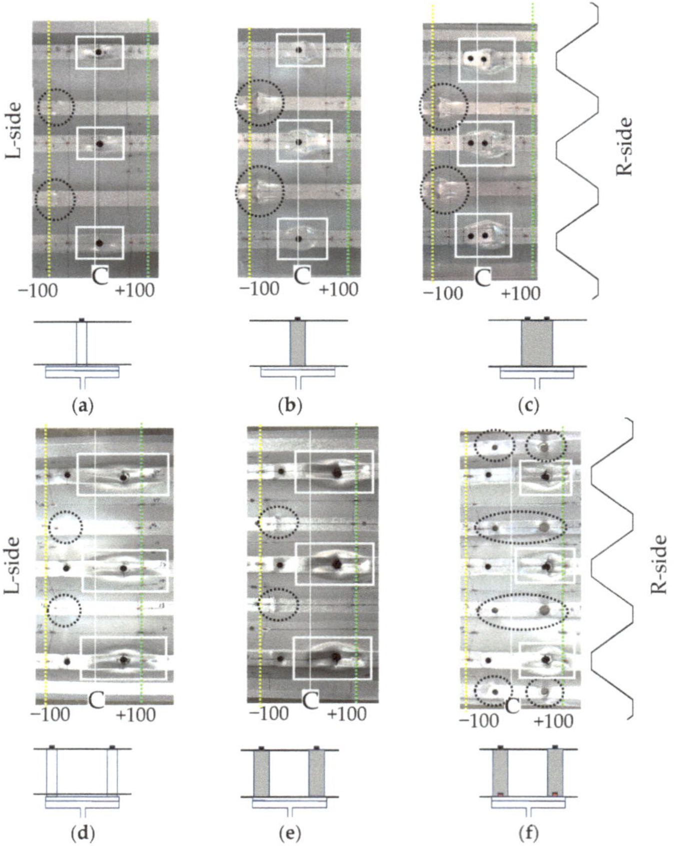

Figure 10. Ultimate statement of specimens (+100 mm). (**a**) TF11-0.8, (**b**) RA11-0.8, (**c**) RB12−0.8, (**d**) TF21-0.8, (**e**) RA21-1.0, (**f**) RA22-1.0.

Figure 9b illustrates the experimental results for specimens with various types of connectors. In the TF11 specimen, $M_{\theta,i}$ starts to increase after $\theta/\theta_{y,r} = 15$, when the bottom flange of a roof folded plate makes contact with the H beam serving as the reaction force beam. As depicted in Figure 7(a-1) for the TF11 specimen, there exists a 7.3 mm gap between the folded plate and the upper flange of the beam. Consequently, the roof folded plate remains unaffected by the torsional deformation of the beam until the lower flange of the roof folded plate contacts the upper flange of the beam. Upon contact, torsional moment is generated in the roof folded plate through two points: the contact surface between the top flange of the beam and the bottom flange of the roof folded plate, and the bolted joint. Consequently, for the RA11 specimen, which is a rigid block with no gap between the connector and the roof folded plate, the torsional moment starts to increase from the initial load. Conversely, for the TF21 specimen with two rows of tight frames welded to the top flange of the loading beam, the torsional moment begins to increase from the beginning of loading. The initial rotational stiffness of TF21, denoted as $k_{\theta,1}$, exhibits approximately 2 to 3% of the theoretical stiffness $k_{\theta,0}$. This is attributed to bending stresses in the top flange of the roof folded plate from the initial load ($\theta/\theta_{y,r} = 0$) through the two bolted joints between the roof folded plate and the tight frame. The initial rotational stiffness of RA11, denoted as $k_{\theta,1}$, exceeds the rotational stiffness of TF21. This is due to the greater stiffness of the roof folded plate–beam connection in RA11 compared with TF21.

Figure 9d displays the results for different numbers of bolts on the top flange of the beam–roof folded plate joints. $M_{\theta,1}$ is defined as the initial torsional moment, indicating the torsional moment at the point of the ▷ plot. The torsional moment of the roof folded plate in RB12 increases from that of RA11, but the initial rotational stiffness $k_{\theta,1}$ is approximately 4.1% of the theoretical stiffness $k_{\theta,0}$ and remains nearly the same for both specimens. On the other hand, the torsional moment and initial rotational stiffness of RA22 are lower than those of RA11, despite the higher theoretical stiffness assumed for rigid connections between the roof folded plate and the beam in RA21 compared with RA11. This is attributed to the fact that the specimen in RA21 experienced plate bending deformation in the mountain flange of the thin roof folded plate from the beginning of loading, as discussed later in Figures 10 and 11.

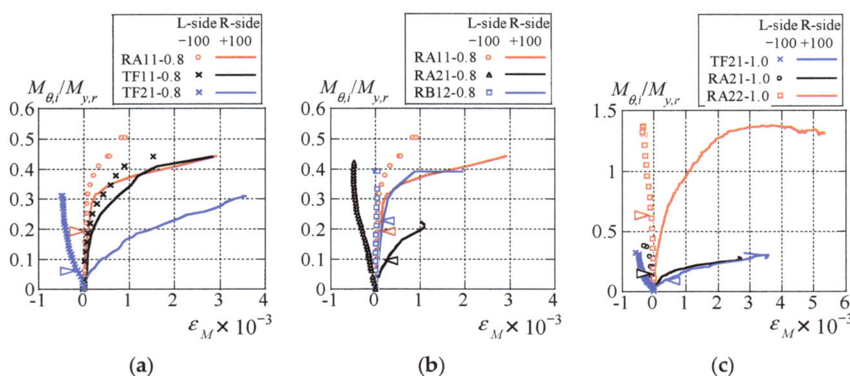

Figure 11. Bending strain of top flange for CM ($z = \pm100$ mm). (**a**) Tight frame and connector, (**b**) Number of bolts, and (**c**) number of joints.

Figure 9e illustrates the comparison between the connector and the lower flange of the roof folded plate when jointed and unjointed. The initial rotational stiffness $k_{\theta,1}$ of RA22 with the upper and lower flanges joined is approximately 12% of $k_{\theta,0}$. The initial rotational stiffness of RA22 is about three times that of RA21. Despite the increased number of joint points from RA21 and the increased rotational stiffness, the local deformation at the joints of the thin roof folded plates and the bending deformation of the roof folded plates at the contact surface of the connectors reduce the rotational rigidity of the roof folded plates. This results in the experimental rotational stiffness becoming only about 12% of the theoretical stiffness assumed for rigid connections between the roof folded plate and the beam.

3.2. Deformation Mechanism of Roof folded Plate

Figure 10 illustrates the deformation of the roof folded plate post-experiment. In specimens TF11, RA11, and RB12, connectors are welded at the midpoint of the top flange, leading to plate local deformation near the bolted joint indicated by the white box. Additionally, bending deformation is observed at the joint where the bottom flange meets the top flange of the loading beam, marked by the black circle. Notably, the bending deformation in RA11 surpasses that of TF11, likely due to the absence of clearance between the roof folded plate and the connector.

In specimens with two rows of connectors welded to the top flange, highlighted by white boxes (TF21 and RA21), significant local deformation, terminal loss, and rupture were observed at the bolted joint of the top flange on the R side. Particularly in RA22, where the bottom flange of the roof folded plate was also bolted together, the bottom flange encircled in black exhibited a wider hole at the joint on the R side than on the L side, resulting in damage. This discrepancy arises because the bolted joint of the lower flange on the L side is pushed up against the plate in a plane, while only the bolt is pulled downward on the R side.

3.3. Stress State of Roof Folded Plate

Figure 11 illustrates the correlation between the plate bending strain ε_M at the top flange of the roof folded plate and the torsional moment at the beam−roof folded plate connection. The strain is calculated according to the subsequent equation.

$$\varepsilon_M = \left(\varepsilon_f - \varepsilon_b\right)/2 \tag{6}$$

where ε_f signifies the strain on the surface of the measurement point and ε_b denotes the strain on the back surface of the measurement point. The strain is measured at the top flange of the roof folded plate, indicated by "CM" in Figure 8 at $z = \pm 100$ mm.

In Figure 11a,b, the plate bending strain of TF21 and RA21 emerges from the initial loading. The plate bending strain on the right side ($z = +100$ mm, solid line) surpasses that on the left side ($z = -100$ mm, dashed line). When the loading beam undergoes torsion, the connectors affixed to the beam rotate, as illustrated in Figure 12b. As the connector rotates, the bolted joint on the right side connecting the connector and the roof folded plate is subjected to downward pulling along the y-axis, as indicated by the downward arrow in Figure 12b, and offers resistance. Consequently, plate bending deformation concentrates on the bolted joint on the right side earlier than on the left side, resulting in higher plate bending stresses. Therefore, the initial rotational stiffness of RA21 is presumed to be lower than that of RA11 in Figure 9b.

Figure 13 illustrates the axial strain ε_{Nz} of the central top flange (CM) on the R side of the roof folded plate. At $z = +400$ mm and $z = +600$ mm, the axial strain of RA22 (specimen with top and bottom flanges joined) surpasses that of other joining methods (specimen with only top flanges joined). When the bottom flange of the roof folded plate is attached to the beam, the bolted joint is subjected to downward pulling along the y-axis due to the torsion of the loading beam, as indicated by the lower arrow in Figure 12c. Consequently, bending stresses arise in both the top and bottom flanges of the roof folded plate, particularly on the right side of the roof folded plate. Therefore, the cross-section at the joint location is constrained, resulting in bending deformation along the length of the roof folded plate. The axial strain of RA22 rapidly decreases near the maximum proof stress. As depicted in Figure 9e, the RA22 specimen, where the bottom flange of the roof folded plate is joined at the bolted joint, exhibits a smaller rotation angle needed to reach the maximum torsional moment when compared with the RA21 specimen where only the top flange of the roof folded plate is joined. This difference leads to early end failure.

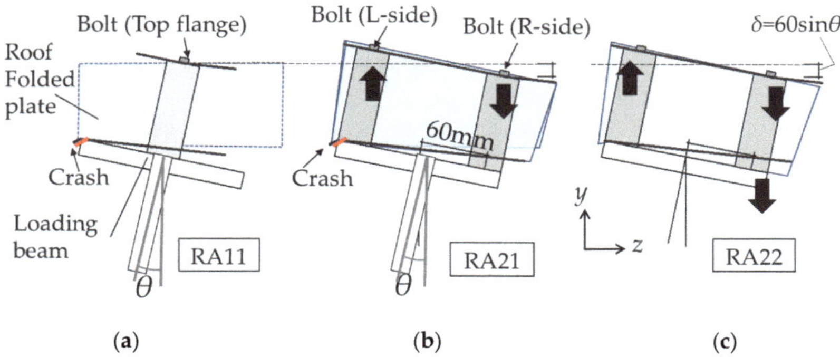

Figure 12. Deformation mechanism of bolted joints due to different connectors. (**a**) RA11, (**b**) RA21, (**c**) RA22.

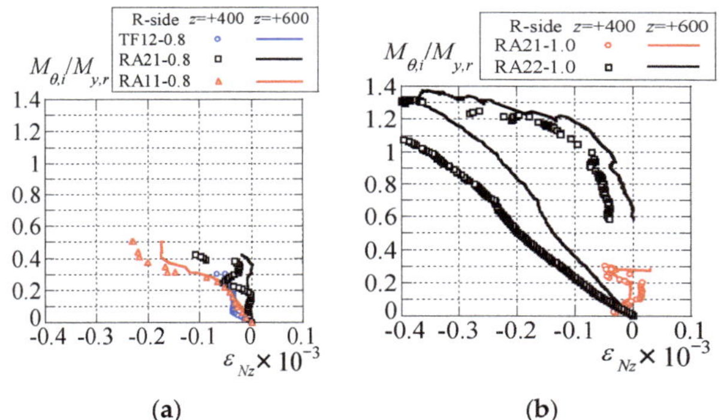

(**a**)   (**b**)

Figure 13. Axial strain of the roof folded plate's top flange ($z = +400$ mm, $+600$ mm). (**a**) Tight frame and connector and (**b**) number of joints.

Figure 14 illustrates the axial strain ε_{Nz} distribution of the central top flange CM in the z-axis direction at 0.1 $M_{y,r}$. The gray dotted line represents the theoretically elastic strain when the roof folded plate is held flat. The theoretical elastic strain ε_{N0} is calculated using the subsequent equation.

$$\varepsilon_{N0} = \frac{\sigma}{E_r} = \frac{M_{\theta,i}}{2E_r Z_r} \tag{7}$$

where Z_r denotes the section modulus of the roof folded plate, as delineated in the 600 mm wide segment illustrated in Figure 5.

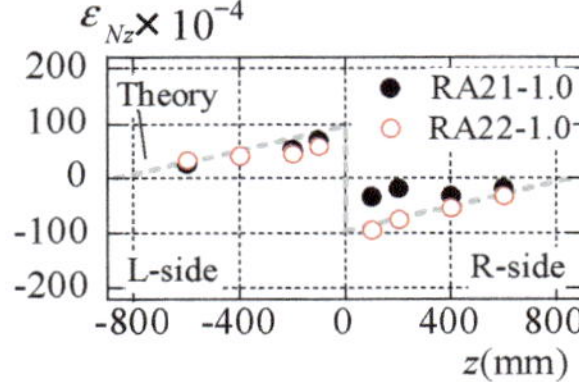

Figure 14. Axial strain in z direction (at 0.1 $M_{y,r}$).

The axial strain on the R side of RA21, where only the top flange of the roof folded plate is connected, is nearly zero. The axial strain of RA22, with the attachment of the top and bottom flanges of the roof folded plate, corresponds with the elastic theoretical strain stipulated in Equation (7). Upon affixing the bottom flange of the roof folded plate to the beam, the entire cross-sectional area of the joint undergoes deformation when subjected to torsion by the loading beam. Therefore, the torsional moment induced in the roof folded plate due to the beam's torsion leads to the bending moment distribution depicted by the gray dashed line in Figure 14 and is transmitted along the length directions (L and R sides) of the roof folded plate.

Figure 15 delineates the distribution of axial strain ε_{Nx} along the x-axis for a roof folded plate. The axial strain ε_{Nx} is gauged at $z = \pm 100$ mm. The gray dotted line is deduced from Equation (7). The axial strain on the left side of the RA21 specimen conforms to the elastic theoretical strain expounded in Equation (7). The axial strain on the right side deviates considerably from the theoretical value across the entire cross-section. In the RA21 specimen, the torsional moment on the right (R) side of the roof folded plate is small. In the RA22 specimen, the axial strain on the left (L) and the right (R) sides are close to the elastic theoretical strain. The roof folded plates of the RA22 specimen exhibit a more uniform cross-sectional deformation when compared with those of the RA21 specimen.

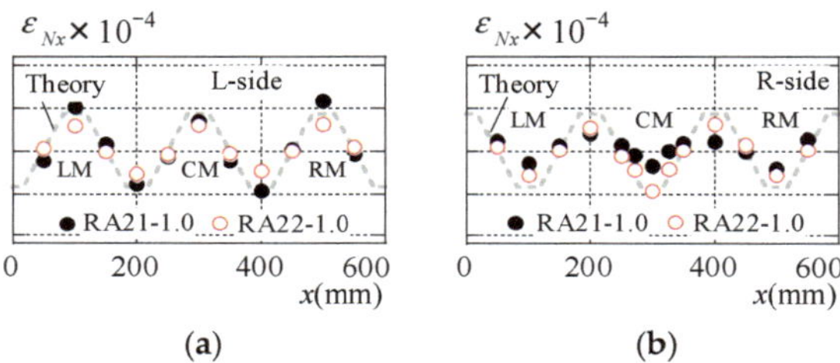

Figure 15. Axial strain in *x* direction. (**a**) L side (−100 mm) and (**b**) R side (+100 mm).

3.4. Summary of Torsion Experiment Results

Table 3 exhibits the rotational stiffness ratio $k_{\theta,i}/k_{\theta,0}$ and torsional moment ratio $M_{\theta,i}/M_{y,r}$ of each specimen at the initiation of stiffness reduction and at maximum load. The torsional moment $M_{\theta,1}$ corresponds to moment at the ▷ plot in Figure 9, while the maximum torsional moment $M_{\theta,2}$ corresponds with the moment at the ▼ plot in Figure 9. The rotational stiffness ratio $k_{\theta,i}/k_{\theta,0}$ is calculated by dividing the initial rotational stiffness $k_{\theta,1}$ or the secant rotational stiffness $k_{\theta,2}$ by the theoretical rotational stiffness $k_{\theta,0}$ in Equation (5). The torsional moment ratio $M_{\theta,i}/M_{y,r}$ is calculated by dividing the torsional moment $M_{\theta,1}$ at the initiation of stiffness reduction or at the maximum moment $M_{\theta,2}$ by the yield moment $M_{y,r}$ of the roof folded plate.

The initial rotational stiffness ratio $k_{\theta,1}/k_{\theta,0}$ and torsional moment ratio $M_{\theta,1}/M_{y,r}$ at the initiation of stiffness reduction increased from 2.85% to 12.4% and from 0.07 to 0.57, respectively, depending on the joining method. Theoretical rotational stiffness assumes that the roof folded plate and beam connections are rigidly connected, but the rotational stiffness of the joint in this experiment corresponds with the sum of the stiffness of the bolted joint and the bending stiffness of the roof folded plate.

The magnitude of the moment transmitted to the roof folded plate depends on the stiffness of the bolted joint. It can be inferred that the initial rotational stiffness in this experiment was lower than the theoretical rotational stiffness due to the lower stiffness of the bolted joints, in turn caused by local deformation and the lower bending stiffness of the roof folded plate due to the out-of-plane deformation of the thin roof folded plate.

Table 3. Summary of results.

Specimen	Rotational Stiffness					Bracing Moment		
	Theory	Initial	Secant	Initial	Secant	Yield	Initial	Maximum
	Equation (5)	Equation (4)		Equation (4)/Equation (5)		Equation (2)	Equation (1)/Equation (2)	
	$k_{\theta,0}$	$k_{\theta,1}$	$k_{\theta,2}$	$k_{\theta,1}/k_{\theta,0}$	$k_{\theta,2}/k_{\theta,0}$	$M_{y,r}$	$M_{\theta,1}/M_{y,r}$	$M_{\theta,2}/M_{y,r}$
	$\times 10^2$ kN/rad			%	%	$\times 10^2$ kN/rad		
TF11-0.8	37.75	0	0	–	–	1271	–	–
TF11-1.0	45.3	0	0	–	–	1528	–	–
TF21-0.8	37.75	1.08	0.36	2.85	0.95	1271	0.07	0.33
TF21-1.0	45.3	1.47	0.47	3.25	1.04	1528	0.11	0.39
RA11-0.8	37.75	1.51	0.54	4.01	1.42	1271	0.18	0.5
RB12-0.8	37.75	1.60	0.57	4.24	1.5	1528	0.22	0.43
RB12-1.0	45.3	2.14	0.72	4.72	1.6	1271	0.24	0.52
RA21-0.8	37.75	1.15	0.53	3.04	1.4	1528	0.11	0.42
RA21-1.0	45.3	1.85	0.66	4.09	1.45	1271	0.2	0.44
RA22-1.0	45.3	5.60	3.85	12.36	8.5	1528	0.57	1.38

Figure 16 elucidates the correlation between the rotational stiffness ratio and torsional moment ratio. Both the rotational stiffness ratio and torsional moment ratio demonstrate a proportional augmentation.

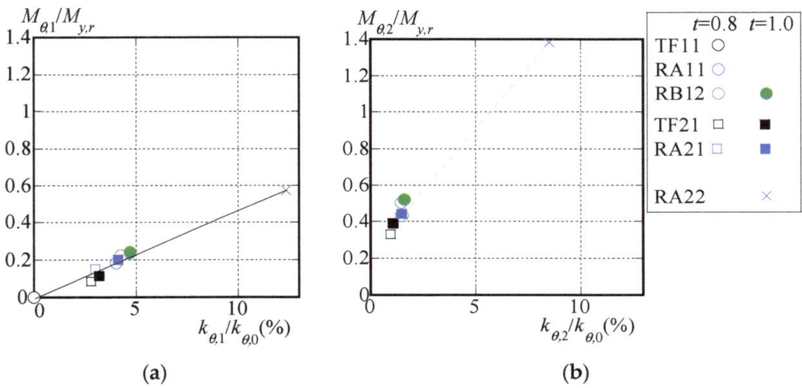

Figure 16. Rotational stiffness and torsional moment at (**a**) initial stiffness and at (**b**) secant stiffness.

4. Effect of Roof Folded Plates on Lateral Buckling of H Beams in Steel Structures

Yoshino et al. undertook an extensive investigation of 1345 large-span steel edifices within Japan, of which 1065 structures, inclusive of gymnasiums, were surveyed to ascertain the cross-sectional configurations and varieties of main beams, small beams, purlins, and roofing members. This examination focuses on 606 of the large-span steel structures, as delineated in Table 4, where the main beams are composed of H beams. Then, the relationship between the demand values against the restraining of the lateral buckling of the main beam in the large-span steel structures and the experimentally determined bearing capacity of the roof folded plate will be elucidated. The aim of this is to verify the feasibility of utilizing a roof folded plate as a continuous brace.

Table 4. Survey of steel structures in Japan with H beams as main beams.

Slenderness Ratio of Main Beams	Percentage of Total (%)	Number of Buildings
$67 \leq \lambda_b \leq 200$	34.0	206
$200 < \lambda_b \leq 300$	45.9	278
$300 < \lambda_b$	20.1	122
Types of Small Beams	Percentage of Total (%)	Number of Buildings
H	94.4	572
Angle and channel	5.0	30
Not listed	0.6	4
Types of Purlin Members	Percentage of Total (%)	Number of Buildings
Channel (C,2C)	89.3	541
Angle (L,2L)	1.0	6
T	0.5	3
Not listed	9.2	56
Types of Roofing Members	Percentage of Total (%)	Number of Buildings
Roof folded plate	2.8	17
Colored steel plate	78.4	476
Not listed	18.8	113

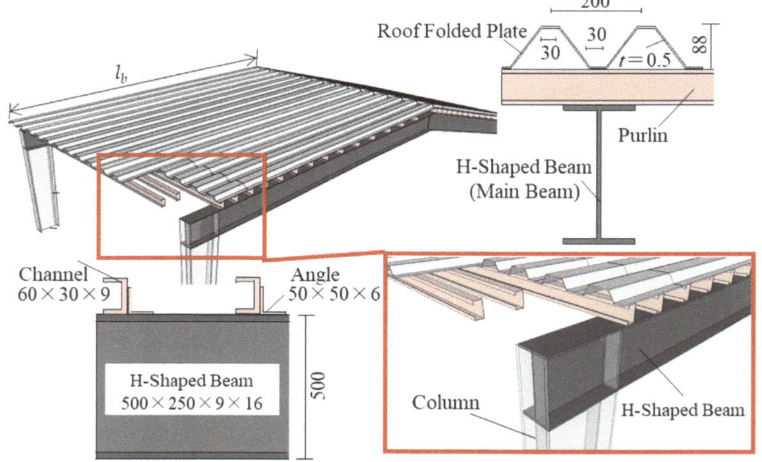

Figure 17 portrays the correlation between the required stiffness ratio of the roof folded plate and the bracing moment ratio engendered within the roof folded plate. The required stiffness ratio is obtained by dividing the rotational stiffness of the roof folded plate, as determined from the experimental data, by the required rotational stiffness, in order to constrain the lateral buckling deformation of the H beam in the investigated actual structure. The rotational stiffness of the roof folded plate corresponds with the value measured for test specimen TF21-0.8, which demonstrates the lowest initial rotational stiffness among the experiments. The bracing moment ratio is calculated by dividing the torsional moment in the roof folded plate, derived from experiment, by the bracing moment when the H beam in the actual structures is laterally buckled. The required rotational stiffness is obtained from the elastic buckling load equation for H beam obtained in [38].

$$P_{cr,n} = -\left\{ \frac{k_u}{2}\left(\frac{L_b}{n\pi}\right)^2 \right\}$$

$$+\frac{1}{2}\sqrt{\left\{k_u\left(\frac{L_b}{n\pi}\right)^2\right\}^2 + 4\left[\begin{array}{c} EI_f\left(\frac{n\pi}{L_b}\right)^2 \left\langle EI_f\left(\frac{n\pi}{L_b}\right)^2 + 2\left\{\frac{GK_w}{d_b^2} + \frac{GK_f}{d_b^2}(\tau_1+\tau_2) + \frac{k_\theta}{d_b^2}\left(\frac{L_b}{n\pi}\right)^2\tau_1\right\} + k_u\left(\frac{L_b}{n\pi}\right)^2\right\rangle \\[2mm] + k_u\left(\frac{L_b}{n\pi}\right)^2\left\{\frac{GK_w}{d_b^2} + \frac{GK_f}{d_b^2}(\tau_1+\tau_2) + \frac{k_\theta}{d_b^2}\left(\frac{L_b}{n\pi}\right)^2\tau_1\right\} \end{array}\right]}$$

$$(8)$$

Here, k_u represents the lateral stiffness of the roof members. L_b denotes the length of the H beam, EI_f signifies the flexural stiffness of each flange for the H beam, GK_f indicates the torsional stiffness of each flange, GK_w refers to the torsional stiffness of the web, d_b stands for the distance between both flanges of an H beam, k_β represents the rotational stiffness of the roof members, τ_1 denotes the reduction of rotational stiffness of the brace and the torsional stiffness of the top flange, and τ_2 is the reduction of the torsional stiffness of the bottom flange.

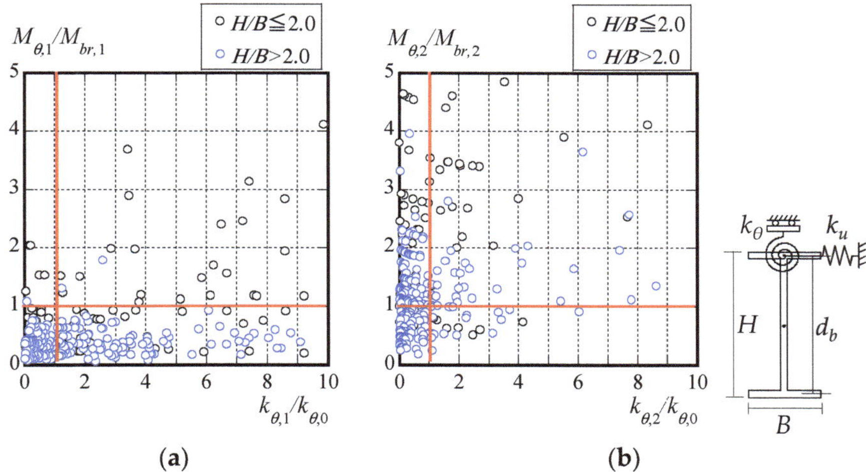

Figure 17. Required stiffness ratio and bracing moment ratio of roof folded plate at (**a**) initial stiffness and at (**b**) secant stiffness.

The elastic lateral buckling load $P_{cr,i}$ of an H beam continuously stiffened by a roof folded plate is determined by replacing $k_{\theta,1}$ and $k_{\theta,2}$ for k_θ in Equation (8). There are two instances of the symbol i in $P_{cr,i}$ (where $i = 1,2$), with $P_{cr,1}$ being the value obtained by substituting $k_{\theta,1}$ into Equation (8), and $P_{cr,2}$ being the value obtained by substituting $k_{\theta,2}$ into Equation (8). While Equation (8) includes a horizontal stiffness term, this paper assumes that no horizontal deformation occurs. Therefore, the horizontal stiffness $k_u = \infty$. When $n = 1$ or 2 in Equation (8), the elastic buckling load $P_{cr,1}$ or $P_{cr,2}$ of the first-order or second-order mode is obtained. The rotational stiffness at the point where $P_{cr,1} = P_{cr,2}$ is delineated as the required rotational stiffness $k_{\theta,0}$, and represents the minimal rotational stiffness at which this transition occurs, as depicted in Figure 18. The lateral buckling mode of the beam transfer from the first order to the second order depends on the magnitude of the rotational stiffness, as illustrated in Figure 18.

In [38–43], it is affirmed that lateral buckling deformation can be mitigated when the rotational stiffnesses $k_{\theta,1}$ and $k_{\theta,2}$ of the roof folded plates exceeds the required rotational stiffness $k_{\theta,0}$.

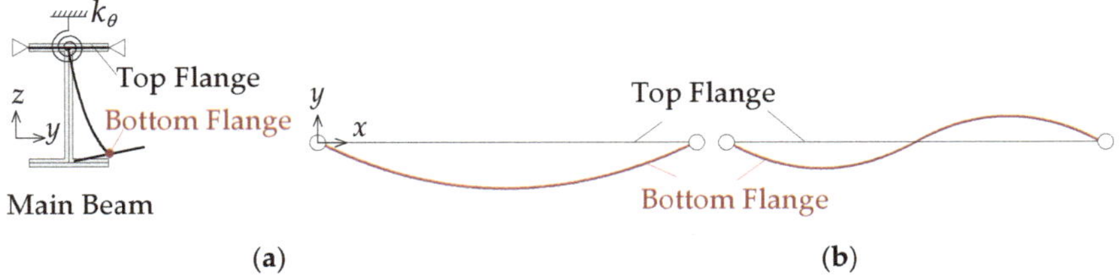

(**a**)   (**b**)

Figure 18. Lateral buckling mode. (**a**) First Buckling mode and (**b**) second buckling mode.

In [38], the required bracing moment $M_{br,i}$ is defined as the moment arising in the roof folded plate during lateral buckling of the H beam, and $M_{br,i}$ is derived from the following equation.

$$M_{br,i} = \left\{ 0.004 \left(0.8 + \frac{1}{\sqrt{K'}} \right) \le 0.008 \right\} M_{p,b} \tag{9}$$

$$K' = k_u k_\theta / \sqrt{\left(\frac{2EI_f \pi^2}{L_b^2} \right) \Big/ \left(\frac{GK}{d_b^2} \right)} \tag{10}$$

where $M_{p,b}$ represents the full plastic bending moment of the main beam across each investigated structure. GK indicates the torsional stiffness of the main beam. $M_{br,1}$ or $M_{br,2}$ is derived by substituting $k_\theta = k_{\theta,1}$ or $k_\theta = k_{\theta,2}$ into Equations (9) and (10), respectively.

In the $\triangleright$ plot of Figure 9, when the torsional moment $M_{\theta,1}$ is considered to be the bearing capacity of the roof folded plate, approximately 16% of $M_{\theta,1}$ of the roof folded plates that are attached to all of the structures does not exceed the required bracing moment $M_{br,1}$ generated in the roof folded plate when the beam buckles laterally ($M_{\theta,1}/M_{br,1} \ge 1.0$). Therefore, the roof folded plates possess the initial rotational stiffness $k_{\theta,1}$ at the torsional moment $M_{\theta,1}$. Furthermore, among the 16% of structures with initial rotational stiffness $k_{\theta,1}$, approximately 90% of these exceed the rotational stiffness $k_{\theta,0}$ required to restrain the lateral buckling deformation of the beam ($k_{\theta,1}/k_{\theta,0} \ge 1.0$).

On the contrary, if the maximum torsional moment $M_{\theta,2}$, shown in in Figure 9, is considered as the bearing capacity of the roof folded plate, approximately 62% of the structures have $M_{\theta,2}$ of roof folded plates that does not surpass the required bracing moment $M_{br,2}$, which arises in the roof folded plate when the beam undergoes lateral buckling ($M_{\theta,2}/M_{br,2} \ge 1.0$). Additionally, the rotational stiffness of the roof folded plate at the maximum torsional moment is lower than that at the initial torsional stiffness $k_{\theta,1}$, resulting in the roof folded plate possessing a secant rotational stiffness $k_{\theta,2}$. Incidentally, among the approximately 62% of structures that possess secant rotational stiffness $k_{\theta,2}$, 50% have a secant rotational stiffness exceeding the required rotational stiffness $k_{\theta,0}$ to restrain the lateral buckling deformation of the beam ($k_{\theta,2}/k_{\theta,0} \ge 1.0$).

From the above, it can be deduced that the number of structures in which the roof folded plate possesses the required performance against lateral buckling of the beams can be increased by utilizing the maximum torsional moment and the secant stiffness, rather than by relying on the initial stiffness and the bearing capacity at the onset of stiffness reduction to act as metrics of the holding performance of the roof folded plate.

Subsequently, the impact of the continuous reinforcement of the roof folded plates during the lateral buckling of the main beams, referring to Table 4, is elucidated. Currently, there exists no globally recognized design formula for the lateral buckling capacity of beams. In the Japanese design guideline, namely the "Limit State Design Guidelines and Commentary" [46], the design capacity of beams, denoted as $M_{b,AIJ}$ is derived from the ensuing equation:

$$(\lambda_b < {}_p\lambda_b = 0.9) \qquad \frac{M_{b,AIJ}}{M_{p,b}} = 1$$

$$({}_p\lambda_b \leq \lambda_b < {}_e\lambda_b \approx 1.29) \qquad \frac{M_{b,AIJ}}{M_{p,b}} = \left(1 - 0.4\frac{\lambda_b - {}_p\lambda_b}{{}_e\lambda_b - {}_p\lambda_b}\right) \tag{11}$$

$$({}_e\lambda_b < \lambda_b) \qquad \frac{M_{b,AIJ}}{M_{p,b}} = \frac{1}{\lambda_b^2}$$

$$\lambda_b = \sqrt{\frac{M_{p,b}}{{}_mM_{cr0,n}}} \tag{12}$$

where ${}_mM_{cr0}$ denotes the elastic lateral buckling moment, which is determined by multiplying the elastic buckling load ${}_mP_{cr0,n}$ of an H beam subjected to the inverse symmetrical bending moment by the distance d_b between flanges. ${}_mP_{cr0,n}$ is obtained from the following equation:

$$_mP_{cr0,n} = P_{cr0,n} \cdot C_m \tag{13}$$

$$C_m = 1.75 + 1.05m + 0.3m^2 \leq 2.3 \tag{14}$$

where $P_{cr0,n}$ represents the value when $k_u = k_\theta = 0$ is substituted into Equation (8). Additionally, when subjected to an inversely symmetrical bending moment, the moment gradient $m = 1.0$ (as indicated by the relationship between M_1 and M_2 in Figure 19), thus $C_m = 2.3$ according to Equation (14).

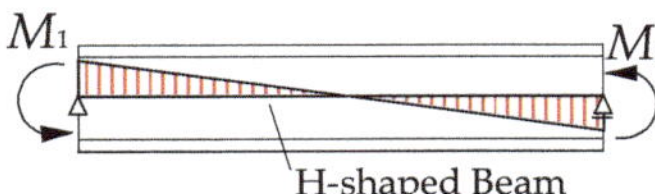

Figure 19. Image of the moment gradient generated in a beam.

The design bearing capacity $M_{b,EC}$ of the beam according to the Eurocode [47] is calculated from the following equation.

$$\frac{M_{b,EC}}{M_{p,b}} = \chi_{LT} = \min\left[1, \quad \frac{1}{\Phi_{LT} + \sqrt{\Phi_{LT}^2 - 0.75\lambda_b^2}}, \quad \frac{1}{\lambda_b^2}\right] \tag{15}$$

Here,

$$\Phi_{LT} = 0.5\left[1 + \alpha_{LT}(\lambda_b - 0.4) + 0.75\lambda_b^2\right] \tag{16}$$

$$(H/B \leq 2.0): \alpha_{LT} = 0.34, \quad (H/B > 2.0): \alpha_{LT} = 0.49$$

where H is the beam height and B is the width of the beam.

Figure 20 illustrates the lateral buckling strength calculated per the Japan code [46] and the Eurocode [47]. Notably, the lateral buckling strength $M_{b,AIJ}$ and $M_{b,EC}$ differ when dimensionless slenderness ratio $\lambda_{b,AIJ} = \lambda_{b,EC}$. Specifically, when $M_{b,AIJ}/M_{p,b} \leq 1.0$, the lateral buckling strength according to the Eurocode varies with H/B. Thus, we present the $\lambda_{b,EC}$ and $\lambda_{b,AIJ}$ when the design capacities of the Japan code [46] and the Eurocode [47] are equivalent. As depicted in the ($\bigcirc$, circular plot), while the lateral buckling strength at $\overline{\lambda}_{b,EC}$ is equivalent to that at $l_{b,AIJ}$, $\overline{\lambda}_{b,EC}$ is smaller than $\lambda_{b,AIJ}$.

Subsequently, if the beams employed in the examined steel structures are affixed to roof folded plates, the design bearing capacity M'_b of the continuously stiffened H beam is considered. This M'_b is determined using Equation (11) or Equation (15), wherein M_{cr0} in Equation (12) replaces the elastic buckling moment ${}_mM_{cr,n}$ of the continuously stiffened beam.

$$_mP_{cr,n} = P_{cr0,n} \cdot C_m + P_{cr,n} - P_{cr0,n} \tag{17}$$

The design bearing capacity M'_b of a continuously braced large beam surpasses the design bearing capacity M_b of a large beam without bracing.

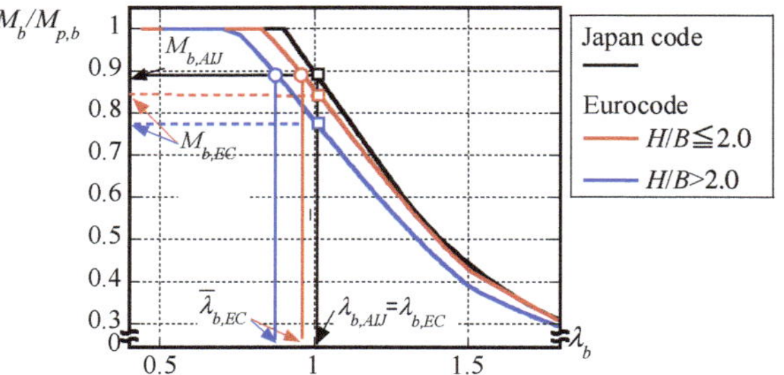

Figure 20. Lateral buckling strength curve [46,47].

Figure 21 presents the procedure for computing the lateral buckling design capacity according to the Japanese code [46] and the Eurocode [47]. By utilizing the rotational stiffness $k_{\theta,1}$ or $k_{\theta,2}$ derived from the experimental results and applying it to Equation (17), the lateral buckling design capacity of an H beam continuously stiffened by a roof folded plate can be determined.

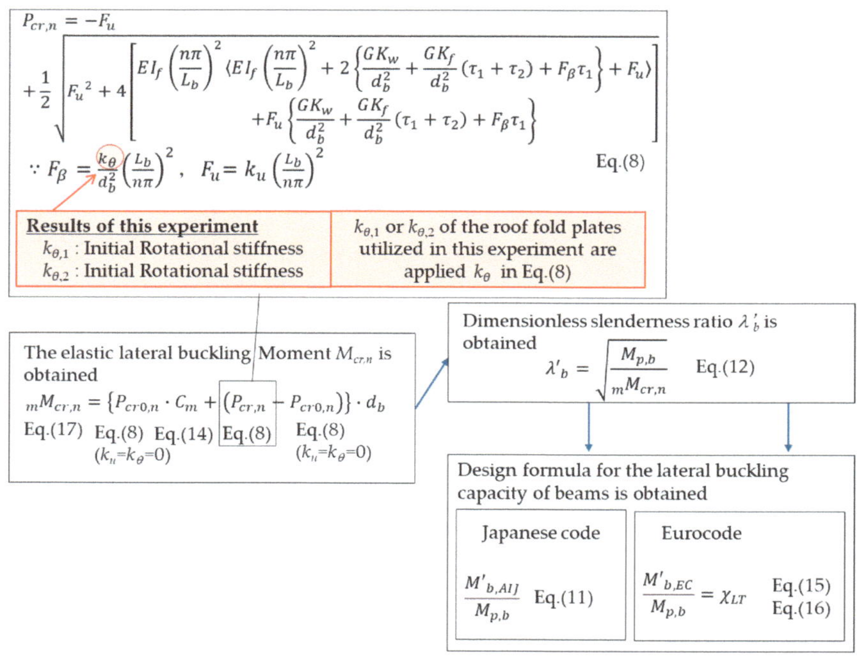

Figure 21. The procedure for computing the lateral buckling design capacity [38,46,47].

Figure 22 illustrates the rate of increase for design bearing capacity in lateral buckling due to rotational stiffness, denoted by 'μ'. The value of μ is determined by the ratio of the design bearing capacity M'_b of the continuously stiffened large beam to the design bearing capacity M_b of the unbraced large beam, where $\mu = M'_b/M_b$. The plot in Figure 22a displays data for surveyed structures that meet the criterion $M_{\theta,1}/M_{br,1} \geq 1.0$ in Figure 17a. The plot in Figure 22b presents data for surveyed structures that meet the criterion $M_{\theta,2}/M_{br,2} \geq 1.0$ in Figure 17b. Despite the initial rotational stiffness $k_{\theta,1}$ and secant rotational stiffness

$k_{\theta,2}$ of the roof folded plate in this experiment being smaller than the required rotational stiffness, the design bearing capacity increases. Moreover, a larger required stiffness ratio correlates with a greater proof stress increase ratio μ.

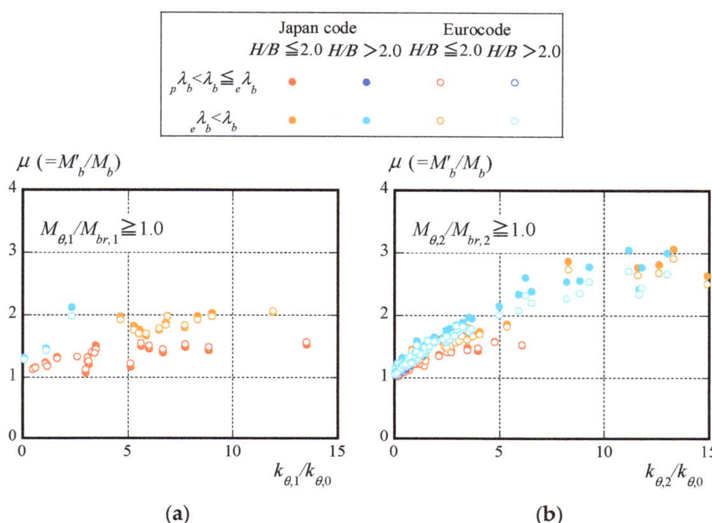

Figure 22. The rate of increase for design bearing capacity in lateral buckling due to rotational stiffness at (**a**) initial stiffness and at (**b**) secant stiffness [46,47].

5. Conclusions

In this paper, torsional tests were conducted on roof folded plates to examine the influence of H beam–roof folded plate joints on the rotational stiffness of roof folded plates, leveraging the findings of a structural survey of steel structures in Japan. Furthermore, the correlation between the rotational stiffness of the roof folded plate and its design bearing capacity was elucidated. The findings are delineated below.

(1) When the top flange of the beam is pressed against the bottom flange of the roof folded plate during torsional deformation of the H beam, the small thickness of the roof folded plate causes bending deformation of the plate where it contacts the top flange of the H beam. Consequently, the rotational stiffness of the roof folded plate falls notably short of the theoretical stiffness posited for a rigid connection between the beam and the roof folded plate.

(2) In the actual structural connection method (where tight frames are welded in a row along the length direction of the beam at the center of the top flange), the gap between the roof folded plate and the tight frame welded to the beam prevents torsional moments in the roof folded plate until the bottom flange of the roof folded plate contacts the top flange of the H steel beam, resulting in the non-exhibition of initial rotational stiffness $k_{\theta,1}$.

(3) For specimens in which the tight frames are welded in two rows along the length direction at the top flange of the beam, a torsional moment is induced in the roof folded plate from the onset of loading, resulting in the initial rotational stiffness $k_{\theta,1}$ of the roof folded plate, being 2–3% of the theoretical rotational stiffness $k_{\theta,0}$.

(4) For specimens in which connectors with stiffened blocks are utilized and both the top and bottom flanges of the roof folded plate are constrained, the initial rotational stiffness surpasses that of the other specimens. Specifically, the initial rotational stiffness $k_{\theta,1}$ of the roof folded plate amounts to approximately 12% of the theoretical rotational stiffness $k_{\theta,0}$.

(5) When evaluating the lateral buckling capacity of continuously stiffened H beams using the buckling design capacities specified by the Japan code [46] and the Eurocode [47], it was observed that the design capacity outlined by the Japan code exceeded that of the Eurocode, even when the dimensionless slenderness ratio was equivalent.

(6) Upon attaching the roof folded plates used in the experiments to the surveyed structures, it was observed that, in 16% of all of the structures, the torsional moment $M_{br,1}$ occurring in the roof folded plates during the lateral buckling of the main beams was smaller than the torsional moment $M_{\theta,1}$ at the onset of stiffness reduction, at which the initial rotational stiffness $k_{\theta,1}$ of the roof folded plates used in this study can be achieved. In contrast, in 62% of all structures, the maximum torsional moment $M_{\theta,2}$, for which the secant rotational stiffness $k_{\theta,2}$ of the roof folded plates used in this study can be demonstrated, exceeds the torsional moment $M_{br,2}$ occurring in the roof folded plates during the lateral buckling of the main beams. Thus, it can be inferred that the number of structures in which the roof folded plate exhibits the required performance against lateral buckling of the beams can be increased by utilizing the maximum torsional moment and the secant stiffness, rather than by relying on the initial stiffness and the bearing capacity at the onset of stiffness reduction, as the metrics for the holding performance of a roof folded plate.

Author Contributions: Conceptualization, Y.Y. and Y.K.; methodology, Y.Y. and Y.K.; software, Y.Y.; validation, Y.Y. and Y.K.; formal analysis, Y.Y. and Y.K.; investigation, Y.Y. and Y.K.; resources, Y.Y. and Y.K.; data curation, Y.Y.; writing—original draft preparation, Y.Y. and Y.K.; writing—review and editing, Y.Y. and Y.K.; visualization, Y.Y. and Y.K.; supervision, Y.Y. and Y.K.; project administration, Y.Y. and Y.K.; funding acquisition, Y.Y. and Y.K. All authors have read and agreed to the published version of the manuscript.

Funding: This research was funded by The Kajima Foundation's Support Program for International Joint Research Activities (principal investigator: Yoshihiro Kimura) and JSPS KAKENHI (grant number 22K14369) (principal investigator: Yuki Yoshino). We express our deepest gratitude for their sincere support.

Data Availability Statement: The raw/processed data necessary to reproduce these findings cannot be shared at this time because the data also form part of an ongoing study.

Conflicts of Interest: The authors declare that they have no known competing financial interest or personal relationship that could have appeared to influence the work reported in this paper.

References

1. Ishida, T.; Iyama, J.; Kishiki, S.; Shimada, Y.; Yamada, S.; Seike, T. Structural Damage to School Gymnasiums due to the 2016 Kumamoto Earthquake. *J. Technol. Des. (Trans. AIJ)* **2018**, *58*, 1313–1318. (In Japanese) [CrossRef]
2. Suzuki, A.; Fujita, T.; Kimura, Y. Identifying damage mechanisms of gymnasium structure damaged by the 2011 Tohoku earthquake based on biaxial excitation. *Structures* **2022**, *35*, 1321–1338. [CrossRef]
3. Kobayashi, F.; Sato, E.; Tomokiyo, E.; Noda, M.; Gavanski, E.; Takadate, Y.; Takamori, K.; Kimura, K.; Nakato, S.; Moriyama, H.; et al. Immediate Report on Wind Disasters Occurred by Typhoon 1915. *J. Wind Eng. (JAWE)* **2020**, *45*, 30–39. (In Japanese) [CrossRef]
4. Akatsuka, T.; Tomokiyo, E. Investigation of Damage to Houses Caused by Typhoon JEBI and the Influence of Surrounding Environments on Its Occurrence. *J. Wind Eng. (JAWE)* **2020**, *26*, 96–101. (In Japanese) [CrossRef]
5. Timoshenko, S.P. *Beam without Lateral Support*; American Society of Civil Engineers (ASCE): Reston, VA, USA, 1924; Volume 87.
6. Bleich, F. *Buckling Strength of Metal Structures*; McGraw-Hill: New York, NY, USA, 1952; p. 405.
7. Nethercot, D.A.; Rockey, K.C. A Unified Approach to the Elastic Lateral Buckling of Beams. *Struct. Eng.* **1971**, *49*, 96–107. [CrossRef]
8. Suzuki, T. Lateral Buckling of Open Web. *Trans. Archit. Inst. Jpn. (Trans. AIJ)* **1960**, *166*, 537–540. (In Japanese)
9. Wakabayashi, M.; Nakamura, T. Numerical Analysis of Lateral Buckling Strength of H-Shaped Steel Beams Subjected to End Moments and Uniformly Distributed Load. *Trans. Archit. Inst. Jpn. (Trans. AIJ)* **1973**, *208*, 7–13. (In Japanese) [CrossRef]
10. Narayanan, R. *Beam and Beam Columns-Stability and Strength*; Applied Science Publisher: Boca Raton, FL, USA, 1983.
11. Salvador, T.G. Lateral Buckling of Eccentrically Loaded I–Columns. *Am. Soc. Civ. Eng. (ASCE)* **1956**, *121*, 1163–1178. [CrossRef]
12. Trahair, N.S. Laterally unsupported beams. *Eng. Struct.* **1996**, *18*, 759–768. [CrossRef]
13. Lebastard, M.; Couchaux, M.; Bureau, A.; Hjiaj, M. Lateral-torsional buckling of uniform and tapered welded I-section beams. *Eng. Struct.* **2024**, *303*, 117301. [CrossRef]
14. Yura, J.A. Fundamentals of beam bracing. *Eng. J.-Am. Inst. Steel Constr.* **2001**, *38*, 11–26. [CrossRef]
15. Gengshu, T.; Shaofan, C. The elastic buckling of interbraced girders. *J. Constr. Steel Res.* **1989**, *14*, 87–105. [CrossRef]

16. Kimura, Y.; Sugita, Y.; Yoshino, Y. Effect of lateral-rotational restraint of continuous braces on lateral buckling strength for H-shaped steel beams with compressive axial force and flexural moment. *J. Struct. Constr. Eng. (Trans. AIJ)* **2016**, *81*, 1321–1331. (In Japanese) [CrossRef]
17. Kimura, Y.; Sato, Y. Effect of warping restraint of beams to column joint on lateral buckling behavior for H-shaped beams with continuous braces under gradient flexural moment. *J. Struct. Constr. Eng. (Trans. AIJ)* **2021**, *86*, 145–155. (In Japanese) [CrossRef]
18. Kimura, Y.; Yoshino, Y. Required Bracing Capacity on Lateral Buckling Strength for H-Shaped Beams with Bracings. *J. Struct. Constr. Eng. (Trans. AIJ)* **2011**, *76*, 2143–2152. (In Japanese) [CrossRef]
19. Kimura, Y.; Matsuo, T.; Yoshino, Y. Estimation of Elasto-Plastic Lateral Buckling Stress for H-Shaped Beams with Lateral-Rotational Braces on Subjected to Axial Force and Flexural Moment. *J. Struct. Constr. Eng. (Trans. AIJ)* **2014**, *79*, 1299–1308. (In Japanese) [CrossRef]
20. Zhang, Z.; Kawai, R.; Kanao, I. Numerical Studies on Lateral Bracing Method with Effective Restraint on Out-of-Plane Deformation of Wide Flange Beam. *J. Struct. Constr. Eng. (Trans. AIJ)* **2014**, *79*, 323–329. (In Japanese) [CrossRef]
21. Igawa, N.; Ikarashi, K.; Mitsui, K. Buckling Behavior of H-shaped Beam with Continuous Restraint Installed Transverse Stiffeners and Method for Deciding Stiffening Position. *J. Struct. Constr. Eng. (Trans. AIJ)* **2020**, *85*, 1491–1501. (In Japanese) [CrossRef]
22. Nguyen, C.T.; Moon, J.; Lee, H.E. Lateral–torsional buckling of I-girders with discrete torsional bracings. *J. Constr. Steel Res.* **2010**, *66*, 170–177. [CrossRef]
23. Ji, X.L.D.; Twizell, S.C.; Driver, R.G.; Imanpour, A. Lateral Torsional Buckling Response of Compact I-Shaped Welded Steel Girders. *J. Struct. Eng.* **2022**, *148*, 04022149. [CrossRef]
24. Mohammadi, E.; Hosseini, S.S.; Rohanimanesh, M.S. Elastic lateral-torsional buckling strength and torsional bracing stiffness requirement for monosymmetric I-beams. *Eng. Struct.* **2016**, *104*, 116–125. [CrossRef]
25. Nguyen, C.T.; Joo, H.S.; Moon, J.; Lee, H.E. Flexural-torsional buckling strength of I-girders with discrete torsional braces under various loading conditions. *Eng. Struct.* **2022**, *36*, 337–350. [CrossRef]
26. Fukunaga, I.; Todaka, T.; Zhang, Z.; Kanao, I. Deformation Capacity of Wide Flange Beam with Lateral Bracings. *J. Struct. Constr. Eng. (Trans. AIJ)* **2022**, *87*, 372–380. (In Japanese) [CrossRef]
27. Matsui, R.; Yamaura, Y.; Takeuchi, T. Lateral Bracing Requirements for H-Section Beams with Supports Attached to Top Flange Subjected to Cyclic Antisymmetric Moment. *J. Struct. Constr. Eng. (Trans. AIJ)* **2013**, *78*, 1485–1492. (In Japanese) [CrossRef]
28. Idota, H.; Kato, Y.; Ono, T. Lateral Bracing Spacing and Strength of H- Shaped Steel Beams under Cyclic Loading. *J. Struct. Constr. Eng. (Trans. AIJ)* **2013**, *78*, 1989–1998. (In Japanese) [CrossRef]
29. Ikarashi, K.; Ohnishi, Y.; Sano, T. Collapse Mode and Plastic Deformation Capacity of H-Shaped Beams with Continuous Restraint on Upper Flange. *J. Struct. Constr. Eng. (Trans. AIJ)* **2018**, *83*, 491–501. (In Japanese) [CrossRef]
30. Kanao, I.; Nakashima, M.; Liu, D. Lateral Buckling Behavior and Lateral Bracing of Wide-Flange Beams Subjected Cyclic Loading. *J. Struct. Constr. Eng. (Trans. AIJ)* **2001**, *66*, 147–154. (In Japanese) [CrossRef] [PubMed]
31. Bradford, M.A.; Gao, Z. Distortional buckling solutions for continuous composite beams. *J. Struct. Eng.* **1992**, *118*, 73–89. [CrossRef]
32. Rossi, A.; Nicoletti, R.S.; de Souza, A.S.C.; Martins, C.H. Lateral distortional buckling in steel-concrete composite beams: A review. *Structures* **2020**, *27*, 1299–1312. [CrossRef]
33. Egilmez, O.O.; Herman, R.S.; Helwig, T.A. Lateral stiffness of steel bridge I-girders braced by metal deck forms. *J. Bridge Eng.* **2009**, *14*, 17–25. [CrossRef]
34. Suzuki, A.; Kimura, Y. Rotation capacity of I-shaped beam failed by local buckling in buckling-restrained braced frames with rigid beam-column connections. *J. Struct. Eng.* **2023**, *149*, 04022243. [CrossRef]
35. Suzuki, A.; Suzuki, K.; Kimura, Y. Ultimate shear strength of perfobond shear connectors subjected to fully reversed cyclic loading. *Eng. Struct.* **2021**, *248*, 113240. [CrossRef]
36. Tsuda, K.; Kido, M. Buckling Strength of Columns Braced Continuously with Varying Axial Force -Flexural Buckling of Compressive Flange of H-shaped Steel Restrained by Web or Slab-. *J. Struct. Constr. Eng. (Trans. AIJ)* **2013**, *78*, 1513–1521. (In Japanese) [CrossRef]
37. Ikarashi, K.; Sano, T. Influence of Upper Flange Restraint Condition on Lateral Buckling Behavior of H-Shaped Beam. *J. Struct. Constr. Eng. (Trans. AIJ)* **2018**, *83*, 1063–1073. (In Japanese) [CrossRef]
38. Kimura, Y.; Yoshino, Y.; Ogawa, J. Effect of Lateral-Rotational Restraint and Strength of Continuously Braces on Lateral Buckling Load for H-Shaped Beams. *J. Struct. Constr. Eng. (Trans. AIJ)* **2013**, *78*, 193–201. (In Japanese) [CrossRef]
39. Kimura, Y.; Sugita, Y. Effect of Lateral-Rotational Restraint by Continuous Braces on Lateral Buckling Strength for H-Shaped Beams with Gradient Flexural Moment and Compressive Axial Force. *J. Struct. Constr. Eng. (Trans. AIJ)* **2017**, *82*, 1799–1809. (In Japanese) [CrossRef]
40. Kimura, Y.; Yoshino, Y. Effect of Lateral and Rotational Restraint for Bracings on Lateral Buckling Load for H-Shaped Beams under Moment Gradient. *J. Struct. Constr. Eng. (Trans. AIJ)* **2014**, *79*, 761–770. (In Japanese) [CrossRef]
41. Kimura, Y.; Yoshino, Y. Effect of Lateral-Rotational Restraint of Continuous Braces on Lateral Buckling Strength for H-Shaped Beams with Flexural Moment Gradient. *J. Struct. Constr. Eng. (Trans. AIJ)* **2016**, *81*, 1309–1319. (In Japanese) [CrossRef]
42. Kimura, Y.; Miya, M.; Liao, W. Effect of Restraint for Continuous Braces on Lateral Buckling Load for H-Shaped Beams with Warping Restraint of Beams to Column Joint. *J. Struct. Constr. Eng. (Trans. AIJ)* **2018**, *83*, 1355–1363. (In Japanese) [CrossRef]
43. Kimura, Y.; Miya, M. Effect of Warping Restraint of Beams to Column Joint on Lateral Buckling Behavior for H-Shaped Beams with Continuous Braces under Gradient Flexural Moment. *J. Struct. Constr. Eng. (Trans. AIJ)* **2019**, *84*, 1601–1611. (In Japanese) [CrossRef]

44. Architectural Institute of Japan (AIJ). *Recommendations for Stability Design of Steel Structures*; Maruzen Publishing Co., Ltd.: Tokyo, Japan, 2012.
45. Architectural Institute of Japan (AIJ). *Design Standard for Structures–Based on Allowable Stress Concept*; Maruzen Publishing Co., Ltd.: Tokyo, Japan, 2012.
46. Architectural Institute of Japan (AIJ). *Recommendation for Limit State Design of Steel Structures*; Maruzen Publishing Co., Ltd.: Tokyo, Japan, 2011.
47. *Eurocode-3: Design of Steel Structures Part 1–1,1-3,1-5: General Rules and Rules for Buildings*; European Committee for Standardization: Brussels, Belgium, 2005.
48. Japan Metal Roofing Association. *Standard of Steel Roofing (SSR)*; Japan Metal Roofing Association: Tokyo, Japan, 2007.
49. JIS-A6514; Component for Metal Roof-Decks. Japanese Industrial Standards Committee: Tokyo, Japan, 2019. (In Japanese)
50. Suzuki, A.; Abe, K.; Suzuki, K.; Kimura, Y. Cyclic behavior of perfobond-shear connectors subjected to fully reversed cyclic loading. *J. Struct. Eng.* **2021**, *147*, 04020355. [CrossRef]
51. Yoshino, Y.; Liao, W.; Kimura, Y. Restraint effect on lateral buckling load of continuous braced H-shaped beams based on partial frame loading tests. *J. Struct. Constr. Eng. (Trans. AIJ)* **2022**, *87*, 634–645. (In Japanese) [CrossRef]
52. Ikarashi, K.; Fujisawa, M.; Shimizu, N. Evaluation of Elastic Buckling Strength of Rectangular Corrugate Plate under Shear Loading. *J. Struct. Constr. Eng. (Trans. AIJ)* **2008**, *73*, 1883–1890. (In Japanese) [CrossRef]
53. Ikarashi, K.; Nakano, S.; Shimizu, N. Evaluation of Shear Stiffness of Rectangular Corrugate Plate under Shear Loading. *J. Struct. Constr. Eng. (Trans. AIJ)* **2009**, *74*, 2327–2334. (In Japanese) [CrossRef]
54. Shimizu, N.; Okada, T.; Ikarashi, K. A Study on Post Shear Buckling Behavior in Corrugated Steel Plate. *J. Struct. Constr. Eng. (Trans. AIJ)* **2010**, *75*, 1013–1020. (In Japanese) [CrossRef]
55. Wei, X.; Li, G.; Xiao, L.; Zhou, L.; He, K.; Han, B. Shear strength reduction of trapezoidal corrugated steel plates with artificial corrosion pits. *J. Constr. Steel Res.* **2021**, *180*, 106583. [CrossRef]
56. Wen, C.-B.; Guo, Y.-L.; Zuo, J.-Q.; Zhao, X.-Y. Strength design of prefabricated corrugated steel plate shear walls under combined compression and shear loads. *J. Build. Eng.* **2023**, *65*, 105790. [CrossRef]
57. Elamary, A.S.; Saddek, A.B.; Alwetaishi, M. Effect of corrugated web on flexural capacity of steel beams. *Int. J. Appl. Eng. Res* **2017**, *12*, 470–481.
58. Kubo, M.; Watanabe, K. Lateral-Torsional Buckling Capacity of Steel Girders with Corrugated Web Plates. *Doboku Gakkai Ronbunshu A* **2007**, *63*, 179–193. (In Japanese) [CrossRef]
59. Watanabe, K.; Uchida, S.; Kubo, M. Shear buckling capacity of steel girders with corrugated webs. *J. Struct. Eng. A* **2007**, *53A*, 13–24. (In Japanese)
60. Elkawas, A.A.; Hassanein, M.F.; Elchalakani, M. Lateral-torsional buckling strength and behaviour of high-strength steel corrugated web girders for bridge construction. *Thin-Walled Struct.* **2018**, *122*, 112–123. [CrossRef]
61. Yi, J.; Gil, H.; Youm, K.; Lee, H. Interactive shear buckling behavior of trapezoidally corrugated steel webs. *Eng. Struct.* **2008**, *30*, 1659–1666. [CrossRef]
62. Suzuki, A.; Liao, W.; Shibata, D.; Yoshino, Y.; Kimura, Y.; Shimoi, N. Structural damage detection technique of secondary building components using piezoelectric sensors. *Build.* **2023**, *13*, 2368. [CrossRef]
63. Degtyarev, V.V. Flexural strength of steel decks with square and rectangular holes: Numerical studies. *J. Constr. Steel Res.* **2020**, *172*, 106241. [CrossRef]
64. Degtyarev, V.V. Concentrated load distribution in corrugated steel decks: A parametric finite element study. *Eng. Struct.* **2020**, *206*, 110158. [CrossRef]
65. Cho, H.; Ozaki, F.; Sato, Y.; Miyabayashi, K. Collapse Temperature of a Simple Supported Roof Member Using a Corrugated Steel Thin Plate. *Steel Constr. Eng.* **2024**, *30*, 63–74. (In Japanese) [CrossRef]
66. Steel Deck Institute. *Standard for Steel Roof Deck*; Steel Deck Institute: Allison Park, PA, USA, 2017.
67. Steel Deck Institute. *Roof Deck Design Manual*; Steel Deck Institute: Allison Park, PA, USA, 2012.
68. Steel Deck Institute. *Manual of Construction with Steel Deck*; Steel Deck Institute: Allison Park, PA, USA, 2016.
69. Otsu, T.; Takagi, J.; Okada, T.; Sato, Y.; Shoji, Y. In-Plane Shear Performance Evaluation of Roof Deck Plates Connected with Puddle Welding. *J. Technol. Des. (Trans. AIJ)* **2019**, *60*, 691–696. (In Japanese) [CrossRef]
70. Takagi, J.; Shoji, Y.; Okada, T.; Sato, Y.; Otsu, T. Material Property of Roof Deck Steel Plates and Shear Behavior of Connections with Puddle Welding. *J. Technol. Des. (Trans. AIJ)* **2019**, *59*, 177–182. (In Japanese) [CrossRef]
71. Kikitsu, H.; Ohkuma, T.; Okuda, Y.; Nishimura, H. Vulnerability of V-beam Steel Roofing Subjected to High Wind Hazard. *J. Wind Eng. (JAWE)* **2008**, *20*, 223–228. (In Japanese) [CrossRef]
72. Coni, N.; Gipiela, M.L.; D'Oliveira, A.S.C.M.; Marcondes, P.V.P. Study of the mechanical properties of the hot dip galvanized steel and galvalume. *J. Braz. Soc. Mech. Sci. Eng.* **2009**, *32*, 319–326. [CrossRef]
73. JIS Z 2241; Metallic Materials—Tensile Testing Method of Test at Room Temperature. Japan Industrial Standards (JIS): Tokyo, Japan, 2011.

 *buildings*

Article

A Study on the Mechanical Characteristics and Wheel–Rail Contact Simulation of a Welded Joint for a Large Radio Telescope Azimuth Track

Xiao Chen [1], Ruihua Yin [1], Zaitun Yang [1], Huiqing Lan [1,*] and Qian Xu [2,3]

[1] Key Laboratory of Vehicle Advanced Manufacturing, Measuring and Control Technology (Ministry of Education), Beijing Jiaotong University, Beijing 100044, China; 21125985@bjtu.edu.cn (X.C.); 23121388@bjtu.edu.cn (R.Y.); 22125985@bjtu.edu.cn (Z.Y.)

[2] Xinjiang Astronomical Observation, Chinese Academy of Sciences, Urumqi 830001, China; xuqian@xao.ac.cn

[3] Key Laboratory of Radio Astronomy, Chinese Academy of Sciences, Urumqi 830011, China

* Correspondence: hqlan@bjtu.edu.cn

Abstract: The azimuth track is an important component of the radio telescope wheel–rail system. During operation, the azimuth track is inevitably subject to phenomena such as track wear, track fatigue cracks, and impact damage to welded joints, which can affect observation accuracy. The 110 m QiTai radio telescope (QTT) studied in this paper is the world's largest fully steerable radio telescope at present, and its track will bear the largest load ever. Since the welded joint of an azimuth track is the weakest part, an innovative welding method (multi-layer and multi-pass weld) is adopted for the thick welding section. Therefore, it is necessary to study the contact mechanical properties between the wheel and the azimuth track in this welded joint. In this study, tensile tests based on digital image correlation technology (DIC) and Vickers hardness tests are carried out in the metal zone (BM), heat-affected zone (HAZ), modified layer, and weld zone (WZ) of the welded joint, and the measured data are used to fit the elastic–plastic constitutive model for the different zones of the welded joint in the azimuth track. Based on the constitutive model established, a nonlinear finite element model is built and used to simulate the rolling mechanical performance between the wheel and azimuth track. Through the analysis of simulated data, we obtained the stress distribution of the track under different pre-designed loads and identified the locations most susceptible to damage during ordinary working conditions, braking conditions, and start-up conditions. The result can provide a significant theoretical basis for future research and for the monitoring of large track damage.

Keywords: azimuth track; welded joint; multi-layer and multi-pass; wheel–rail contact; mechanical property; finite element model

Citation: Chen, X.; Yin, R.; Yang, Z.; Lan, H.; Xu, Q. A Study on the Mechanical Characteristics and Wheel–Rail Contact Simulation of a Welded Joint for a Large Radio Telescope Azimuth Track. *Buildings* **2024**, *14*, 1300. https://doi.org/10.3390/buildings14051300

Academic Editors: Atsushi Suzuki and Dinil Pushpalal

Received: 7 April 2024
Revised: 30 April 2024
Accepted: 2 May 2024
Published: 5 May 2024

1. Introduction

During the operation of radio telescopes, a series of problems affecting the pointing accuracy of the azimuth tracks [1], such as track wear, fatigue cracks, and impact damage to track welded joints, occur in the wheel–rail contact system. A lot of design optimizations and improvements to tracks have already been conducted [2–4]. For the QiTai radio telescope (QTT) studied in this paper, the joints of the azimuth track are all welded. Since the welded joint is the weakest part of the entire track, most of the damage occurs in this zone. The welded joint of the azimuth track adopts an innovative multi-pass welding method of multi-layer and multi-pass due to the thick track. Therefore, it is significant to study the contact mechanical characteristics in the welded joint.

In the research of welded joints, there are four methods commonly used to obtain the local mechanical properties. The first method is based on hardness testing, characterizing the local constitutive behavior of the welded joint by the correlation between hardness values and strength, but the empirical hardness–strength relationship proposed by different researchers is only valid for the materials studied and lacks universality. The second method is to obtain the stress–strain constitutive relationship by cutting microscopic specimens and conducting tensile, shear, and impact tests. This method can provide a relatively accurate stress–strain constitutive relationship as long as the material in the specimen is uniform. However, it is difficult to ensure the uniformity of the material in the joint due to the presence of welding heat, especially in the heat-affected zone where material property gradients exist. The third method is to simulate the welding process through thermodynamic finite element analysis, but it is difficult to accurately reproduce the temperature history of the welded joint. The fourth method is the DIC technology approach. This method has made good progress. According to these methods, many scholars have conducted all kinds of experiments. Microscopically, Kumar et al. [5] conducted micro-hardness, tensile, and impact tests on the welded joints and analyzed the microstructural properties of the multi-pass welded joints. Han [6] explained the evolution of the microstructure and mechanical properties of welded joints. These studies focused on the microstructure of materials and achieved good results. Macroscopically, Saranath et al. [7] conducted research on the welded joints by cutting microscopic samples and subjecting them to tension, shear, and impact tests, but it is difficult to ensure the uniformity of the materials due to the influence of thermal processes in the welded joint. Zhang et al. [8] studied the welded joints by simulating the weld process through thermodynamic finite element analysis, but it is difficult to accurately reproduce the temperature history of the welded joint. However, with the widespread application of digital image correlation (DIC) technology, the aforementioned problems have been solved. Zhang et al. [9] used DIC technology to analyze the strain field of the welded joint, which can effectively obtain the local constitutive relation of the nonuniform welded joint, but it also has a drawback in that it cannot obtain the complete stress–strain curve of the nonfracture position.

In the research on wheel–rail contact, there are already a sufficient theoretical foundation and technical methods for determining the wheel–rail contact of track vehicles. For the mechanical analysis and damage analysis of wheel–rail contact, the mainstream methods currently are finite element simulation analysis and numerical simulation analysis. Tan and Shi [10] used numerical simulation to solve wheel–rail fatigue damage. Moreover, they [11] also used the multi-finite-element coupling method and the multi-time-step solution method to solve the dynamics of wheel–rail contact. Many wheel–rail contact methods have been applied in azimuth tracks such as the finite element method. Arslan [12] and Twllsikivi [13] found that the track is more prone to significant plastic deformation during wheel–rail contact. Kuminek and Aniolek [14] investigated the influence of computational accuracy. Tian et al. [15] analyzed the deformation patterns during wheel–rail contact. Fu et al. [16] obtained the maximum contact stress and contact width of the wheel–rail based on the Hertz elastic contact theory. Yang et al. [17] discussed wheel–rail frictional rolling contact using the explicit finite element analysis method. However, these methods lack an experimental basis for material parameters, material properties, and track division. Moreover, the models established above are finite element models of the form "base metal–welded joint–base metal". Although this method saves a lot of computation time, it ignores the influence of the heat-affected zone of the welded joint, leading to significant discrepancies between the simulation results and the actual test results.

In this paper, we conducted experiments at a macro level. We combined and improved existing methods to analyze the welded joint of the azimuth track by conducting tensile tests and Vickers hardness measurements based on digital image technology. Then, we fitted constitutive models of the welded joint by comparing and analyzing the experimental data, and a real finite element model of the azimuth track was established. Based on these, the Mises stress of welded joints during smooth operation under different loads and the distribution of stress on the welded joint surface during start-up and braking conditions were obtained by simulation.

2. Experiment

2.1. Tensile Experiment Based on DIC

The test material was selected from the 110 m radio telescope overall track weld sample provided by 39th Research Institute of China Electronics Technology Group Corporation. The track material is 42CrMo steel and adopts a method of multi-pass, single-groove welding for the track. The welding process first implements the welding of the modified layer, followed by the welding of the filling layer, and finally implements the welding of the surface overlay layer. Figure 1 shows the overall track weld sample and joint form of the 110 m radio telescope.

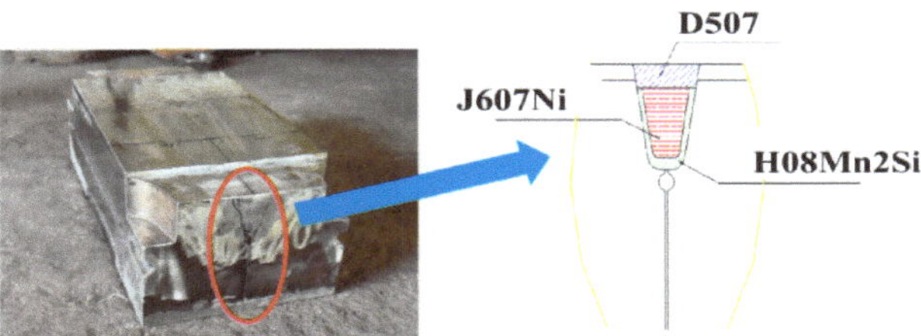

Figure 1. A 110 m radio telescope overall azimuth track weld sample and joint form.

According to the tensile test method for metallic materials in ISO 6892-1:2019 [18], a quasi-static tensile specimen was cut and designed at the welded joint of the 110 m radio azimuth track using wire electrical discharge machining. Considering the use of a layered weld process for the azimuth track welded joint, a cut was carried out along the depth direction of the welded joint based on the weld morphology. There were two types of test specimens, namely specimen 1 and specimen 2. They both consisted of the metal zone (BM), the heat-affected zone (HAZ), the modified layer, and the weld zone (WZ), but the difference between them was that they use different weld methods for the WZ. The dumbbell specimen was sampled longitudinally along the track. The sampling position and the physical object are shown in Figure 2.

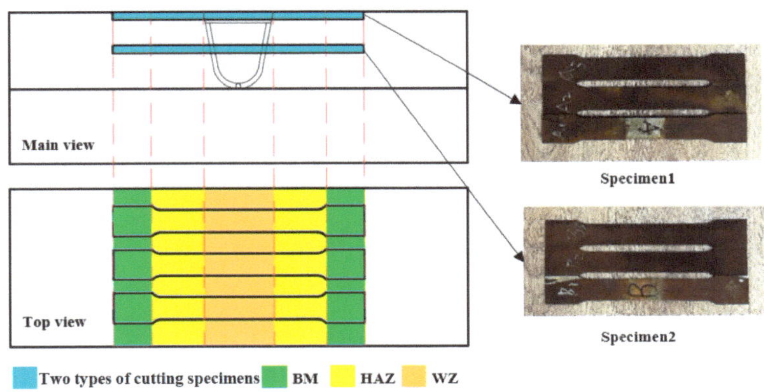

Figure 2. Schematic diagram of specimen sampling location.

The uniaxial tensile test was conducted on an INSTRON 5982 composite material testing machine and a VIC-3D measurement system, as shown in Figure 3a. The stretching experiment machine uses displacement control. The VIC-3D strain measurement system includes two highly integrated high-speed cameras that can track the displacement and strain of each pixel within the field of view. They were placed at an angle of 30° approximately in front of the specimen to capture the speckle changes on the surface of the specimen during the stretching process. DIC technology can obtain the strain of any point on the predetermined path at any time. The optical measurement points mainly serve as virtual extensometers to measure the strain distribution at the ROI zone of the welded joint. The optical measurement path combined with the stretching sample is shown in Figure 3b, with an optical measurement range of 160 mm, including the BM, HAZ, and WZ.

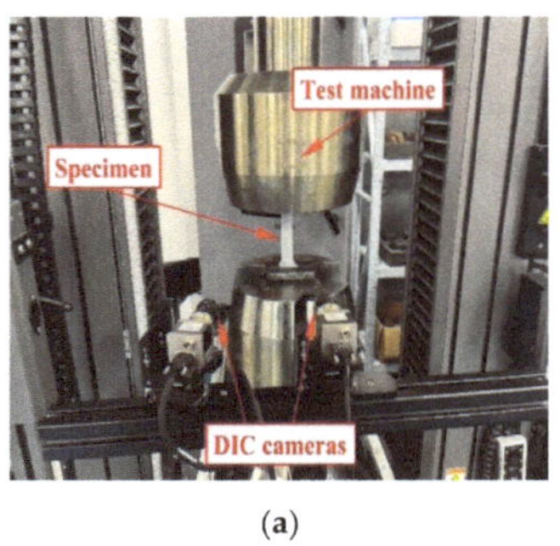

(a)

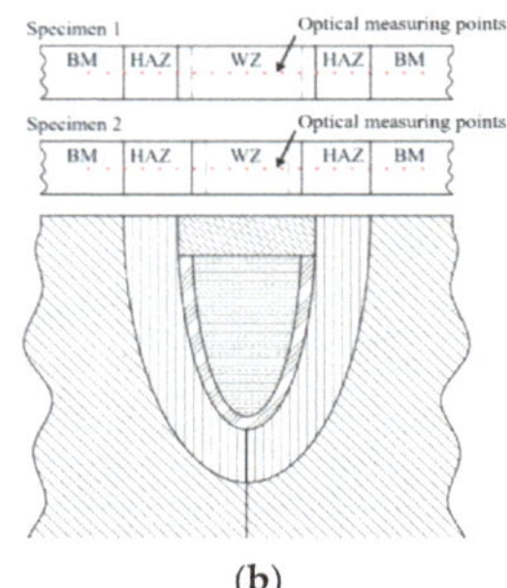

(b)

Figure 3. Tensile test with DIC. (**a**) Test measurement equipment; (**b**) schematic diagram of the location of optical measurement points.

2.2. Hardness Experiment

According to standard ISO 6507-1:2005 [19], the Vickers hardness test was conducted on the welded joint, as shown in Figure 4a. In this test, a small force was applied perpendicular to the test surface for Vickers hardness, with a maximum loading time of 8 s and a head descent speed not exceeding 0.3 mm/s. The lengths of the two diagonal lines of the indentation were measured, and the Vickers hardness value was calculated according to the relevant standard. As shown in Figure 4b, the hardness measurement was centered on the welded joint, spanning four zones from position A to B, namely the WZ, the modified layer, the HAZ, and the BM. The total length of AB was approximately 75 mm, with a spacing of 5 mm, and a total of 16 hardness points were tested. The test force was 0.98 N, with a duration of 8 s. To improve the accuracy of the hardness test and eliminate anomalies, the welded joint underwent three hardness tests, and the average value was taken as the final result of the Vickers hardness test for that welded joint.

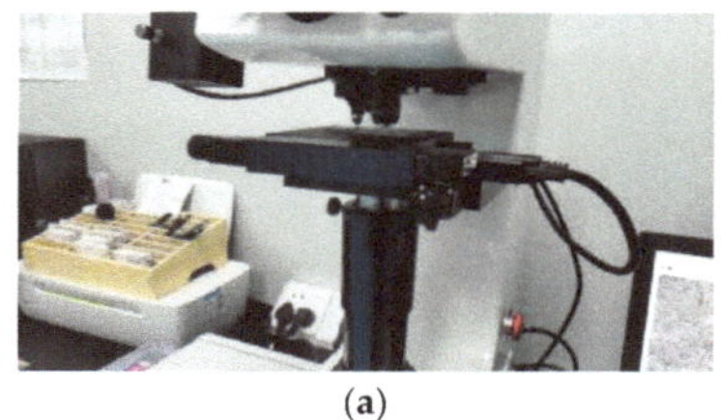

(a)

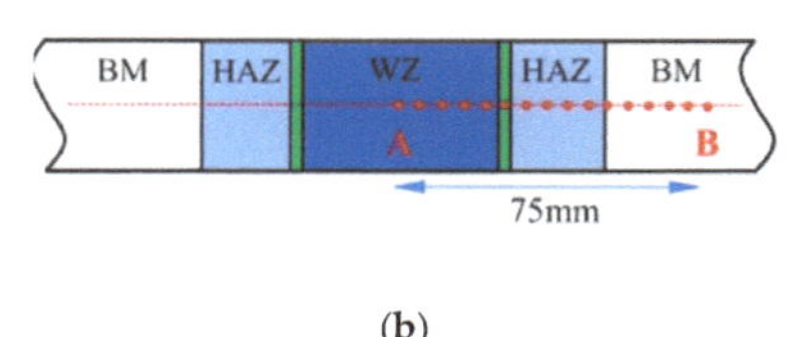

(b)

Figure 4. Hardness test. (**a**) Instrument for Vickers hardness test; (**b**) schematic diagram of the location of Vickers hardness measurement points.

2.3. Experimental Results and Data Analysis

Figure 5 shows the strain contours of the welded joint DIC at different times. It can be seen that the strain distribution on both sides of the WZ remained symmetrical. In specimen 1, the strain was mainly concentrated in the HAZ, with smaller strain values in the BM and WZ. As the load increased, necking occurred in the HAZ. In specimen 2, the strain was also concentrated in the zone close to the welded joint, namely the HAZ and the modified layer zone, while the strain in the BM and WZ remained relatively stable. With the increase in load, a fracture occurred in the HAZ position.

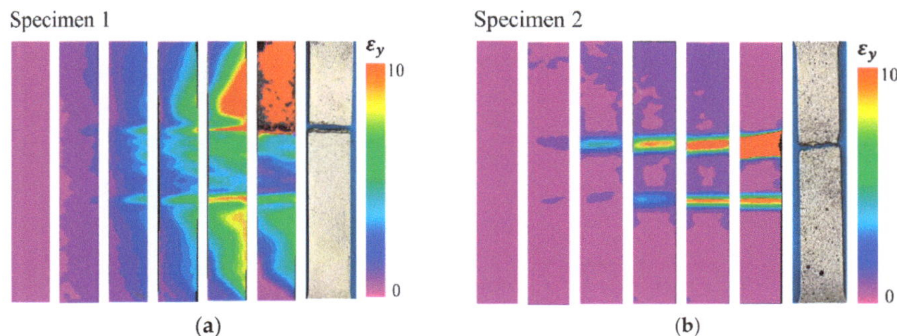

Figure 5. The strain contours of specimens. (**a**) Specimen 1; (**b**) specimen 2.

From Figure 6, it can be seen that only the HAZ and the modified layer exhibited relatively complete stress–strain curves, indicating that, the weaker the mechanical performance of the stress–strain parameters, the more prone the weld is to fracture. Additionally, the weld material showed significant heterogeneity, with noticeable differences in the mechanical performance of the BM, HAZ, overlay layer, modified layer, and filler layer.

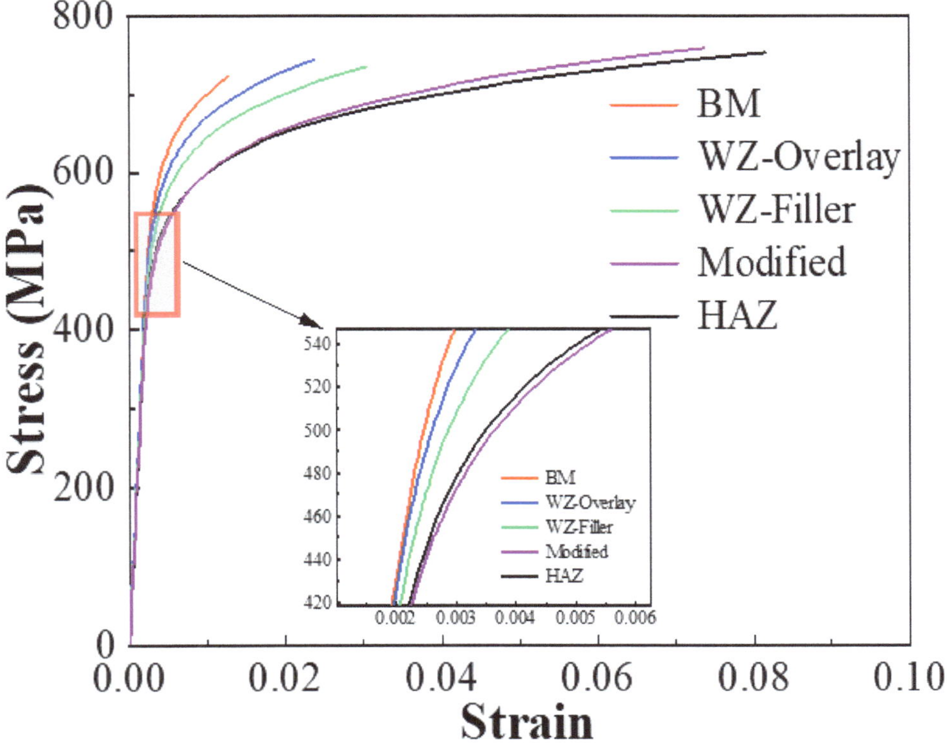

Figure 6. Stress–strain curve of welded joints.

The characteristic points were extracted from the BM, HAZ, overlay layer, modified layer, and filler layer. The strain–time curves of each zone's characteristic points were plotted on the same graph, as shown in Figure 7. It can be seen that the strain rate of the HAZ and the modified layer was significantly higher than that of the overlay and the BM. As the time increased, the changes in strain in the HAZ were most evident.

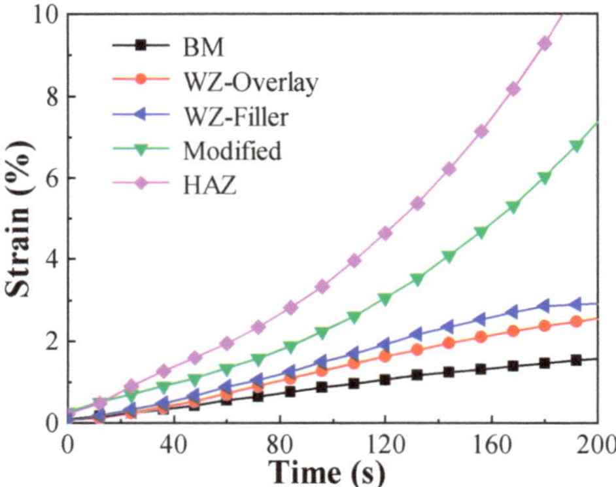

Figure 7. Strain–time variation curve.

Figure 8 shows the distribution curve of strain on the surface of the welded joint under different tensile loads. It can be observed that, as the tensile load increased, the strain distribution on the surface of the welded joint exhibited approximately symmetrical characteristics with respect to the WZ. In addition, the strain variation in the BM and WZ was relatively small, while it was more significant in the HAZ.

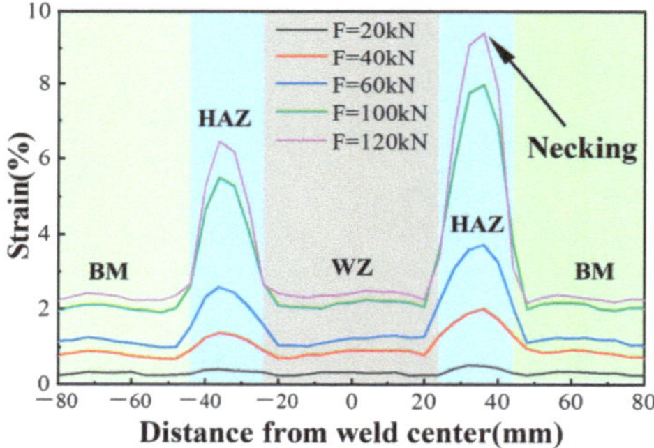

Figure 8. Changes in strain of welded joints under different loads.

From Figure 9, it can be seen that the distribution pattern of the elastic modulus in the welded joint showed a W-shaped distribution. The elastic modulus of the BM was the highest, followed by the WZ, and the HAZ and the modified layer had the lowest elastic modulus. This can explain the reason why plastic deformation first occurred at the HAZ and modified layer in the DIC strain contour. Figure 10 shows the distribution pattern of yield strength in the welded joint: The order of magnitude for the yield strength and tensile strength of specimen 1 was BM > WZ-Overlay > modified layer > HAZ. The yield strength of specimen 2 was in the following order: BM >WZ-Filler> modified layer > HAZ. Although the HAZ and the modified layer were more prone to plastic deformation, their elongation after fracture was higher than that for the BM and WZ.

Figure 11 shows that the hardness of the WZ-Overlay of specimen 1 fluctuated within the range of 310 HV to 340 HV, while the hardness of the BM fluctuated in the range 325 HV~375 HV; the minimum hardness of the welded joint was located at the HAZ. As the distance from the center of the welded joint increased, the hardness of the welded joint gradually decreased. At a distance of 35 mm from the center of the welded joint, there was a low hardness point, with hardness values fluctuating in the range of 200 HV~225 HV. The hardness of the WZ-Filler in specimen 2 fluctuated in the range of 300 HV~337 HV, and the hardness of the BM fluctuated in the range of 325 HV~375 HV. The minimum hardness of the welded joint was located at the HAZ.

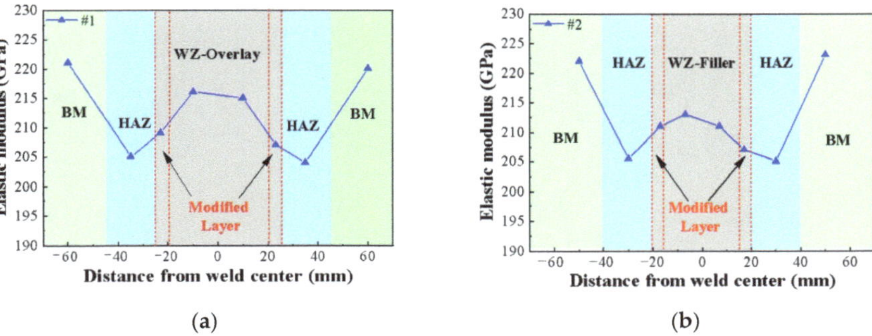

Figure 9. Elastic modulus contours of specimens.(**a**) Sample 1; (**b**) Sample 2.

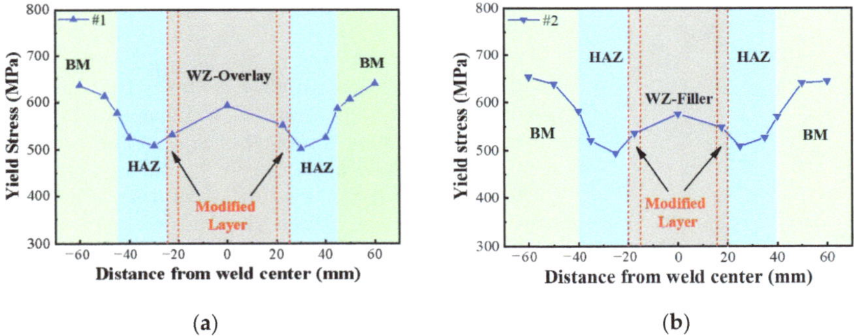

Figure 10. The yield stress contours of specimens. (**a**) Sample 1; (**b**) Sample 2.

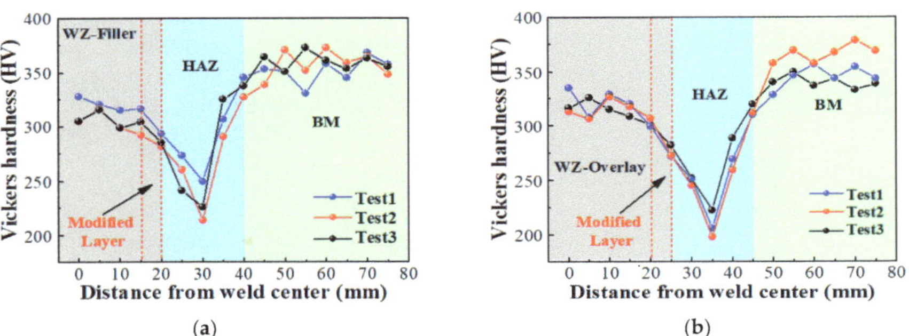

Figure 11. The Vickers hardness contours of specimens. (**a**) Sample 1; (**b**) Sample 2.

The hardness values of WZ and BM changed relatively little, while the hardness of HAZ fluctuated greatly. This is because the HAZ undergoes heating and cooling cycles, which cause changes in its microstructure and properties. The most severe change is grain growth, which leads to the most significant softening and affects the hardness value. As the distance from the center of the welded joint increases, the influence of heat input weakens, and the hardness gradually recovers to the BM, which determines that the weakest zone of the welded joint is located at the HAZ.

Figure 12 compares the distribution of Vickers hardness and yield strength in the welded joint, it was found that there was a certain correlation between them. Similarly, the hardness value fluctuated significantly at the HAZ.

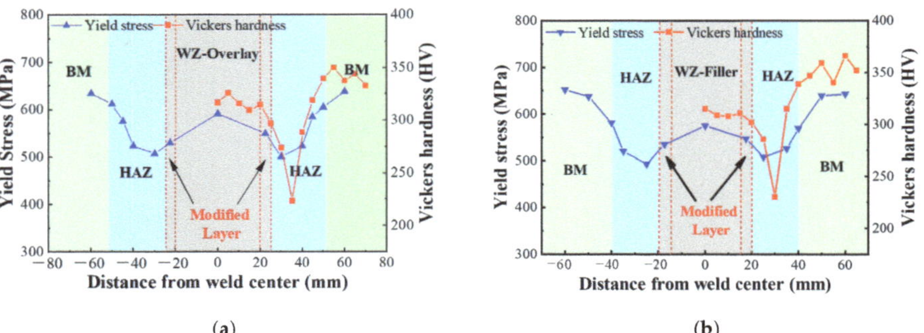

Figure 12. Yield stress and Vickers hardness contours of specimens. (**a**) Sample 1; (**b**) Sample 2.

3. Model

3.1. Elastic–Plastic Constitutive Model

From the mechanical curves, it is known that there are differences in the mechanical properties of the BM, HAZ, and WZ, and their strain relationships all exhibited significant nonlinear characteristics. Therefore, the use of linear elastic models or bilinear constitutive models cannot accurately describe the hardening of the welded joint.

The Ramberg–Osgood model [20], as an excellent method for fitting material characteristic curves considering strain hardening effects, is widely used to establish and describe the elastic–plastic and the stress–strain relationships of metal materials. This model was initially proposed by C. G. Ramberg and W. R. Osgood in 1943. The basic form of the model is as follows:

$$\varepsilon = \frac{\sigma}{E} + \varepsilon_0 \left(\frac{\sigma}{\sigma_{\varepsilon_0}} \right)^n \tag{1}$$

where E is the elastic modulus, σ_{ε_0} is the conventional elastic limit, ε is the residual strain corresponding to σ_{ε_0}, and n is the strain hardening parameter.

As shown in Figure 6, it can be seen from the experimental curve that both the base material and the weld material did not have a clear yield. Therefore, the constitutive parameters of the R-O constitutive relationship fitting model were chosen. The stress at 0.2% plastic strain was taken as the conventional elastic limit.

$$\sigma_{\varepsilon_0} = \sigma_{0.2} \tag{2}$$

Correspondingly, if $\sigma_{\varepsilon_0} = 0.002$, the equation can be transformed into the following:

$$\varepsilon = \frac{\sigma}{E} + 0.002 \left(\frac{\sigma}{\sigma_{0.2}} \right)^n \tag{3}$$

Equation (3) represents the stress–strain constitutive equation of the track welded joint under uniaxial tensile load. By conducting uniaxial tensile tests, the nominal stress–strain curves of each zone's optical measurement points were obtained. After converting them into true stress–strain curves, we can obtain the elastic modulus and yield limit of the material. Further fitting the experimental data using the R-O constitutive model yields the strain hardening exponent, thus completing the determination of the material's R-O constitutive parameters. The constitutive parameters are shown in Table 1 and the comparison between the experiment and the fitting model is given in Figure 13. It shows that the R-O fitting model was a good fit.

Table 1. Constitutive parameters of welded joint material for azimuth track.

Zone	E/GPa	$\sigma_{0.2}$/MPa	n
BM	221	623.7	10.55
HAZ	205	530.6	9.469
WZ-Modified	207	542.2	9.672
WZ-Overlay	215	594.6	10.16
WZ-Filler	212	574.8	10.245

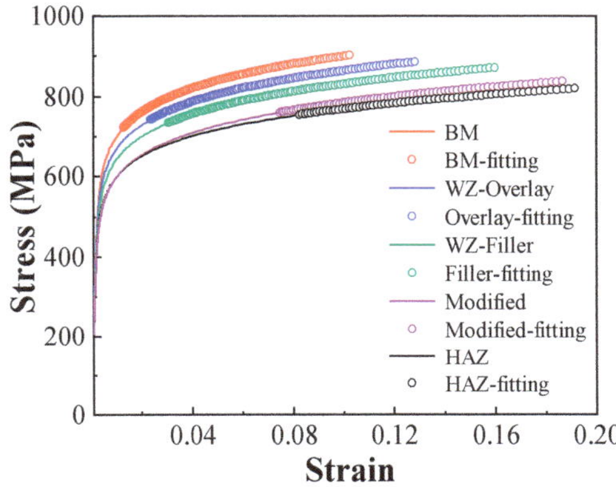

Figure 13. Experimental stress–strain curve and fitting.

3.2. Finite Element Model

3.2.1. Structure and Mesh

A simulation model of the track welded joint was built with five parts: BM, HAZ, WZ-Modified layer, WZ-Overlay, and WZ-Filler. Figure 14a shows the different material properties assigned to the five parts for subsequent simulation analysis. Based on the results of the mechanical experiment, the width of the welded joint on the track surface was 50 mm, with a depth of 80 mm. The ends of the welded joint were closely attached to the BM, with a transition layer width of 5 mm. The width of the overlay layer on the top of the welded joint was 15 mm, with the center part being the filler layer, and the width of the HAZ was 20 mm. The material properties were set as 42CrMo for the track and wheel base material, H08Mn2Si for the modified layer, D507 for the surfacing layer, and J607Ni for the filling layer, with specific plastic parameters obtained from the experiments.

(a)　　　　　　　　　　　　　　　　(b)

Figure 14. Wheel-track model. (**a**) Solid model, (**b**) Finite element model.

To ensure calculation accuracy, hexahedral elements were used for discretization of the antenna track, wheel, and welding area in the model, and C3D8R elements suitable for contact calculation were selected. Compared to regular fully integrated elements, one integration point was reduced in each direction. Linear reduced integration elements have only one integration point at the center of the element, making it easy to view the analysis results at the integration point. Figure 14b shows the local mesh refinement in the welding seam area and the wheel–rail contact area.

3.2.2. Boundary Conditions

To simulate fixed boundary conditions, all degrees of freedom of the bottom nodes of the track were constrained. The contact properties between the wheel and the track can be divided into normal action and tangential action. For normal action, the system defaulted to using a "hard contact" model, where the contact pressure between contact surfaces can reach infinity, and the gap can be a negative value. As for tangential action, the system used the Coulomb friction model to describe the friction characteristics between contact surfaces, where the friction coefficient is used to measure the resistance of the contact surfaces in the tangential direction.

3.2.3. Stress Initialization

There were three steps for the analysis. The purpose of the first analysis step was to move the wheel downwards to make it come into contact with the track with an interference fit, thus establishing a stable contact between the track and the wheel. The purpose of the second analysis step was to release gravity to allow the wheel track to naturally establish steady contact. The purpose of the third analysis step was to establish steady rolling of the wheel. Figure 15 shows the cross distribution of contact stress on the track surface after steady rolling.

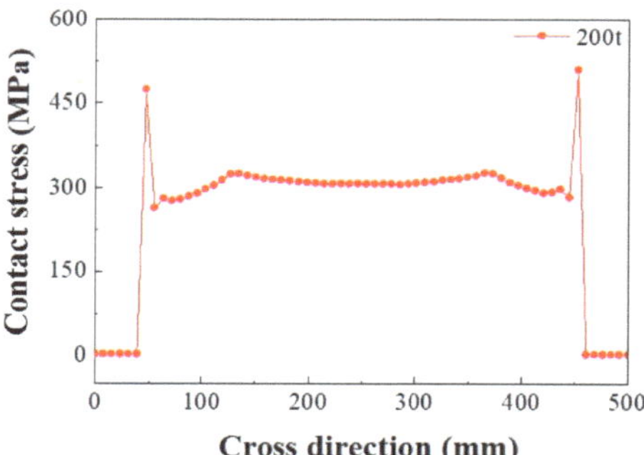

Figure 15. Contact stress distribution on the surface of track.

The purpose of the second analysis step was to apply gravity to make the wheel press down and rebound on the track, ultimately establishing a stable natural contact state between the wheel and the track. This process is to gradually reduce the gap between the wheel and the track, achieving a tighter wheel–rail contact effect. As shown in Figure 16, it can be seen that as the incremental step increased, the contact stress fluctuated violently before 0.4 s, repeatedly pressing down and rebounding. The stress will increase when pressed down and decrease when rebounding. Finally, after 0.4 s, the contact stress of the track gradually tended to a constant value, indicating that a steady contact state between the wheel and track was established at this time.

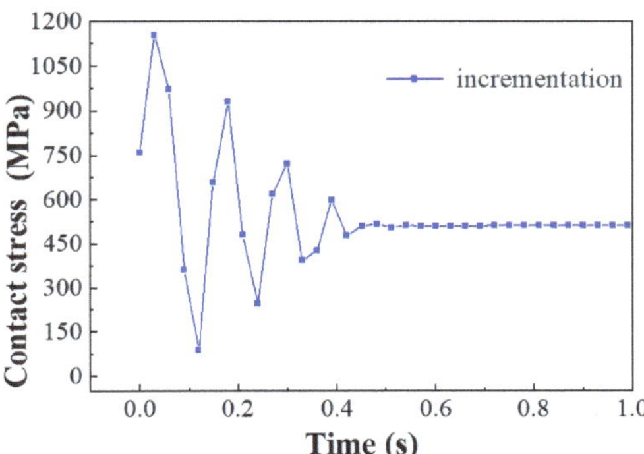

Figure 16. Curve of contact stress with analysis step time.

4. Simulation and Discussion

4.1. The Mises Stress of Welded Joints during Ordinary Working Condition under Different Loads

The overall weight of the QTT is about 6000–7000 tons. It is supported by approximately 30 or more bearing wheels, with each wheel bearing a load of about 200–240 tons. In order to study the influence of the self-weight of the structure on the mechanical characteristics of the azimuth wheel–rail rolling contact, a simulation study for ordinary working condition was conducted to examine the variation in contact stress under five different loads: 200 t, 210 t, 220 t, 230 t, and 240 t.

Figure 17a–c shows the contour of Mises stress under partial loads at different depths of the azimuth track in the BM, HAZ, and WZ. Extracting the Mises stress along the depth direction, the trend of the Mises stress in the BM, HAZ, and WZ with different loads can be obtained in Figure 17d–f. It can be seen that the Mises stress showed a trend of increasing first and then decreasing with the change in the contact surface depth, and the values of the Mises stress all reached the maximum at a subsurface depth of about 6 mm on the contact surface. For example, the Mises contour under 210 t at 6 mm depth in the HAZ was 375.2 MPa, and it was obviously higher than that at other depths in the same conditions.

Figure 18a shows a stress contour example of the Mises stress distribution in the BM under 200 t. Figure 18b shows the situation of the maximum Mises stress in the different zones under different loads. It can be observed that the Mises stress on the wheel–rail gradually increased with the increase in load. Meanwhile, it also showed a trend of the stress field with "HAZ > WZ > BM"; it was fully and completely consistent with the distribution pattern of yield strength in the experiment. Figure 19 shows the vertical deformation distribution of the azimuth track welding joints under different loads, with the vertical deformation of the HAZ being the largest, followed by the WZ, and that of the BM being the smallest.

Through simulation analysis in this part, we first predicted the overall stress field distribution of the QTT under the ordinary working condition with different pre-loads, and this verified the correctness of the experiment by the distribution of the stress in each zone. Secondly, the zone with the highest stress was located at 25–27 mm on both sides of the center of the weld in the track welding area, while the areas with lower stress were at 0–24 mm on both sides and 50 mm away. Thirdly, it was found that the most perilous points for the future operation of the QTT were 22 mm inward from both sides of the cross track and 6 mm vertically from the track. Finally, it was concluded that the maximum vertical deformation occurred in the HAZ zone when the design axle load was 230 t.

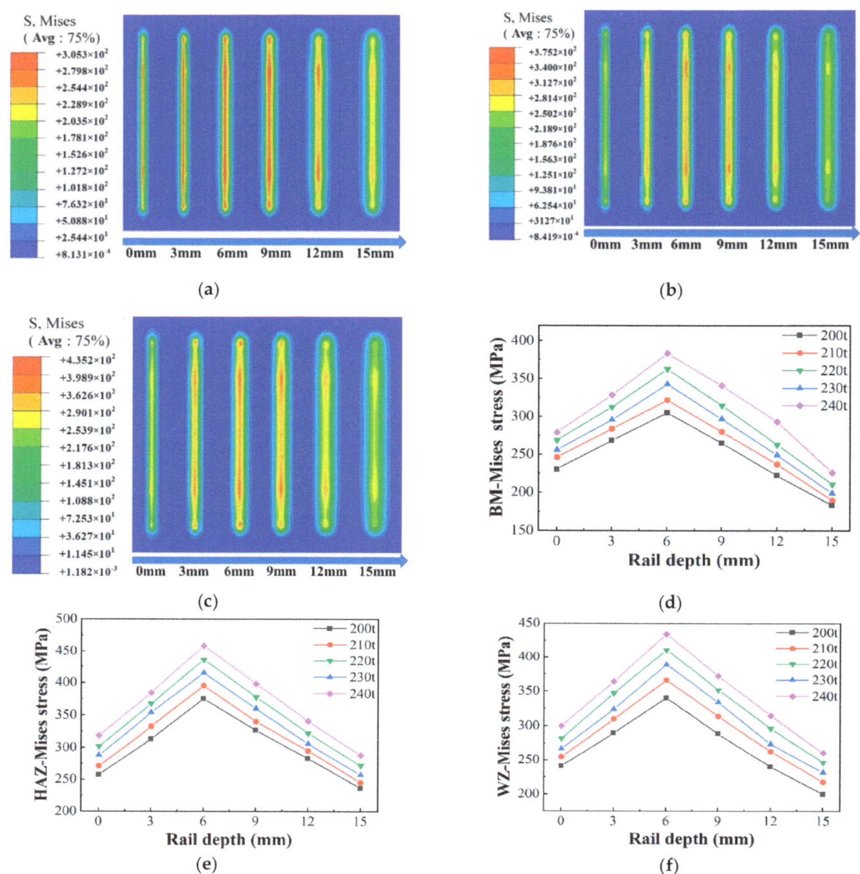

Figure 17. Mises stress distribution in the weld zone at different depths. (**a**) Mises contour under 200 t at different depths in the BM; (**b**) Mises contour under 210 t at different depths in the HAZ; (**c**) Mises contour under 240 t at different depths in the WZ; (**d**) Mises contour under different loads at different depths in the BM; (**e**) Mises contour under different loads at different depths in the HAZ; (**f**) Mises contour under different loads at different depths in the WZ.

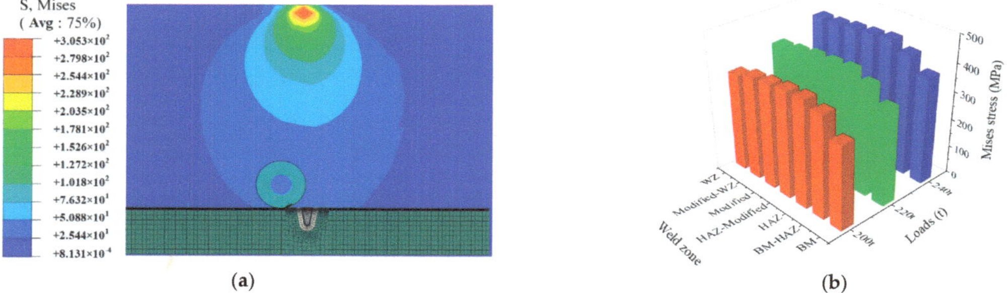

Figure 18. Mises stress distribution in the weld zone under different loads. (**a**) Mises stress distribution in the BM under 200 t. (**b**) Mises stress distribution in the welded joint under different loads.

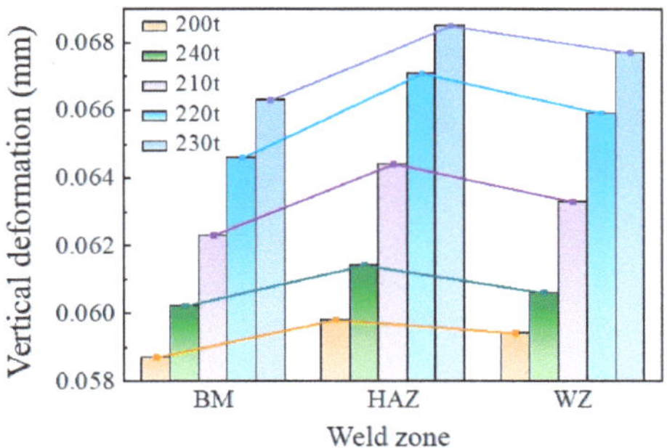

Figure 19. Vertical deformation distribution of weld zone under different loads.

4.2. Formatting of Mathematical Components

This section conducts finite element simulations of the start-up and rapid braking of the large radio telescope to better understand the force conditions on the track. Based on the results, appropriate measures can be taken, such as limiting acceleration and velocity, to reduce damage to the track and ensure its safety and reliability.

4.2.1. Start-Up Working Condition

Simulation analysis of the mechanical response of the welded joint at various zones during the telescope start-up process under a 200 t load. The start-up speed is defined using the smooth step amplitude curve provided in ABAQUS, with an acceleration to 1 rad/s within 1 s. The start-up speed loading method is shown in Figure 20a.

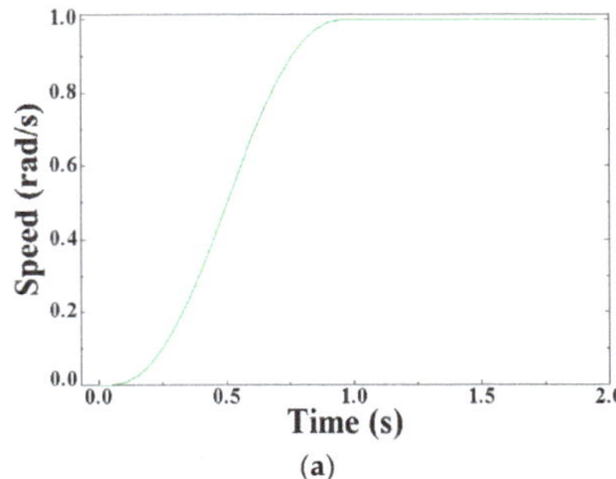

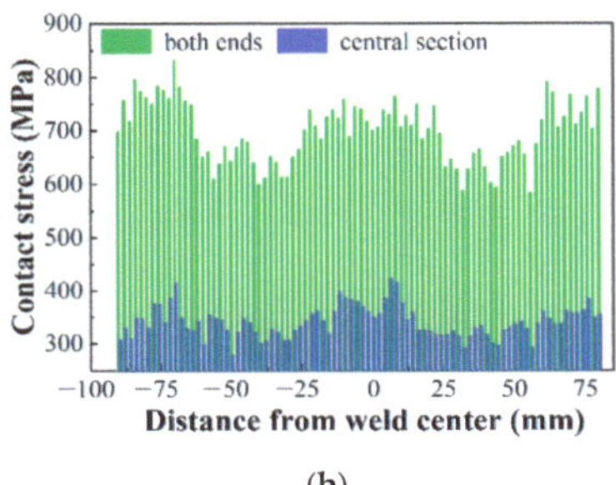

Figure 20. Start-up condition. (**a**) Speed loading method; (**b**) distribution of contact stress on the surface of the welded joint.

Figure 20b depicts the contact stress distribution on the surface of the welded joint under start-up conditions. It can be observed that the BM, HAZ, and WZ exhibited significant differences in contact stress, and there was also considerable fluctuation in the distribution of contact stress values within the same zone. However, the stress in each zone fluctuated within a fixed range. From a global perspective, under the operating conditions, the distribution of contact stress on the surface of the welded joint showed a W-like pattern.

Therefore, it can be seen that the contact stress amplification in the HAZ was the highest among the three different zones. This indicates that the start-up condition has the greatest impact on the contact stress of the HAZ, which may have a significant effect on the durability of the joint.

4.2.2. Braking Condition

To test the braking condition, we simulated the mechanical response of the welded joint in different zones under a 200 t load during the rapid braking process. In the rapid braking condition, the smooth step amplitude curve provided by ABAQUS was used to define the braking process, as shown in Figure 21a, and it was set to decelerate to 0 rad/s within 1 s.

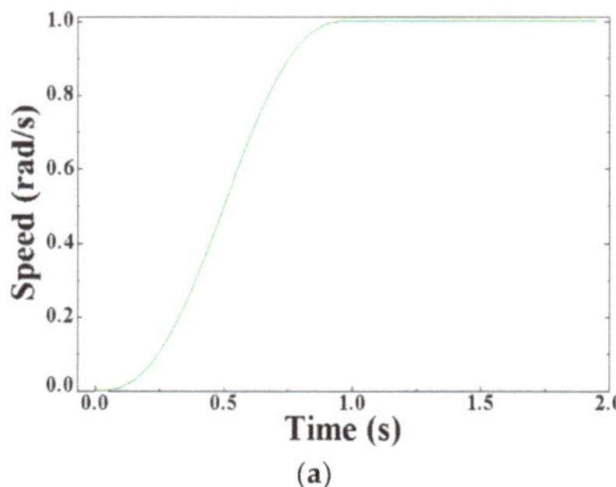
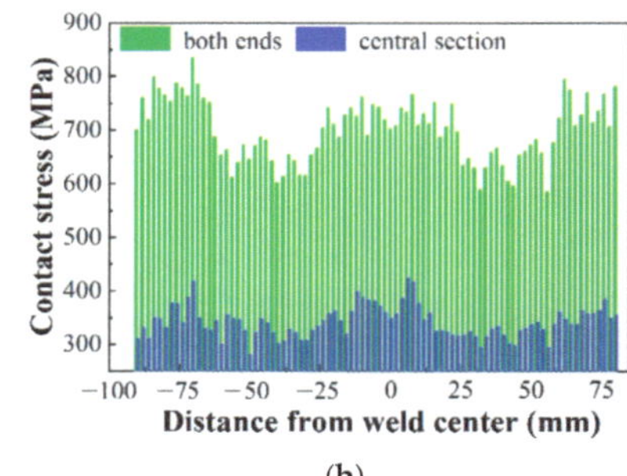

(a) (b)

Figure 21. Braking condition. (**a**) Speed loading method; (**b**) distribution of contact stress on welded joint surface.

According to Figure 21b, the contact stress distribution on the surface of the welded joint under braking conditions also showed a W-shaped distribution.

Under the start-up condition, the average contact stress at both ends of the BM contact area was approximately 750 MPa, while the average contact stress in the middle of the contact area was approximately 460 MPa; the average contact stress at both ends of the HAZ contact area was approximately 630 MPa, with the average stress in the middle of the contact area at approximately 430 MPa; the average contact stress at both ends of the WZ contact area was approximately 720 MPa, with the average stress in the middle of the contact area at approximately 450 MPa. A comparison with the mechanical parameters of the joint under steady conditions in the previous section revealed that the contact stress in the rail joint area under the start-up conditions was significantly higher than that under stable contact conditions. Specifically, in the HAZ contact area, the contact stress at both ends and in the middle increased by 161 MPa and 102 MPa. It can be seen that, among the three different areas, the HAZ contact area experienced the greatest increase in contact stress. This indicates that the start-up has the greatest impact on the contact stress of the HAZ, which may have a significant impact on the durability of the joint. Therefore, greater attention should be paid to the health of the HAZ. Meanwhile, compared to the distribution of contact stress on the surface of the welded joint in the start-up condition, the braking condition had a similar trend, and its HAZ was also weak.

5. Conclusions

This article experimentally investigated the position distribution and mechanical properties of the BM, HAZ, modified layer, overlay layer, and filling layer in the welded joint of a large azimuth track for the first time through the combination of image recognition with tensile and Vickers hardness tests macroscopically and explored the stress–strain relationship in each region.

By obtaining the tensile curve, image recognition strain field, Vickers hardness curve, and distribution of yield strength and tensile strength from the experiments, a constitutive equation for the elastic–plastic nonlinearity of the welded joint was fitted with the Ramberg–Osgood equation.

By combining experimental data and welding properties, a finite element model of the wheel–rail contact in the welded joint of the large azimuth track was established. The global stress field distribution under different design loads was obtained through simulation, and the weak points under the ordinary working condition, start-up conditions, and braking conditions were determined, providing a theoretical basis for the health monitoring of the track and research on wear and damage.

Author Contributions: Conceptualization, X.C., H.L. and Q.X.; methodology, X.C.; software, R.Y.; validation, X.C., R.Y. and Z.Y.; formal analysis, Z.Y.; investigation, X.C.; resources, H.L.; data curation, Z.Y.; writing—original draft preparation, R.Y.; writing—review and editing, H.L.; visualization, R.Y.; supervision, Q.X.; project administration, H.L. and Q.X.; funding acquisition, H.L. and Q.X. All authors have read and agreed to the published version of the manuscript.

Funding: This research was funded by the National Key Research and Development Program of China, grant number 2021YFC2203600.

Data Availability Statement: The data presented in this study are available on request from the corresponding author. The data is not publicly available.

Acknowledgments: We thank the 39th Research Institute of China Electronics Technology Group Corporation for providing the weldment.

Conflicts of Interest: The authors declare that they have no conflicting financial interests.

References

1. Antebi, J.; Kan, F.W. Precision continuous high-strength azimuth track for large telescopes. In *Future Giant Telescopes*; SPIE: Avenue de l'Entreprise, France, 2002; pp. 612–623.
2. Anderson, R.; Symmes, A.; Egan, D. *Replacement of the Green Bank Telescope Azimuth Track*; International Society for Optics and Photonics: San Francisco, CA, USA, 2008; p. 784807.
3. Symmes, A.; Robert, A.; Dennis, E. Improving the service life of the 100-meter green bank telescope azimuth track. In *Ground-Based and Airborne Telescopes II*; SPIE: Avenue de l'Entreprise, France, 2008; p. 701238.
4. Juneja, G.; Kan, F.W.; Antebi, J. Update on slip and wear in multi-layer azimuth track systems. In *Optomechanical Technologies for Astronomy*; SPIE: Avenue de l'Entreprise, France, 2006; p. 627318.
5. Kumar, N.; Arora, N.; Goel, S.K. Weld joint properties of nitrogen-alloyed austenitic stainless steel using multi-pass GMA weld. *Archiv. Civ. Mech.* **2020**, *20*, 82. [CrossRef]
6. Han, P.; Wang, K.; Wang, W.; Ni, L.; Lin, J.; Xiang, Y.; Liu, Q.; Qiao, K.; Qiang, F.; Cai, J. Microstructure evolution and mechanical properties of Ti-15-3 alloy joint fabricated by submerged friction stir weld. *Archiv. Civ. Mech.* **2024**, *54*, 1–15.
7. Saranath, K.M.; Ramji, M. Local zone wise elastic and plastic properties of electron beam welded Ti-6Al-4V alloy using digital image correlation technique. a comparative study between uniform stress and virtual fields method. *Opt. Lasers Eng.* **2015**, *68*, 222–234. [CrossRef]
8. Zhang, Z.; Pan, B.; Grédiac, M.; Song, W. Accuracy-enhanced constitutive parameter identification using virtual fields method and special stereo-digital image correlation. *Opt. Lasers Eng.* **2018**, *103*, 55–64. [CrossRef]
9. Zhang, L.; Min, J.; Wang, B.; Lin, J.; Li, F.; Liu, J. Constitutive model of friction stir weld with consideration of its inhomogeneous mechanical properties. *Chin. J. Mech. Eng.* **2016**, *29*, 357–364. [CrossRef]
10. Tan, Y.; Shi, J.; Liu, P.; Tao, J.; Zhao, Y. Research on the Mechanical Performance of a Mountainous Long-Span Steel Truss Arch Bridge with High and Low Arch Seats. *Buildings* **2023**, *13*, 3037. [CrossRef]
11. Xu, L.; Li, Z.; Zhao, Y.; Yu, Z.; Wang, K. Modelling of vehicle-track related dynamics: A development of multi-finite-element coupling method and multi-time-step solution method. *Veh. Syst. Dyn.* **2022**, *60*, 1097–1124. [CrossRef]
12. Arslan, M.A.; Kayabasi, O. 3-D Rail–Wheel contact analysis using FEA. *Adv. Eng. Softw.* **2012**, *45*, 325–331. [CrossRef]
13. Tellikivi, T.; Olofsson, U. Contact mechanics analysis of measured wheel-rail profiles using the finite element method. *Proc. Inst. Mech. Engineers. Part F J. Rail Rapid Transit* **2001**, *215*, 65–72. [CrossRef]
14. Kuminek, T.; Anioek, K. Methodology and verification of calculations for contact stresses in a wheel-rail system. *Veh. Syst. Dyn.* **2014**, *52*, 111–124. [CrossRef]
15. Tian, J.H.; Jing, G.Q.; Lu, X.X.; Xiao, K.; Zhang, H.R. The rolling contact research of three dimensional wheel-rail based on finite element analysis. *Mater. Sci. Eng.* **2019**, *657*, 012072. [CrossRef]
16. Fu, L.; Qian, H.L.; Fan, F.; Zhong, J.; Liu, G.X. Analysis on Wheel-Rail Contact of 65-m Radio Telescope. In Proceedings of the 2011 International Conference on Electric Technology and Civil Engineering (ICETCE), Lushan, China, 22–24 April 2011; pp. 107–110.
17. Yang, Z.; Deng, X.; Li, Z. Numerical modeling of dynamic frictional rolling contact with an explicit finite element method. *Tribol. Int.* **2019**, *129*, 214–231. [CrossRef]
18. *ISO 6892-1:2019*; Metallic Materials—Tensile Tests—Part 1: Test Method at Room Temperature. International Organization for Standardization: Geneva, Switzerland, 2019.
19. *ISO 6507-1:2005*; Metallic Materials-Vickers Hardness test—Part1: Test Method. International Organization for Standardization: Geneva, Switzerland, 2015.
20. Ramberg, W. *Description of Stress-Strain Curves by Three Parameters*; National Advisory Committee for Aeronautics: Washington, DC, USA, 1943.

buildings

MDPI

Article

Evaluation of Rotation Capacity and Bauschinger Effect Coefficient of I-Shaped Beams Considering Loading Protocol Influences

Yoshihiro Kimura

Graduate School of Engineering, Tohoku University, Sendai 980-8579, Japan; kimura@tohoku.ac.jp; Tel.: +81-22-795-7865

Abstract: Recent catastrophic earthquake events have reinforced the necessity of evaluating the seismic performance of buildings. Notably, the buildings can go into the plastic phase during a striking earthquake disaster. Under this condition, the current design codes assume seismic response reduction by virtue of the energy dissipation capacity of the structural members. In the strong-column–weak-beam design, which involves I-shaped beams and boxed columns, the mechanism is defined as a standard design scheme to prevent the building from collapsing. Therefore, energy dissipation relies highly on the I-shaped beam performance. However, the I-shaped beam performance can differ depending on the loading history experienced, whereas this effect is untouched in the prevailing evaluation equation. Hence, this study first performs cyclic loading tests of 11 specimens using different loading protocols. The experimental results clarify the fluctuation in the structural performance of I-shaped beams depending on the applied loading hysteresis, proving the necessity of considering the stress history for proper assessment. Furthermore, the database of experimental results is constructed based on the previous experimental studies. Ultimately, the novel evaluation equation is proposed to reflect the influences of the loading protocol. This equation is demonstrated to effectively assess the member performance retrieved from the experiment of 65 specimens, comprising 11 specimens from this investigation and 54 specimens from the database. The width–thickness ratio, shear span-to-depth ratio, and loading protocols are utilized as the evaluation parameters. Moreover, the prediction equation of the Bauschinger effect coefficient is newly established to convert the energy dissipation capacity under monotonically applied force into hysteretic energy dissipation under the cyclic forces.

Keywords: I-beam; rotation capacity; energy dissipation capacity; Bauschinger effect coefficient; loading protocol

Citation: Kimura, Y. Evaluation of Rotation Capacity and Bauschinger Effect Coefficient of I-Shaped Beams Considering Loading Protocol Influences. *Buildings* **2024**, *14*, 1376. https://doi.org/10.3390/buildings14051376

Academic Editor: Hugo Rodrigues

Received: 28 March 2024
Revised: 30 April 2024
Accepted: 9 May 2024
Published: 11 May 2024

1. Introduction

I-shaped steel beams (I-beams) are widely utilized in steel moment-resisting frames (MRFs) due to their high in-plane bending stiffness and strength against applied bending moments. However, in addition to dead and live loads, seismic loads increase the bending moment, leading to instability phenomena such as lateral buckling and local buckling. The research on I-beams initiated by elucidating these buckling phenomena through classical theory [1,2] and then expanded to evaluate lateral buckling resistance and rotation capacity [3–25]. Since main girders in frames are typically connected to restraining beams or concrete slabs, studies on buckling stiffening and stress transfer mechanisms between these components [26–33] have improved the performance evaluation of beams in frames. Additionally, numerous analytical and experimental studies have assessed the width–thickness ratio, ultimate bending strength, and rotation capacity to avoid plate buckling of the flange and web. Through these efforts, the beam can exhibit sufficient plastic rotation during a major earthquake by limiting the width-to-thickness ratio [34–52]. Regarding the width–thickness ratio, the prevailing design codes [53,54] specify their values for the member classifications. Even though recent provisions by AIJ [55] included the index with the web–flange interactions, the loading protocol influence was not reflected in the evaluation equation.

The mechanical performance of I-beams has been extensively studied in previous experimental research, such as the work by Kato et al. and Lignos, D. G. et al. [56–61]. Furthermore, the loading protocol and slenderness ratio influences are addressed in previous studies [62,63], whereas the evaluation scheme has not been established in an explicit manner. The evaluation criteria typically involve the rotation capacity for unidirectional loading (monotonic loading) and the energy dissipation capacity for cyclic loading. Parameters such as the width–thickness ratio of the member, shear span-to-depth ratio, or loading conditions are considered. The energy dissipation capacity refers to the dimensionless load–displacement relationship (skeleton curve) obtained by connecting the skeleton curves of each loop when the member is subjected to cyclic loading. It has been considered equivalent to the rotation capacity of the member under monotonic loading. Therefore, it is used as an index for evaluating the performance of members in the static incremental analysis (pushover analysis) of steel frame structures subjected to static seismic forces [64].

Conversely, in time–history response analyses, the energy dissipation of the hinge parts of columns and beams in the structure is used as an evaluation criterion [65]. The relationship between the energy dissipation capacity under monotonic loading and hysteretic energy dissipation is evaluated using the Bausinger effect coefficient, and different methods for assessing the performance of other members are used in seismic design analysis methods. The Bausinger effect coefficient is the ratio of the energy absorption in the softening zone during cyclic loading, which is not evaluated in the skeleton curve, to the energy absorption in the skeleton curve of the load–deformation history curve of the member subjected to cyclic loading. In the case of beams, a coefficient of 2.0 is set according to the recommendation of the Architectural Institute of Japan (AIJ). However, since the Bausinger effect varies with the loading amplitude and number of cycles [66,67], it is necessary to clarify the relationship between the energy dissipation capacity obtained from the skeleton curve and the hysteretic energy dissipation of the entire loading history. Otherwise, differences in mechanical performance may occur for the same member when different analysis methods are used. Therefore, it is necessary to ascertain the rotation and energy dissipation capacity of members constituting the main structure and clarify their relationship.

On the other hand, although experimental and analytical data on I-beams subjected to monotonic loading and cyclic loading have been collected in previous studies [68–77], the influence of the loading history on the rotation capacity, energy dissipation capacity, hysteretic energy dissipation, and Bausinger effect coefficient has not been clarified yet.

This paper aims to establish primary data on the history characteristics of I-beams subjected to monotonic loading and cyclic incremental loading by conducting loading experiments on I-beams with different loading histories, using the parameters of the flange width–thickness ratio and shear span-to-depth ratio. It collects experimental data on the structural performances at the maximum flexural strength time and stores the experimental results in a database. Then, based on the data extracted from earlier experimental studies [78–97], where the load–deformation relationship can be extracted as a cantilever beam type, the experimental results of I-beams which failed due to local buckling under monotonic loading and cyclic incremental loading are collected, and the data are databased. Differences in the width–thickness ratio, shear span-to-depth ratio, loading history, etc., are examined to clarify their effects on the ultimate strength ratio and energy dissipation capacity. Furthermore, for cyclic incremental loading, an evaluation equation for the Bausinger effect coefficient with the loading amplitude and number of loadings as parameters is proposed, and a performance evaluation method considering differences in loading history characteristics is presented.

The novel evaluation equation proposed and validated in this study realizes the proper reflection of the impact of various loading protocols (monotonic loading, repeated loading, number of cycles) on the Bauschinger effect coefficient, even for beams with identical cross-sections and lengths. This enables the specification of the behavior of beams within an MRF, reaching their ultimate flexural strength contingent upon the stress history encountered during earthquakes.

2. Cyclic Loading Tests on I-Shaped Steel Beams with Different Flange Width-to-Thickness Ratios, Shear Span Ratios, and Loading Histories

2.1. Outline of Experiment on I-Beams

Figure 1 shows the test specimen for the loading experiment. A 2000 kN universal testing machine was used for the experiment, employing a three-point bending method. As shown in the figure, the left side of the loading point is reinforced with cover plates welded to the upper and lower flanges, and the loading beam is designed to maintain elasticity even when the test specimen reaches its maximum load. Table 1 lists the test specimens. In their designation, the first letter stands for the loading protocols (A: monotonic, B: cyclic, one loading cycle with a specific displacement, and C: cyclic, three loading cycles with a specific displacement), the second number denotes the flange width–thickness ratio (1: $b/t_f = 6.25$, 2: $b/t_f = 4.17$, and 3: $b/t_f = 8.33$), and the third number indicates the shear span-to-depth ratio (1: $L/H = 5$, $L/H = 4$, and $L/H = 6$).

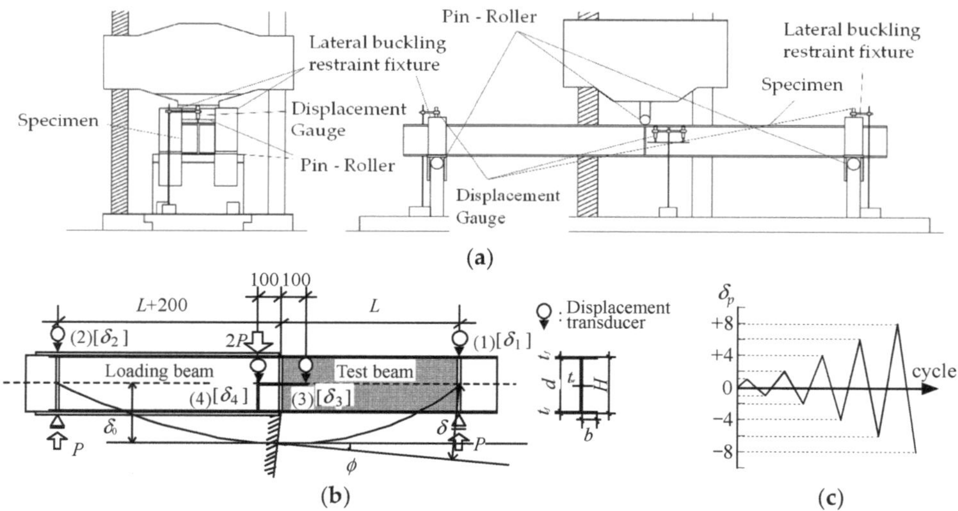

Figure 1. Outline of experiment and specimen: (**a**) testing apparatus; (**b**) measurements; (**c**) loading protocol (cyclic, one loading cycle with a specific displacement).

There are a total of 11 specimens, and SN400 steel was used for the specimens. The cross-section of the specimens has a web depth of 200 mm, a web thickness of 8 mm, and a flange thickness of 12 mm, with varying flange width-to-thickness ratios and shear span-to-depth ratios. This experiment selected beams with a small web depth-to-thickness ratio, for which there are few existing experimental results. The loading methods include monotonic and cyclic incremental loading, with the cyclic loading program designed based on the assumption of beams subjected to seismic forces. During an earthquake, when horizontal forces act on a moment-resisting frame, bending moments resulting from shear forces acting on the columns are transmitted to the beams. Given that the entire beam experiences an unsymmetric bending moment, the loading condition for half of the beams in the moment-resisting frame was adopted for this experiment. Figure 1c depicts the loading protocol for the cyclic loading experiments. The deformation δ_p corresponding to the full plastic bending moment M_p of the test specimen divided by the beam length L was set as the criterion for cyclic loading, and one cycle of loading was repeated at each step of $\pm\delta_p$, $\pm2\delta_p$, $\pm4\delta_p$, and $\pm6\delta_p$ of the incremental loading, except for specimen C-1-1. In contrast, for specimen C-1-1, three loading cycles were applied at each step. The M_p was calculated using the measured values shown in Table 2. As shown in Figure 1, one displacement transducer was installed at the position 100 mm from the center of the beam on both sides, and one displacement transducer was installed at each end support of the test specimen. The controlled displacement δ was calculated as a cantilever beam based on the displacement δ_0 at the position of the transducer and the rotation angle φ of the transducer.

$$\delta = \delta_0 + \varphi L = (\delta_3 + \delta_4)\frac{100}{200} - (\delta_1 + \delta_2)\frac{L}{2L+200} + \left(\frac{\delta_3 - \delta_4}{200} - \frac{\delta_1 - \delta_2}{2L+200}\right)L \tag{1}$$

Table 1. Test configurations.

Designation	L [mm]	b/t_f [mm]	d/t_w [mm]	L/H [-]	Slenderness Ratio [-]	Protocol	No. of Cycles
A-1-1				4.95		Monotonic	-
B-1-1		6.25		4.95	24.55	Cyclic	1
C-1-1				4.98		Cyclic	3
A-2-1	1000			4.98		Monotonic	1
B-2-1		4.17		4.95	37.88	Cyclic	1
A-3-1			22	4.98		Monotonic	1
B-3-1		8.33		4.95	18.15	Cyclic	1
A-1-2						Monotonic	1
B-1-2	800			3.96	19.64	Cyclic	1
A-1-3		6.25				Monotonic	1
B-1-3	1200			5.97	29.46	Cyclic	1

Table 2. Measurement of section dimensions.

Specimen	H [mm]	B [mm]	t_f [mm]	t_w [mm]
A-1-1	202	150	11.9	8.20
B-1-1	202	150	12.1	8.10
C-1-1	201	150	12.0	8.25
A-2-1	201	101	12.0	8.25
B-2-1	202	100	12.1	8.25
A-3-1	201	200	11.8	8.20
B-3-1	202	201	11.8	8.20
A-1-2	202	150	12.0	8.25
B-1-2	202	150	11.9	8.25
A-1-3	201	150	11.9	8.35
B-1-3	201	151	12.0	8.25

Here, δ_1 to δ_4 represent the vertical displacements of displacement transducers numbered (1) to (4) in Figure 1. Additionally, δ_p is calculated from the following equation as the sum of the displacements δ_{mp} due to bending deformation and δ_{sp} due to shear deformation as referenced in the previous literature [69]:

$$\delta_p = \delta_{mp} + \delta_{sp}, \delta_{mp} = \frac{P_p L^3}{3EI}, \delta_{sp} = \frac{\kappa P_p L}{GA} \tag{2}$$

Here, δ_{mp} represents the displacement of the simply supported beam due to bending deformation when the plastic moment is reached at the loaded side beam end, and δ_{sp} represents the displacement of the simply supported beam due to shear deformation. Additionally, E is Young's modulus, I is the moment of inertia, A is the cross-sectional area, G is the shear modulus, and κ is the shape factor.

The test specimens were made of SN400B steel in Japan. Two specimens were extracted from a 12 mm thick flange and an 8 mm thick web for tensile testing. Stress–strain curves are illustrated in Figure 2, and Table 3 presents the characteristics of the tensile test results. It is noted that the yield stress level of the 12 mm thick plate is lower than that of the 8 mm thick plate, while their tensile strengths are nearly equal. The material test of the steel was conducted in accordance with JIS Z 2241 [98].

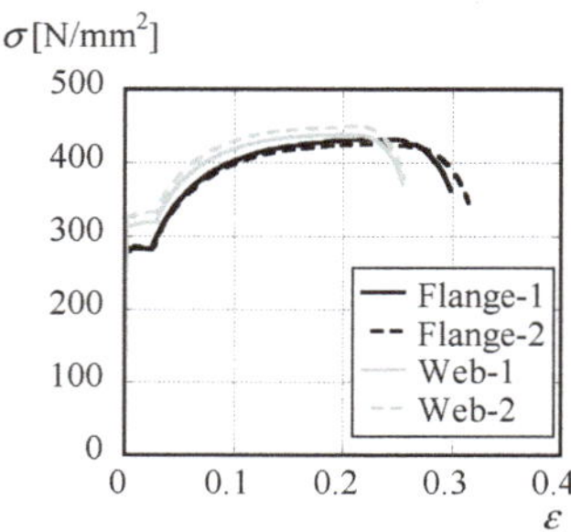

Figure 2. Stress–strain curves.

Table 3. Material test results.

	Thickness [mm]	E [N/mm^2]	σ_y [N/mm^2]	σ_u [N/mm^2]	Y.R.	Elongation [%]
Flange	8	1.99×10^6	323	444	0.727	21
Web	12	2.00×10^6	283	428	0.661	24

2.2. Plastic Deformation Characteristics of I-Beam with Different Loading Protocols

Figure 3a shows the load–displacement relationship of specimen C-1-1 subjected to cyclic loading. C-1-1 is a specimen subjected to loading three times with the same amplitude. The vertical axis represents the shear force acting on the beam P normalized by the ultimate plastic load P_p ($=M_p/L$), and the horizontal axis represents the displacement δ of the beam normalized by the displacement δ_p at the ultimate plastic strength.

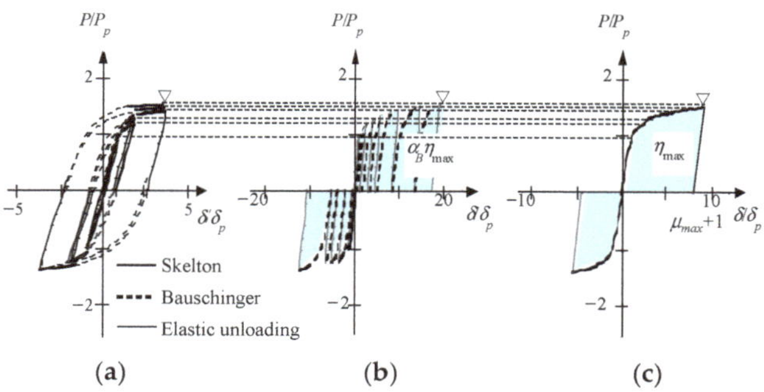

Figure 3. Procedure used to draw skeleton curve (C-1-1): (**a**) hysteresis curve; (**b**) cumulative hysteresis curve; (**c**) skeleton curve.

The cyclic load–displacement curve is divided into the skeleton curve, the Bauschinger effect region, and the elastic unloading region. The skeleton curve part is represented by thick solid lines corresponding to the load level experienced by the member for the first time. The Bauschinger effect region is defined by dashed lines corresponding to load levels previously experienced by the member. The elastic unloading region is represented by thin solid lines corresponding to the load level during unloading. The $\triangledown$ in the figure indicates the point of maximum strength.

Figure 3b shows the cumulative hysteresis curve of specimen C-1-1. The cumulative hysteresis curve decomposes the loading history curve into each loop. It connects the final displacement at the unloading of the previous loop and the initial displacement at the beginning of the next loop at $P/P_p = 0$ for each positive and negative loop. $\alpha_B \eta_{max}$ is the value obtained by subtracting the elastic deformation part from the total area up to the maximum strength in the cumulative history curve. It represents the dimensionless accumulated hysteretic energy dissipation.

Figure 3c shows the skeleton curve of specimen C-1-1. The skeleton curve connects only the skeleton curve part indicated by thick solid lines in the hysteresis curve of Figure 3a. In this paper, the value obtained by subtracting the elastic deformation part from δ_{max}/δ_p at the maximum strength is defined as the rotation capacity μ_{max} in the skeleton curve. The energy dissipation capacity η_{max} at the maximum strength in the skeleton curve is obtained by subtracting the elastic deformation part from the total area up to the maximum load and represents the dimensionless cumulative energy dissipation.

Here, α_B is the ratio of the energy dissipation capacity of the skeleton curve in Figure 3b to the hysteretic energy dissipation of the cumulative hysteresis curve in Figure 3c, which is the Bauschinger effect coefficient. The Bauschinger effect coefficient α_B is always greater than 1.

Figure 4 illustrates the load–displacement relationship for I-beams with equal cross-sections and lengths but different loading histories. Figure 4a shows the results of the cyclic loading tests, while Figure 4b shows the skeleton curves and results of the monotonic loading tests from Figure 4a. For the monotonic loading test A-1-1, the same history curve is drawn for both positive and negative sides. Even for specimens with the same shape, the skeleton curve of cyclic loading has a higher maximum strength and a smaller rotation capacity at the maximum strength compared to the load–displacement relationship of monotonic loading. This indicates that the skeleton curve obtained from cyclic loading is not necessarily equivalent to the load–displacement relationship of monotonic loading.

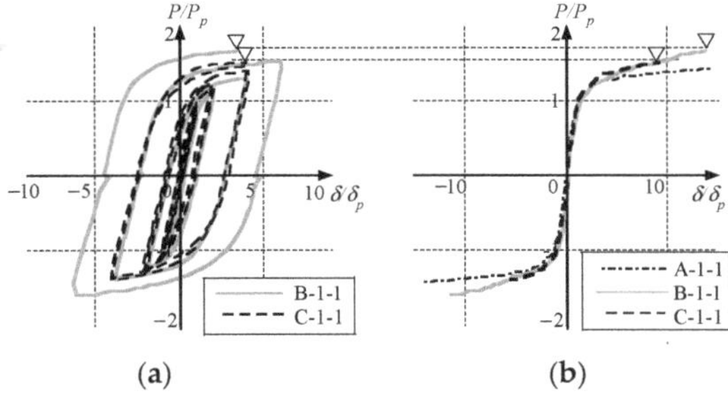

(a)   (b)

Figure 4. Comparison of test results of behavior under monotonic and cyclic loading: (**a**) hysteresis curve; (**b**) skeleton curve.

Figure 5 illustrates the strain distribution in the beam flange and web of specimen C-1-1 for the first, second, and third loading cycles with a loading amplitude $\delta/\delta_p = 3$. Figure 5a shows the strain distribution in the flange, with the upper side representing the compressive side and the lower side representing the tensile side. Figure 5b,c show the strain distribution in the web, with Figure 5b indicating the position 100 mm from the load point side of the specimen end and Figure 5c indicating the position 200 mm from the load point side. At $\delta/\delta_p = 3$, local flange buckling has already occurred near the beam end during the first loading cycle, with some variation in strain values between the compressive and tensile flanges, but overall, they are nearly matched. As the loading cycles increase for the second and third cycles, the strain values increase in the plasticized region within 400 mm ($2H$) from the beam end. In other words, even at the same dimensionless displacement level, as the number of loading cycles increase, the strain values increase, and due to strain hardening, the stress also increases. Therefore, as shown in Figure 4, before local buckling occurs, the load on the skeleton curve of cyclic loading is higher at the same displacement compared to the load–displacement relationship of monotonic loading, and with more loading cycles, the load at the same displacement is slightly higher on the positive side.

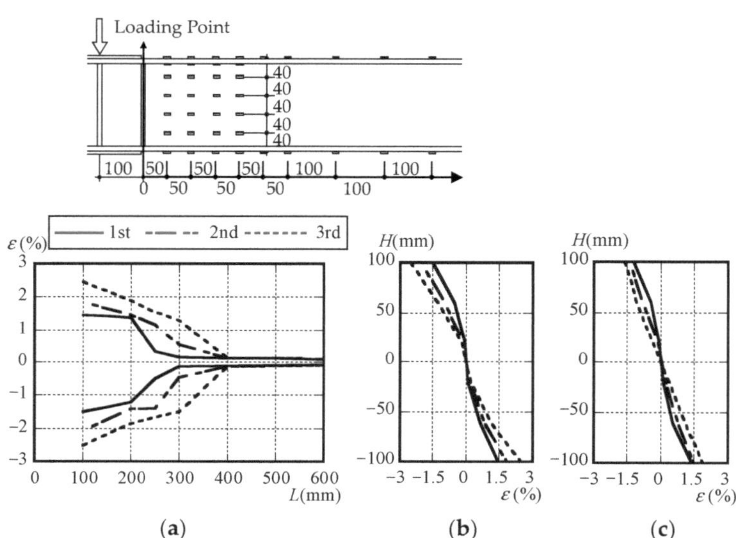

Figure 5. Strain distribution: (**a**) flange; (**b**) web (100 mm apart from fixed end); (**c**) web (200 mm apart from fixed end).

2.3. Influence of Loading Protocols on Cyclic Behavior of I-Beams

Table 4 presents a summary of the experimental results for all specimens. It includes the failure mechanism, the number of cycles at which local buckling occurred based on strain data, the rotation capacity μ_{max} in the skeleton curve, the energy dissipation capacity η_{max} in the skeleton curve, and the Bauschinger effect α_B. In cases where the flange width is narrow, such as specimens A-2-1 subjected to monotonic loading and B-2-1 subjected to cyclic loading, combined buckling occurred shortly after local buckling, leading to rapid strength degradation. Specimen B-3-1 experienced fracture due to welding defects at the welded joint between the longitudinal stiffener near the load point side and the tension flange after local buckling. On the other hand, in cases where the shear span-to-depth ratio is small, specimen A-1-2 reached its maximum strength due to local buckling, while specimen B-1-2 subjected to cyclic loading experienced flange local buckling due to bending moments, showing more prominent shear deformation compared to other specimens. Even for the specimens of the same section and length, different loading histories can result in different failure mechanisms.

Table 4. Summary of experimental results.

Specimen	Failure Modes	Cycle at Failure	μ_{max}	η_{max}	α_b
A-1-1	Local buckling		13.27	15.96	
B-1-1	Local buckling	$+2\delta_p$	12.65	16.16	2.06
C-1-1	Local buckling	$+2\delta_p$ (1st)	8.17	9.45	2.48
A-2-1	Combined buckling (local and lateral buckling)		8.34	7.96	
B-2-1	Combined buckling (local and lateral buckling)	$-4\delta_p$	8.64	13.84	1.63
A-3-1	Local buckling		11.86	13.33	
B-3-1	Local buckling and flange failure	$+2\delta_p$	5.46	10.41	1.61
A-1-2	Local buckling		20.90	26.59	
B-1-2	Combined buckling (local and shear buckling)	$+6\delta_p$	8.88	17.10	2.02
A-1-3	Local buckling		12.84	15.47	
B-1-3	Local buckling	$+2\delta_p$	7.85	14.09	1.82

Figure 6 illustrates the relationship between rotation capacity and energy dissipation capacity for I-beams with different flange width-to-thickness ratios and shear span ratios. Figure 6a,b depict the relationship between rotation capacity, energy dissipation capacity, and flange width-to-thickness ratio for monotonic and cyclic loading, while Figure 6c,d show the relationship between rotation capacity, energy dissipation capacity, and shear span-to-depth ratio. In Figure 6a,b, specimens with smaller width-to-thickness ratios (A-2-1 and B-2-1) experienced lateral buckling shortly after local buckling, resulting in a lower rotation capacity and energy dissipation capacity compared to specimens with larger width-to-thickness ratios (A-1-1 and B-1-1). However, for other specimens, the rotation capacity and energy dissipation capacity decreased as the width-to-thickness ratio increased. In Figure 6c,d, for specimens subjected to cyclic loading, specimen B-1-2 exhibited a lower rotation capacity due to significant shear deformation along with flange local buckling. However, for other specimens, the rotation capacity and energy dissipation capacity decreased as the shear span-to-depth ratio increased. Different buckling mechanisms between monotonic and cyclic loading indicate that cyclic loading induces local buckling in both flanges, leading to more advanced section deformation at the same loading displacement compared to monotonic loading, making it more susceptible to other buckling modes.

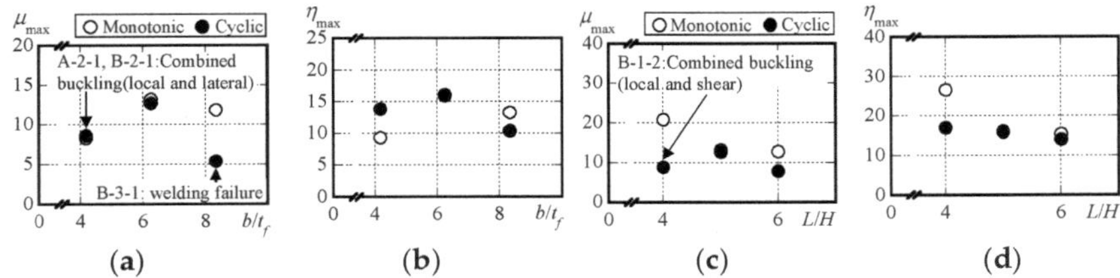

(a)   **(b)**   **(c)**   **(d)**

Figure 6. Relation between section dimensions and rotation capacity and energy dissipation capacity: (a) μ_{max} vs. b/t_f; (b) η_{max} vs. b/t_f; (c) μ_{max} vs. L/H; (d) η_{max} vs. L/H.

Figure 7 compares the experimental results of monotonic and cyclic loading for specimens with the same cross-section and length. Figure 7a–c show the ultimate strength ratio, rotation capacity, and energy dissipation capacity, respectively. Although the ultimate strength ratio is higher for cyclic loading in all specimens, the rotation capacity and energy dissipation capacity are higher for monotonic loading, except for specimens A-2-1 and B-2-1, where the buckling mode was combined with lateral buckling. In the case of monotonic loading, local buckling occurs in the flanges subjected to compressive stress, while the flanges subjected to tensile stress experience plasticization and stretching. Under cyclic loading, the flange initially undergoes local buckling under compressive stress, followed by tension, resulting in extended local buckling deformation. Flanges subjected to tensile stress undergo plasticization and stretching, followed by local buckling under compressive stress. As a result, the stress state changes within the same loading amplitude, leading to pinching due to the alternating expansion and contraction of deformations. Consequently, even with the same displacement, cyclic loading causes a reduction in bearing capacity in a smaller deformation.

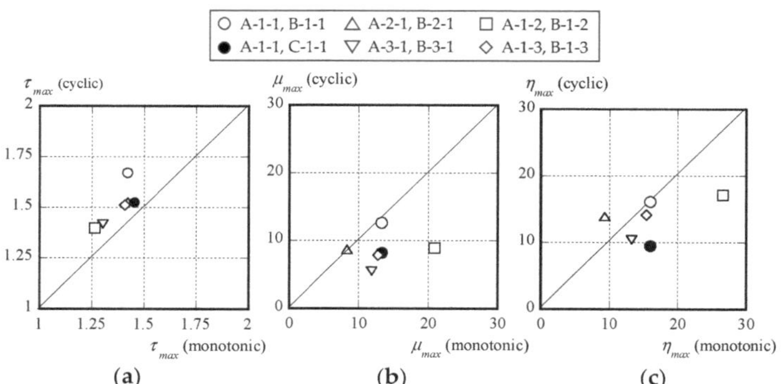

Figure 7. Comparison of experimental results under monotonic and cyclic loading: (**a**) τ_{max}; (**b**) μ_{max}; (**c**) η_{max}.

3. Rotation Capacity and Energy Dissipation Capacity of I-Beams That Failed Due to Local Buckling under Different Loading Protocols

3.1. Parameters of Experiment

Figure 8 illustrates the relationship between the width-to-thickness ratio of the test specimens from previous experiments [78–97] and the width-to-thickness ratio classification according to AISC, EC8-1, and AIJ codes [53,55,99,100]. The symbol ○ represents specimens from monotonic loading experiments, while ● represents specimens from the cyclic loading experiments. The test specimens are made of ordinary steel equivalent to either the 400 N/mm^2 class or the 490 N/mm^2 class. The figure shows 41 specimens from cyclic loading experiments for which history curves could be extracted, represented as ●, from previous articles [78–92] and 38 specimens from monotonic loading experiments, shown as ○, from earlier reports [93–97]. The relationship between each classification and rotation capacity is shown in Figure 8b. Additionally, the test specimens from this paper are denoted with a plus sign (+) for monotonic loading and a cross (×) for cyclic loading. Specimens where the maximum strength was reached due to lateral buckling (A-2-1 and B-2-1) and those fractured at welded joints (B-3-1) are excluded.

For instance, when comparing the width-to-thickness ratio of each beam with the Eurocode, both the flanges and webs display a broad spectrum of width-to-thickness ratios, spanning from Class 1 to Class 3, with the majority falling into Class 1. The web width-to-thickness ratios range approximately from 20 to 85, while the flange width-to-thickness ratios range from about 5 to 15. Notably, there was only one specimen for cyclic loading and two specimens for monotonic loading in the Class 4 equivalent rank.

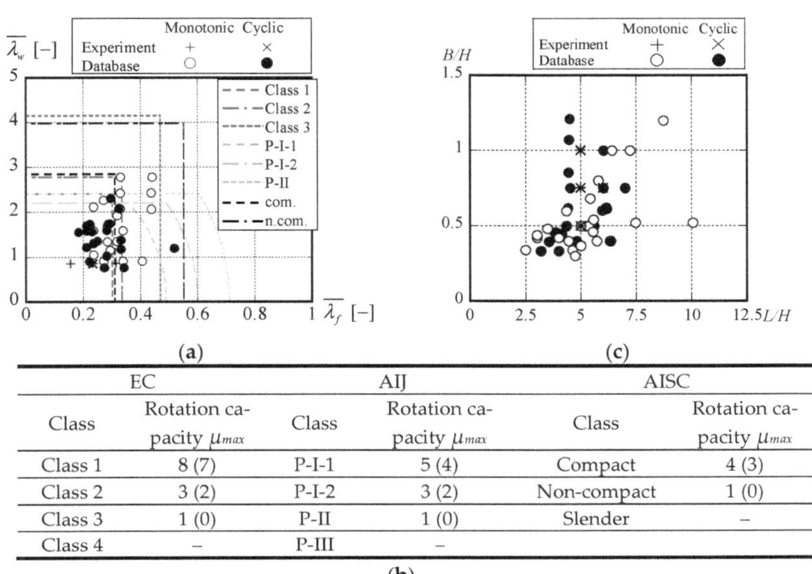

	EC		AIJ		AISC
Class	Rotation capacity μ_{max}	Class	Rotation capacity μ_{max}	Class	Rotation capacity μ_{max}
Class 1	8 (7)	P-I-1	5 (4)	Compact	4 (3)
Class 2	3 (2)	P-I-2	3 (2)	Non-compact	1 (0)
Class 3	1 (0)	P-II	1 (0)	Slender	–
Class 4	–	P-III	–		

(b)

Figure 8. Distribution of specimen configuration: (**a**) width–thickness ratio; (**b**) rotation capacities depending on cross-section class; (**c**) section aspect ratio and shear span-to-depth ratio.

According to AISC [54], cross-sections classified as compact members must possess an inelastic rotation capacity (μ_{max}) equal to or greater than 3 (thus, a rotation capacity μ greater than 4). In contrast, EC 8-1 [99] mandates that cross-sections fall into Class 1, 2, or 3 for seismic design. Each class is associated with a specific behavior factor (q), with Class 1 requiring $q > 4$, Class 2 needing $q > 2$, and Class 3 stipulating $q > 1.5$. Conversely, EC 8-3 [100] specifies rotation capacity values of $\mu_{max} > 8$ for Class 1 and $\mu_{max} > 3$ for Class 2. Among the design codes, only the Japanese classification system outlined in AIJ [53] accounts for segment interaction. It comprises four classes (P-I-1, P-I-2, P-II, and P-III), with the required μ_{max} defined as follows: $\mu_{max} \geq 4$ for P-I-1, $\mu_{max} \geq 2$ for P-I-2, $\mu_{max} \geq 0$ for P-II, and $\mu_{max} < 0$ for P-III.

Figure 8c illustrates the relationship between the sectional aspect ratio and the shear span-to-depth ratio of I-beams. Here, the sectional aspect ratio refers to the ratio of beam width to height (B/H), and the shear span-to-depth ratio refers to the ratio of beam length to height (L/H). For hot-rolled members, these two indices can generally capture the section performance of I-beams and are used as indicators for section selection during design. Therefore, the figure shows the distribution of both indices. The sectional aspect ratio ranges from approximately 0.4 to 1.2, while the shear span-to-depth ratio ranges from about 2.5 to 10, within the range of hot-rolled I-beams used in actual structures.

3.2. Rotation Capacity, Ultimate Strength Ratio, and Energy Dissipation Capacity of I-Beams with Different Loading Protocols

Figure 9 organizes the experimental results of I-beams in terms of equivalent width-to-thickness ratio. The legend is the same as in Figure 8. Figure 9a illustrates the ultimate strength ratio, denoted as τ_{max} ($\tau_{max} = P_{max}/P_p$), while Figure 9b shows the rotation capacity, denoted as μ_{max}. The equivalent width-to-thickness ratio of I-beams is defined in Kadono et al. [101] by the following equation:

$$\left(\frac{b}{t_f}\right) = \sqrt{\frac{\sigma_{yf}}{E}\left(\frac{b}{t_f}\right)^2 + \frac{\sigma_{yw}}{41E}\left(\frac{d}{t_w}\right)^2} \tag{3}$$

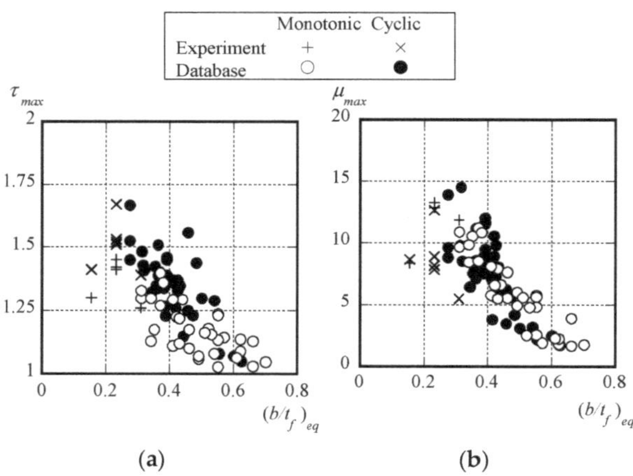

(a)

(b)

Figure 9. Width–thickness ratio and structural performance of I-beams; (**a**) ultimate strength ratio; (**b**) rotation capacity.

Here, σ_{yf} and σ_{yw} are the yield stresses of the flange and web, respectively, E is Young's modulus, and b/t_f and d/t_w are the flange width-to-thickness ratio and web width-to-thickness ratio, respectively.

Figure 9a,b show that as the equivalent width-to-thickness ratio decreases, the ultimate strength ratio τ_{max} and rotation capacity μ_{max} tend to increase. However, even for equal width-to-thickness ratios, the ultimate strength ratio is higher in the case of cyclic load experiments compared to monotonic load experiments, and the rotation capacity shows variability in the range from $(b/t_f)_{eq} = 0.3$ to 0.6.

Here are the experimental equations shown in a previous study [52], presented as Equations (4)–(6):

$$\tau_{max} = 1.46 - \left[0.63\frac{b}{t_f} + 0.053\frac{d}{t_w} + 0.02(\lambda_y - 50)\right]\sqrt{\varepsilon_y} \tag{4}$$

$$\mu_{max} = \frac{(\tau_{max} - 1)}{0.03} \tag{5}$$

$$\eta_{max} = \frac{(\tau_{max} + 1)(\mu_{max} + 1 - \tau_{max})}{2} \tag{6}$$

This paper considers the parameters of previous experimental equations and uses the least squares method to create experimental equations based on the results of 65 specimens from this experiment and the database outlined in Appendix A. Furthermore, considering the differences in loading methods, the number of repetitions and the repetition amplitude are added as parameters. The experimental equations proposed in this paper are shown in Equations (7)–(9):

$$\tau_{max} = \left[1.5 - 0.57\left(b/t_f\right)_{eq} - 0.010L/H\right]\left(1.1 - 0.1\frac{N}{\sum \mu_{xi}}\right) \tag{7}$$

$$\mu_{max} = \left[-7.2 + \frac{4.5}{\left(b/t_f\right)_{eq}} + \frac{8}{L/H}\right]\left(0.55 + 0.08\frac{\sum \mu_{xi}}{N}\right) - 1 \tag{8}$$

$$\eta_{max} = \frac{(\tau_{max} + 1)(\mu_{max} + 1 - \tau_{max})}{2} \tag{9}$$

In this context, N denotes the cumulative number of loading cycles until the attainment of the maximum load, where each loading cycle, whether positive or negative, contributes a value of 1. However, for instances of monotonic loading, N is stipulated to be 1. The symbol μ_{xi} represents the dimensionless amplitude of loading displacement in the i-th cycle.

In Equation (4), the flange-to-web thickness ratio, web thickness ratio, and yield strain are evaluated using the equivalent thickness ratio in Equation (7), with the aspect ratio being considered the shear span ratio.

Equation (5) previously comprised only the ultimate strength and axial force ratios (assumed as 0 here) to determine the plastic strain amplification. However, recognizing that rotation capacity is influenced not only by the thickness ratio but also by the proportion of the plasticization region [102], Equation (8) accounts for the effects of the equivalent thickness ratio and shear span-to-depth ratio. The rationale behind employing the shear span-to-depth ratio as an indicator of moment gradient lies in the disparity in the bending moments experienced by the flanges, even under equivalent moments. In cases of localized buckling failure, the local buckling of the flanges can impact the structural performance of the beam.

Figure 10 illustrates the correspondence between the experimental results from previous studies [78–97], the experimental results presented in this paper, as well as the experimental equations from both the earlier studies and the proposed equations in this paper. In Figure 10a, the comparison between the experimental results and the empirical equation (Equation (4)) from Kato's research [52] regarding the ultimate strength ratio is demonstrated. This paper aims to evaluate the mechanical performance of I-beams under cyclic loading by comparing them with the results of monotonic loading experiments and considering factors such as the number of cycles and load amplitudes, thereby enabling a consistent evaluation regardless of the loading conditions. It is noteworthy that compared to the experimental equations from previous studies, the experimental results generally exhibit larger values, particularly noticeable in the case of cyclic loading experiments. This discrepancy can be attributed to the fact that the experimental equations from previous studies do not consider the parameters of the cyclic loading history, thus not necessarily aligning well with the cyclic loading experimental results collected in this paper.

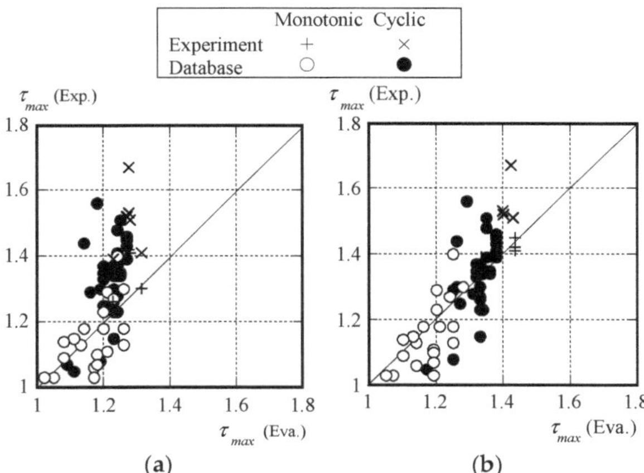

Figure 10. Comparison of ultimate strength ratio: (**a**) previous evaluation equation; (**b**) proposed evaluation equation.

Figure 10b presents a comparison between the experimental results of the ultimate strength ratio and the modified experimental equation (Equation (7)). Compared to the experimental equations from previous studies, the modified experimental equation proposed in this paper demonstrates better correspondence with the experimental results.

The error rates for the experimental equations from previous studies are 10.7% (12.6%) for Figure 10a and 6.1% (5.7%) for Figure 10b, while those for the modified experimental equations proposed in this paper are 6.1% (5.7%) and 6.1% (5.7%), respectively. This indicates a relatively good approximation of the experimental results for both monotonic and cyclic loading conditions. It is worth noting that the error rate is calculated by summing the differences between the experimental values and the values obtained from the experimental equations, divided by the total number of experimental data points, and expressed as a percentage. The first number in the error rate represents the case where both monotonic and cyclic loading experimental results are combined. In contrast, the number in parentheses represents the case where only cyclic loading experimental results are considered.

Figure 11 compares the experimental results of the rotation capacity, μ_{max}, with the experimental equations. Figure 11a utilizes the experimental equation (Equation (5)) from a previous study [52], while Figure 11b employs Equation (8). The shaded areas in the figures indicate the range of a 30% error for reference. Although there are no significant disparities between the experimental equations and the experimental results, the values of the experimental results vary widely around μ_{max} = 8 for the experimental equation (Equation (8)). On the other hand, Equation (8) demonstrates a generally good correspondence with the experimental results. The error rates for the experimental equation (Equation (5)) and Equation (8) are 31.3% (28.9%) and 23.1% (25.8%), respectively, indicating that using the experimental equations based on the equivalent thickness ratio and aspect ratio proposed in this paper can reduce the variability.

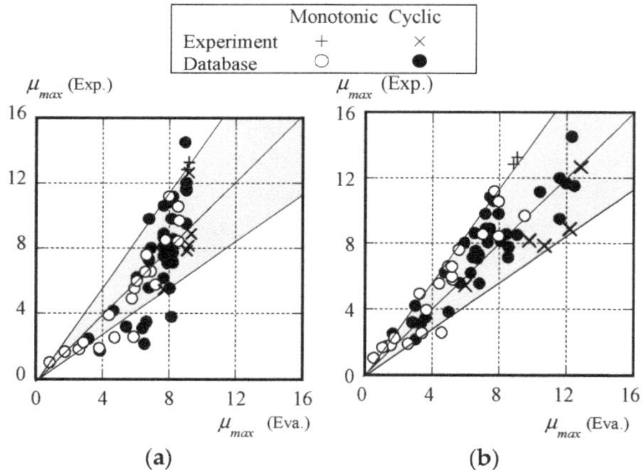

Figure 11. Comparison of rotation capacity: (**a**) previous evaluation equation; (**b**) proposed evaluation equation.

Figure 12 compares the experimental results of the energy dissipation capacity, η_{max}, with the experimental equations. Figure 12a,b employ the experimental equation (Equation (6)) from Kato's research [52] and Equation (9), respectively. The shaded areas in the figures indicate a 30% error range, similar to that in Figure 12. Meanwhile, the experimental equations from previous studies represent the gradient up to the maximum load, with a linear approximation. However, they may not correspond well with the experimental results obtained from Equations (4) and (5). Hence, several data points for η_{max} exceed the range of a 30% error, resulting in an error rate of 35.5% (36.1%). In contrast, the modified experimental equation proposed in this paper, Equation (9), demonstrates a better fit with many experimental results, falling within the 30% error range, resulting in an error rate of 25.2% (27.0%).

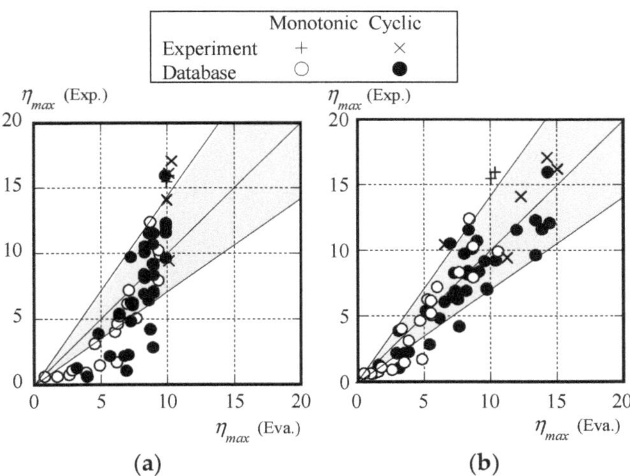

Figure 12. Comparison of energy dissipation capacity: (**a**) previous evaluation equation; (**b**) proposed evaluation equation.

In summary, by proposing experimental equations that consider parameters such as the equivalent thickness ratio, shear span-to-depth ratio, number of cycles up to the maximum load, and cumulative normalized displacement based on the parameters of previous experimental equations, this paper demonstrates that the experimental results can be more accurately represented.

3.3. Influence of Bauschinger Effect Coefficient of I-Beams That Failed Due to Local Buckling under Cyclic Loading

Figure 13 illustrates the relationship between the Bauschinger effect coefficient, the number of loading cycles, and the loading amplitude. The crosses represent the results of the cyclic loading experiments in this paper. Figure 13a depicts the relationship with the number of loading cycles N. In contrast, Figure 13b shows the relationship between the total dimensionless loading displacement amplitude $\sum \mu_{xi}$ and the Bauschinger effect coefficient α_B. The Bauschinger effect coefficient α_B is defined as the ratio of cumulative hysteretic energy dissipation in the cumulative hysteresis curve to the energy dissipation capacity in the skeleton curve, as described in earlier research [64], which is set to 2.0 for beams. While α_B shows an almost linear relationship with N and correlates with $\sum \mu_{xi}$, some variability is observed. Therefore, Figure 14 illustrates the relationship between both of the coefficients N and $\sum \mu_{xi}$ and the Bauschinger effect coefficient. The vertical axis represents the Bauschinger effect coefficient α_B, while the horizontal axis represents the variable used in the second term on the right-hand side of the equation.

$$\alpha_b = 1 + 0.5(N - 1)\log \sqrt{\sum\nolimits_{i=1} \mu_{xi}} \tag{10}$$

However, in the case of $N \leq 1$ (monotonic loading), N is set to 1. When the loading history remains within the elastic range during cyclic loading, only the first occurrence is counted.

From this paper and previous experimental results, it has been shown that this enhancement ratio depends on both the number of loading cycles and their amplitudes. Therefore, using both indicators in this paper, the evaluation equation is set as $\alpha_b = 1$ for monotonic loading and the above equation is formed through a trial-and-error method.

The plots in the figure represent the experimental results, while the dashed line corresponds to Equation (10). There is a tendency for the Bauschinger effect coefficient to increase with higher values of the total number of loading cycles and loading displacement amplitudes, and the experimental results generally agree with Equation (10).

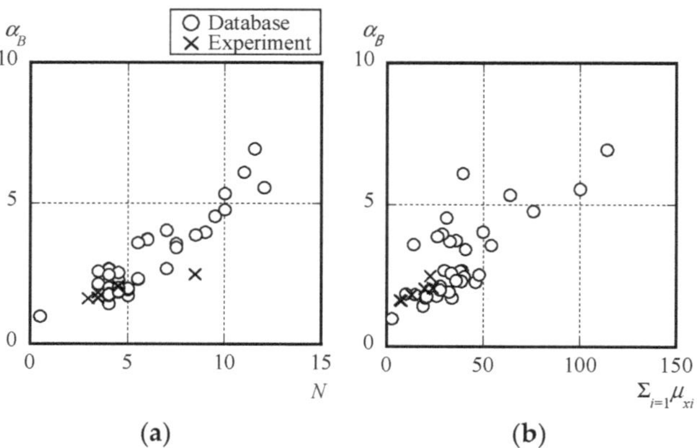

Figure 13. Relationship between Bauschinger effect coefficient and loading histories: (**a**) number of cycles; (**b**) cumulative plastic deformation.

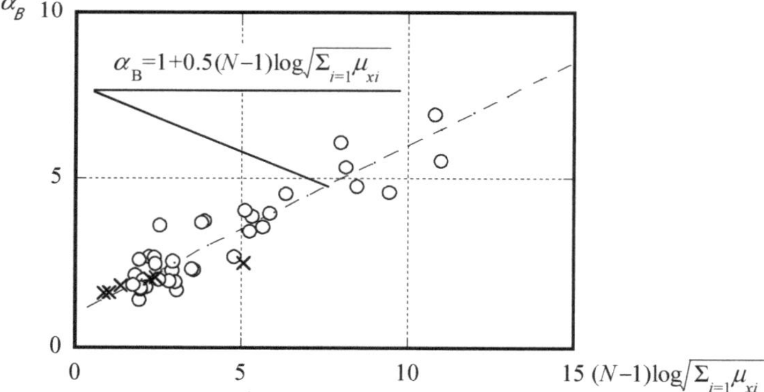

Figure 14. Evaluation of Bauschinger effect coefficient.

Thus, the hysteretic energy dissipation of an I-beam subjected to cyclic loading can be obtained by multiplying η_{max} obtained from Equation (9) by Equation (10).

In the field of structural engineering, numerous studies have focused on evaluating the performance of composite beams, where steel beams are attached to concrete slabs as composite members [27–33]. However, this paper focuses on the failure mechanism of steel beams, explicitly targeting the phenomenon of local buckling, and aims to evaluate the inherent performance of steel beams that failed due to local buckling. The performance evaluation of the composite beam will be clarified in a future study.

4. Conclusion

This paper elucidated the differences in the hysteresis curves and buckling mechanisms of locally buckling I-beams through monotonic and cyclic loading experiments. Subsequently, a simplified and accurate experimental equation was proposed based on the previous experimental equation as a unified evaluation method independent of loading history differences. It is noteworthy that Table A1 comprises several specimens sharing identical cross-sectional shapes and lengths, yet lacks variation in loading protocols as a parameter. Therefore, concluding remarks (1) and (2) stem from the findings retrieved from the experiment in this study. The conclusions obtained are as follows:

(1)   When subjected to cyclic loading at the same amplitude in the plastic region where local buckling occurs in I-beams, it was confirmed that the strain values increase and plasticization progresses as the number of cycles increase. When subjected to cyclic loading, the rotation capacity and energy dissipation capacity obtained from the experimental results decrease with increasing cycles at the same amplitude in the skeleton curve.

(2)   For specimens of the same cross-section and length subjected to monotonic and cyclic loading, the rotation capacity and energy dissipation capacity are higher under monotonic loading if local buckling occurs.

(3)   In the previous experimental equation, particularly in the case of cyclic loading, there are significant errors in the maximum load, rotation capacity, and energy dissipation capacity. On the other hand, the experimental equations proposed in this paper (Equations (7)–(9)) can generally evaluate the ultimate strength ratio, rotation capacity, and energy dissipation capacity regardless of the loading type.

(4)   The Bauschinger effect coefficient increases with a more significant number of loading cycles and larger loading amplitudes, and its value can be evaluated using Equation (10). Moreover, hysteretic energy dissipation under cyclic loading can be assessed using Equations (9) and (10).

It should be noted that the application range of the experimental equations in this paper covers web-to-thickness ratios from 20 to 85, flange-to-thickness ratios from 5 to 15, section shape ratios from 0.4 to 1.2, and shear span-to-depth ratios from 2.5 to 10, based on the previous literature and the specimens tested in this paper.

This paper only targets the incremental amplitude type, and evaluating the performance of I-beams under decremental amplitude or random amplitude types is a future task.

Funding: This research study was funded by the Research Project of Co-Creation for Disaster Resilience (Principal Investigator: Yoshihiro Kimura). I express my deepest gratitude for their sincere support.

Data Availability Statement: The raw/processed data necessary to reproduce these findings cannot be shared at this time because the data also form part of an ongoing study.

Conflicts of Interest: The author declares that they have no known competing financial interests or personal relationships that could have appeared to influence the work reported in this paper.

Appendix A

The summary of the database obtained from previous articles is listed below. The hysteresis curves and residual deformation of all 11 specimens are presented below. All specimens except for A-2-1 and B-2-1 failed due to local buckling, while specimens A-2-1 and B-2-1 experienced coupled buckling induced by lateral buckling after local buckling, leading to ultimate failure. Specimen B-1-2 exhibited significant shear deformation in addition to bending deformation.

Table A1. Summary of database.

Section	L	b/t_f	d/t_w	L/H	τ_{max}	μ_{max}	η_{max}	N	$\Sigma\mu_{xi}$	Refs.
H-330 × 150 × 4.5 × 9	1260	8.3	69.3	3.8	1.29	3.21	2.24	11	39.07	[78]
H-330 × 150 × 4.5 × 9	1266	8.3	69.3	3.8	1.05	1.76	0.68	5.5	13.34	[78]
H-250 × 150 × 4.5 × 9	1496	8.3	51.6	6.0	1.25	3.5	2.33	9.5	30.49	[80]
H-250 × 150 × 4.5 × 12	1492	6.3	50.2	6.0	1.28	3.83	2.89	9	28.53	[80]
H-250 × 150 × 4.5 × 9	1497	8.3	51.6	6.0	1.3	3.13	2.22	8.5	25.91	[80]
H-250 × 150 × 4.5 × 9	1499	8.3	51.6	6.0	1.08	2.2	1.11	4.5	9.76	[80]
H-200 × 150 × 6 × 9	900	8.3	30.3	4.5	1.35	7.16	7.12	7	49.89	[81]
H-200 × 150 × 6 × 9	900	8.3	30.3	4.5	1.34	11.17	11.62	11.5	113.84	[81]
H-200 × 150 × 6 × 9	900	8.3	30.3	4.5	1.51	8.58	9.18	5.5	38.24	[81]
H-200 × 150 × 6 × 9	900	8.3	30.3	4.5	1.39	8.17	8.4	5.5	35.4	[81]
H-500 × 200 × 9 × 19	2400	5.3	51.3	4.8	1.23	5.6	4.23	27	130.71	[82]
H-450 × 200 × 9 × 12	1850	8.3	47.3	4.1	1.37	5.63	4.87	7.5	40.66	[83]
H-450 × 200 × 9 × 12	1850	8.3	47.3	4.1	1.35	9.82	9.78	12	99.8	[83]
H-450 × 200 × 9 × 12	1850	8.3	47.3	4.1	1.33	7.31	6.3	10	75.72	[83]
H-506 × 201 × 11 × 19	1800	5.3	42.5	3.6	1.35	7.77	7.05	4	29.42	[84]
H-300 × 130 × 6 × 12	1200	5.4	46.0	4.0	1.45	12.03	12.35	4	37.78	[85]
H-300 × 130 × 6 × 12	1200	5.4	46.0	4.0	1.45	11.54	12.15	4.5	47.62	[85]
H-300 × 130 × 6 × 12	1200	5.4	46.0	4.0	1.39	9.5	9.69	3.5	33.05	[85]
H-300 × 130 × 6 × 12	1200	5.4	46.0	4.0	1.46	11.69	11.66	4	39.85	[85]
H-300 × 100 × 9 × 9	1200	5.6	31.3	4.0	1.41	8.52	9.24	7	38.84	[86]
H-300 × 100 × 6 × 6	1200	8.3	48.0	4.0	1.44	4.2	3.9	4.5	14.18	[86]
H-488 × 300 × 11 × 18	2150	8.3	41.1	4.4	1.56	6.26	5.41	10	63.67	[87]
H-600 × 300 × 12 × 22	3125	6.8	46.3	5.2	1.36	7.18	6.48	5	25.28	[88]
H-600 × 300 × 12 × 22	3125	6.8	46.3	5.2	1.37	7.12	6.64	4	22.16	[88]
H-450 × 150 × 9 × 12	1425	6.3	47.3	3.2	1.35	8.06	6.08	6	32.4	[89]
H-250 × 125 × 6 × 9	1375	6.9	38.7	5.5	1.48	9.82	10.76	6	35.5	[90]
H-500 × 200 × 10 × 16	2168	6.3	46.8	4.3	1.34	10.85	11.59	7.5	53.87	[91]
H-500 × 200 × 10 × 16	3150	6.3	46.8	6.3	1.3	10.59	10.15	5	33.56	[92]
H-500 × 200 × 10 × 16	3150	6.3	46.8	6.3	1.26	8.9	8.15	5	31.85	[92]
H-500 × 200 × 10 × 16	3150	6.3	46.8	6.3	1.15	6.24	10.57	4	18.45	[92]
H-500 × 200 × 10 × 16	3175	6.3	46.8	6.4	1.37	8.7	8.44	4	25.51	[93]
H-500 × 200 × 10 × 16	3175	6.3	46.8	6.4	1.23	8.92	8.28	4.5	27.27	[93]
H-500 × 200 × 10 × 16	3175	6.3	46.8	6.4	1.27	8.65	8.3	4	19.91	[93]
H-500 × 200 × 10 × 16	3175	6.3	46.8	6.4	1.27	7.56	6.96	4	20.38	[93]
H-300 × 125 × 4.5 × 9	1200	6.9	62.7	4.0	1.29	6.65	6.3	-	-	[94]
H-300 × 125 × 9 × 9	1200	6.9	31.3	4.0	1.3	9.69	9.94	-	-	[94]
H-300 × 125 × 6 × 9	900	6.9	47	3.0	1.4	11.21	12.45	-	-	[94]
H-300 × 125 × 6 × 9	1200	6.9	47	4.0	1.27	8.55	8.33	-	-	[94]
H-180 × 144 × 6 × 9	1040	8.0	27	5.8	1.18	10.56	10.31	-	-	[95]
H-180 × 180 × 6 × 9	1300	10.0	27	7.2	1.11	5.81	5.11	-	-	[95]
H-180 × 216 × 6 × 9	1570	12.0	27	8.7	1.06	4.96	4.07	-	-	[95]
H-180 × 144 × 6 × 9	1040	8.0	27	5.8	1.13	8.46	7.98	-	-	[95]
H-300 × 180 × 6 × 9	1300	10.0	47	4.3	1.1	5.59	4.69	-	-	[95]
H-420 × 144 × 6 × 9	1040	8.0	67	2.5	1.07	6.03	5.23	-	-	[95]
H-420 × 144 × 6 × 9	1040	8.0	67	2.5	1.13	3.93	3.15	-	-	[95]
H-300 × 144 × 6 × 9	1040	8.0	47	3.5	1.03	2.6	1.77	-	-	[95]
H-270.5 × 108.5 × 4.23 × 5.57	1200	9.7	61.3	4.4	1.18	2.56	1.53	-	-	[96]
H-269.6 × 144.3 × 4.23 × 5.57	1500	13.0	61.1	5.6	1.09	1.86	0.81	-	-	[96]
H-314.4 × 108.3 × 4.23 × 5.57	1450	9.7	71.7	4.6	1.15	1.93	0.96	-	-	[96]
H-315.5 × 143.8 × 4.23 × 5.57	1750	12.9	72.0	5.5	1.03	1.69	0.66	-	-	[96]
H-359.7 × 108.8 × 4.23 × 5.57	1700	9.8	82.4	4.7	1.14	2.26	1.13	-	-	[96]
H-359.5 × 144.8 × 4.23 × 5.57	2050	13.0	82.4	5.7	1.03	1.78	0.68	-	-	[96]
H-250 × 125 × 5.8 × 8.5	1250	7.4	40.2	5.0	1.23	6.62	6.2	-	-	[97]
H-450 × 200 × 9 × 14	1350	7.1	46.9	3.0	1.18	7.62	7.22	-	-	[98]

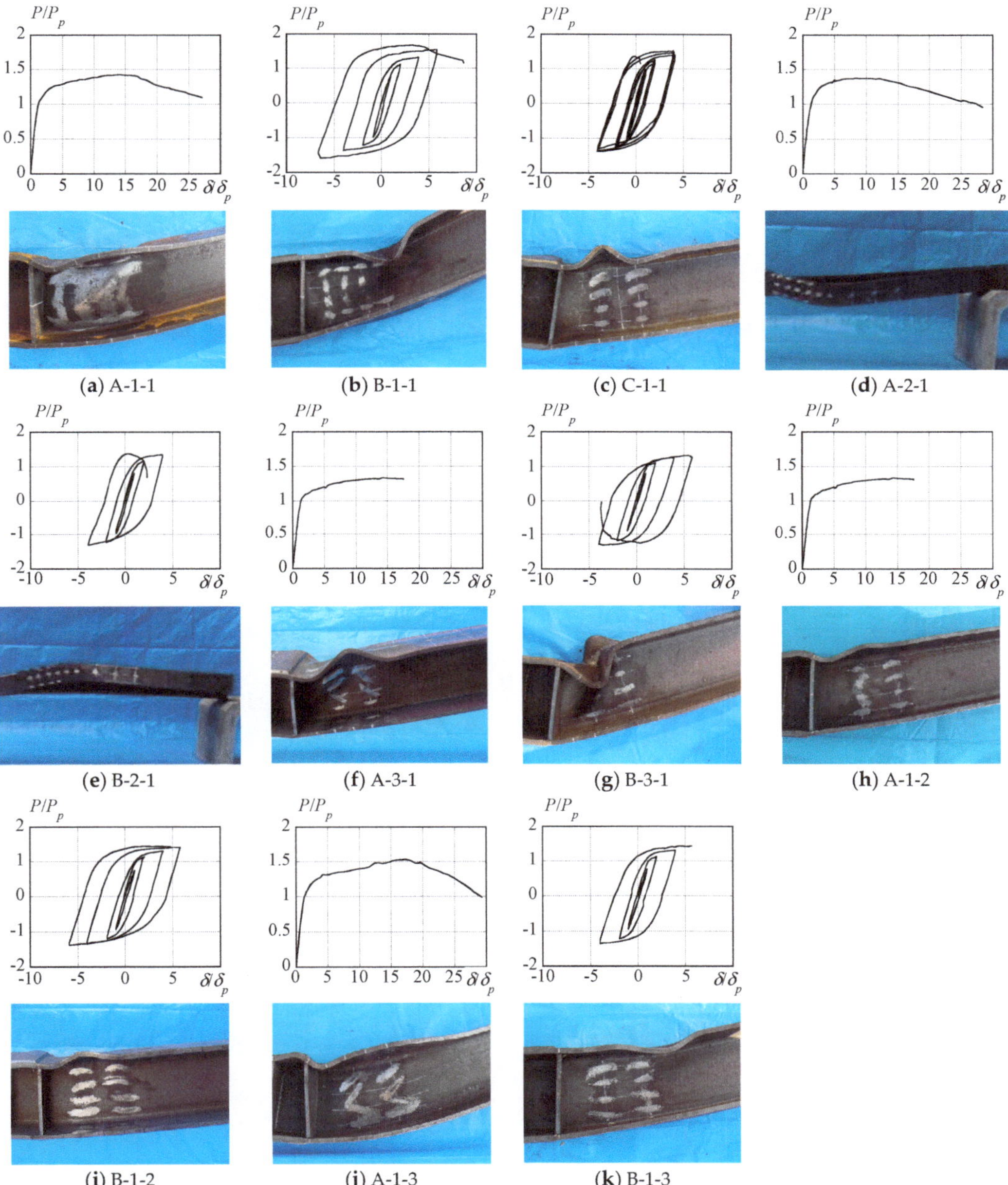

Figure A1. Load–displacement relationship and ultimate deformation state.

The list of symbols is presented below.

Table A2. List of symbols.

b/t_f	Flange width–thickness ratio
L/H	Shear span-to-depth ratio
L	Beam length
d/t_w	Web width-to-thickness ratio
H	Hight of beam
B	Width of beam
t_f	Flange thickness
t_w	Web thickness
δ_p	Displacement corresponding to full plastic bending moment
M_p	Full plastic bending moment
δ	Controlled displacement
δ_0	Displacement at position of transducer
φ	Rotation angle measured by the transducers
δ_{mp}	Displacements due to bending deformation of beam
δ_{sp}	Displacements due to shear deformation of beam
E	Young's modulus
I	Moment of inertia
A	Cross-sectional area
G	Shear modulus
κ	Shape factor
σ_y	Yield stress
σ_u	Tensile strength
Y.R.	Yield ratio (σ_y / σ_u)
P	Shear force acting on beam
P_p	Shear force imposing full plastic bending moment of beam
$\alpha_B \eta_{max}$	Hysteretic energy dissipation up to ultimate strength
α_B	Bauschinger effect coefficient
μ_{max}	Rotation capacity
η_{max}	Energy dissipation capacity
$\frac{b}{t_f}\sqrt{\frac{235}{\sigma_{yf}}}$	Generalized flange width-to-thickness ratio
$\frac{d}{t_w}\sqrt{\frac{235}{\sigma_{yw}}}$	Generalized web width-to-thickness ratio
B/H	Section aspect ratio
τ_{max}	Ultimate strength ratio
σ_{yf}	Yield stress of flange plate
σ_{yw}	Yield stress of web plate
N	Cumulative number of loading cycles until attainment of maximum load
μ_{xi}	Dimensionless amplitude of loading displacement in i-th cycle

References

1. Bleich, F. *Buckling Strength of Metal Structures*; McGraw Hill: New York, NY, USA, 1952.
2. Timoshenko, S.P. *Theory of Elastic Stability*; McGraw Hill: New York, NY, USA, 1963.
3. Nethercot, D.A.; Rockey, K.C. A Unified Approach to the Elastic Lateral Buckling of Beams. *Struct. Eng.* **1971**, *7*, 96–107. [CrossRef]
4. Narayanan, R. *Beam and Beam Columns-Stability and Strength*; Applied Science Publishers: Amsterdam, The Netherlands, 1983.
5. Salvador, T.G. Lateral Buckling of Eccentrically Loaded I—Columns. *Am. Soc. Civ. Eng.* **1956**, *121*, 1163–1179. [CrossRef]
6. Trahair, N.S. laterally unsupported beams. *Eng. Struct.* **1996**, *18*, 759–768. [CrossRef]
7. Nguyen, C.T.; Moon, J.; Lee, H.E. Lateral–torsional buckling of I-girders with discrete torsional bracings. *J. Constr. Steel Res.* **2010**, *66*, 170–177. [CrossRef]
8. Ji, X.L.D.; Twizell, S.C.; Driver, R.G.; Imanpour, A. Lateral Torsional Buckling Response of Compact I-Shaped Welded Steel Girders. *Eng. Struct.* **2022**, *148*, 04022149. [CrossRef]
9. Mohammadi, E.; Hosseini, S.S.; Rohanimanesh, M.S. Elastic lateral-torsional buckling strength and torsional bracing stiffness requirement for monosymmetric I-beams. *Eng. Struct.* **2016**, *104*, 116–125. [CrossRef]
10. Nguyen, C.T.; Joo, H.S.; Moon, J.; Lee, H.E. Flexural-torsional buckling strength of I-girders with discrete torsional braces under various loading conditions. *Eng. Struct.* **2022**, *36*, 337–350. [CrossRef]
11. Fukunaga, I.; Todaka, T.; Zhang, Z.; Kanao, I. Deformation Capacity of Wide Flange Beam with Lateral Bracings. *J. Struct. Constr. Eng. (Trans. AIJ)* **2022**, *87*, 372–380. (In Japanese) [CrossRef]
12. Bradford, M.A.; Gao, Z. Distortional buckling solutions for continuous composite beams. *J. Struct. Eng.* **1992**, *118*, 73–89. [CrossRef]

13. Kimura, Y.; Yoshino, Y. Required Bracing Capacity on Lateral Buckling Strength for H-Shaped Beams with Bracings. *J. Struct. Constr. Eng. (Trans. AIJ)* **2011**, *76*, 2143–2152. (In Japanese) [CrossRef]
14. Kimura, Y.; Yoshino, Y.; Ogawa, J. Effect of Lateral-Rotational Restraint and Strength of Continuously Braces on Lateral Buckling Load for H-Shaped Beams. *J. Struct. Constr. Eng. (Trans. AIJ)* **2013**, *78*, 193–201. (In Japanese) [CrossRef]
15. Kimura, Y.; Yoshino, Y. Effect of Lateral and Rotational Restraint for Bracings on Lateral Buckling Load for H-Shaped Beams under Moment Gradient. *J. Struct. Constr. Eng. (Trans. AIJ)* **2014**, *79*, 761–770. (In Japanese) [CrossRef]
16. Kimura, Y.; Matsuo, T.; Yoshino, Y. Estimation of Elasto-Plastic Lateral Buckling Stress for H-shaped Beams with Lateral Rotational Braces on subjected to Axial Force and Flexural Moment. *J. Struct. Constr. Eng. (Trans. AIJ)* **2014**, *79*, 1299–1308. (In Japanese) [CrossRef]
17. Kimura, Y.; Yoshino, Y. Effect of Lateral-Rotational Restraint of Continuous Braces on Lateral Buckling Strength for H-Shaped Beams with Flexural Moment Gradient. *J. Struct. Constr. Eng. (Trans. AIJ)* **2016**, *81*, 1309–1319. (In Japanese) [CrossRef]
18. Kimura, Y.; Sugita, Y.; Yoshino, Y. Effect of Lateral-Rotational Restraint of Continuous Braces on Lateral Buckling Strength for H-shaped Beams with Compressive Axial Force and Flexural Moment. *J. Struct. Constr. Eng. (Trans. AIJ)* **2016**, *81*, 1321–1331. (In Japanese) [CrossRef]
19. Kimura, Y.; Sugita, Y. Effect of Lateral-Rotational Restraint of Continuous Braces on Lateral Buckling Strength for H-shaped Beams with Gradient Flexural Moment and Compressive Axial Force. *J. Struct. Constr. Eng. (Trans. AIJ)* **2016**, *82*, 1799–1809. (In Japanese) [CrossRef]
20. Kimura, Y.; Miya, M.; Liao, W. Effect of Restraint for Continuous Braces on Lateral Buckling Load for H-shaped Beams with Warping Restraint of Beams to Column Joint. *J. Struct. Constr. Eng. (Trans. AIJ)* **2018**, *81*, 1353–1363. (In Japanese) [CrossRef]
21. Kimura, Y.; Miya, M. Effect of Warping Restraint of Beams to Column Joint on Lateral Buckling Behavior for H-Shaped Beams with Continuous Braces under Gradient Flexural Moment. *J. Struct. Constr. Eng. (Trans. AIJ)* **2019**, *84*, 1601–1611. (In Japanese) [CrossRef]
22. Kimura, Y.; Sato, Y. Effect of Warping Restraint of Beams to Column Joint on Lateral Buckling Behavior for H-Shaped Beams with Continuous Braces under Gradient Flexural Moment. *J. Struct. Constr. Eng. (Trans. AIJ)* **2021**, *86*, 145–155. (In Japanese) [CrossRef]
23. Yoshino, Y.; Liao, W.; Kimura, Y. Restraint effect on lateral buckling load of continuous braced H-shaped beams based on partial frame loading tests. *J. Struct. Constr. Eng. (Trans. AIJ)* **2022**, *87*, 634–645. (In Japanese) [CrossRef]
24. Kimura, Y.; Sato, Y.; Suzuki, A. Effect of Fork Restraint of Column on Lateral Buckling Behavior for H-Shaped Beams with Continuous Braces under Flexural Moment Gradient. *J. Struct. Constr. Eng. (Trans. AIJ)* **2022**, *87*, 316–327. (In Japanese) [CrossRef]
25. Yoshino, Y.; Kimura, Y. Rotational Stiffening Performance of Roof Folded Plates in Torsion Tests and the Stiffening Effect of Roof Folded Plates on the Lateral Buckling of H Beams in Steel Structures. *Buildings* **2024**, *14*, 1158. [CrossRef]
26. Rossi, A.; Nicoletti, R.S.; de Souza, A.S.C.; Martins, C.H. Lateral distortional buckling in steel-concrete composite beams: A review. *Structures* **2020**, *27*, 1299–1312. [CrossRef]
27. Suzuki, A.; Kimura, Y. Cyclic behavior of component model of composite beam subjected to fully reversed cyclic loading. *J. Struct. Eng.* **2019**, *145*, 04019015. [CrossRef]
28. Suzuki, A.; Abe, K.; Kimura, Y. Restraint performance of stud connection during lateral-torsional buckling under synchronized in-plane displacement and out-of-plane rotation. *J. Struct. Eng.* **2020**, *146*, 04020029. [CrossRef]
29. Suzuki, A.; Abe, K.; Suzuki, K.; Kimura, Y. Cyclic behavior of perfobond-shear connectors subjected to fully reversed cyclic loading. *J. Struct. Eng.* **2021**, *147*, 04020355. [CrossRef]
30. Suzuki, A.; Suzuki, K.; Kimura, Y. Ultimate shear strength of perfobond shear connectors subjected to fully reversed cyclic loading. *Eng. Struct.* **2021**, *248*, 113240. [CrossRef]
31. Suzuki, A.; Kimura, K. Mechanical performance of stud connection in steel-concrete composite beam under reversed stress. *Eng. Struct.* **2021**, *249*, 113338. [CrossRef]
32. Suzuki, A.; Hiraga, K.; Kimura, Y. Mechanical performance of puzzle-shaped shear connectors subjected to fully reversed cyclic stress. *J. Struct. Eng.* **2023**, *149*, 04023087. [CrossRef]
33. Suzuki, A.; Hiraga, K.; Kimura, Y. Cyclic behavior of steel-concrete composite dowel by clothoid-shaped shear connectors under fully reversed cyclic stress. *J. Adv. Concr. Technol.* **2023**, *21*, 76–91. [CrossRef]
34. Bedair, O. Stability of web plates in W-shape columns accounting for flange/web interaction. *Thin-Walled Struct.* **2009**, *47*, 768–775. [CrossRef]
35. Roberts, T.M.; Jhita, P.S. Lateral, local and distortional buckling of I-beams. *Thin-Walled Struct.* **1983**, *1*, 289–308. [CrossRef]
36. Bradford, M.A.; Hancock, G.J. Elastic interactions of local and lateral buckling in beams. *Thin-Walled Struct.* **1984**, *2*, 1–25. [CrossRef]
37. Chin, C.K.; Al-Bermani, F.G.A.; Kitipornchai, S. Stability of thin-walled members having arbitrary flange shape and flexible web. *Eng. Struct.* **1990**, *141*, 121–132. [CrossRef]
38. Seif, M.; Schafer, B.W. Local buckling of structural steel shapes. *J. Construct. Steel Res.* **2010**, *66*, 1232–1247. [CrossRef]
39. Han, K.H.; Lee, C.H. Elastic flange local buckling of I-shaped beams considering effect of web restraint. *Thin-Walled Struct.* **2016**, *105*, 101–111. [CrossRef]
40. Gardner, L.; Fieber, A.; Macorini, L. Formulae for calculating elastic local buckling stresses of full structural cross-sections. *Structures* **2019**, *17*, 2–20. [CrossRef]

41. Tankova, T.; da Silva, L.S.; Marques, L.; Tankova, T.; da Silva, L.S.; Marques, L. Buckling resistance of non-uniform steel members based on stress utilization: General formulation. *J. Construct. Steel Res.* **2019**, *149*, 239–256. [CrossRef]
42. Ragheb, W.F. Local buckling of welded steel I-beams considering flange–web interaction. *Thin-Walled Struct.* **2015**, *97*, 241–249. [CrossRef]
43. Gioncu, V. Framed structures. Ductility and seismic response: General Report. *J. Constr. Steel Res.* **2000**, *55*, 125–154. [CrossRef]
44. Gioncu, V.; Mosoarca, M.; Anastasiadis, A. Prediction of available rotation capacity and ductility of wide-flange beams: Part 1: DUCTROT-M computer program. *J. Constr. Steel Res.* **2011**, *69*, 8–19. [CrossRef]
45. Anastasiadis, A.; Mosoarca, M.; Gioncu, V. Prediction of available rotation capacity and ductility of wide-flange beams: Part 2: Applications. *J. Constr. Steel Res.* **2011**, *68*, 176–191. [CrossRef]
46. Shokouhian, M.; Shi, Y. Classification of I-section flexural members based on member ductility. *J. Constr. Steel Res.* **2014**, *95*, 198–210. [CrossRef]
47. Araujo, M.; Macedo, L.; Castro, J. Evaluation of the rotation capacity limits of steel members defined in EC8-3. *J. Constr. Steel Res.* **2017**, *135*, 11–29. [CrossRef]
48. Cardoso, D.; Vieira, J. Comprehensive local buckling equations for FRP I-sections in pure bending or compression. *Compos. Struct.* **2017**, *182*, 301–310. [CrossRef]
49. Ascione, F.; Feo, L.; Lamberti, M.; Minghini, F.; Tullini, N. A closed-form equation for the local buckling moment of pultruded FRP I-beams in major-axis bending. *Compos. Part B* **2016**, *97*, 292–299. [CrossRef]
50. Qiao, P.; Davalos, J.; Wang, J. Local Buckling of Composite FRP Shapes by Discrete Plate Analysis. *J. Struct. Engineering. Am. Soc. Civ. Eng.* **2001**, *127*, 245–255. [CrossRef]
51. Torabian, S.; Schafer, B. Role of local slenderness in the rotation capacity of structural steel members. *J. Constr. Steel Res.* **2013**, *95*, 32–43. [CrossRef]
52. Kato, B.; Akiyama, H.; Obi, Y. Deformation characteristics of H-shaped steel members influenced by local buckling. *J. Struct. Constr. Eng. (Trans. AIJ)* **1997**, *257*, 49–57. (In Japanese)
53. Architectural Institute of Japan (AIJ). *Recommendation for Limit State Design of Steel Structures*; Maruzen Publishing Co., Ltd.: Tokyo, Japan, 2010. (In Japanese)
54. American Institute of Steel Construction (AISC). *Seismic Provisions for Structural Steel Buildings*; AISC: Chicago, IL, USA, 2016.
55. Architectural Institute of Japan (AIJ). *AIJ Recommendations for Plastic Design of Steel Structures*; Maruzen Publishing Co., Ltd.: Tokyo, Japan, 2017. (In Japanese)
56. Lignos, D.G.; Krawinkler, H. *Sidesway Collapse of Deteriorating Structural Systems under Seismic Excitations*; Report No. TB 172; The John A. Blume Earthquake Engineering Center, Stanford University: Stanford, CA, USA, 2009.
57. Lignos, D.G.; Krawinkler, H. A steel database for component deterioration of tubular hollow square steel columns under varying axial load for collapse assessment of steel structures under earthquakes. In Proceedings of the 7th International Conf. on Urban Earthquake Engineering (7CUEE), Tokyo, Japan, 3–5 March 2010.
58. Lignos, D.G.; Laura Eads Krawinkler, H. Effect of Composite Action on the Dynamic Stability of Special Steel Moment Resisting Frames Designed in Seismic Regions. In Proceedings of the ASCE Structures Congress, Budapest, Hungary, 31 August–2 September 2011.
59. Lignos, D.G.; Krawinkler, H. Deterioration modeling of steel components in support of collapse prediction of steel moment frames under earthquake loading. *J. Struct. Eng.* **2011**, *137*, 1291–1302. [CrossRef]
60. Lignos, D.G.; Krawinkler, H.; Whittaker, A. Prediction and validation of sidesway collapse of two scale models of a 4-story steel moment frame. *J. Earthq. Eng. Struct.* **2011**, *40*, 807–825. [CrossRef]
61. Lignos, D.G.; Hikino, T.; Matsuoka, Y.; Nakashima, M. Collapse Assessment of Steel Moment Frames Based on E-Defense Full-Scale Shake Table Collapse Tests. *J. Struct. Eng.* **2013**, *139*, 120–132. [CrossRef]
62. Kanao, I.; Nakashima, M.; Takehara, S. Braced frame model considering buckling and fracture and its responses under near-fault strong motions. *J. Struct. Constr. Eng. (Trans. AIJ)* **2004**, *577*, 117–122. (In Japanese) [CrossRef] [PubMed]
63. Uang, C.M.; Yu, Q.S.; Gilton, C.S. Effects of loading history on cyclic performance of steel RBS moment connections. In Proceedings of the 12th World Conference on Earthquake Engineering, Auckland, New Zealand, 30 January–4 February 2000.
64. Architectural Institute of Japan (AIJ). *Ultimate Strength and Deformation Capacity of Building in Seismic Design (1990)*; Maruzen Publishing Co., Ltd.: Tokyo, Japan, 1990. (In Japanese)
65. Architectural Institute of Japan (AIJ). *Evaluation Procedure for Performance-Based Design of Buildings-Calculation of Response and Limit Strength, Energy Balance-Based Seismic Resistant Design, Time History Response Analysis*; AIJ: Tokyo, Japan, 2005. (In Japanese)
66. Shimazu, M.; Kimura, Y. Correlation between plastic deformation capacity and energy absorption capacity for H-shaped beams with local buckling. In *Summaries of Technical Papers of Annual Meeting Architectural Institute of Japan*; Japan Society for Finishings Technology: Tokyo, Japan, 2006; pp. 845–846. (In Japanese)
67. Shimazu, M.; Kimura, Y. Effects of flange width-to-thickness ratio and shear span ratio of plastic deformation capacity and energy absorption capacity for H-shaped beams. In *Summaries of Technical Papers of Annual Meeting Architectural Institute of Japan*; Japan Society for Finishings Technology: Tokyo, Japan, 2007; pp. 627–628. (In Japanese)
68. Qi, H.; Tada, M.; Tani, N.; Shirai, Y. Monotonic and hysteretic model for H-shaped beams incorporating deterioration behavior owing to the local buckling. *Thin-Walled Struct.* **2020**, *157*, 107016. [CrossRef]

69. Kimura, Y.; Yamanishi, T.; Kasai, K. Cyclic Hysteresis Behavior and Plastic Deformation Capacity for H-shaped Beams on Local Buckling under Compressive and Tensile Forces. *J. Struct. Constr. Eng. (Trans. AIJ)* **2013**, *78*, 1307–1316. (In Japanese) [CrossRef]
70. Kimura, Y.; Suzuki, A.; Kasai, K. Estimation of plastic deformation capacity for H-shaped beams on local buckling under compressive and tensile forces. *J. Struct. Constr. Eng. (Trans. AIJ)* **2016**, *81*, 2133–2142. (In Japanese) [CrossRef]
71. Suzuki, A.; Kimura, Y.; Kasai, K. Plastic deformation capacity of H-shaped beams collapsed with combined buckling under reversed axial forces. *J. Struct. Constr. Eng. (Trans. AIJ)* **2018**, *83*, 297–307. (In Japanese) [CrossRef]
72. Suzuki, A.; Kimura, Y.; Kasai, K. Rotation capacity of I-shaped beams under alternating axial forces based on buckling-mode transitions. *J. Struct. Eng.* **2020**, *146*, 04020089. (In Japanese) [CrossRef]
73. Kimura, Y.; Fujak, S.M.; Suzuki, A. Elastic local buckling strength of I-beam cantilevers subjected to bending moment and shear force based on flange-web interaction. *Thin-Walled Struct.* **2021**, *162*, 107633. [CrossRef]
74. Fujak, S.M.; Kimura, Y.; Suzuki, A. Estimation of elastoplastic local buckling capacities and novel classification of I-beams based on beam's elastic local buckling strength. *Structures* **2022**, *39*, 765–781. [CrossRef]
75. Fujak, S.M.; Suzuki, A.; Kimura, Y. Estimation of ultimate capacities of Moment-Resisting Frame's subassemblies with Mid-Storey Pin connection based on elastoplastic local buckling. *Structures* **2023**, *48*, 410–426. [CrossRef]
76. Suzuki, A.; Kimura, Y. Rotation capacity of I-shaped beam failed by local buckling in buckling-restrained braced frames with rigid beam-column connections. *J. Struct. Eng.* **2023**, *149*, 04022243. [CrossRef]
77. Suzuki, A.; Kimura, Y.; Matsuda, Y.; Kasai, K. Rotation capacity of I-shaped beams with concrete slab in buckling-restrained braced frames. *J. Struct. Eng.* **2024**, *150*, 04023204. [CrossRef]
78. Fujikawa, T.; Fujiwara, K. Research on strength and deformation capacity of H-shaped beam under cyclic loading. In *Proceedings of Architectural Institute of Japan Kinki Chapter Research Meeting*; AIJ: Tokyo, Japan, 1984; pp. 337–340. (In Japanese)
79. Fujiwara, K.; Kato, S. A study on deformation capacity of H-shaped steel beams with relative large ratio of web depth to thickness under cyclic loading. In *Proceedings of Architectural Institute of Japan Kinki Chapter Research Meeting*; AIJ: Tokyo, Japan, 1985; pp. 425–428. (In Japanese)
80. Sakai, J.; Matsui, C.; Kuno, T. Effects of the difference of the value of yield ratio of steel and the collapse-types of frame on the ductility of steel frame. In *Summaries of Technical Papers of Annual Meeting Architectural Institute of Japan*; Japan Society for Finishings Technology: Tokyo, Japan, 1990; pp. 1351–1352. (In Japanese)
81. Fujita, T.; Nakagomi, T.; Mizusaki, Y. The experimental research concern with structural behaviors of welded beam-to-column connections using cold-formed square steel tube column. In *Summaries of Technical Papers of Annual Meeting Architectural Institute of Japan*; Japan Society for Finishings Technology: Tokyo, Japan, 1992; pp. 1537–1538. (In Japanese)
82. Kai, S.; Sera, K.; Abukawa, T.; Yabe, Y.; Terada, G. Structural behavior of H-shaped beam-end connected to RHS-column. In *Summaries of Technical Papers of Annual Meeting Architectural Institute of Japan*; Japan Society for Finishings Technology: Tokyo, Japan, 1992; pp. 1541–1542. (In Japanese)
83. Makishi, T.; Yamamoto, N.; Tsutsui, S.; Fujisawa, K.; Uemori, H.; Ishii, T.; Morita, K. Experimental study on the deformation capacity of welded beam end connection without weld access hole. In *Summaries of Technical Papers of Annual Meeting Architectural Institute of Japan*; Japan Society for Finishings Technology: Tokyo, Japan, 1993; pp. 1257–1260. (In Japanese)
84. Ito, H.; Kimura, M.; Kaneko, H.; Yagi, T.; Ishii, H. Experimental study on beam-to-column welding connection without beam scallops. In *Summaries of Technical Papers of Annual Meeting Architectural Institute of Japan*; Japan Society for Finishings Technology: Tokyo, Japan, 1993; pp. 1261–1262. (In Japanese)
85. Suzuki, T.; Motoyui, S.; Fukazawa, T.; Uchikoshi, T. Study on effect of width-thickness ratio on deformation capacity of beams with scallops. *J. Struct. Constr. Eng. (Trans. AIJ)* **1996**, *486*, 107–114. (In Japanese) [CrossRef] [PubMed]
86. KMinami; Fujita, T.; Sasaki, Y. Experimental study on effect of non-scallops on deformation capacities of beam-to-column welded joints. In *Summaries of Technical Papers of Annual Meeting Architectural Institute of Japan*; Japan Society for Finishings Technology: Tokyo, Japan, 1996; pp. 631–632. (In Japanese)
87. Ishimaru, R.; Kagami, S.; Tanaka, A.; Masuda, K. Experimental study on the statical characteristics of the WBFW type beam-to-column connections. In *Summaries of Technical Papers of Annual Meeting Architectural Institute of Japan*; Japan Society for Finishings Technology: Tokyo, Japan, 1997; pp. 429–430. (In Japanese)
88. Sawamoto, Y.; Yoshida, H.; Kihiara, H.; Torii, S.; Tanaka, N. Experimental study on detail of H-shaped beam-to-box column connection. In *Summaries of Technical Papers of Annual Meeting Architectural Institute of Japan*; Japan Society for Finishings Technology: Tokyo, Japan, 1998; pp. 375–376. (In Japanese)
89. Yasuno, T.; Tsujioka, S.; Tada, M. Alternate loading test of haunch-beams made from rolled wide-flange shape. In *Summaries of Technical Papers of Annual Meeting Architectural Institute of Japan*; Japan Society for Finishings Technology: Tokyo, Japan, 2001; pp. 583–586. (In Japanese)
90. Kamaga, H.; Makino, Y.; Kuroba, K.; Tanaka, M.; Fukudome, Y.; Kobukuro, Y. Testing of beam-to-RHS column connections with-out weld access holes for field welding. In *Summaries of Technical Papers of Annual Meeting Architectural Institute of Japan*; Japan Society for Finishings Technology: Tokyo, Japan, 2001; pp. 919–922. (In Japanese)
91. Nakano, T.; Masuda, H.; Sasaji, T.; Tanaka, A. Experimental study on behavior of WBFW type beam-to-column connection. *J. Struct. Constr. Eng. (Trans. AIJ)* **2002**, *556*, 139–144. (In Japanese) [CrossRef] [PubMed]
92. Nakano, T.; Masuda, H.; Tanaka, A. Experimental study on effect of reinforcements at web connected parts of WF beam to SHS column connections. *J. Struct. Constr. Eng. (Trans. AIJ)* **2003**, *566*, 145–152. (In Japanese) [CrossRef] [PubMed]

93. Suzuki, T.; Ono, T.; Kanebako, Y. The local buckling and inelastic deformation capacity of steel beams under shear bending. *J. Struct. Constr. Eng. (Trans. AIJ)* **1977**, *260*, 91–98. (In Japanese)
94. Kato, T.; Akiyama, H.; Obi, Y. Experimental study of H-shaped beam with relatively large width-thickness ratio on local buckling characteristics. In *Summaries of Technical Papers of Annual Meeting Architectural Institute of Japan*; Japan Society for Finishings Technology: Tokyo, Japan, 1976; pp. 1075–1076. (In Japanese)
95. Yoda, K.; Imai, K.; Kuroba, K.; Ogawa, K.; Kimura, K.; Motoyui, S. Bending capacity of thin-walled welding H-section beams. *J. Struct. Constr. Eng. (Trans. AIJ)* **1989**, *397*, 60–72. (In Japanese) [CrossRef] [PubMed]
96. Yoshikawa, M.; Suzuki, T.; Ogawa, T.; Kimura, K.; Motoyui, S.; Matsuoka, T. Investigation on estimation of inelastic behavior of H-shaped beam with circular hole. In *Summaries of Technical Papers of Annual Meeting Architectural Institute of Japan*; Japan Society for Finishings Technology: Tokyo, Japan, 1993; pp. 1473–1474. (In Japanese)
97. Ikarashi, K.; Suzuki, T. A study on plastic deformation capacity of H-shaped stiffened beams with web opening. In *Summaries of Technical Papers of Annual Meeting Architectural Institute of Japan*; Japan Society for Finishings Technology: Tokyo, Japan, 2000; pp. 505–506. (In Japanese)
98. *JIS Z 2241*; Metallic Materials-Tensile Testing Method of Test at Room Temperature. Japan Industrial Standards (JIS): Tokyo, Japan, 2011.
99. *EN 1998-1*; Eurocode 8: Design of Structures for Earthquake Resistance—Part 1: General Rules, Seismic Actions and Rules for Buildings. European Committee for Standardization: Brussels, Belgium, 2004.
100. *EN 1998-3*; Eurocode 8: Design of Structures for Earthquake Resistance—Part 3: Assessment and Retrofitting of Buildings. European Committee for Standardization: Brussels, Belgium, 2005.
101. Kadono, A.; Sasaki, M.; Okamoto, K.; Akiyama, H.; Matsui, C.; Inoue, K. Experimental study on the effect of yield ratio on the bending strength increasing ratio and the ductility of steel structures' members. *J. Struct. Eng.* **1994**, *40B*, 673–682. (In Japanese)
102. Suzuki, T.; Ikarashi, K.; Satsukawa, K. Estimate for plastic deformation capacity of steel members with moment gradient reflected material properties. *J. Struct. Constr. Eng. (Trans. AIJ)* **2000**, *537*, 129–134. (In Japanese) [CrossRef]

 *buildings*

Article

Egg White and Eggshell Mortar Reinforcing a Masonry Stone Bridge: Experiments on Mortar and 3D Full-Scale Bridge Discrete Simulations

Murat Cavuslu * and Emrah Dagli

Department of Civil Engineering, Zonguldak Bülent Ecevit University, 67100 Zonguldak, Türkiye; emrahdagli@beun.edu.tr
* Correspondence: murat.cavusli@beun.edu.tr; Tel.: +90-543-8793515

Citation: Cavuslu, M.; Dagli, E. Egg White and Eggshell Mortar Reinforcing a Masonry Stone Bridge: Experiments on Mortar and 3D Full-Scale Bridge Discrete Simulations. *Buildings* **2024**, *14*, 1672. https://doi.org/10.3390/buildings14061672

Academic Editors: Atsushi Suzuki and Dinil Pushpalal

Received: 24 April 2024
Revised: 29 May 2024
Accepted: 31 May 2024
Published: 5 June 2024

Abstract: In this study, experimental and numerical investigations were conducted to examine the time-dependent creep and earthquake performance of the historical Plaka stone bridge, which was constructed in 1866 in Arta, Greece. During the original construction of the bridge in 1866, Khorasan mortar with an egg white additive was used between the stone elements. Furthermore, when the bridge underwent restoration in 2015, Khorasan mortar with an eggshell additive was employed between the stone elements. Consequently, two distinct 3D finite-difference models were developed for this study. In the first bridge model, egg white was used in the Khorasan mortar, replacing water at various proportions of 0%, 25%, 50%, 75%, and 100%. In contrast, for the second model, eggshell was incorporated into the Khorasan mixture at percentages of 25%, 50%, 75%, and 100%, relative to the lime amount. Subsequently, the mortars were subjected to curing periods of 1 day, 7 days, and 28 days, and their mechanical properties were determined through unconfined compression strength experiments. Taking into account the determined strengths of the mortars, the kn and ks stiffness values of the interface elements between the stone elements and Khorasan mortar were calculated. In the 3D model, each stone element was individually represented, resulting in a total of 1,849,274 stone elements being utilized. Non-reflecting boundary conditions were applied to the edge boundaries of the bridge model, and the Burger creep and Mohr–Coulomb material models was employed for time-dependent creep and seismic analyses, respectively. Subsequently, time-dependent creep analyses were conducted on the bridge, and seismic events that occurred in the region where the bridge was located were simulated to assess their impact. Based on the results of the time-dependent creep and seismic analyses, we observed that the use of 50% eggshell-mixed Khorasan mortar between the stone elements had a positive influence on the earthquake and creep behaviors of both restored and yet-to-be-restored historical bridges.

Keywords: eggshell; egg white; finite-difference method; interface element; Khorasan mortar; masonry bridge; seismic damage

1. Introduction

Historical structures play a crucial role in shedding light on the history of mankind. Preserving and transmitting these structures to future generations holds immense significance. Therefore, safeguarding historical buildings under significant external loads such as earthquakes and floods, and implementing earthquake retrofitting measures, are of vital importance. Historical structures were shaped according to the needs of their inhabitants, with each possessing unique architectural characteristics. One of the most important of these structures that shed light on our day is the historical stone arch bridges. These bridges were constructed in regions where communities relied on river transportation. They were frequently used by the local population during the Ottoman Empire and have managed to endure until the present day. Typically, these structures were built using stone elements that were compatible with the geological makeup of the construction area. The length of the stone elements was generally similar, and special mortars were applied between each stone to enable interaction. The Ottoman Empire utilized a specific type of mortar known as Khorasan mortar to facilitate interaction between the stone elements. This mortar, composed of special materials, provided substantial bonding between the stone elements.

Investigating the composition of this mortar and its utilization in the restoration and reinforcement of historical bridges today is of utmost importance. Recently, studies on historical buildings have increased. However, there is not much information and study in the literature about Khorasan mortar, which is used among the stone elements in historical buildings. Bouzas et al. [1] examined the impact of variations in the thickness dimensions of arches on the assessment of a masonry bridge's load-carrying capacity. The study began with an experimental campaign utilizing geomatic techniques to characterize the bridge's geometry. Subsequently, a limit analysis model was developed based on the outer dimensions of the arches. The study highlighted the importance of considering both the outer and inner dimensions of the arches in the probabilistic model, as relying solely on the outer dimensions can lead to an overestimation of the load capacity. Majtan et al. [2] examined the structural behavior of masonry arch bridges under extreme flood flows and impact forces from flood-borne debris. The study employed a validated numerical modeling approach, utilizing the smoothed particle hydrodynamics (SPH) method to simulate fluid behavior and determine pressure distributions on a single-span arch bridge. The findings demonstrated that debris impact significantly increased stress levels, particularly when the abutment was fully submerged and the debris had a side-on orientation. Silva et al. [3] introduced two nonlinear finite element modeling approaches for the assessment of damaged masonry arch bridges. These strategies were applied to the Leça railway bridge, serving as a case study, considering different damage scenarios and railway traffic loading. The proposed numerical strategies demonstrated potential in representing localized damage, particularly longitudinal cracks in various opening conditions, in masonry bridges. Furthermore, these calibrated numerical strategies proved to be computationally efficient for nonlinear dynamic analysis under service loading. Ozakgul et al. [4] focused on assessing the structural behavior of a 94-year-old reinforced concrete open-spandrel arch bridge. Dynamic tests, including acceleration measurements, were performed to understand the bridge's in situ behavior. A 3D finite element model was created based on the original constructional drawings and updated using the Response Surface Method. Structural assessment and evaluation were conducted under UIC-71 live loading, obtaining rating factors and reliability indices for each bridge member using the Load and Resistance Factor Rating approach. The results indicated that the deck of the bridge was the most critical member, suggesting potential strengthening and retrofitting requirements for higher live loads. Chen et al. [5] determined the damage on a historical bridge using different methods. Sonar techniques were used to detect underwater foundation damage in Gongchen Bridge. The results showed that NDT technology improved work efficiency and accurately identified the damage situation of the foundation. Then, it was recommended to reinforce the foundation, seal exposed wooden piles, and remove underwater obstacles for the bridge's stability. Gönen and Soyöz [6] performed a seismic assessment of masonry arch bridges and highlighted the importance of considering their nonlinear behavior. Initial testing was conducted to gather information on the bridge's geometry and material properties, and dynamic identification using ambient vibrations was performed to enhance the accuracy of the finite element model. Nonlinear static (NSA), nonlinear dynamic (NDA), and incremental dynamic analyses (IDA) were carried out to assess the bridge's seismic response. NSA provided conservative displacement results but fell short in capturing the behavior and damage mechanisms associated with higher vibration modes. Moreover, NDA yielded more reliable results but required significant computational resources. Saygılı and Lemos [7] investigated the seismic behavior of two historical stone masonry bridges in Turkey using finite- and discrete-element approaches. The Kazan Bridge exhibited the capacity to withstand seismic excitations, although some damage was predicted due to excessive stresses. Moderate damages, such as slight openings, were observed at the mid-span of the arch. On the other hand, the Şenyuva Bridge was found to be significantly vulnerable and prone to strong damage or collapse during earthquakes with a long return period. It was seen that bridge geometry, including arch span and deck width, influenced the seismic response, with longer spans increasing vulnerability. Papa et al. [8] investigated the arch-backfill interaction problem in masonry bridges using an adaptive NURBS-based limit analysis approach. Comparisons with experimental results and other analysis approaches demonstrated the significance of direct modeling in accurately representing the failure response of the bridge. Zani et al. [9] investigated the response of the Azzone Visconti bridge, a historical masonry arch bridge in northern Italy, to vertical static loads. A detailed 3D nonlinear finite element model was created to assess the effect of soil–structure interaction (SSI) on the bridge's mechanical behavior. The results showed that the model with SSI and soil nonlinearity accurately reproduces the bridge's vertical settlements under load tests. Chen et al. [10] examined the weathering of stone ashlars used in the construction of Guyue Bridge in Yiwu, Zhejiang. It was concluded that stone weathering in historic structures is primarily influenced by atmospheric conditions and internal stress caused by loading. Sánchez-Aparicio et al. [11] proposed a non-destructive multidisciplinary approach for the structural diagnosis of masonry arch bridges. The proposed methodology was validated on a case study of a Roman bridge in Spain, showing promising results. Simos et al. [12] focused on earthquake-induced damage on a stone arch bridge (Konitsa Bridge) in an earthquake-prone zone. The bridge, located on an active fault, survived a near-field earthquake in 1996 with no damage. Field studies and laboratory tests were conducted to establish the bridge's dynamic characteristics and material properties. The study compared the damage caused by near-field and far-field earthquakes and far-field earthquakes were found to be more destructive. Aydin and Özkaya [13] focused on non-linear analyses of arched structures, examining their behavior and the occurrence of cracks and fractures under different loads. It was seen that the load distribution before collapse exhibited a decreasing tendency near the $L/2$ section, while variations after the $L/4$ section towards the left and right ends differed, albeit slightly, from the filler and side walls. Conde et al. [14] proposed a multidisciplinary approach to analyze masonry arch bridges, with the Vilanova Bridge in Allariz, Spain, serving as a case study. The results indicated that the cohesion of fill materials and the non-linear properties of masonry significantly influenced the bridge's structural behavior. Based on numerous numerical simulations, the Vilanova Bridge demonstrated an acceptable safety level, but lower service loads were recommended for preserving the structure's life expectancy. Karaton et al. [15] investigated the nonlinear seismic responses of the Malabadi Bridge, an Artuqid structure constructed in 1147 on the Batman River in Turkey. Seismic acceleration data were generated for three different levels (D1, D2, and D3) based on the seismic characteristics of the region where the bridge is located. Under D1 and D2 earthquake loadings, no damage was observed in the arches and spandrel walls of the Malabadi Bridge. However, plastic deformations were observed in the backfill material under D2 earthquake loadings. Conde et al. [16] focused on the structural analysis of a masonry arch bridge situated in Galicia, Spain. The bridge's geometric characterization was conducted using terrestrial laser scanning, providing an accurate and detailed description of the arches. The results indicated that without precise data, caution must be exercised when making idealizations about the geometry of stone arches for numerical calculations, as it often leads to an overestimation of the predicted collapse load. Costa et al. [17] presented an overview of numerical strategies employed to assess the load-carrying capacity of stone masonry arch bridges. Detailed numerical models were developed using the finite element method (FEM), discrete element method (DEM), and rigid block analysis. The results demonstrated a strong correlation between the numerical response of the bridge models and the maximum load, as well as the corresponding hinge mechanism. Bayraktar et al. [18] focused on the operational modal analysis of eight historical masonry arch bridges in Turkey, namely the Aspendos, Pehlivanlı, Mikron, Osmanlı, Şenyuva, Şahruh, Osmanbaba, and Torul bridges. Ambient vibration data were utilized to determine the experimental frequencies, damping ratios, and mode shapes of these bridges. It was seen that the first and second mode shapes of the bridges exhibit bending behavior, while the third mode shape predominantly shows a vertical form. Moreover, the first, second, and third frequencies range between 4–8 Hz, 6–10 Hz, and 8–12 Hz, respectively. Ozmen and Sayin [19] observed the effect of soil structure interaction (SSI) on the seismic behavior of masonry bridges and found significant differences when comparing the model of fixed base and SSI. Shabani and Kioumarsi [20] investigated the effect of three different strengthening models on the bridge. Two methods include using polyparaphenylene benzobisoxazole and carbon fiber-reinforced concrete mortar layers to cover the pier of the bridge and the other one is improving all mechanical properties of indefensible part of the bridges. Effects of the usage of different additives (egg white and eggshell) on Khorasan mortar were not examined in those studies [19,20]. Vuoto et al. [21] explored digital twin concepts for the conservation of cultural heritage buildings. Their study highlights the transformative potential of the digital twin paradigm in preserving Built Cultural Heritage (BCH) assets. Addressing the fragmented nature of current conservation efforts, the utilization of digital twin technology offers a means to enhance management processes through precise asset modeling and advanced analytics. Furthermore, the study emphasizes the importance of collaborative research and inter-disciplinary cooperation to fully realize the benefits of this paradigm. Supported by EU and global policies advocating for sustainable BCH utilization, digital twin technology emerges as a crucial tool in safeguarding cultural heritage for future generations. Ferretti and Pascale [22] analyzed the strengthening technics for masonry buildings. The CAM system improved the seismic performance of masonry buildings. Hafner et al. [23] compared the methods for strengthening masonry walls and piers. Angiolilli et al. [24] utilized a reinforced cementitious matrix for strengthening brick and stone masonry walls, reporting a remarkable 420% increase in shear strength. Ferretti et al. [25] examined the combined effect of fiber-reinforced cementitious matrix composite reinforced mortar on structural strength, yielding an average increase of 53.61% in strength. Triantafillou [26] conducted analyses on the impact of textile-reinforced mortar, demonstrating its ability to enhance the strength of unreinforced masonry walls. Kouris and Triantafillou [27] proposed a design methodology for strengthening textile-reinforced mortar, validating it through experiments and establishing a good correlation for masonry structures. Carozzi and Poggi [28] employed fabric-reinforced cementitious matrix to strengthen masonry struc-tures, with polyparaphenylene benzobisoxazole fiber exhibiting the highest strength and increase in performance among the fibers studied. Grande and Milani [29] investigated the fiber-reinforced cementitious matrix strengthening system's impact on the tensile strength of masonry, proposing a model and validating it through experimental tests. Lemos [30] explored the discrete element method on various masonry structures, affirming its suit-ability for determining the dynamic behavior of masonry monuments. Gobbin et al. [31] introduced new algorithms to define the seismic behavior of masonry structures using the discrete element method, finding the rigid body method suitable for simple problems. Masi et al. [32] developed a model for testing masonry structures through experiments, with the discrete element model proving successful in modeling the dynamic response of the structures.

As can be seen from the studies in the literature, there is no study on the use of Khorasan mortar with egg white and eggshell to examine the structural damage behavior of historical buildings. For this reason, this study adds new information to the literature about the structural behavior of historical bridges and the use of Khorasan mortar in historical bridges.

2. General Information about the Content of the Study

The name "Khorasan" is known as a district located in Iran. It is also used instead of concrete in Saudi Arabia. In this study, we evaluated the effects of using Khorasan mortar with different additives as an interface element between the discrete stone elements in his-torical bridges on time-dependent creep and seismic behavior. The bridge was constructed in 1866 and later destroyed during the 2015 flood. During its original construction in 1866, Khorasan mortar with egg white was employed to ensure the interaction between the stone elements. Therefore, in the initial phase of the study, Khorasan mortars with egg white additives were prepared in the laboratory, and the strengths of the 1-, 7-, and 28-day Khorasan mortars were comprehensively assessed. The mechanical properties obtained from these experiments were used to calculate the properties of the interface elements between the discrete stone elements in the 3D finite-difference model of the bridge. The Burger creep material model was applied to the stone elements and foundation while modeling the bridge, and non-reflecting boundary conditions were defined for the bridge's boundaries. Each stone element of the bridge was individually modeled. Since the bridge experienced a total of 5 flood disasters, the water level was taken into account, matching the bridge's height during these flood events. Furthermore, excluding floods, the water level was considered to be 1/6 of the bridge's height. Taking these water levels into consideration, it was investigated the time-dependent creep behavior of the bridge from 1866 to 2015. As a result of the creep analyses, significant failures were observed around the auxiliary arch parts of the bridge. In the subsequent phase of the study, it was examined the long-term creep and seismic behaviors of the bridge, which was reconstructed following its collapse in the 2015 flood, after using Khorasan mortar with eggshell additives between the stone elements. Khorasan mortars with eggshell additives were prepared in the laboratory, and 1, 7, and 28-day Khorasan mortars were subjected to unconfined compression strength tests. The experiments allowed us to determine the mechanical properties of the eggshell-added Khorasan mortar, which were then employed for the mechanical properties of the interface elements, namely kn and ks stiffness elements, between the discrete stone elements in the bridge model. Subsequently, the bridge model was subjected to the 2023 Kahramanmaraş earthquakes. The seismic analyses revealed that there was no significant seismic damage observed during the earthquake in the historical bridges using Khorasan mortar with eggshell additives between discrete stone elements. Additionally, it was concluded that the use of Khorasan mortar with eggshell additives between stone elements during the reconstruction or restoration of historical bridges provides significant positive contributions to the seismic behavior of the bridge.

3. Burger Creep Material Model

The Burger creep model is a viscoelastic material model that combines elastic and viscous behavior to simulate the time-dependent deformation of materials. It is based on the assumption that materials exhibit both instantaneous elastic strains and time-dependent creep strains when subjected to stress. The model assumes non-linear elasticity, meaning that the stress–strain relationship follows Hooke's Law [33]. Burger's model consists of a series connection of a Kelvin model and a Maxwell model.

According to Hooke's Law, the stress (σ) in a material is proportional to the strain (ε) it experiences, with the proportionality constant being the elastic modulus (E) of the material. The elastic modulus represents the stiffness of the material and determines how it deforms under load. In addition to the instantaneous elastic strain, the Burger creep model incorporates creep, which refers to the time-dependent deformation that occurs in materials under constant stress [33]. The creep strain in the model is described by a creep law, which relates the applied stress to the resulting creep strain over time. The creep behavior is typically characterized by a creep compliance function, which defines the relationship between stress and strain rate. The creep compliance function can take different forms depending on the specific material and its behavior. It can be defined using experimental data or mathematical models that represent the creep behavior of the material accurately. When using the Burger creep model in FLAC3D, the material properties such as Young's modulus (E) and Poisson's ratio (ν) are specified to define the elastic behavior of the material [33]. These properties determine the material's response to instantaneous loading. To capture the time-dependent creep behavior, the parameters provided are related to the creep compliance function. These parameters can include the initial creep strain, the characteristic time constant, and the shape of the creep compliance function. These values are typically determined through laboratory tests or obtained from material property databases. The Burger creep model in FLAC3D allows researchers to simulate the long-term behavior of materials subjected to sustained loading or changing stress conditions over time. It is particularly useful in geotechnical engineering applications, such as analyzing the settlement of soil under a structure or the behavior of rock masses in underground excavations. It is important to note that the specific details and implementation of the Burger creep model may vary depending on the version of FLAC3D and the specific requirements of the analysis to be performed [33]. The Burger creep material model is vital for analyzing historical structures because it accurately simulates how these structures deform over time due to various loads. It helps engineers predict long-term performance issues and plan maintenance accordingly, ensuring the preservation of these valuable assets. This model also assists in assessing risks related to time-dependent effects, enabling the development of strategies to safeguard historical structures.

4. Free-Field and Quiet Non-Reflecting Boundary Conditions

In FLAC3D, non-reflective boundary conditions are employed to model boundaries that allow waves or stresses passing through them to be fully absorbed without reflecting into the model domain. These boundary conditions are particularly crucial when dealing with semi-infinite or unbounded domains, where minimizing the influence of boundary reflections on the results is essential. Non-reflective boundary conditions are commonly used in wave propagation analyses and other dynamic simulations. They are implemented as absorbing boundaries, effectively dissipating incident waves to prevent reflections back into the model domain [33]. These boundaries are especially valuable in modeling wave propagation phenomena, such as seismic waves or acoustic waves, as they ensure a more accurate representation of wave behavior within the model. Two significant non-reflective boundary conditions in FLAC3D are the free-field (FF) boundary condition and the quiet boundary condition. The free-field boundary condition is a fundamental feature in FLAC3D, a powerful numerical modeling tool designed for simulating geo-mechanical problems in three-dimensional space. It represents the behavior of the model's exterior domain, assuming that the displacements and stresses exerted by the surrounding free-field are negligible. In this approach, the analysis domain is represented with a finite element mesh, and the free-field boundary condition defines the behavior beyond this mesh, assuming uniform and known material properties and displacements [33]. This assumption holds when the model's boundaries are positioned far enough from the area of interest, ensuring minimal influence on the region of interest. Users specify the external boundaries where the free-field boundary condition is imposed, typically selecting specific nodes or elements at the outer limits of the mesh and designating them as free-field boundary nodes/elements. Material properties of these nodes/elements are set to represent the properties of the surrounding free-field medium, often with zero displacements (both translational and rotational) to reflect the assumption of negligible external influence [33]. The lateral boundaries of the main grid are connected to the free-field grid through the use of viscous dashpots, creating a simulation of a quiet boundary (Figure 1). The unbalanced forces originating from the free-field grid are then imposed on the main-grid boundary. These conditions are mathematically described by the following three equations, which apply to the free-field boundary along one side of the boundary plane, with its normal aligned in the x-axis direction. Similar formulations can be derived for the other side and for the corner boundaries (Equations (1)–(3)) [33]. The density of the material along the vertical model boundary is denoted by the symbol ρ. The P-wave speed at the side boundary is represented by the variable C_p. The S-wave speed at the side boundary is given by the parameter C_s. The area of influence of the free-field grid-point is defined as A. The velocity of the grid-point in the main grid at the side boundary in the x-direction is designated as V_x^m. V_y^m signifies the velocity of the grid-point in the main grid at the side boundary in the y-direction. V_z^m corresponds to the velocity of the grid-point in the main grid at the side boundary in the z-direction. V_x^{ff} stands for the velocity of the grid-point in the side free field in the x-direction. Similarly, V_y^{ff} represents the velocity of the gridpoint in the side free field in the y-direction. Lastly, V_z^{ff} denotes the velocity of the gridpoint in the side free field in the z-direction. The free-field grid-point force with contributions from the stresses of the free-field zones around the grid-point in the x-direction is expressed by the variable F_x^{ff}. Likewise, the free-field grid-point force with contributions from the stresses of the free-field zones around the grid-point in the y-direction is denoted by F_y^{ff}. Lastly, F_z^{ff} stands for the free-field grid-point force with contributions from the stresses of the free-field zones around the grid-point in the z-direction [33].

$$F_x = -\rho C_p \left(V_x^m - V_x^{ff} \right) A + F_x^{ff} \tag{1}$$

$$F_y = -\rho C_s \left(V_y^m - V_y^{ff} \right) A + F_y^{ff} \tag{2}$$

$$F_z = -\rho C_s \left(V_z^m - V_z^{ff} \right) A + F_z^{ff} \tag{3}$$

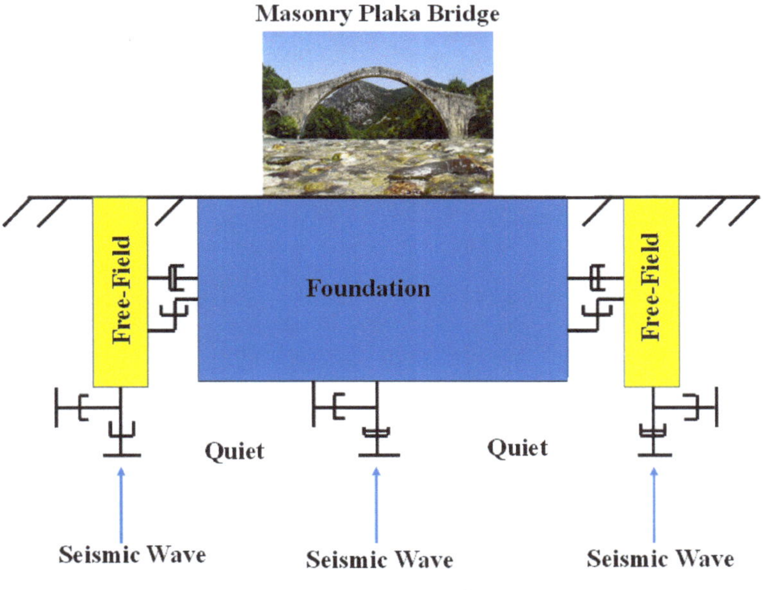

Figure 1. View of non-reflective boundary conditions [33].

The quiet boundary condition, on the other hand, is a specialized feature in FLAC3D, designed to simulate boundaries where waves and stresses passing through are entirely absorbed without reflection. It is particularly useful when dealing with semi-infinite or unbounded domains, ensuring that unwanted boundary reflections do not affect the accuracy of the simulation results. Like the free-field boundary condition, it is strategically applied to specific regions on the model's exterior boundaries, absorbing waves, stress waves, and displacements entirely, to simulate an infinitely distant boundary [33]. Its implementation is crucial for accurately simulating wave propagation and dynamic problems, effectively minimizing artificial reflections that could compromise the accuracy and stability of the simulation. Moreover, the quiet boundary condition excels in accurately analyzing wave propagation, making it particularly valuable in simulations involving seismic waves or similar dynamic phenomena, where precise representation without boundary reflections is essential [33]. Additionally, eliminating reflections significantly contributes to maintaining numerical stability in dynamic simulations, reducing artificial oscillations, and ensuring more reliable and consistent results. A notable advantage of the quiet boundary condition is its ability to simulate unbounded or semi-infinite domains without the need to explicitly model the entire surrounding space, resulting in substantial computational resource savings [33].

5. General Information about Plaka Masonry Bridge

The Plaka Stone Bridge, located in the city of Arta, Greece, is an iconic historical structure with significant cultural and architectural value. Originally constructed during the Ottoman Empire in 1866, this bridge served as a vital transportation link for the local community, facilitating river crossings and trade routes. Over the years, the Plaka Stone Bridge faced numerous challenges, including natural disasters and the test of time. In 2015, a devastating flood caused the collapse of the bridge, necessitating extensive reconstruction and restoration efforts. From the mechanical point of view, the size effect is the reason why the masonry bridge collapses or is heavily damaged during floods [34]. Recognizing the importance of preserving this remarkable piece of history, a comprehensive renovation project was initiated to revive the bridge and ensure its continued existence for future generations. The reconstruction process involved meticulous planning, advanced engineering techniques, and the use of traditional construction methods to retain the bridge's original character and architectural integrity. Skilled craftsmen and engineers collaborated to painstakingly replicate the intricate stonework and structural elements that defined the bridge's unique charm. In addition to physical restoration, comprehensive studies and assessments were conducted to enhance the bridge's resilience and ability to withstand external forces, such as earthquakes and heavy rainfall [35]. The renovated Plaka Stone Bridge stands as a testament to the harmonious blend of historical preservation and modern engineering. It not only re-establishes the vital transportation link it once provided but also symbolizes the cultural heritage and resilience of the local community. The restored bridge serves as a captivating landmark, attracting visitors from around the world, while also honoring the legacy of those who originally built and traversed this remarkable structure. In conclusion, the Plaka Stone Bridge in Arta, Greece, stands as a remarkable example of historical architecture and engineering prowess. The meticulous reconstruction and incorporation of innovative materials ensure its longevity and continued significance as a cherished historical monument. The mechanical properties of the stone elements of the bridge are shown in Table 1. Szabo et al. [36] delineated six distinct patterns for masonry buildings, and the Plaka bridge was characterized as Type D (regular masonry from soft stones) in this study. Furthermore, Szabo et al. [36] classified typologies into two groups: Class 1 and Class 2, with the subject bridge falling under Class 2, denoting "good quality masonry". Additionally, the construction technique of the Plaka Bridge was meticulously detailed in the study conducted by Giannelos et al. [37]. The examination revealed that the central arch of the bridge, initially perceived as semi-circular, exhibits a more intricate geometry upon closer inspection. A topographic survey unveiled a dual curvature within the arch structure: from its base to a $60°$ angle, it conforms to an incircle with a radius (R1) of 20.05 m, while transitioning to a smaller incircle with a radius (R2) of 17.41 m towards its apex [37]. The center of the smaller incircle is situated 3.18 m above that of the larger one, resulting in a 0.53-meter elevation of the apex. This curvature variation significantly influences the distribution of structural stress, as elaborated in Section 6. Furthermore, the examination of collapsed fragments and failure surfaces facilitated the identification of construction typology across various bridge segments [37]. The western and eastern approaches feature solid masonry, with approximately 0.40-meter-thick spandrel walls filled with rubble stone masonry and mortar. Conversely, the central $60°$ segment comprises exclusively arches, while the area between the solid approaches and the central arch segment employs a distinct filling material: tufa stones combined with mortar. This deliberate selection aims to reduce the bridge's self-weight and alleviate lateral pressure on the spandrel walls [37]. Furthermore, Khorasan mortar has been employed amidst the discrete stones of the Plaka Bridge. The term "Khorasan" was initially coined during the Sassanid era, signifying "the land of the sun" [38]. Throughout the periods of the Ottoman and Roman Empires, mortars composed of tile, lime, and brick were referred to as Khorasan mortar [39]. This study extensively investigated Khorasan mortar and its effects on the seismic behavior of the bridge's mortar.

Table 1. Material properties of the stone.

Item	Density (kg/m^3)	Modulus of Elasticity (Pa)	Poisson Ratio	Cohesion (kPa)	Friction Angle (Degree)	Dilatancy Angle (Degree)
The Lowest Main Arch	2628	7.05×10^7	0.28	3970	47.1	46.2
The Middle Main Arch	2583	6.52×10^7	0.26	3842	47.0	46.1
Supporter Arch (Right)	2280	5.86×10^7	0.21	3370	47.5	47.3
Supporter Arch (Left)	2152	5.24×10^7	0.19	2975	42.6	41.6
Rock Material	2300	4.75×10^7	0.25	1750	36.5	36.2
Foundation	3120	9.00×10^7	0.38	4817	49.6	48.5

6. Experimental Background

In this section of the study, we present detailed information about the experiments and test results. Khorasan mortar was prepared by using a mixture of brick powder, hydrated lime, standard sand, eggshell, and egg white to examine the structural failure behavior of historical bridges with more realistic data. The utilization of egg white and eggshell in the production of Khorasan mortar during the Ottoman and Roman Empires was widespread. However, there are no published works documenting the use of eggshell and egg white specifically for producing Khorasan mortar in bridge construction. It is worth noting that the cementing effect of eggshell and egg white surpasses that of fiber, which is why they were chosen for this study. The egg white used in the study was sourced from commercially purchased white eggs. The standard sand was prepared following the TS EN 196-1 [40] standard. Hydrated lime and brick were obtained from a private company. Figure 2 illustrates all the materials used in the experiments. To achieve the maximum Khorasan mortar mixture, egg white and eggshell were used separately as additives, in varying proportions. Nine different mixtures were formulated, including a control sample without any egg white or eggshell. The mixtures containing egg white had ratios of 4:4:3:9 for "water + egg white", lime, standard sand, and brick powder, respectively. Mixtures prepared with eggshell had the same ratio (4:4:3:9) of materials; only "water + egg white" was replaced by eggshell. While the proportions of water and egg white were altered in these mixtures, the other components remained constant. Specifically, the ratios of "water + egg white" were selected as 0%, 25%, 50%, 75%, and 100%, in the mixtures. The total combined weight of egg white and water was set at 36 g, and the percentage of "water + egg white" in each mixture was determined. A 0% egg white additive ratio means that there is 0 g of egg white and 36 g of water, while a 100% additive is defined as 0 g of water and 36 g of egg white in the mixture. Eggshell additive ratios were used as 25%, 50%, 75%, and 100% of the lime amount. The amount of eggshell was increased while keeping the amounts of water and other materials constant for mixtures involving eggshell. Additionally, the effect of curing time was investigated using different periods (1, 7, and 28 days). The prepared mixtures were assigned codes for easy identification. For instance, AC1 denoted the control sample, C represented curing, and 1 indicated a curing period of 1 day. Similarly, EW25W75C7 stood for a mixture containing egg white (EW) with a 25% percentage, water

(W) with a 75% percentage, and subjected to curing (C) for 7 days. Furthermore, mixture ES25C1 was labeled as such, with ES referring to eggshell, 25 representing the eggshell percentage, C indicating curing, and 1 representing the curing time. Table 2 provides comprehensive details regarding the composition of each mixture.

Figure 2. Materials used for Khorasan mortar preparation.

Table 2. Mixture ratios and curing time.

Mixture Code	Water (g)	Egg White (g)	Eggshell (g)	Lime (g)	Standard Sand (g)	Brick Powder (g)	Curing Time (Day)
AC1	36	0	0	36	27	81	1
AC7	36	0	0	36	27	81	7
AC28	36	0	0	36	27	81	28
EW25W75C1	27	9	0	36	27	81	1
EW25W75C7	27	9	0	36	27	81	7
EW25W75C28	27	9	0	36	27	81	28
EW50W50C1	18	18	0	36	27	81	1
EW50W50C7	18	18	0	36	27	81	7
EW50W50C28	18	18	0	36	27	81	28
EW75W25C1	9	27	0	36	27	81	1
EW75W25C7	9	27	0	36	27	81	7
EW75W25C28	9	27	0	36	27	81	28
EW100W0C1	0	27	0	36	27	81	1
EW100W0C7	0	36	0	36	27	81	7
EW100W0C28	0	36	0	36	27	81	28
ES25C1	36	0	9	36	27	81	1
ES25C7	36	0	9	36	27	81	7
ES25C28	36	0	9	36	27	81	28
ES50C1	36	0	18	36	27	81	1
ES50C7	36	0	18	36	27	81	7
ES50C28	36	0	18	36	27	81	7
ES75C1	36	0	27	36	27	81	1
ES75C7	36	0	27	36	27	81	7
ES75C28	36	0	27	36	27	81	28
ES100C1	36	0	36	36	27	81	1
ES100C7	36	0	36	36	27	81	7
ES100C28	36	0	36	36	27	81	28

6.1. Preparation of Mortar Samples

Before the mixtures were prepared, the brick ballast was pulverized through 1900 cycles in the Los Angeles abrasion device to obtain brick powder, as illustrated in Figure 3.

The Los Angeles abrasion instrument is constructed from steel and has a spacious interior for sample placement. It uses balls to break down the material into finer dimensions. Each full turn of the device is considered one cycle during the experiment. Subsequently, the egg white was vigorously whisked in a porcelain bowl. All the materials were gathered in a container and thoroughly mixed until a uniform blend was achieved. The resulting mixture was then compacted using the Harvard mini compactor mold to create a cylindrical test sample, as shown in Figure 4. The compaction process consisted of 5 layers, with each layer subjected to 10 strokes. An equal amount of the sample was evenly distributed on each layer. Afterward, the samples were carefully removed from the mold using a sample extractor. The diameters and heights of the extracted samples were measured using a caliper, resulting in mean diameter and height values of 33 mm and 71 mm, respectively. To ensure proper preservation, the measured samples were meticulously wrapped with stretch film. Following this, identifying labels were attached, and the samples were placed inside a desiccator, as depicted in Figure 5. The desiccators were stored in a dark and cool environment, away from direct sunlight. The primary goal of using desiccators was to prevent any loss of water content from the samples. Consequently, the samples were kept within the desiccator for the designated curing periods (1, 7, and 28 days).

Figure 3. Powdering of brick with a Los Angeles abrasion test device.

Figure 4. Specimen preparation with Harvard miniature compaction equipment.

Figure 5. Wrapping the sample with stretch film and placing it in the desiccator.

6.2. Test Setup

The experiments were conducted following the ASTM D2166/D2166M-16 [41] standard. The prepared samples for the test were placed on the bottom table of the automatic triaxial strength test device, as shown in Figure 6. Subsequently, relevant data concerning the sample, such as the mixing code, sample diameter, and height, were input into the program. The test specimens were set at a height of 71 mm, and the test speed was applied at a 10% strain rate (0.71 mm/min) over 10 min. The data recording rate was selected at 1 data point per second during the test. Before starting the experiment, the lower platform was raised to ensure that the sample was adequately loaded. Once the sample made contact with the upper platform and the load value exhibited a change, the load and deformation values were reset in the program, and the test was initiated. Throughout the experiments, real-time monitoring was conducted for both the graphical representation and the physical sample. Upon completion of the experiment, the stress–strain graph was generated. In this graphical representation, the maximum stress value provides the unconfined compression strength of the sample, while half of this value corresponds to the cohesion of the sample.

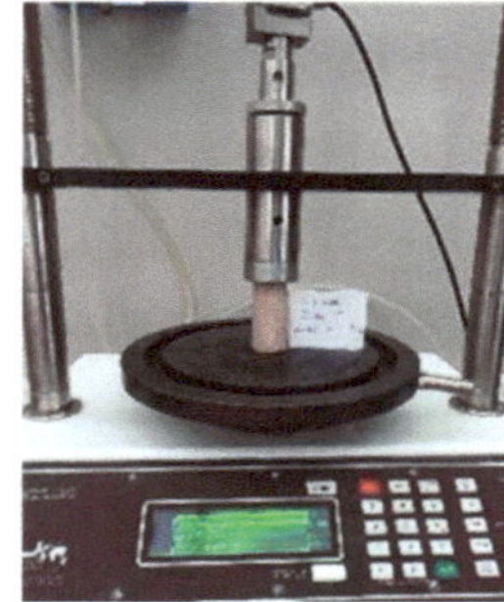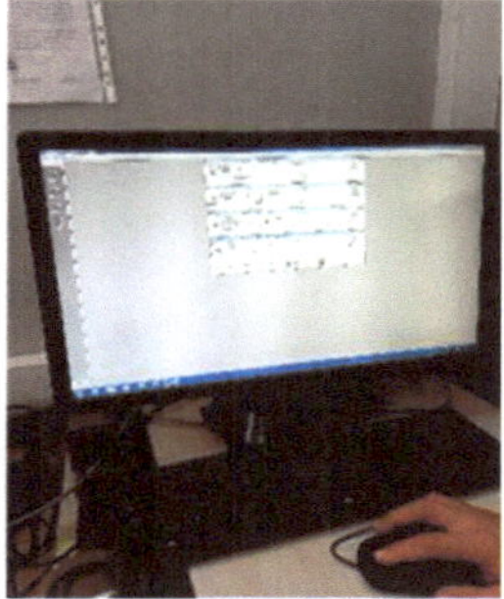

Figure 6. Triaxial strength test device and placing the sample on the device.

7. Experimental Results

7.1. UCS Results for Control Mortars

In this section of the study, the results of the unconfined compression strength test of the control sample (mortar) are presented in detail. Stress–strain graphs for the 1-day cured control sample (AC1) are provided in Figure 7a. According to the test results of the AC1 sample, the actual fracture pattern indicated a brittle fracture. When examining the stress (σ)–axial strain (ε) graph obtained from the program, it can be observed that the unconfined compression strength (UCS) of the AC1 sample was 143.22 kPa, and the sample reached the maximum stress value when the strain reached 3.03%. The stress–strain graph of the AC7 sample is presented in Figure 7b. The actual fracture pattern indicated ductile fracture, unlike the AC1 sample. The unconfined compression strength of the AC7 sample was 210.24 kPa. The axial strain value obtained at maximum stress was 2.95% and very close to the ε value of the AC1 sample. The unconfined compression strength value of the AC7 sample was 46.80% larger than AC1, demonstrating the strength-enhancing effect of the lime during the 7-day reaction period. The σ–ε graph for the AC28 sample is provided in Figure 7c. The fracture pattern was similar to AC7 but different from the AC1 sample. The unconfined compression strength of the sample was 338.79 kPa. The ε value obtained at this strength was 4.22%, slightly larger than the ε values of the AC1 and AC7 samples. The unconfined compression strength of the AC28 sample increased by 136.55% compared to the AC1, indicating that the chemical bond between lime and other materials strengthened as a result of the 28-day curing period.

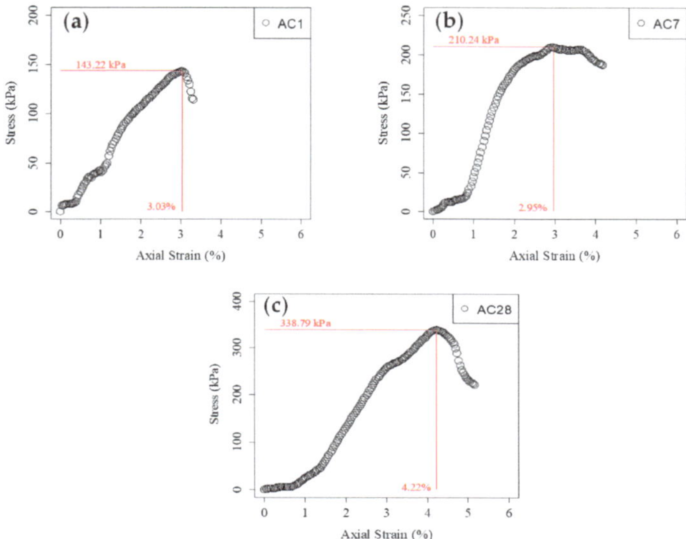

Figure 7. Stress–strain curves for control sample: (**a**) 1 day of curing (**b**) 7 days of curing (**c**) 28 days of curing.

7.2. UCS Results for Mortars Containing Egg White and Eggshell Stabilized Samples

Unconfined compression strength results of the egg white and eggshell stabilized samples are presented in this section. The effect of the test on the failure type of the samples and the σ–ε graphs of stabilized samples for different curing times (1 day, 7 days, and 28 days) are provided. EW25W75C1 showed the worst strength performance compared to all mixtures for all curing times even though its strength was 28.58% larger than the AC1, as shown in Figure 8a. Axial strain for reaching maximum stress (unconfined compression strength) for egg white-stabilized samples cured for 1 day tended to decrease when egg white in the mixture increased, as shown in Figure 8c,e,g. The maximum value for unconfined compression strength (UCS) of the egg white mixture stabilized and cured for 1 day was EW50W50C1 and the strength of it was 195.76% higher than the AC1. The eggshell stabilized mixtures show a similar stress–strain behavior to that of the egg white stabilized mixtures, for a curing time of 1 day, according to Figure 8b,d,f,h. Although the UCS value of ES25C1 is less than other eggshell stabilized mixtures, it is still 95.93% higher than the AC1. A maximum increase in unconfined compression strength was obtained at ES50C1 when compared to all mixtures cured for 1 day, and it is 282.62%.

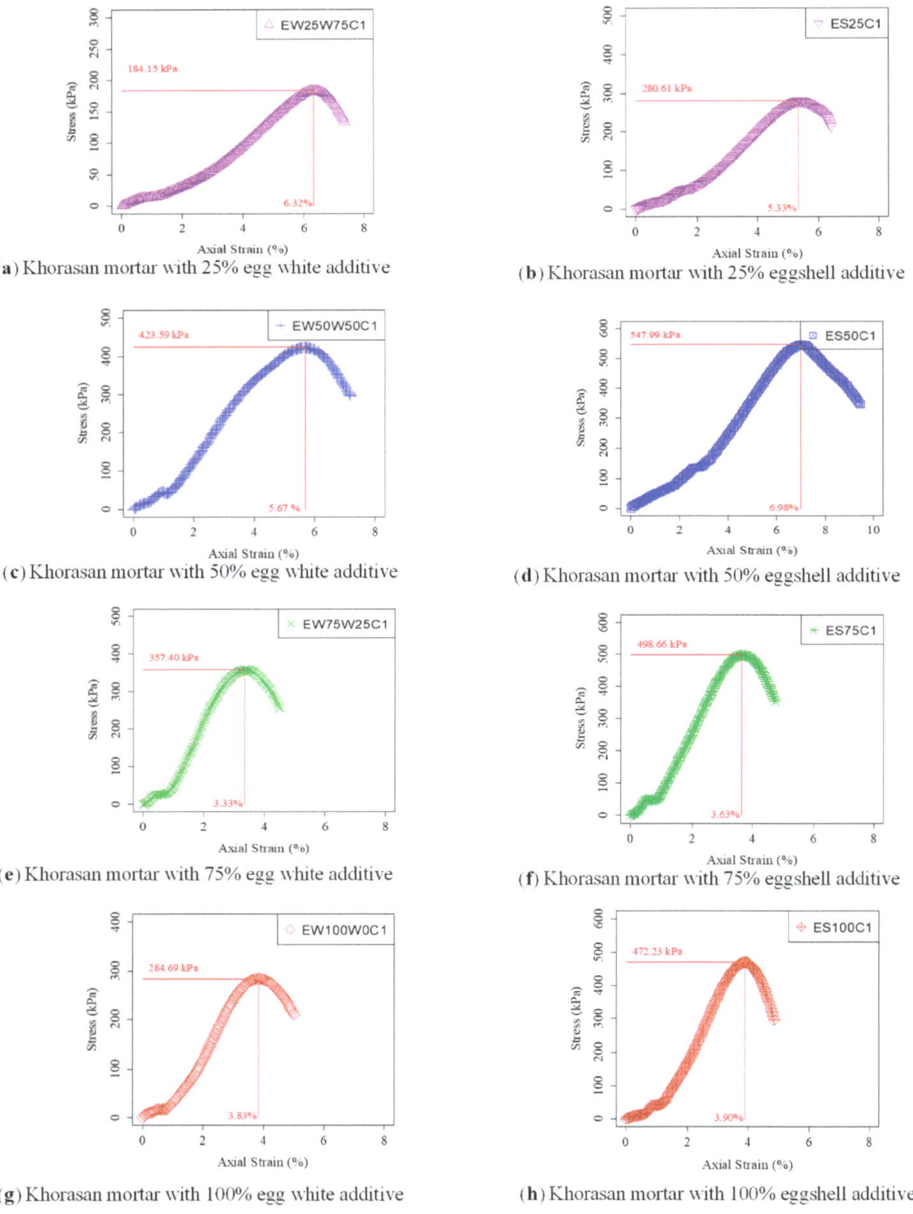

(**a**) Khorasan mortar with 25% egg white additive

(**b**) Khorasan mortar with 25% eggshell additive

(**c**) Khorasan mortar with 50% egg white additive

(**d**) Khorasan mortar with 50% eggshell additive

(**e**) Khorasan mortar with 75% egg white additive

(**f**) Khorasan mortar with 75% eggshell additive

(**g**) Khorasan mortar with 100% egg white additive

(**h**) Khorasan mortar with 100% eggshell additive

Figure 8. Experimental results of egg white/eggshell added to Khorasan mortars at the end of 1 day.

The stress–strain graph of mixtures cured for 7 days is shown in Figure 9. UCS values of all mixtures increased significantly. EW25W75C7 was the mixture having the lowest UCS value when compared to all samples cured for 7 days, as presented in Figure 9a. The behavior of the stress–strain indicates that increasing the egg white content causes a sharp reduction in the axial strain value at the point of maximum stress occurring. This value decreased suddenly after adding egg white, by more than 50%, and it is seen in Figure 9c,e,g. However, increasing egg shell content in stabilized samples caused almost no change in the axial stress value reached at maximum stress after 50%, as shown in Figure 9b,d,f,h. ES50C7 was the mixture having the highest UCS value when compared to all mixtures cured for 7 days and the increase in strength was 571.45%, according to the AC1.

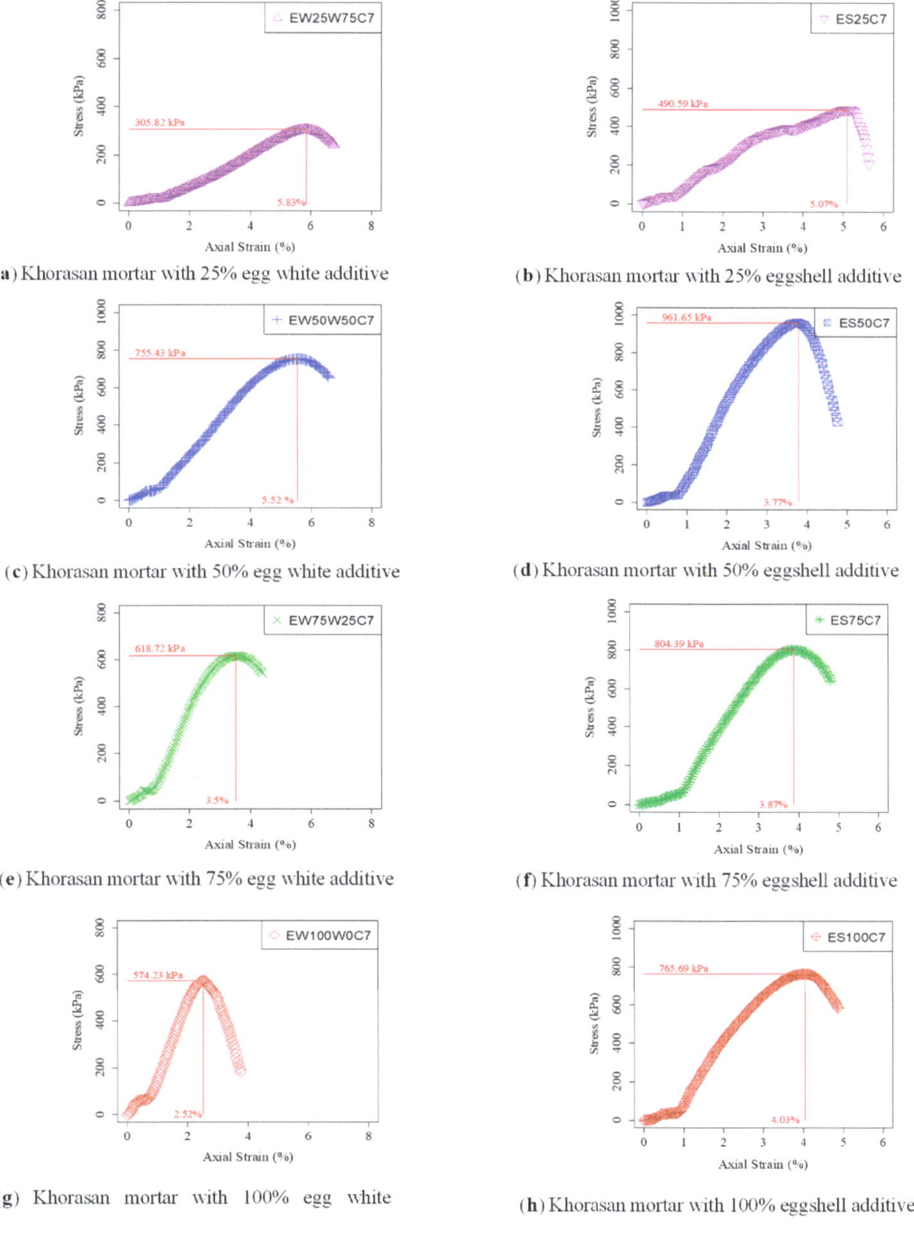

(**a**) Khorasan mortar with 25% egg white additive

(**b**) Khorasan mortar with 25% eggshell additive

(**c**) Khorasan mortar with 50% egg white additive

(**d**) Khorasan mortar with 50% eggshell additive

(**e**) Khorasan mortar with 75% egg white additive

(**f**) Khorasan mortar with 75% eggshell additive

(**g**) Khorasan mortar with 100% egg white

(**h**) Khorasan mortar with 100% eggshell additive

Figure 9. Experimental results of egg white/eggshell added Khorasan mortars at the end of 7 days.

Egg white stabilized samples cured for 28 days behave a little bit differently compared to other curing times, since the addition of egg white after 50% has no significant change at axial strain value occurred at maximum stress as observed in Figure 10a,c,e,g. EW25W75C28 is again the worst mixture compared to other mixtures cured for 28 days just as 1 and 7 days. The unconfined compression strength value of ES50C28 is the highest one compared to all mixtures and curing times since it is 1056.44% higher than AC1. Egg shell stabilized samples cured for 28 days behave more brittle compared to other curing times as shown in Figure 10b,d,f,h. Stress-strain behavior of ES50C28 and ES75C28 are similar when compared with lime-stabilized samples as presented in Figure 10f,h. Those results are strong evidence that as the curing time increased, the chemical reaction became stronger, and a lime-like behavior emerged as the amount of CaO in the sample increased.

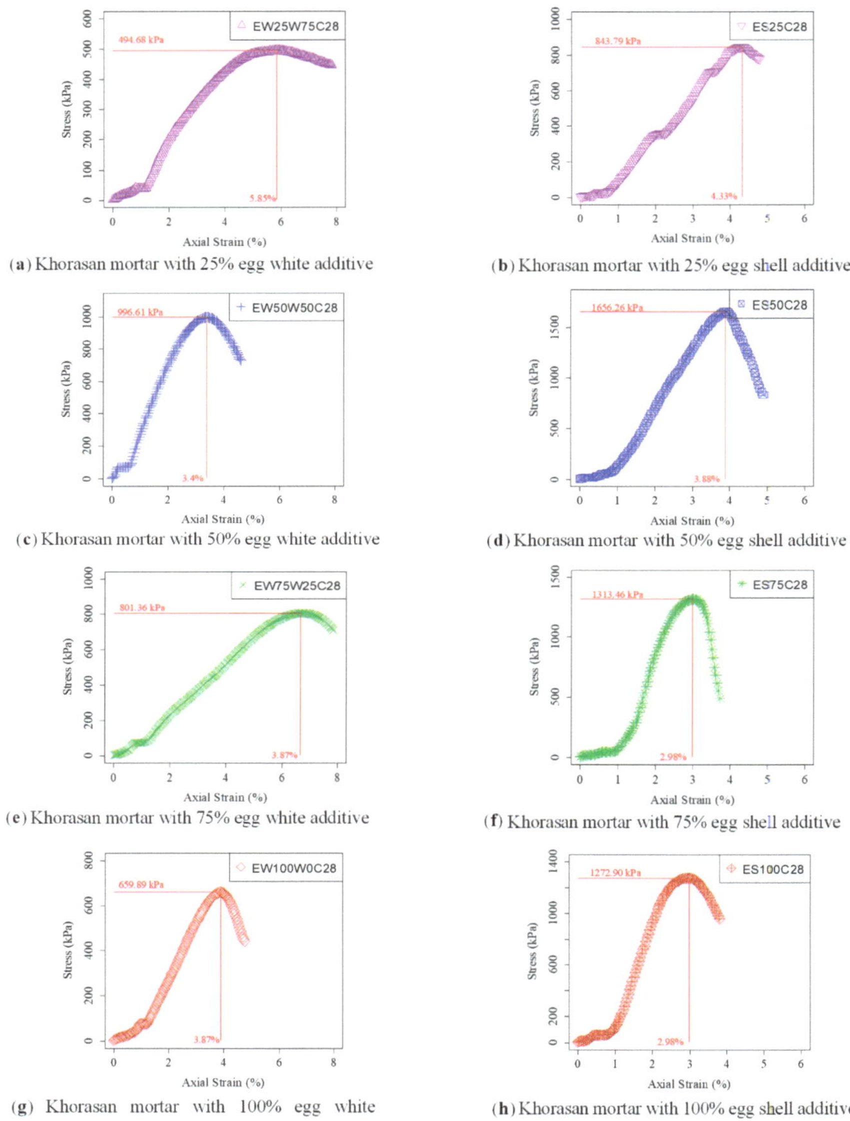

(**a**) Khorasan mortar with 25% egg white additive

(**b**) Khorasan mortar with 25% egg shell additive

(**c**) Khorasan mortar with 50% egg white additive

(**d**) Khorasan mortar with 50% egg shell additive

(**e**) Khorasan mortar with 75% egg white additive

(**f**) Khorasan mortar with 75% egg shell additive

(**g**) Khorasan mortar with 100% egg white

(**h**) Khorasan mortar with 100% egg shell additive

Figure 10. Experimental results of egg white/eggshell added to Khorasan mortars at the end of 28 days.

Figure 11 illustrates the unconfined compression strength (UCS) of all mixtures, including the control sample. The UCS performance of the control sample for all curing times has been depicted on the diagrams to facilitate comparison with the results of all mixtures. In Figure 11a, curing times of 1 day for all mixtures are displayed. The unconfined compression strengths of the EW50WC1, EW75WC1, ES50C1, ES75C1, and ES100C1 samples surpass those of the AC1, AC7, and AC28 samples, even though they were cured for just 1 day. Results for samples cured for 7 days are presented in Figure 11b. With the exception of EW25W75C7, all samples exhibit UCS values higher than the control sample. In Figure 11c, the UCS values of egg white-treated samples after 28 days of curing are compared. The lowest UCS among these samples is 3.45 times, 2.35 times, and 1.46 times that of AC1, AC7, and AC28, respectively, as inferred from the figure.

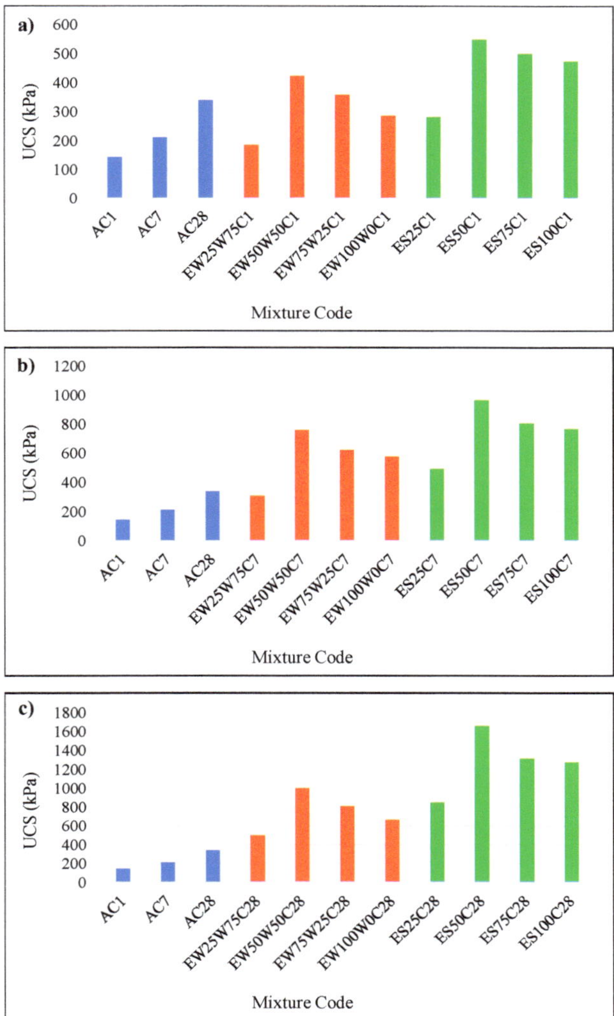

Figure 11. UCS results of all mixtures including control sample: (**a**) 1 day of curing (**b**) 7 days of curing (**c**) 28 days of curing.

While creating mortar, slump tests are performed and the consistency of mortar is determined based on these slump tests [42]. Figure 12 illustrates the slump values of both the control sample and the mixtures utilized in the analyses. The slump value serves as a reliable indicator of workability. A notable decrease in slump was observed when a mixture with a 50% addition of eggshell was used as an additive, followed by an increase with higher ratios. A similar trend was observed in mixtures containing egg white + water. The best workability was achieved with the EW50W50 and ES50 mixtures, compared to the control sample.

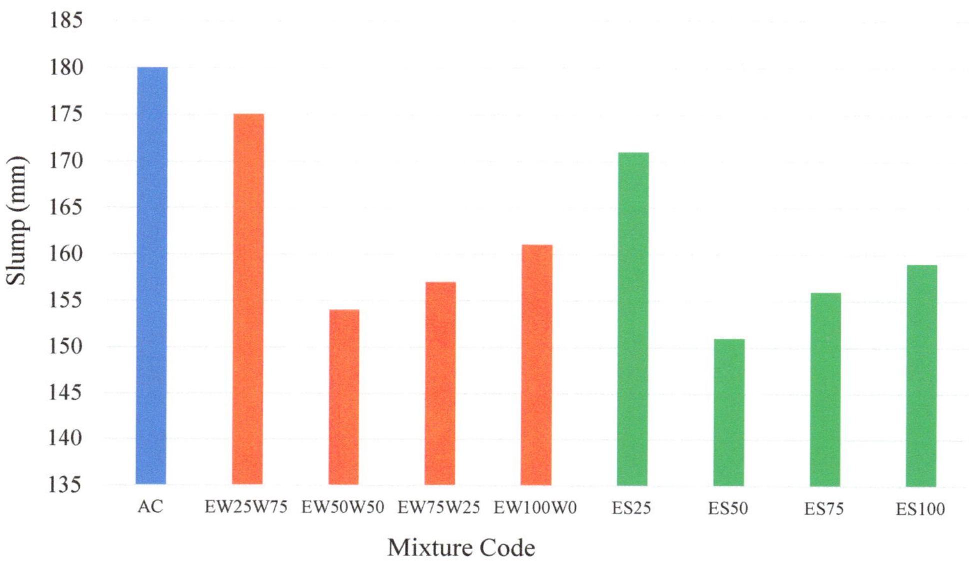

Figure 12. Slump results of all mixtures including control sample.

8. Three-Dimensional Modeling of Plaka Bridge

The finite-difference modeling technique offers a range of advantages that make it a valuable tool in various fields. One of its key strengths is its versatility, as it can be applied to a wide array of problems, including differential equations and partial differential equations [33]. Additionally, it is known for its simplicity, making it relatively easy to understand and implement compared to other numerical methods. Despite its simplicity, finite-difference modeling can yield accurate results when properly utilized, enabling fine-grained resolution and the capture of intricate details within the modeled system [33]. Furthermore, the method is efficient and can be deployed on modern computing architectures, resulting in high-performance simulations and reduced computational time. In this study, the FLAC3D program was employed, which is based on the finite-difference method, to develop the bridge model timeline. Each stone element was individually modeled using a separate FLAC3D code, with brick or wedge elements utilized for each stone. Essentially, our approach utilized the finite-difference method instead of the discrete-element method or finite-element method. This choice was made because the FLAC3D program and the finite-difference method offer a crucial alternative for revealing the time-dependent creep and seismic behavior of stone elements. The Plaka bridge had been destroyed and subsequently rebuilt after a flood disaster in Greece in 2015. Two distinct bridge models were developed. The first model considered the Plaka bridge, which had endured from 1866 to 2015, and utilized Khorasan mortar with egg white additives between the stone elements. On the other hand, the second model accounted for the use of eggshell-added Khorasan mortar between the discrete stone elements. In the creation of both models, each stone element was individually represented. Brick and wedge elements were employed for modeling the stone elements. Brick elements, also known as hexahedral elements, are three-dimensional elements with six faces, eight nodes, and twelve edges. They are widely used in FLAC3D to discretize the geotechnical domain and represent solid materials, such as soil, rock, or concrete [33]. In addition, wedge elements, specialized three-dimensional elements available in FLAC3D, are particularly useful for modeling geological features such as faults, joints, or stratigraphic interfaces. These elements possess a triangular base and taper to a point, resembling the shape of a wedge. Wedge elements are advantageous when modeling inclined or irregular surfaces, as they can conform well to complex geological geometries. Brick and wedge elements used in the bridge modeling process are illustrated in Figure 13.

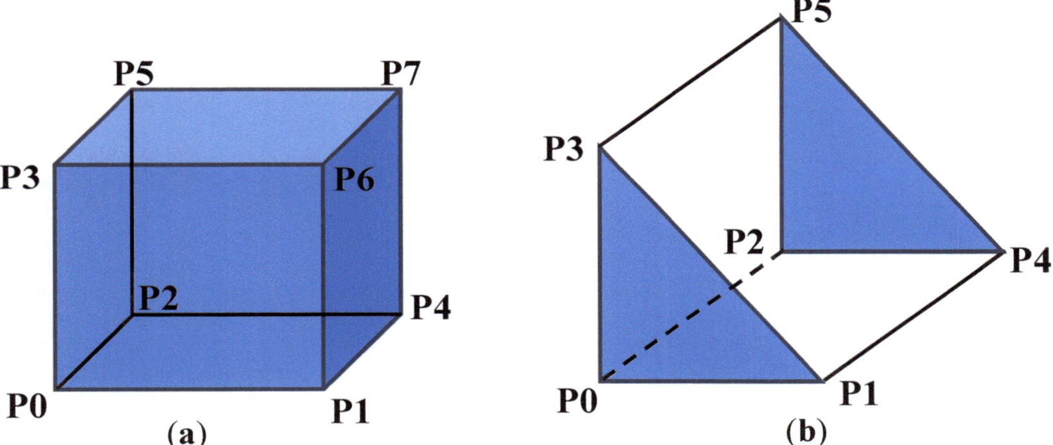

Figure 13. Modeling of discrete stone elements: (**a**) brick element (**b**) wedge element [33].

Khorasan mortar was simulated between the discrete stone elements by considering specialized interface elements. Utilizing the mechanical properties of Khorasan mortar (Figures 7–11), the mechanical characteristics (kn and ks stiffness) of the interface elements between Khorasan mortar and stone elements were computed (Equation (4)) [33]. Subsequently, the calculated kn and ks stiffness values, representing Khorasan mortar between the stone elements in the bridge model, were defined to realistically model the interaction between the stones. The most straightforward approach to modeling Khorasan mortar within a three-dimensional model is to introduce interface elements into the bridge model instead of Khorasan mortar. The mechanical properties of the stiffness elements established between the stone elements instead of Khorasan mortar were entirely determined by considering Equation (4). Therefore, in this study, interface elements, as depicted in Figure 14, were defined between distinct surfaces in place of Khorasan mortar. One interface element was computed between two different stones with identical mechanical properties. However, during the bridge modeling process, it was necessary to define interface elements between stones with varying mechanical properties in certain sections of the bridge. As illustrated in Figure 14, two distinct interface elements were computed between stones with different mechanical properties, and the larger interface element between two dissimilar stones with varying mechanical properties was considered in the analyses.

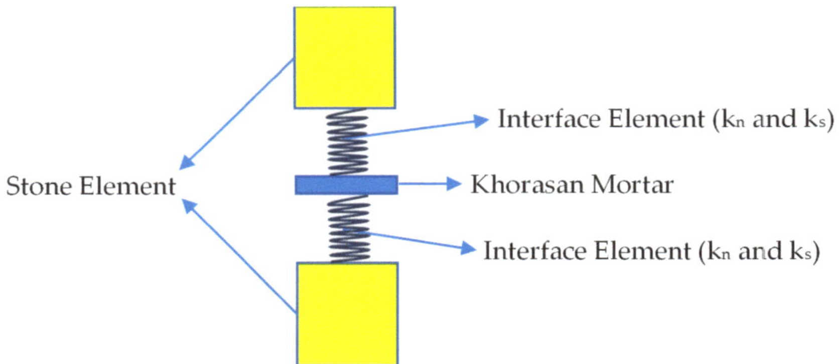

Figure 14. Khorasan mortar and interface elements in the model.

The mechanical properties of the stone elements and Khorasan mortar were utilized in the calculation of interface elements. The 'Interface' functionality in FLAC3D is employed to establish and simulate interactions between two distinct sub-grids within a geotechnical model. It represents the boundary or contact surface between different materials or regions. By specifying properties such as bulk and shear moduli, the interface facilitates the transfer of forces and displacements between the interconnected sub-grids, allowing for the modeling of intricate geotechnical phenomena such as soil–structure interaction and fault behavior. The interface feature in FLAC3D offers a versatile approach to simulating the interaction of separate regions, contributing to the comprehension and prediction of the system's behavior. In this study, the kn (normal) and ks (shear) stiffness coefficients of the interface elements, defined in place of Khorasan mortar between the discrete stone elements, were determined using the following formulation [33].

$$k_n \text{ and } k_s = \frac{K_{average} + {}^4/_3 G_{average}}{Z_{average}} \tag{4}$$

In Equation (4) [33], K and G represent the bulk modulus and shear modulus, respectively. Additionally, the symbol Z corresponds to the mesh lengths of the stone element and Khorasan mortar. Mesh size plays a crucial role in finite-difference analysis, significantly impacting the results of structural analyses. In this study, the mesh size was determined based on the creep analysis of the bridge. Specifically, 10 different mesh ranges (0.1 m, 0.2 m, 0.3 m, 0.4 m, 0.5 m, 0.6 m, 0.7 m, 0.8 m, 0.9 m, and 1 m) were selected for displacements and stresses and it was observed that no significant changes occurred in the range from 0.1 m to 0.3 m mesh size. Consequently, a mesh interval of 0.3 m was chosen for the bridge model. It is important to note that the mesh intervals in this study were not arbitrarily selected but were based on the results of the creep analysis. For this study, the mesh lengths of the stone elements and Khorasan mortars were set to 0.4 m and 0.03 m, respectively. The mechanical properties of mortars and stones are shown in Table 3. When computing the kn and ks values for the interface elements, the averages of the K and G values of the stone elements and Khorasan mortar are considered, as depicted in Table 4, to obtain the kn and ks interface coefficients. The calculated kn and ks values are presented in Table 4. The variable "K" denotes the bulk modulus, while "kn" and "ks" represent stiffness values in the normal and shear directions, respectively. In the context provided, "EW50W50C28" signifies that the mixture contains a 50% ratio of egg white to water and was cured for 28 days. When referencing "between the lowest main arch and EW50W50C28", this indicates that Khorasan mortar with the composition EW50W50C28 was utilized in the Lowest Main Arch section of the bridge.

Table 3. Mechanical properties of mortars and stones.

Mixture Code	E (Pa)	Poisson Ratio	G (Pa)	K (Pa)
EW50W50C28 (Mortar)	3.47×10^7	0.34	1.29×10^7	3.61×10^7
ES50C28 (Mortar)	6.01×10^7	0.28	2.35×10^7	4.55×10^7
The Lowest Main Arch	7.05×10^7	0.28	2.75×10^7	5.34×10^7
The Middle Main Arch	6.52×10^7	0.26	2.59×10^7	4.53×10^7
Supporter Arch Right	5.86×10^7	0.21	2.42×10^7	3.37×10^7
Supporter Arch Left	5.24×10^7	0.19	2.20×10^7	2.82×10^7
Rock Material	4.75×10^7	0.25	1.90×10^7	3.17×10^7
Foundation	9.00×10^7	0.38	3.26×10^7	1.25×10^8

Table 4. Stiffness parameters between the discrete surfaces.

Location of Interface Element	G (Average) (Pa)	K (Average) (Pa)	k_n and k_s Stiffness (Pa/mm)
Between The Lowest Main Arch and EW50W50C28	2.02×10^7	4.48×10^7	4.35×10^5
Between The Middle Main Arch and EW50W50C28	1.94×10^7	4.07×10^7	4.03×10^5
Between Supporter Arch Right and EW50W50C28	1.86×10^7	3.49×10^7	3.62×10^5
Between Supporter Arch Left and EW50W50C28	1.75×10^7	3.22×10^7	3.37×10^5
Between Rock Material and EW50W50C28	1.60×10^7	3.39×10^7	3.35×10^5
Between Foundation and EW50W50C28	2.28×10^7	8.06×10^7	6.73×10^5
Between The Lowest Main Arch and ES50C28	2.55×10^7	4.95×10^7	5.06×10^5
Between The Middle Main Arch and ES50C28	2.47×10^7	4.54×10^7	4.75×10^5
Between Supporter Arch Right and ES50C28	2.39×10^7	3.96×10^7	4.33×10^5
Between Supporter Arch Left and ES50C28	2.28×10^7	3.69×10^7	4.08×10^5
Between Rock Material and ES50C28	2.13×10^7	3.86×10^7	4.06×10^5
Between Foundation and ES50C28	2.81×10^7	8.53×10^7	7.44×10^5

While examining historical bridges in the literature, the influence of Khorasan mortars is often disregarded. However, it is crucial to accurately model Khorasan mortars within the bridge model to yield more realistic results. Therefore, these values contribute significantly to the literature regarding the modeling of historical stone arch bridges. Subsequently, the Burger creep material model was employed for the bridge's rockfill, arch, and foundation materials. This material model allowed for an accurate representation of the time-dependent failure behavior of the bridge material. When using this material model, the following material properties were incorporated: shear modulus, bulk modulus, density, cohesion, friction angle, and dilatancy angle values. The bridge model was constructed with a non-linear approach, employing the Burger model to characterize the non-linear behavior of materials. Additionally, free-field and quiet-boundary conditions were applied to the lateral boundaries of the three-dimensional bridge model. These boundary conditions serve to minimize the reflection of earthquake waves within the model. Furthermore, a reflecting boundary condition was specified for the base boundaries of the bridge model. The bridge model comprised a total of 1,849,274 individual brick elements. Each brick element was individually modeled within the overall structure. An overview of the three-dimensional bridge model is presented in detail in Figure 15. The height of the primary arch of the bridge was modeled as 18.7 m. Subsequently, two auxiliary arches were created, measuring 7.8 m and 8.9 m, respectively. The thickness of the primary arch section in the model was established at 0.57 m. To incorporate structure–soil interaction into the analysis of significant structures like bridges, soil was modeled beneath the bridge. Furthermore, soil was included in the bridge model to simulate a more realistic effect of earthquake waves on the bridge. Additionally, the foundation section extended downward by four times the height of the bridge (Figure 15).

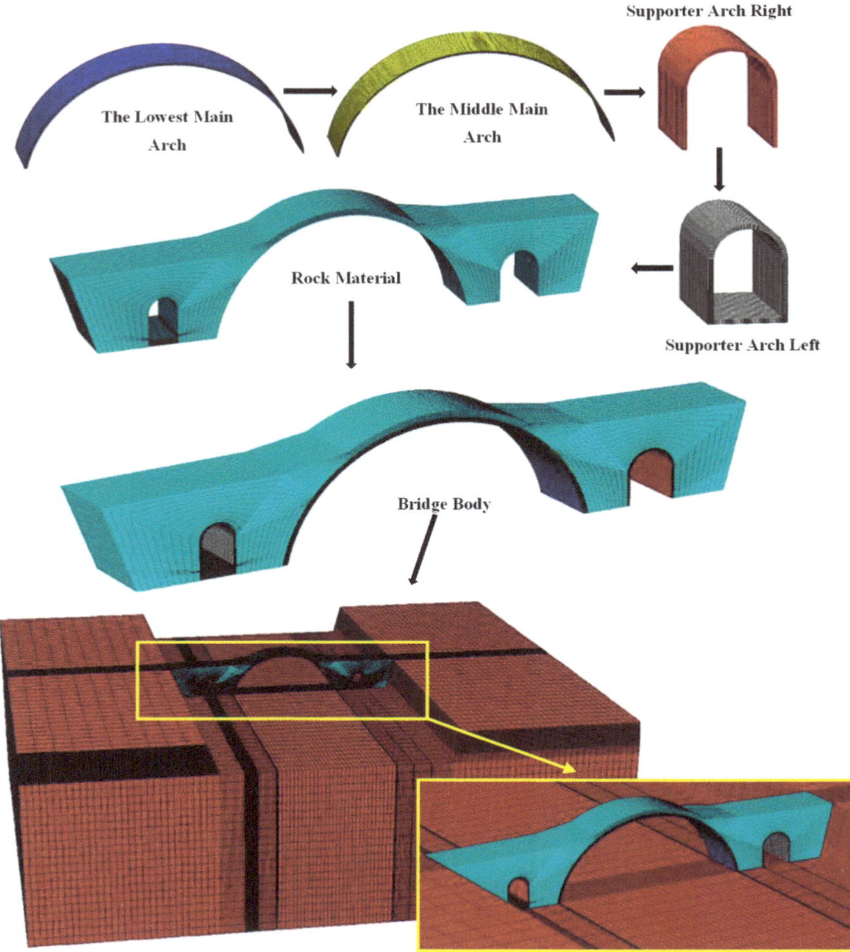

Figure 15. Three-dimensional view of the bridge model.

9. Three-Dimensional Numerical Analysis Results

9.1. Long-Term Creep Analysis Results beween 1866 and 2015 for Bridge Using Khorasan Mortar with Egg White Additive

Historical bridges represent one of the most significant ways of illuminating our history. These bridges provide insights into the lifestyles of people who resided in the past. Therefore, it is essential to scrutinize the prospects of these bridges, considering the genuine structural behavior of historical bridges that have managed to endure to this day. Restoration efforts should be carried out without compromising the historical fabric. This section of the study gives a comprehensive examination of the time-dependent creep analyses of the Plaka bridge. This bridge, constructed in 1866, has withstood a total of five floods over its history. The specific dates of these flood disasters are detailed in Table 5.

Table 5. Dates of the flood disasters.

Symbol	Date of Flood Disaster	Explanation
FD1	1919	53 years after the bridge was built
FD2	1930	64 years after the bridge was built
FD3	1948	82 years after the bridge was built
FD4	1975	109 years after the bridge was built
FD5	2015	149 years after the bridge was built

The water level at the bridge was simulated using hydrostatic water forces and a water table. During the examination of the time-dependent creep analyses, it was considered that Khorasan mortar mixed with egg white should be used between the stone elements. The primary reason for this choice is that when the bridge was constructed in 1866, Khorasan mortar with an egg white additive was employed to bond the discrete stone elements together. As egg white-mixed Khorasan mortar enhances the interaction between these stone elements, the stiffness values (kn and ks) of the interface elements were calculated, taking into account the mechanical properties of Khorasan mortar with egg white additive. The computed kn and ks values are provided in detail in Table 3. Stiffness values were established between individual stone elements, and the bridge was prepared for realistic creep analyses. In the execution of time-dependent creep analyses, specific time intervals were defined within the FLAC3D program. The definition of precise time steps is crucial for conducting more realistic analyses. To carry out the time-dependent creep analyses, we initially commenced with 1-second analyses. After each 1-second creep analysis, all stresses and displacements on the bridge were reset. Subsequently, the next 1-second analysis was conducted, and this analysis cycle was repeated 4,698,864,000 times, covering 149 years of creep analyses. The stages of analysis for the creep analyses are comprehensively presented in Figure 16.

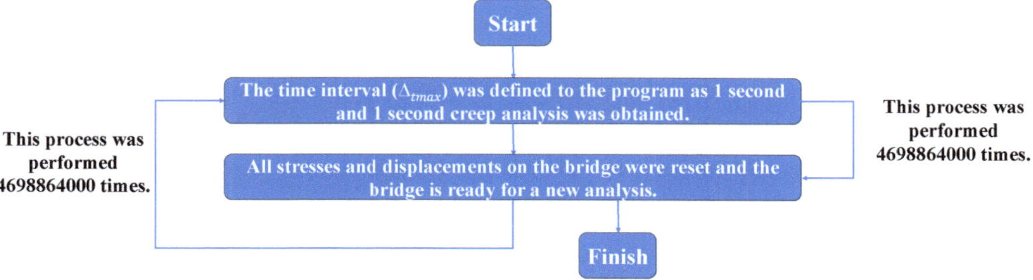

Figure 16. Analysis stages of the creep analyses.

Since the bridge was subjected to five flood disasters in 149 years, the water level of the bridge fluctuated during these flood events. Under normal conditions, except for flood disasters, the water level at the bridge was considered to be one-sixth of the bridge's height. However, during flood events, the water level at the bridge was assumed to be at the same level as the bridge height. Following the analyses, an evaluation was made regarding the principal stress values that occurred on the bridge over 149 years. The flood events that the bridge encountered during this period are indicated on the graphs with the symbol FD (Flood Disaster). Following the results of the time-dependent creep analysis, the time-dependent principal stress values at Point 1, Point 2, and Point 3 are elaborated upon in Figure 17. When examining the time-dependent principal stress values at Point 1, it becomes evident that there were minimal principal stresses during the initial 53 years. After the first flood disaster (FD1), an increase in principal stresses at Point 1 was observed. This rise can be attributed to the water level reaching Point 1, subjecting it to significant hydrostatic forces. During the bridge's second flood disaster (FD2), there was an approximate 0.2 MPa increase in principal stresses at Point 1 (Figure 17a). Notably, significant stress increases occurred at Point 1 between the second and third flood disasters. Furthermore, after the third flood disaster, stress values at Point 1 increased by approximately 0.3 MPa. This result demonstrated the fact that flood disasters profoundly alter the principal stress behavior of historical bridges. The maximum principal stress value observed at Point 1 during the fourth flood disaster was 4.11 MPa. In this study, the maximum capacity value for Point 1 was determined as 6.3 MPa. In Figure 17b, the principal stress values that evolve at Point 2 are detailed. Significant principal stress values were not observed at Point 2 until the first flood disaster. However, the principal stress change at Point 2 after the first flood disaster was 0.23 MPa, significantly altering the time-dependent failure behavior of Point 2. No significant principal stress changes were detected at Point 2 between the first and second flood disasters. The principal stress value just before the second flood disaster at Point 2 was 0.76 MPa. The principal stress value after the second flood disaster at Point 2 was 1.38 MPa. It is evident that the second flood disaster also led to substantial principal stress changes at Point 2. The third flood disaster caused more significant principal stress changes at Point 2 compared to the first and second flood disasters. After the third flood disaster, an approximate principal stress change of 0.24 MPa was observed at Point 2. Similar to other flood disasters, the fourth flood disaster significantly increased the principal stress values at Point 2. It was concluded that the maximum capacity value for Point 2 was 4.8 MPa. In Figure 17c, detailed time-dependent principal stress values occurring at Point 3 over 149 years are presented. It was found that no significant principal stress values occurred at Point 3 during the first 53 years. However, 53 years later, after the first flood, significant changes were observed in the principal stress values at Point 3. Following the first flood disaster, principal stress increased by approximately 0.39 MPa at Point 3. Furthermore, the flood disaster accelerated the rate of principal stress increase at Point 3. This result provides valuable insights into how flood disasters alter the structural behavior of historical stone bridges. After the second flood disaster, there was an approximate 0.17 MPa increase in principal stress at Point 3. Significant principal stress changes were observed at Point 3 between the second and third flood disasters. Following the third flood disaster, critical increases in principal stress values were noted at Point 3, and it was concluded that the maximum principal stress value at Point 3 was 3.98 MPa between the third and fourth floods. Substantial increases in principal stress values at Point 3 were observed between the third and fourth flood disasters. This outcome suggests that the bridge experienced structural fatigue after the third flood. The rate of principal stress increase at Point 3 accelerated significantly after the fourth flood disaster. After the fifth flood, it was observed that structural damage occurred at Point 3 when the maximum principal stress value reached 5.4 MPa.

In Figure 18, the stress-displacement behavior of the historical bridge is presented in detail. According to Figure 18, it was concluded that the displacement values over time at Point 1 were significantly greater than those at Point 2 and Point 3. Additionally, when comparing these three different points, it was observed that the smallest displacement values occurred at Point 2. The largest displacement value recorded at Point 1 over 149 years was approximately 29.7 mm. Furthermore, the largest displacement values observed over 149 years at Point 2 and Point 3 were 15.8 mm and 19.7 mm, respectively. Point 1, Point 2, and Point 3 reached their maximum displacement values at 6.3 MPa, 4.8 MPa, and 5.4 MPa, respectively. This result indicates that the maximum principal stress value for the structural behavior of Point 1, Point 2, and Point 3 on historical stone bridges is 6.3 MPa, 4.8 MPa, and 5.4 MPa, respectively.

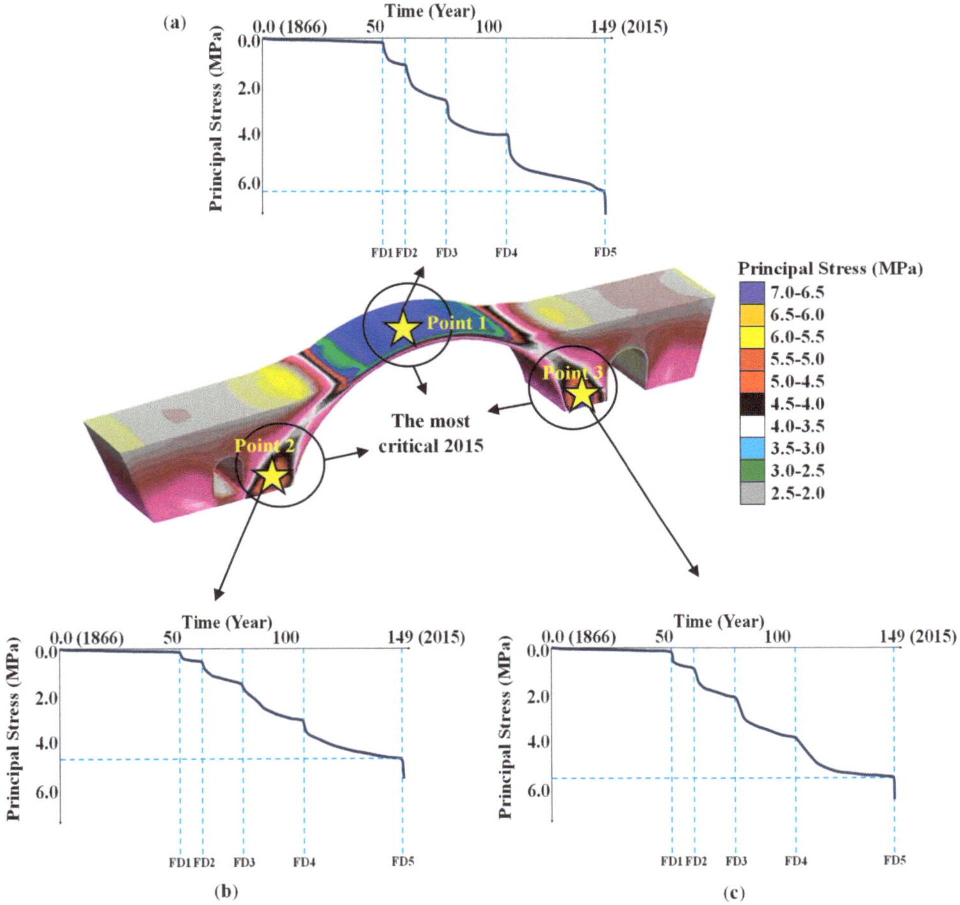

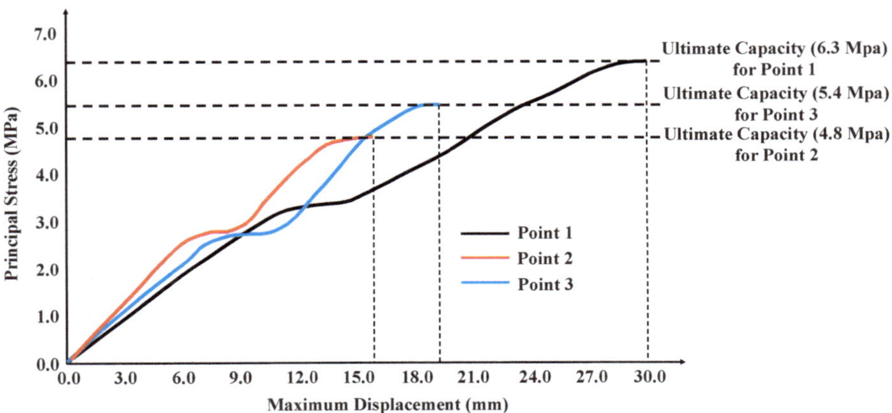

Figure 17. Long-term (1866–2015) principal stress behavior of the bridge containing egg white-added Khorasan mortar: (**a**) Point 1 (**b**) Point 2 (**c**) Point 3.

Figure 18. Stress-displacement behavior of the bridge containing egg white-added Khorasan mortar.

9.2. Long-Term Creep Analysis Results between 2015 and 2164 for Bridge Using Khorasan Mortar with Eggshell Additive

In this section of the study, a comprehensive examination of the time-dependent creep behavior of the bridge following its restoration was conducted. During the 2015 restoration of the bridge, Khorasan mortar with eggshell additives was employed between the individual stone elements. The mechanical properties of these mortars are extensively detailed in Table 2. Furthermore, when eggshell-added Khorasan mortar was used between separate stone elements, the calculated kn and ks stiffness parameters between the stones were calculated, and are presented in Table 3. Once the calculated kn and ks stiffness values were defined for the individual stone elements, the bridge underwent static analysis from 2015 to 2164. These analyses took into consideration the flood disasters the bridge had previously experienced. In essence, time-dependent creep analyses were conducted, assuming that the bridge would encounter floods in 2068, 2079, 2097, 2124, and 2164 (Table 6). This approach allowed us to compare how Khorasan mortar with egg white additive and Khorasan mortar with eggshell additive influenced the time-dependent creep behavior of the bridge. During flood disasters, it was assumed that the water level would rise to match the height of the bridge, and the water level was modeled using water loads and a water table. These analyses provided us with detailed insights into the future structural behavior of the restored Plaka bridge. The time-dependent creep behavior of the bridge from 2015 to 2164 is comprehensively presented in Figure 19, with the 'FD' symbol denoting flood disasters. Based on the principal stress values evolving at Point 1, Point 2, and Point 3, it was observed that the largest principal stress values occurred at Point 1, while the smallest principal stress values were recorded at Point 2. According to the analyses conducted between 1866 and 2015, it was deduced that the optimum capacity values for Point 1, Point 2, and Point 3 were 6.3 MPa, 4.8 MPa, and 5.4 MPa, respectively. However, as per the results of the creep analysis conducted from 2015 to 2164, the maximum principal stress values at Point 1, Point 2, and Point 3 were determined to be 1.92 MPa, 0.91 MPa, and 1.48 MPa, respectively (Figure 19). These findings suggest that Point 1, Point 2, and Point 3 would not experience any structural damage in the event of five flood disasters occurring between 2015 and 2164. Moreover, they indicate that the historical bridge constructed with Khorasan mortar containing eggshells exhibited greater durability than the bridge constructed using egg white, displaying enhanced resistance to floods without collapsing. Consequently, this study strongly recommends the use of eggshell-containing Khorasan mortar between individual stone elements in the restoration of historical bridges.

Table 6. General information about flood disasters that are assumed to occur in the future.

Symbol	Year of Flood Disaster	Explanation
FD1	2068	It is thought that a flood disaster would occur 53 years after the bridge was restored.
FD2	2079	It is thought that a flood disaster would occur 64 years after the bridge was restored.
FD3	2097	It is thought that a flood disaster would occur 82 years after the bridge was restored.
FD4	2124	It is thought that a flood disaster will occur 109 years after the bridge was restored.
FD5	2164	It is thought that a flood disaster would occur 149 years after the bridge was restored.

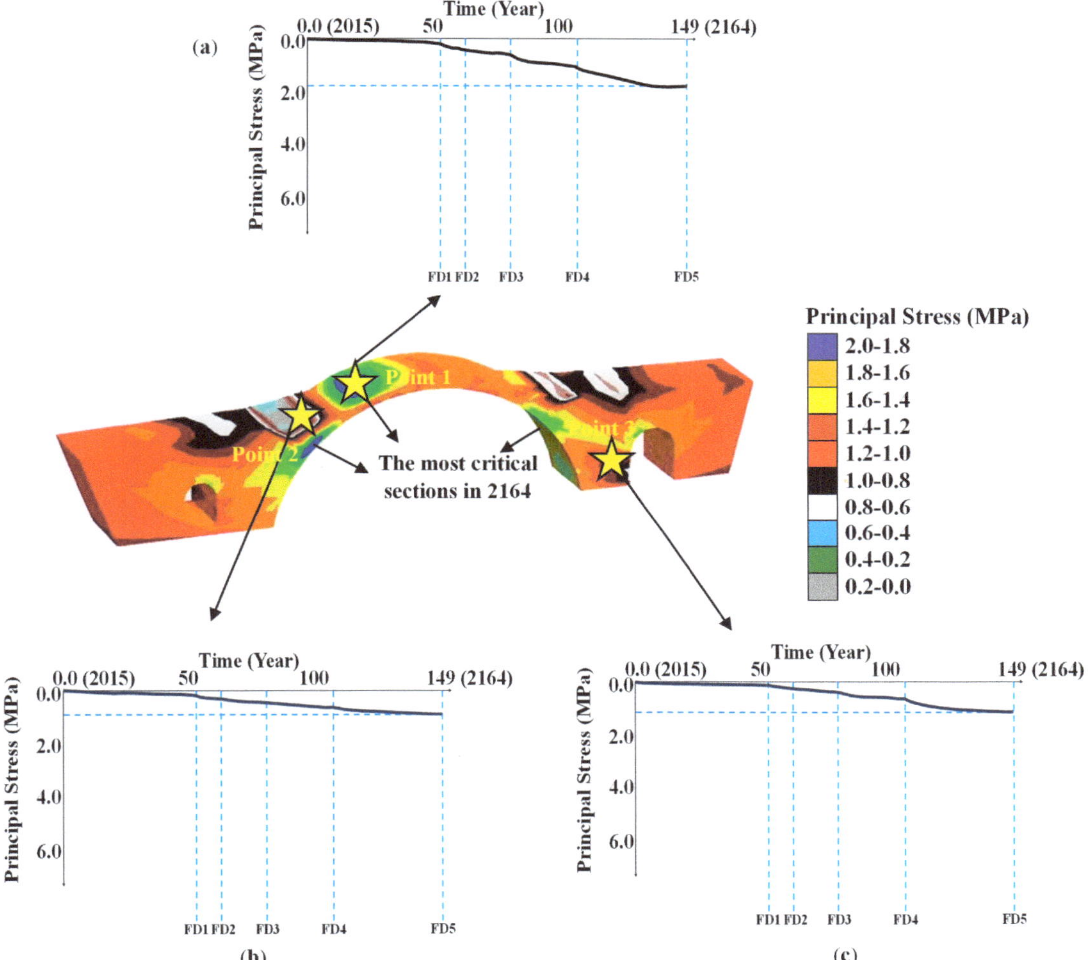

Figure 19. Long-term (2015–2164) principal stress behavior of the bridge containing eggshell-added Khorasan mortar: (**a**) Point 1 (**b**) Point 2 (**c**) Point 3.

9.3. Modal and Seismic Analysis Results for Bridge Using Khorasan Mortar with Eggshell Additive

In this section of the study, detailed modal and seismic analysis results for the bridge that has been restored using Khorasan mortar containing eggshells are presented in detail. In this study, while the Burger creep material model was used for creep analyses, the material model was changed for seismic analyses. Specifically, the Mohr–Coulomb material model was used for the seismic analyses. The material parameters for the Mohr–Coulomb model were obtained from the literature (Table 1) [28]. All assumptions of this material model are included in the FLAC3D program [33]. Figure 20 illustrates the modal behavior of the restored Plaka bridge. According to Figure 20, the first mode of the restored bridge has a frequency of 12.071 Hz. Additionally, the second and third modes of the bridge have frequencies of 14.623 Hz and 21.136 Hz, respectively. The first mode (with a frequency of 12.071 Hz) is characterized by 'bending' deformation dominance, while the second mode (with a frequency of 14.623 Hz) is dominated by 'axial' deformation. These findings offer valuable insights into the modal behavior of historical bridges that have undergone restoration with Khorasan mortar containing eggshells.

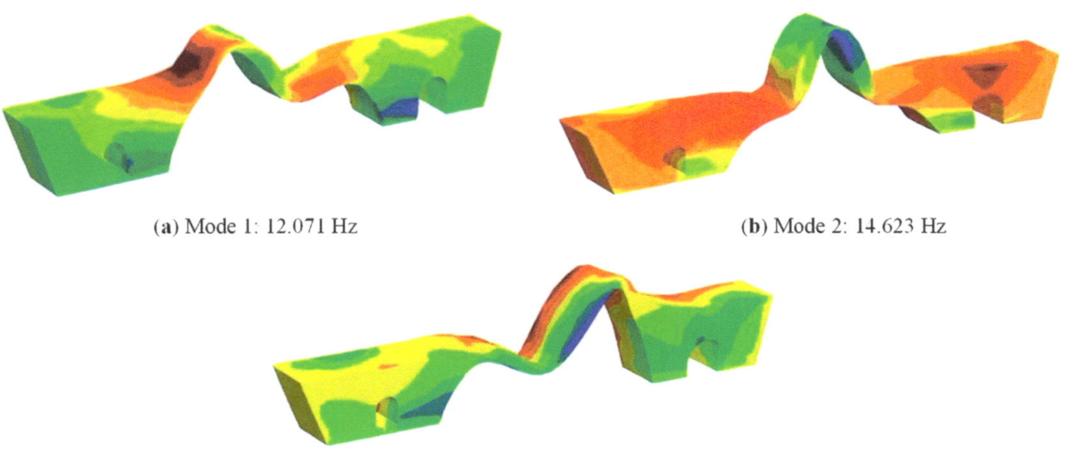

(**a**) Mode 1: 12.071 Hz

(**b**) Mode 2: 14.623 Hz

(**c**) Mode 3: 21.136 Hz

Figure 20. Natural frequencies of three vibration modes for the restored bridge.

In Figures 21–24, the seismic response of the restored bridge to 10 different earthquakes is investigated in detail. When conducting seismic analyses of the bridge, non-reflective boundary conditions were implemented. For lateral boundaries, free-field and quiet-boundary conditions were considered to minimize wave reflection within the model. Hysteresis damping was incorporated into the earthquake analyses, and G/Gmax values of the materials were utilized for calculating the dampings. Earthquake accelerations were defined in the program using the acceleration-time format. Although the seismic events used in these analyses had durations exceeding 40 s, only the most critical 40 s of each earthquake were considered due to the computational length of the analyses. Detailed mechanical properties of the earthquakes employed in these analyses are provided in Table 7.

Table 7. Earthquake characteristics [43].

Symbol	Earthquake	M_w	Distance (km)	PGA (cm²/s)	PGV (cm/s)	PGD (s)
EQ1	Pazarcık (Kahramanmaraş)	5.5	6.87	49.84	2.84	0.55
EQ2	İslahiye (Gaziantep)	5.7	10.46	363.52	13.85	1.02
EQ3	Ekinözü (Kahramanmaraş)	5.5	10.93	79.35	4.26	0.42
EQ4	Pazarcık (Kahramanmaraş)	7.6	8.6	1966.74	186.78	661.9
EQ5	Elbistan (Kahramanmaraş)	7.6	7	635.45	170.79	614.52
EQ6	Yeşilyurt (Malatya)	5.6	6.15	25.23	2.36	12.77
EQ7	Nurdağı (Gaziantep)	6.6	6.2	445.29	40.49	9.27
EQ8	Doğanşehir (Malatya)	5.6	10.23	47.28	2.90	0.40
EQ9	Nurdağı (Gaziantep)	5.6	6.98	44.15	2.91	0.73
EQ10	Defne (Hatay)	6.4	21.7	445.38	75.78	44.90

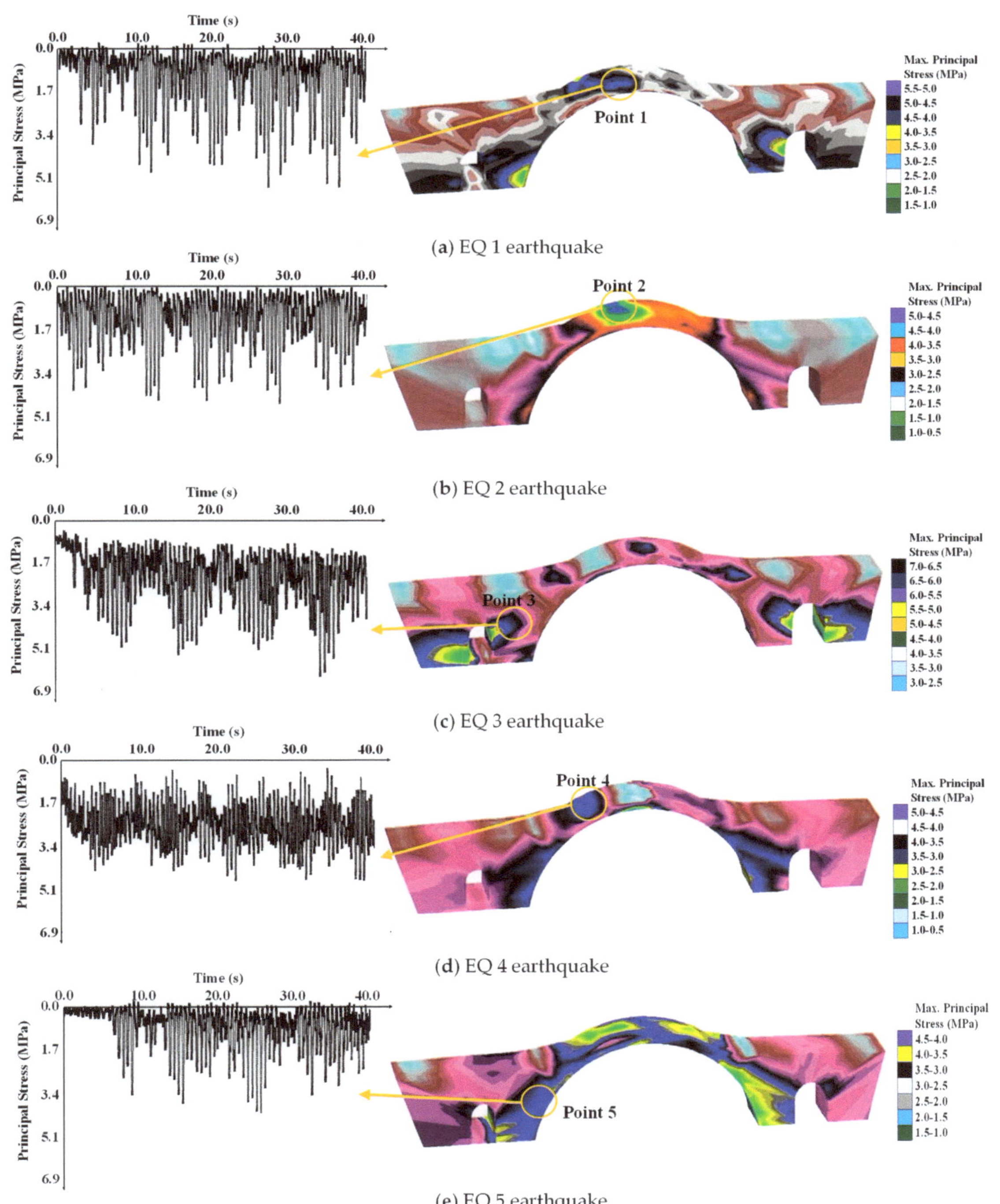

Figure 21. Principal stress behavior of Plaka bridge under EQ 1, EQ 2, EQ 3, EQ 4, and EQ 5 earthquakes.

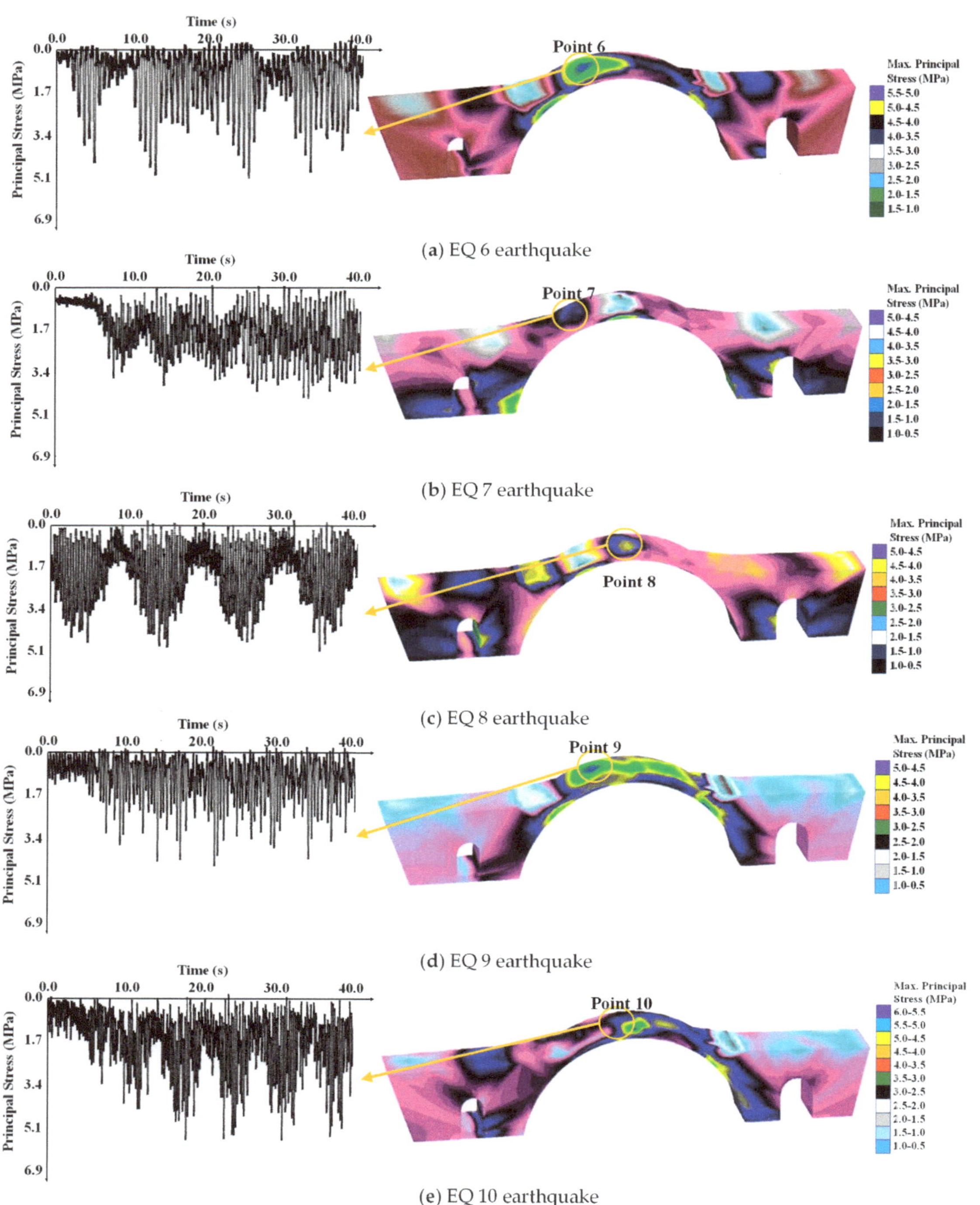

Figure 22. Principal stress behavior of Plaka bridge under EQ 6, EQ 7, EQ 8, EQ 9, and EQ 10 earthquakes.

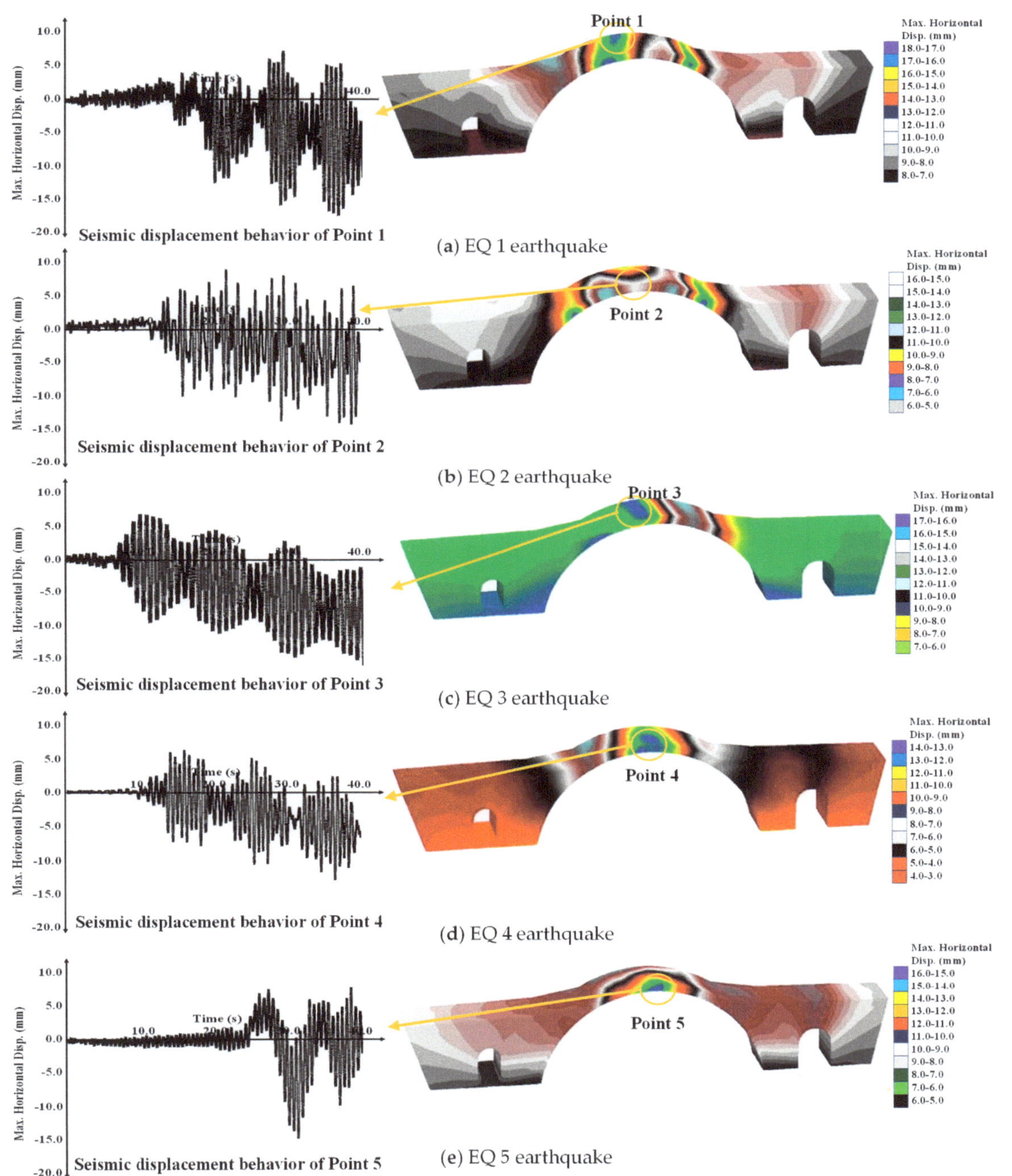

Figure 23. Displacement behavior of Plaka bridge under EQ 1, EQ 2, EQ 3, EQ 4, and EQ 5 earthquakes.

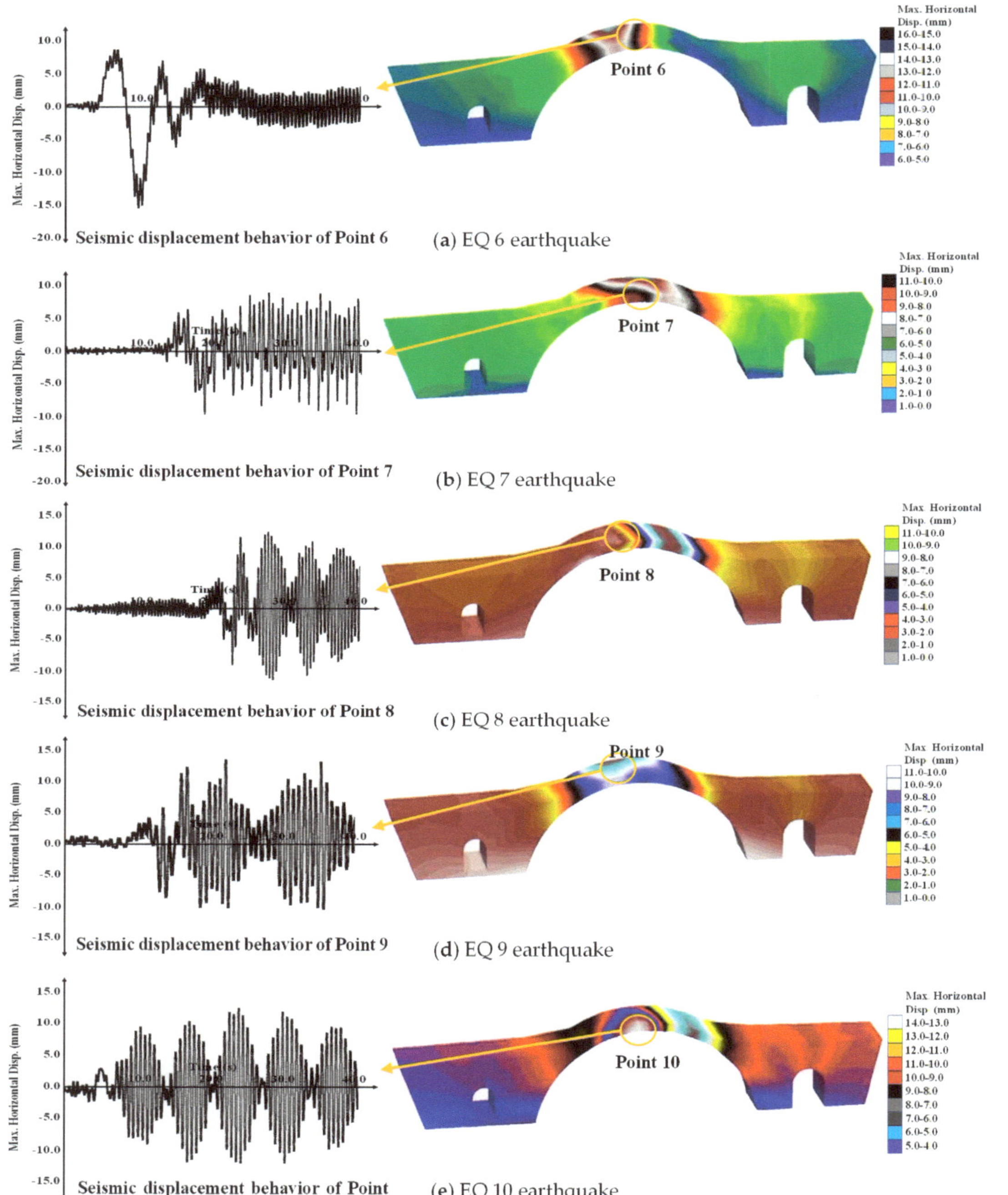

Figure 24. Displacement behavior of Plaka bridge under EQ 6, EQ 7, EQ 8, EQ 9, and EQ 10 earthquakes.

The seismic analyses in this study are based on important ground motions that occurred in Turkey in 2023, resulting in the loss of thousands of lives. Given the proximity of the Plaka bridge to Turkey, utilizing these earthquakes in three-dimensional seismic analyses holds significant importance for assessing the bridge's earthquake response. In Figure 21, the principal stress behavior of the Plaka bridge was investigated under five different earthquakes. During the EQ 1 earthquake, substantial principal stresses were observed in the middle of the main arch section of the bridge and at the foot parts of the auxiliary arches (Figure 21a). At Point 1 on the main arch, a maximum principal stress value of 5.4 MPa was recorded during the EQ 1 earthquake. During the EQ 2 earthquake, significant principal stress values manifested in the middle and along the edges of the main arch of the Plaka bridge. Furthermore, substantial principal stress values were evident at the peripheries of the auxiliary arches (Figure 21b). Under the EQ 4 earthquake, significant principal stress values emerged at the edges of the main arch of the bridge and in the lower sections of the secondary arches. Additionally, substantial principal stresses were observed in the lower regions of the main arch. Nevertheless, no critical principal stress values were noted in the upper portions of the auxiliary arches. The maximum principal stress value recorded at Point 4 during the EQ 4 earthquake was 4.5 MPa (Figure 21d). In Figure 22, the principal stress behavior of the Plaka bridge was analyzed under five distinct earthquakes. The highest principal stress values on the bridge during the EQ 6 earthquake manifested in the upper segments of the main arch (Figure 22a). Furthermore, substantial principal stress values were observed in the lower sections of the auxiliary arches. Conversely, minimal stress values were detected at the far-right and -left extremities of the bridge. The maximum principal stress value recorded at Point 6 during the EQ 6 earthquake was 5.05 MPa. In Figure 22b, the seismic behavior of the Plaka bridge during the EQ 7 earthquake is comprehensively presented. As a result of the earthquake analysis, significant principal stress values were observed in the auxiliary arch segments of the bridge (Figure 22b). During the EQ 9 earthquake, the highest principal stress values manifested in the upper segments of the main arch section of the Plaka bridge. Additionally, notable stress values were evident in the base sections of the auxiliary arches. The maximum principal stress value recorded at Point 9 during the EQ 9 earthquake was 4.7 MPa (Figure 22d). These findings indicate that the principal stress values in the body of the Plaka bridge, restored using Horasan mortar mixed with eggshell, are lower compared to those restored using Horasan mortar mixed with egg white. This suggests that such mortars significantly enhance the seismic resilience of the Plaka bridge.

Figure 23 illustrates the displacement patterns of the Plaka Bridge in response to five different earthquakes. During the EQ 1 earthquake, the most extensive horizontal displacement values were recorded in the main arch section of the Plaka bridge. The maximum horizontal displacement value recorded at Point 1 during the EQ 1 earthquake was 17.8 mm. Additionally, the maximum horizontal displacement in the auxiliary arches was approximately 11 mm (Figure 23a). In Figure 23b, the displacement behavior of the Plaka bridge was assessed during the EQ 2 earthquake. Following the EQ 2 earthquake, the most significant displacements occurred in the central section of the main arch. Conversely, the smallest displacements were observed at the bridge's base. The maximum horizontal displacement recorded at Point 2 during the EQ 2 earthquake was 15.1 mm (Figure 23b). The EQ 3 earthquake resulted in maximum horizontal displacement values within the central regions of the main arch. Additionally, substantial displacements occurred in the lower segments of the secondary arches, while minimal displacements were observed in the upper sections of the auxiliary arches. The maximum horizontal displacement recorded at Point 3 during the EQ 3 earthquake was 16.2 mm (Figure 23c). These findings underscore the criticality of the main arch sections in historical bridges. In Figure 24, the detailed horizontal displacement behavior of the bridge during five different earthquakes is examined. The most significant horizontal displacements observed on the bridge when considering the EQ 6 earthquake analysis were at the edges of the main arch section. During the earthquake, minimal displacement values were registered at the base of the auxiliary arches. Furthermore, the upper parts of the auxiliary arches exhibited maximum displacement values of approximately 10 mm. The largest recorded displacement value at Point 6 during the earthquake was 15.4 mm (Figure 24a). Under the EQ 8 earthquake, maximum displacements were observed at the edges of the main arch section of the bridge. Significant displacements also occurred at the edges of the auxiliary arch components. This finding underscores the primary importance of the main arches in the seismic behavior of masonry arch bridges (Figure 24c). Therefore, when constructing or restoring masonry arch bridges, careful consideration should be given to the middle of the main arch sections. Under the EQ 10 earthquake, significant horizontal displacements occurred both in the middle and at the edges of the main arch section of the Plaka bridge. The maximum displacement value recorded during the earthquake analysis at Point 10 was 13.3 mm. These results highlight the crucial role played by both the main arch and auxiliary arch sections in the seismic behavior of masonry arch bridges. Moreover, it was observed that the incorporation of eggshell-containing Khorasan mortar in masonry arch bridges positively contributes to their seismic displacement behavior. In light of these analysis results, it is recommended to use eggshell-containing Khorasan mortar between the stone elements during the reconstruction and restoration processes of historical bridges.

10. Conclusions

In this study, the long-term creep and seismic behaviors of a historical bridge were investigated using Khorasan mortar with additives of both egg white and eggshell. The mechanical properties of Khorasan mortar were derived from experiments, and the stiffness parameters between the stone elements were computed based on these mechanical properties. Subsequently, three-dimensional creep and seismic analyses of the bridge were conducted to assess how the behavior of the bridge is influenced when either egg white- or eggshell-added Khorasan mortar is used between the stone elements. In the first part of the study, experiments were carried out to obtain the mechanical properties of Khorasan mortar. Based on the experimental findings, it was observed that the unconfined compression strength of the control soil and all other mixtures increased with longer curing times. Notably, the maximum mixture was found to contain 50% additives of either egg white or eggshell. To optimize the use of egg white, water content was reduced, contributing to efficient water usage, a critical global resource. This reduction in water content led to increased unconfined compression strength in the samples. Furthermore, the eggshell additive exhibited superior performance compared to egg white, due to its stronger and faster chemical reaction, attributed to its lime content. Importantly, even the weakest strength value obtained with additives remained higher than that of the control sample, affirming the positive impact of adding Khorasan mortar with these additives. However, it was determined that exceeding a 50% additive ratio reduced the strength compared to samples stabilized at this percentage. Nevertheless, the strength values remained significantly higher than those of the control soil. In the second part of the study, three-dimensional numerical analyses were performed. Historically, during the construction of the bridge in 1866, Khorasan mortar with egg white additives was used between the stone elements. Static analysis of the bridge from 1866 to 2015 (149 years) revealed that significant principal stresses were concentrated in the main-arch and auxiliary-arch sections following five different flood disasters. This analysis suggested that one of the primary causes of the bridge's collapse during the 2015 flood was the use of Khorasan mortar with egg white additives in these critical sections. Analyzing the bridge constructed with Khorasan mortar containing egg white additives from 1866 to 2015 revealed maximum principal stress values of 6.3 MPa, 4.8 MPa, and 5.4 MPa at Point 1, Point 2, and Point 3, respectively. When the bridge was being restored in 2015, Khorasan mortar with eggshell additives was employed between the stone elements. For this reason, the bridge was statically analyzed from 2015 to 2164 considering Khorasan mortar with eggshell additives between stone elements. The results indicated that no structural damage would occur in any part of the bridge, even when subjected to five different hypothetical flood disasters. For the bridge constructed with eggshell-added Khorasan mortar from 2015 to 2164, maximum principal stress values of 1.92 MPa, 0.92 MPa, and 1.48 MPa were observed at Point 1, Point 2, and Point 3, respectively. These results demonstrated that the use of eggshell-mixed

Khorasan mortar between stone elements provided more favorable contributions to the long-term creep behavior of the bridge. Moreover, according to modal analysis results of the bridge using eggshell-added Khorasan mortar, the natural frequencies of Mode 1, Mode 2, and Mode 3 are 12.071 Hz, 14.623 Hz, and 21.136 Hz, respectively. Additionally, the maximum horizontal displacement value observed in the bridge body during these seismic analyses was 17.8 mm. The seismic principal-stress and displacement values obtained from the seismic analyses are very similar to those found in the literature. This result validates the accuracy of the seismic analyses of the Plaka bridge and demonstrates the reliability of using Khorasan mortar for assessing the seismic behavior of masonry bridges. Considering these comprehensive results, it is highly recommended to use eggshell-mixed Khorasan mortar, rather than egg white-mixed Khorasan mortar, for restoring or rebuilding historical bridges.

Author Contributions: Conceptualization, M.C.; methodology, M.C.; software, M.C.; validation, M.C.; formal analysis, M.C.; experiments: M.C. and E.D.; investigation, M.C. and E.D.; resources, M.C.; data curation, M.C.; writing—original draft preparation, M.C. and E.D.; writing—review and editing, M.C. and E.D.; visualization, M.C.; supervision, M.C.; project administration, M.C.; funding acquisition, M.C. and E.D. All authors have read and agreed to the published version of the manuscript.

Funding: This research received no external funding.

Data Availability Statement: Data are contained within the article.

Conflicts of Interest: The authors declare no conflicts of interest.

References

1. Bouzas, O.; Conde, B.; Matos, J.C.; Solla, M.; Cabaleiro, M. Reliability-based structural assessment of historical masonry arch bridges: The case study of Cernadela bridge. *Case Stud. Constr. Mater.* **2023**, *18*, e02003. [CrossRef]
2. Majtan, E.; Cunningham, L.S.; Rogers, B.D. Numerical study on the structural response of a masonry arch bridge subject to flood flow and debris impact. *Structures* **2023**, *48*, 782–797. [CrossRef]
3. Silva, R.; Costa, C.; Arêde, A. Numerical methodologies for the analysis of stone arch bridges with damage under railway loading. *Structures* **2022**, *39*, 573–592. [CrossRef]
4. Ozakgul, K.; Yilmaz, M.F.; Caglayan, B.O. Assessment of an old reinforced concrete open-spandrel arch railway bridge. *Structures* **2022**, *44*, 284–294. [CrossRef]
5. Chen, B.; Yang, Y.; Zhou, J.; Zhuang, Y.; McFarland, M. Damage detection of underwater foundation of a Chinese ancient stone arch bridge via sonar-based techniques. *Measurement* **2021**, *169*, 108283. [CrossRef]
6. Gönen, S.; Soyöz, S. Seismic analysis of a masonry arch bridge using multiple methodologies. *Eng. Struct.* **2021**, *226*, 111354. [CrossRef]
7. Saygılı, Ö.; Lemos, J.V. Seismic vulnerability assessment of masonry arch bridges. *Structures* **2021**, *33*, 3311–3323. [CrossRef]
8. Papa, T.; Grillanda, N.; Milani, G. Three-dimensional adaptive limit analysis of masonry arch bridges interacting with the backfill. *Eng. Struct.* **2021**, *248*, 113189. [CrossRef]
9. Zani, G.; Martinelli, P.; Galli, A.; Prisco, M.d. Three-dimensional modelling of a multi-span masonry arch bridge: Influence of soil compressibility on the structural response under vertical static loads. *Eng. Struct.* **2020**, *220*, 110998. [CrossRef]
10. Chen, X.; Qi, X.-B.; Xu, Z.-Y. Determination of weathered degree and mechanical properties of stone relics with ultrasonic CT: A case study of an ancient stone bridge in China. *J. Cult. Herit.* **2020**, *42*, 131–138. [CrossRef]
11. Sánchez-Aparicio, L.J.; Castro, A.B.-D.; Conde, B.; Carrasco, P.; Ramos, L.F. Non-destructive means and methods for structural diagnosis of masonry arch bridges. *Autom. Constr.* **2019**, *104*, 360–382. [CrossRef]
12. Simos, N.; Manos, G.C.; Kozikopoulos, E. Near- and far-field earthquake damage study of the Konitsa stone arch bridge. *Eng. Struct.* **2018**, *177*, 256–267. [CrossRef]
13. Aydin, A.C.; Özkaya, S.G. The finite element analysis of collapse loads of single-spanned historic masonry arch bridges (Ordu, Sarpdere Bridge). *Eng. Fail. Anal.* **2018**, *84*, 131–138. [CrossRef]
14. Conde, B.; Ramos, L.F.; Oliveira, D.V.; Riveiro, B.; Solla, M. Structural assessment of masonry arch bridges by combination of non-destructive testing techniques and three-dimensional numerical modelling: Application to Vilanova bridge. *Eng. Struct.* **2017**, *148*, 621–638. [CrossRef]
15. Karaton, M.; Aksoy, H.S.; Sayın, E.; Calayır, Y. Nonlinear seismic performance of a 12th century historical masonry bridge under different earthquake levels. *Eng. Fail. Anal.* **2017**, *79*, 408–421. [CrossRef]

16. Conde, B.; Díaz-Vilariño, L.; Lagüela, S.; Arias, P. Structural analysis of Monforte de Lemos masonry arch bridge considering the influence of the geometry of the arches and fill material on the collapse load estimation. *Constr. Build. Mater.* **2016**, *120*, 630–642. [CrossRef]

17. Costa, C.; Arêde, A.; Morais, M.; Aníbal, A. Detailed FE and DE Modelling of Stone Masonry Arch Bridges for the Assessment of Load-carrying Capacity. *Procedia Eng.* **2015**, *114*, 854–861. [CrossRef]

18. Bayraktar, A.; Türker, T.; Altunişik, A.C. Experimental frequencies and damping ratios for historical masonry arch bridges. *Constr. Build. Mater.* **2015**, *75*, 234–241. [CrossRef]

19. Ozmen, A.; Sayin, E. 3D soil structure interaction efects on the seismic behavior of single span historical masonry bridge. *Geotech Geol Eng* **2023**, *41*, 2023–2041. [CrossRef]

20. Shabani, A.; Kioumarsi, M. Seismic assessment and strengthening of a historical masonry bridge considering soil-structure interaction. *Eng. Struct.* **2023**, *293*, 116589. [CrossRef]

21. Vuoto, A.; Funari, M.F.; Lourenço, P.B. Shaping Digital Twin Concept for Built Cultural Heritage Conservation: A Systematic Literature Review. *Int. J. Archit. Herit.* **2023**, 1–34. [CrossRef]

22. Ferretti, E.; Pascale, G. Some of the latest active strengthening techniques for masonry buildings: A critical analysis. *Materials* **2019**, *12*, 1151. [CrossRef]

23. Hafner, I.; Kisicek, T.; Gams, M. Review of methods for seismic strengthening of masonry piers and walls. *Buildings* **2023**, *13*, 1524. [CrossRef]

24. Angiolilli, M.; Gregori, A.; Pathiragi, M.; Cusatis, G. Fiber Reinforced Cementitious Matrix (FRCM) for strengthening historical stone masonry structures: Experiments and computations. *Eng. Struct.* **2020**, *224*, 111102. [CrossRef]

25. Ferretti, F.; Khatiwada, S.; Incerti, A.; Giacomin, G.; Tomaro, F.; De Martino, F.; Mazzotti, C. Structural strengthening of masonry elements by reinforced repointing combined with FRCM and CRM. *Procedia Struct. Integr.* **2023**, *44*, 2254–2261. [CrossRef]

26. Triantafillou, T. Strengthening of existing masonry structures: Concepts and structural behavior. In *Textile Fibre Composites in Civil Engineering*; Springer: Berlin/Heidelberg, Germany, 2016; pp. 361–374.

27. Kouris, L.A.S.; Triantafillou, T.C. Design methods for strengthening masonry buildings usting textile-reinforced mortar. *J. Compos. Constr.* **2019**, *23*, 361–374. [CrossRef]

28. Carozzi, F.G.; Poggi, C. Mechanical properties and debonding strength of Fabric Reinforced Cementitious Matrix (FRCM) systems for masonry strengthening. *Compos. Part B Eng.* **2015**, *70*, 215–230. [CrossRef]

29. Grande, E.; Milani, G. Numerical simulation of the tensile behavior of FRCM strengthening systems. *Compos. Part B Eng.* **2020**, *189*, 107886. [CrossRef]

30. Lemos, J.V. Discrete element modeling of masonry structures. *Int. J. Archit. Herit.* **2007**, *1*, 190–213. [CrossRef]

31. Gobin, F.; de Felice, G.; Lemos, J.V. Numerical procedures for the analysis of collapse mechanisms of masonry structures using discrete element modelling. *Eng. Struct.* **2021**, *246*, 113047. [CrossRef]

32. Masi, F.; Stefanou, I.; Maffi-Berthier, V.; Vannucci, P. A Discrete Element Method based-approach for arched masonry structures under blast loads. *Eng. Struct.* **2020**, *216*, 110721. [CrossRef]

33. Itasca Consulting Group, Inc. *FLAC3D—Fast Lagrangian Analysis of Continua in Three-Dimensions, Ver. 9.0*; Itasca Consulting Group, Inc.: Itasca, MN, USA, 2023.

34. Mercuri, M.; Pathirage, M.; Gregori, A.; Cusatis, G. Influence of self-weight on size effect of quasi-brittle materials: Generalized analytical formulation and application to the failure of irregular masonry arches. *Int. J. Fract.* **2023**, *246*, 117–144. [CrossRef]

35. Yazdani, M.; Habibi, H. Residual capacity evaluation of masonry arch bridges by extended finite element method. *Struct. Eng. Int.* **2021**, *33*, 183–194. [CrossRef]

36. Szabó, S.; Funari, M.F.; Lourenço, P.B. Masonry patterns' influence on the damage assessment of URM walls: Current and future trends. *Dev. Built Environ.* **2023**, *13*, 100119. [CrossRef]

37. Giannelos, C.; Plainis, P.; Vintzileou, E. The historical bridge of Plaka: Interpretation of structural behavior and collapse. *Eng. Fail. Anal.* **2023**, *153*, 107589. [CrossRef]

38. Britannica. Available online: https://www.britannica.com/place/Khorasan-historical-region-Asia (accessed on 3 May 2024).

39. Oğuz, C.; Turker, F.; Kockal, N.U. Construction materials used in the historical roman era bath in myra. *ScientificWorldJournal.* **2014**, *2014*, 536105. [CrossRef]

40. *TS EN 196-1*; Methods of Testing Cement—Part 1: Determination of Strength. Turkish Standard: Ankara, Turkey, 2016.

41. *ASTM D2166/D2166M-16*; Standard Test Method for Unconfined Compressive Strength of Cohesive Soil. ASTM: West Conshohocken, PA, USA, 2016; pp. 1–7.

42. Angiolilli, M.; Gregori, A.; Vailati, M. ime-Based Mortar Reinforced by Randomly Oriented Short Fibers for the Retrofitting of the Historical Masonry Structure. *Materials* **2020**, *13*, 3462. [CrossRef]

43. AFAD. *Disaster and Emergency Management Presidency*; AFAD: Ankara, Turkey, 2023.

Article

A New Methodology to Estimate the Early-Age Compressive Strength of Concrete before Demolding

Bayarjavkhlan Narantogtokh [1], Tomoya Nishiwaki [1,*], Fumiya Takasugi [1], Ken Koyama [1], Timo Lehmann [1,2], Anna Jagiello [1,3], Félix Droin [1,4] and Yao Ding [1]

1   Graduate School of Engineering, Tohoku University, Sendai 980-8579, Japan; javkhaa0911@gmail.com (B.N.); fumiya.takasugi.r5@dc.tohoku.ac.jp (F.T.); koyama.ken.r7@dc.tohoku.ac.jp (K.K.); timo.lema@gmail.com (T.L.); anu.annajagiello@gmail.com (A.J.); felix.droin@outlook.com (F.D.); yao.ding.a5@tohoku.ac.jp (Y.D.)
2   School of Process Engineering, Hamburg University of Technology, 21073 Hamburg, Germany
3   Faculty of Economics, West Pomeranian University of Technology, 71-210 Szczecin, Poland
4   Department of Civil Engineering, National Institute of Applied Sciences of Toulouse, 31077 Toulouse, France
*   Correspondence: tomoya.nishiwaki.e8@tohoku.ac.jp

Abstract: Non-destructive testing has many advantages, such as the ability to obtain a large number of data without destroying existing structures. However, the reliability of the estimation accuracy and the limited range of applicable targets remain an issue. This study proposes a novel pin penetration test method to determine the early-age compressive strength of concrete before demolding. The timing of demolding and initial curing is determined according to the strength development of concrete. Therefore, it is important to determine the compressive strength at an early age before demolding at the actual construction site. The applicability of this strength estimation methodology at actual construction is investigated. Small test holes (12 mm in diameter) are prepared on the mold surface in real construction sites and mock-up specimens in advance. The pin is penetrated into these test holes to obtain the relationship between the compressive strength and the penetration depth. As a result, it is confirmed that the pin penetration test method is suitable for measuring the early-age compressive strength at the actual construction site. This allows the benchmark values for compressive strength, necessary to avoid early frost damage, to be directly verified on the concrete structural members at the construction site. For instance, the compressive strengths of greater than 5 MPa and 10 MPa can be confirmed by the penetration depths benchmark values of 8.0 mm and 6.7 mm or less, respectively.

Keywords: compressive strength estimation; pin penetration test; non-destructive test; on-site test; demolding; early-age compressive strength; cold weather concreting

Citation: Narantogtokh, B.; Nishiwaki, T.; Takasugi, F.; Koyama, K.; Lehmann, T.; Jagiello, A.; Droin, F.; Ding, Y. A New Methodology to Estimate the Early-Age Compressive Strength of Concrete before Demolding. *Buildings* **2024**, *14*, 2099. https://doi.org/10.3390/buildings14072099

Academic Editor: Dan Bompa

Received: 15 June 2024
Revised: 1 July 2024
Accepted: 5 July 2024
Published: 9 July 2024

1. Introduction

It is important to determine the early-age compressive strength at the actual construction site when concrete work is executed in cold weather conditions to prevent early-age frost damage. Most of the international norms and guidelines for cold weather concreting are recommending to obtain at least 5 MPa strength before exposing concrete to early-age freezing. However, there are no suitable methods available for the measurement of low-strength concrete at a very early age, in particular before demolding at the construction site. As well known, cement hydration is quite sensitive to temperature, and strength development would be delayed at low temperatures [1]. The mold provides an important protection function against low temperatures for early-age concrete. Therefore, it is desirable to extend the period during which the concrete mold remains in place as much as possible to ensure sufficient initial curing. Japanese Architectural Standard Specification for Reinforced Concrete Works, JASS 5 [2], provided by the Architectural Institute of Japan, specifies that the period where the mold remains in place shall be controlled by strength.

JASS 5 requires the removal of the mold after confirming that the compressive strength of the structural concrete attains specific criteria depending on the planned service life. In the case of the 'short-term' and 'standard' service life level, the compressive strength should be 5 N/mm^2 or more for demolding. In the case of the 'long-term' and 'extra-long-term' service life level, the compressive strength should be 10 N/mm^2 or more. In the case of cold weather concreting, the recommendation for the practice of cold weather concreting [3,4] and its commentary also require initial curing until the compressive strength exceeds 5 N/mm^2 in order to avoid initial frost damage.

As mentioned above, the timing of demolding and initial curing is determined according to the strength development of concrete. However, because compressive strength is obtained by compressive strength tests using test pieces, it undergoes different curing conditions from the structural concrete placed in the mold. To obtain the compressive strength of the actual structural concrete, it is desirable to take cores from the hardened structure and test them, which is destructive. However, it is challenging to take cores from unmatured concrete members before demolding. In addition, conducting a compression test is not easy to do at the construction site, and it is unavoidably costly in terms of time and economics, such as transportation to the testing location and the testing procedure itself.

There are various non-destructive testing (NDT) methods available for the assessment of the in situ concrete strength. Non-destructive and micro-destructive testing methods for estimating the compressive strength of concrete have been studied extensively, and many methods have been put into practical use and standardized. The most commonly used NDT methods are a rebound hammer test [5–8] and ultrasonic pulse velocity [9–15]. A combination of those methods [16–18] is also widely used to evaluate the existing concrete structures. However, most of them have been applied to the concrete surface to estimate the internal mechanical properties (compressive strength). Any of those methods are only applicable for the existing or hardened concrete structures but not applicable to the unmatured concrete structures before demolding. The target concrete of these NDTs is post-cured concrete or structures that have deteriorated over time, and there are not enough tests available for young concrete, especially before demolding.

Non-destructive testing has many advantages, such as the ability to obtain a large number of data without destroying existing structures on a large scale [19]. However, the reliability of the estimation accuracy remains an issue. For example, the rebound hammer test is an extremely simple test method and has been standardized in Japan as JIS A 1155 [20], which is modified from ISO 1920-7 [21]. However, the method of estimating strength from the degree of rebound is not included in these standards due to a unified calculation method for estimating strength from the degree of rebound has not been obtained [22–25].

There are several studies that investigated the NDT methods to use for alternative assessment of in situ concrete strength. Gunes et al. [26] have studied the drilling-based test methodology for non-destructive estimation of in situ concrete strength and carried out to develop a relationship between the drilling resistance parameter and compressive strength. However, they concluded that the most accurate estimations for strength are obtained when the drilling resistance measurements are combined with rebound hammer or ultrasonic pulse velocity measurements as additional NDT data. Al-Sabah et al. [27] investigated the post-installed screw pull-out test for the assessment of the compressive strength of in situ concrete, and they found that the correlation between the compressive strength of mortar and the peak load was significant. In addition, one of our previous studies [28] investigated the screening method to evaluate the low-strength concrete using the combination of two low-energy non-destructive testing devices, a type L rebound hammer and a scratching test. As a result, a concrete classification chart is proposed based on the boundary values of two NDT methods, and it provides a concrete strength range via a classification chart and a conservative estimate of the compressive strength. Nguyen et al. [29] studied the simple non-destructive method for evaluating the cover concrete quality, and they concluded that the water intentional spray test method could sensitively detect the poor-quality concrete caused by a high water–cement ratio and short curing time. However, all NDT methods mentioned here were performed on the surface of concrete; it is difficult to apply it directly to concrete before demolding when the surface has not yet been exposed.

The mold plays a role in protecting the concrete from external stimuli and ensuring its quality, and it is undesirable to remove even a part of the mold for the purpose of confirming the strength of young-aged concrete. A proposal to estimate the strength of the concrete with mold has been considered. In the BOSS (Broken Off Specimens by Splitting) specimen method [30] shown in Figure 1, which has been standardized in Japan as JIS A 1163 [31], concrete is poured into the mold with the mold for the BOSS specimen installed in advance to obtain a specimen that has hardened in the same environment as the structural concrete. By performing a compressive test on this BOSS specimen, strength estimation can be performed with high accuracy. However, there are some problems with the simplicity of the test, such as the fact that the compressive test cannot be performed at the construction site, the need for repair after demolding, and the limited number of specimens.

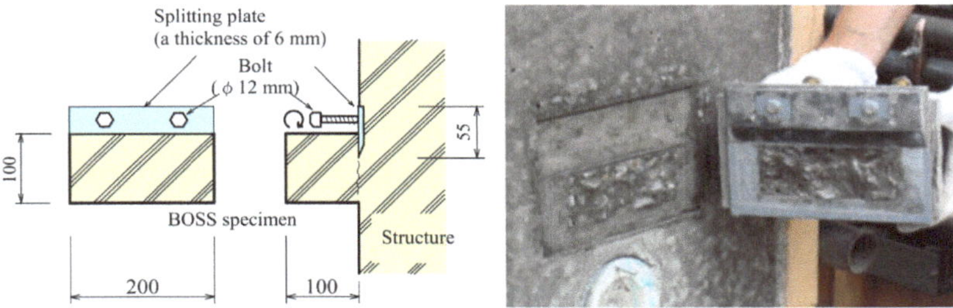

Figure 1. Non-destructive testing method. BOSS specimen [30].

The strength of concrete generally depends on the strength of the hardened cement. Therefore, a non-destructive testing method called the penetration resistance method has been proposed [32–35], in which pins or needles are inserted into the mortar portion of concrete, and the strength is estimated from the penetration depth. Maliha et al. also used the same pin penetration device used in this study; they obtained a correlation between the penetration depth and compressive strength but confirmed that it was affected by coarse aggregate [34]. Conversely, if the influence of coarse aggregate can be eliminated, strength estimation can be performed with higher accuracy. For example, the accuracy of strength estimation for mortar without coarse aggregate is high, and there are also standardized methods for strength estimation, such as shotcrete [35].

In addition, a method called 'smart sensor mold', in which a mold is equipped with a sensor that can measure the surface temperature history of concrete, etc., has been proposed to confirm the development of standard strength at demolding [36,37]. However, this method cannot be applied to conventional mold easily because it requires the use of a mold with special devices.

Based on the above backgrounds, this study proposes to use the pin penetration test method to determine the early-age compressive strength before demolding. There are two main advantages of this method that is reason to use in this study. First, the proposed pin penetration test is applicable to use before demolding. Second, this method is suitable to use for low-strength concrete. However, the previous studies [37] have only shown the effectiveness of this method on laboratory-sized specimens and have not examined it on full-size concrete specimens, which may include different conditions, e.g., compaction conditions and uncontrolled temperature. In this study, the applicability of this strength estimation method at actual construction is investigated. Small test holes (12 mm in diameter) are prepared on the mold surface in real construction sites and mock-up specimens in advance. The pin is penetrated into these test holes to obtain the relationship between the compressive strength and the penetration depth.

2. Testing Method and Materials

2.1. Pin Penetration Testing Device

The pin penetration testing device used in this study is shown in Figure 2, and its specifications are in Table 1. In this device, a metal pin of 2.5 mm in diameter and 60 mm in length is ejected with constant energy (6 Nm) from an internally compressed spring, pushing the pin at the tip up to a predetermined height, and the penetration depth is then measured. The measured penetration depth is digitally displayed on a control unit connected to the main unit of the tester. The measuring principle is similar to that of conventional pin penetration testers [32,33] used for concrete. This device was originally developed to estimate the degree of decay or deterioration of wood based on the penetration depth. In previous studies [34,35], this device was employed to determine the compressive strength of concrete on the cubic mold on a laboratory scale. No result has been verified on full-size mock-up specimens and/or actual construction sites. This study investigated the applicability of this method in actual construction sites to determine the early compressive strength. Two types of pipes with different materials were used to compare the effect of material type on penetration depth, as shown in Figure 2.

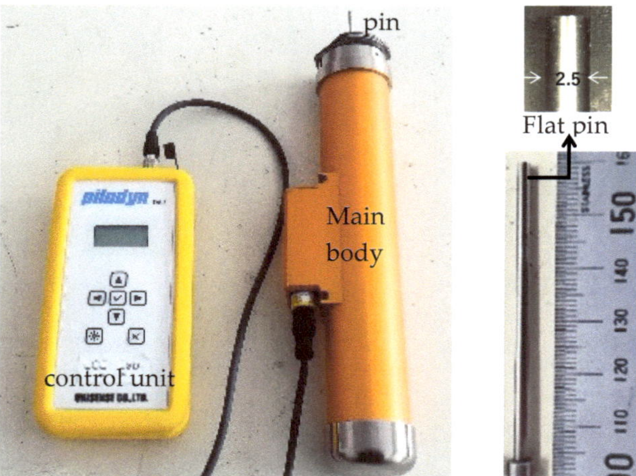

Figure 2. Pin penetration tester.

Table 1. Pin penetration device specifications.

Device Specification	Value Range
Measuring range	0~35 mm
Measurement accuracy	0.1 mm
Dimensions of device	50 × 70 × 335 mm
Weight	~2 kg
Energy	6 J (Nm)

2.2. Technique of Measuring Strength before Demolding

Figure 3 shows the pin penetration test of the mock-up specimen. As shown in this photo, the specimens were not demolded. The pin penetration testing device is compact and lightweight, does not require an external power source other than the built-in dry cell batteries, and can be brought to the construction site easily. The pin penetration tests were performed on mortar filled in the holes, as shown in Figure 4. The measurement results were obtained from each hole.

Figure 3. Pin penetration test of the mock-up mold before demolding.

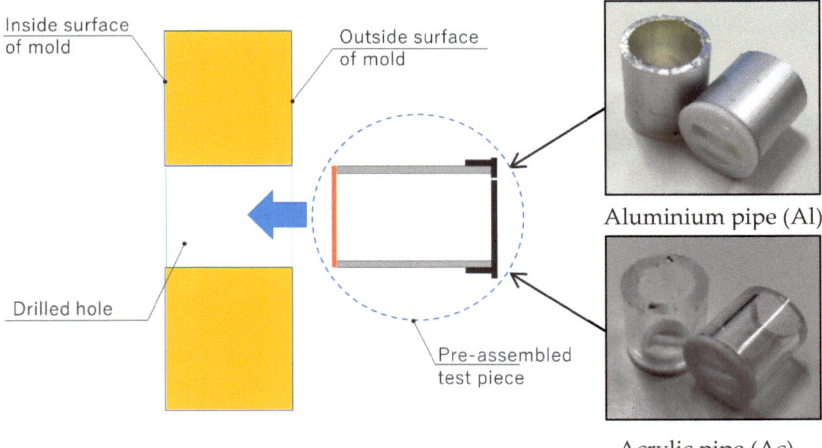

Figure 4. Test hole before inserting the pipe.

Figures 4–6 show the experimental procedure for preparing the test holes on the mold. The pre-assembled test piece consists of an aluminum (Al) or acrylic (Ac) pipe, a plastic cover, and a needle, as shown in Figure 4. The holes in the mold have the same diameter as those used for ordinary separators of the mold, and the test can be conducted simply by placing an aluminum pipe with a coarse aggregate penetration prevention needle and a plastic cap with holes on the side facing the outside of the mold. At first, test holes with a diameter of 12 mm were drilled in the mold before inserting the test piece, as shown in Figure 4. The space between holes was not less than 40 mm. After that, the pre-assembled test pieces were inserted into drilled holes, as shown in Figure 5. The concrete was poured into the mold after finishing the preparation of the mold. The internal and surface vibrators were used to compact the concrete until cement paste leaked from the plastic cap, as shown in Figure 6.

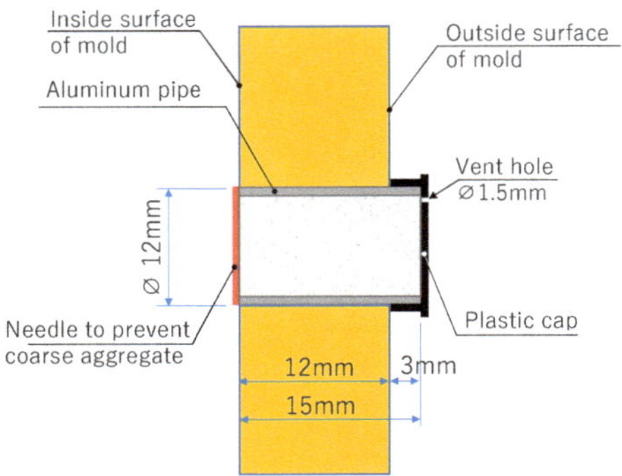

Figure 5. Test piece after inserting the pipe.

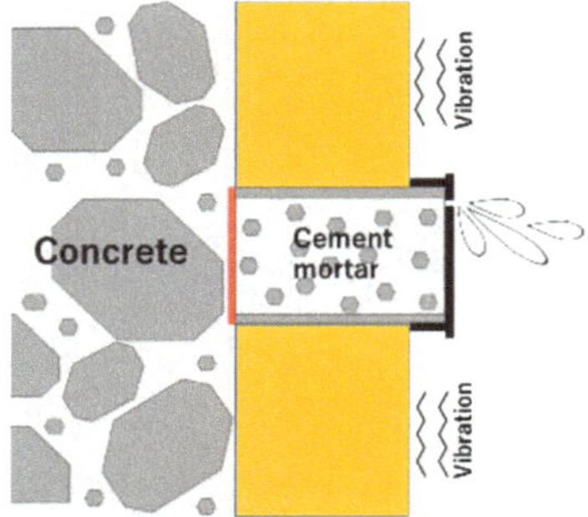

Figure 6. Test hole after concrete casting.

2.3. Pin Penetration Test at Construction Site and Mock-Up Mold

Figure 7 shows the outline of experimental work for the pin penetration test. The pin penetration test was performed at both the laboratory and the actual construction site to investigate the applicability of this method. Also, mock-up specimens were prepared to conduct pin penetration tests to simulate the actual construction site. Cylindrical specimens with a size of 100×200 mm were prepared to determine the compressive strength of concrete. Concrete temperatures and ambient temperatures were measured for each type of specimen and on the construction site using the digital thermometer. The temperature measurement interval was 5 min.

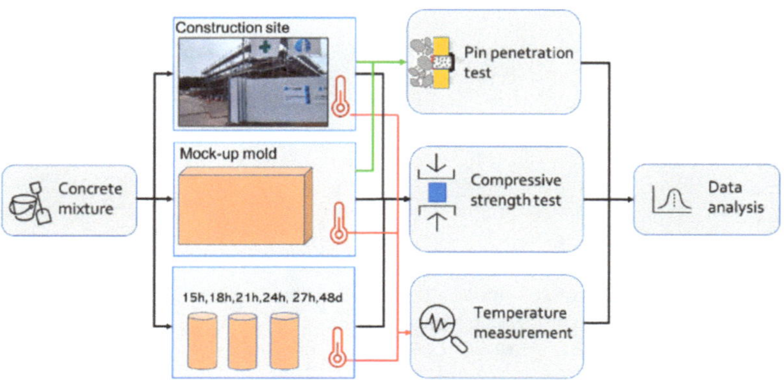

Figure 7. Outline of experimental work.

In the scope of this study, penetration tests were conducted with 30 holes with both Al and Ac pipes. For each series, pin penetration tests were conducted according to the schedule shown in Table 2. At the same accumulated temperature, the compressive strength tests were also conducted to obtain the corresponding compressive strength.

Table 2. Testing schedule.

Determination		Curing Time					
		15 h	18 h	21 h	24 h	27 h	48 h
Mock-up mold	Al pipe	○	○	○	○	○	○
	Ac pipe	○	○	○	○	○	○
Construction site	Al pipe		○	○	○		○
	Ac pipe		○	○	○		○

Figure 8 shows the schematic diagram of the mock-up mold. The size of the mock-up mold was 1800 × 900 × 200 mm each in length, height and thickness. In the case of the mock-up mold, the pin penetration test was performed once 15 h after casting and then repeated every 3 h. Therefore, testing times were 15, 18, 21, 24, 27, and 48 h. The core drilled specimens were taken from mock-up mold at a time interval of 21, 24, 27, and 48 h. The core drilling was conducted with the mold in place, as opposed to the usual situation. Temperatures were measured from the surface of the mock-up mold using the digital thermometer.

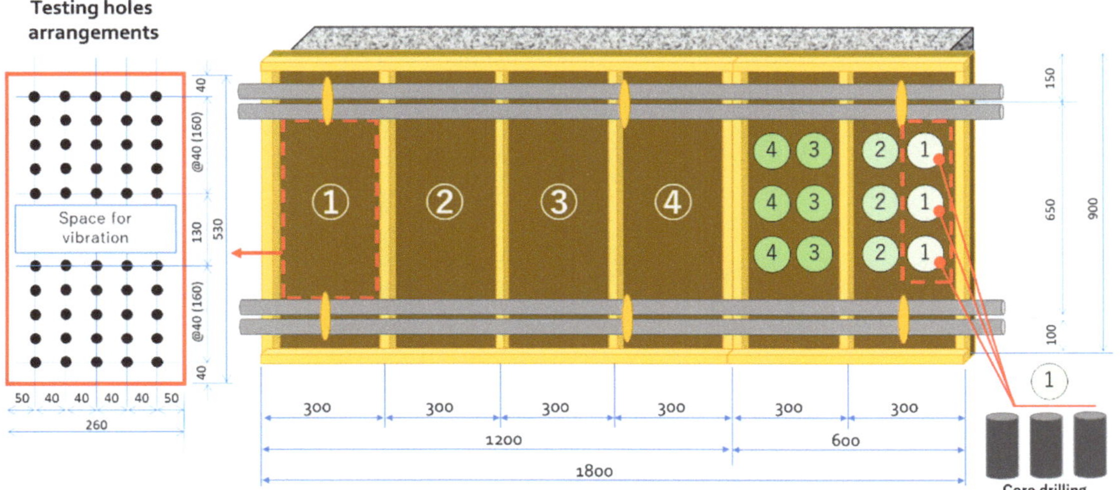

Figure 8. Design of mock-up mold.

On the construction site, it was impossible to perform the test at the construction site before 8 a.m. in the morning and after 5 p.m. in the evening due to the limitation of working time at the construction site. Therefore, pin penetration tests were performed at 18, 21, 24, and 48 h at the construction site. The temperature was also measured from the concrete surface. Figure 9 shows the prepared test holes at the construction site and mock-up mold before conducting the pin penetration test.

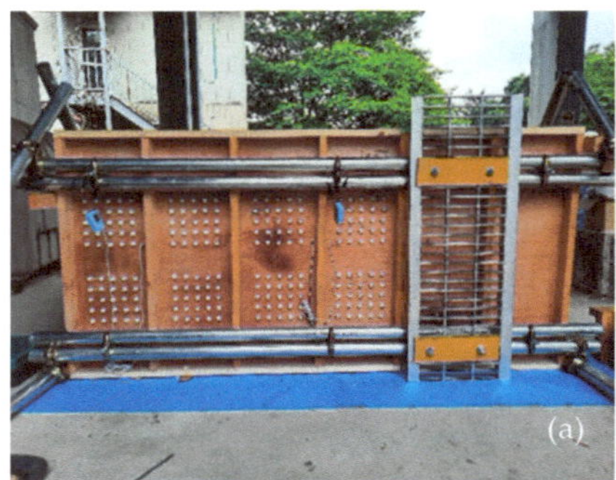

Figure 9. Test holes at mold (**a**) mock-up mold (**b**) construction site.

2.4. Concrete Used in the Experiment

The concrete for the specimens was supplied by a ready-mixed concrete plant near Sendai City, Japan. Table 3 shows its specifications, and Table 4 shows its mix proportions. As shown in these tables, tests were conducted on the different concretes with different nominal strength, slump, and curing conditions also were different. Two different fine aggregates were used for adjustment of particle size distribution. After casting, the cylindrical specimens were sealed and cured at the construction site until testing.

Table 3. Used concrete specifications.

Series	Testing Condition	Nominal Strength [MPa]	Slump [cm]
CS-1	Construction site	36	21
CS-2	(at ambient air temperature)	36	21
MS-1	Mock-up specimens	24	18
MS-2	(at ambient air temperature)	30	18

Table 4. Mix proportions (kg/m^3).

Series	W/C [%]	Cement	Water	Fine Aggregate *	Coarse Aggregate	Admixture
CS-1	41.0	427	175	598, 158	969	6.41
CS-2	41.0	427	175	598, 158	969	6.41
MS-1	54.0	324	175	659, 165	990	3.24
MS-2	46.5	376	175	626, 157	990	3.76

* The fine aggregate contains a combination of two different sources.

2.5. Compressive Strength Test

Compressive strength tests were conducted in accordance with JIS A 1108. For each series of tests, three cylindrical specimens were used for compressive strength tests to confirm the strength. The testing time was determined by temperature measurement, which was reached when the cured specimens obtained the same accumulated temperature as the site concrete. The accumulated temperatures were calculated using Equation (1) [3].

$$M = \sum (10 + T)\Delta t \tag{1}$$

where

M: accumulated temperature (°D·D);
T: temperature at Δt;
Δt: time.

Both ends of the cylindrical specimens were polished. For the young specimens that were not strong enough to withstand polishing, unbonded capping devices with soft rubber were used. A 1000 kN universal testing machine was used for loading.

3. Results and Discussion

3.1. Temperature Difference between Site Concrete and Specimens

Figures 10–13 show the temperature histories of the construction site and mock-up specimens. The temperature measured from the site or mock-up specimen concrete gives the highest temperature profiles, and their temperatures slightly decreased to ambient air temperature. The cylindrical specimens were cured at the same ambient air temperature as mock-up molds and construction sites. However, the highest temperature profile occurred within the first 8 h and then decreased to ambient air temperature. In general, cylindrical specimens' temperatures depend on the ambient air temperature. Therefore, it is confirmed that even if the ambient air temperature is the same, the actual construction site concrete temperature is higher than the temperature of site-cured specimens.

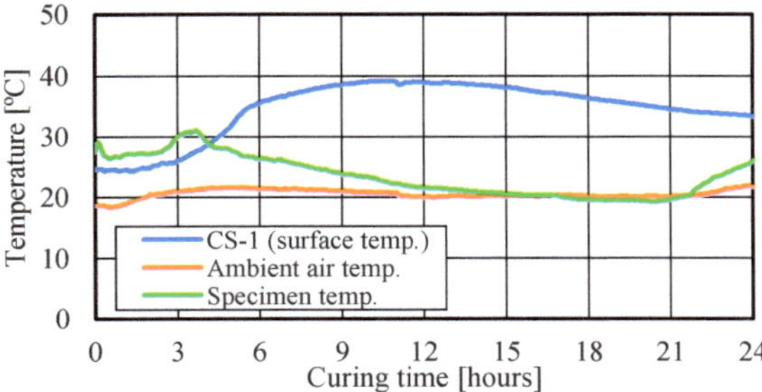

Figure 10. Temperature history of construction site (CS-1).

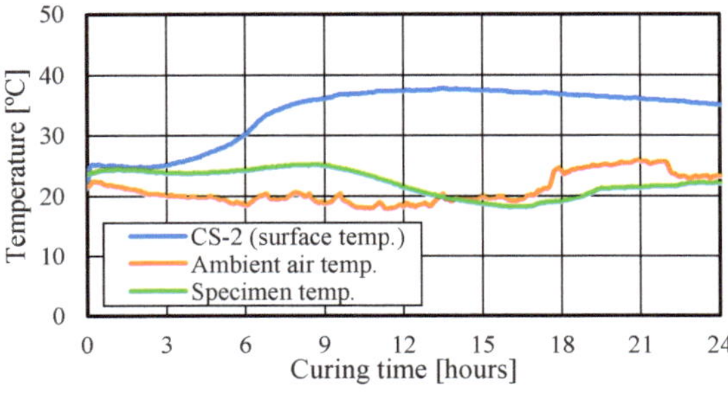

Figure 11. Temperature history of construction site (CS-2).

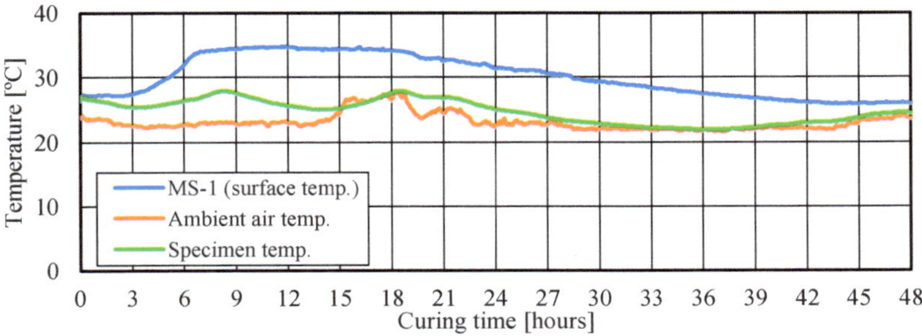

Figure 12. Temperature history of mock-up specimens (MS-1).

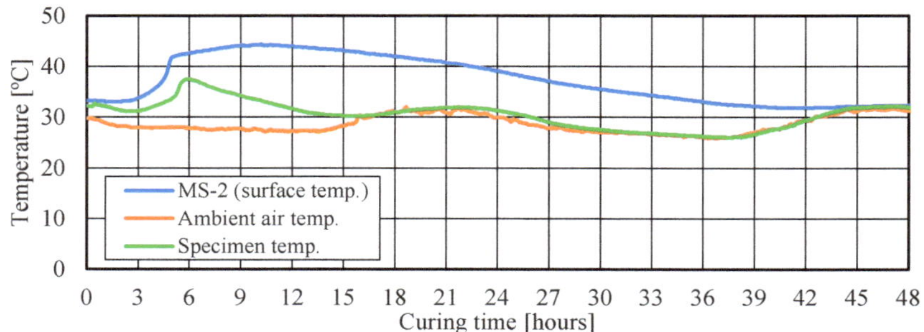

Figure 13. Temperature history of mock-up specimens (MS-2).

3.2. Accumulated Temperature and Strength Delaying Time

As defined by [38], the concrete of the same mix at the same maturity has approximately the same strength, whatever combination of temperature and time that leads to maturity. Therefore, the strength delaying time was determined based on the time to reach the same accumulated temperature for all series. For instance, Figure 14 shows the accumulated temperature and curing time for MS-1. As shown in this figure, the accumulated temperature of the mock-up specimens (MS-1) is 27.5 °D·D, 43.5 °D·D, and 81.2 °D·D after 15 h, 24 h, and 48 h, respectively. However, cylindrical specimens obtained the same accumulated temperature after 17 h 45 min, 29 h 26 min, and 56 h 40 min. Therefore, the strength delaying time is 2 h 45 min, 5 h 26 min, and 8 h 40 min for the cylindrical specimens. It confirms that the on-site cured specimens are not suitable to evaluate the actual compressive strength of concrete at the construction site, and the direct measurement method is important to determine the pre-curing time, especially in cold weather conditions.

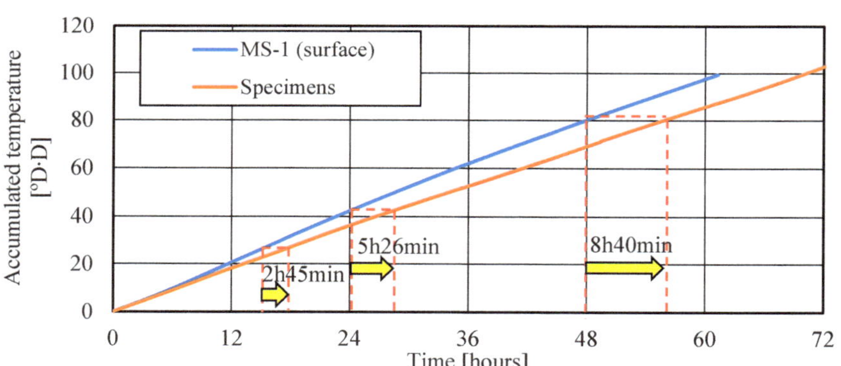

Figure 14. Accumulated temperature and curing time (MS-1).

3.3. Relationship between Compressive Strength and Penetration Depth

The relationship between the penetration depth d and compressive strength is shown in Figure 15. In the case of Al pipe (Figure 15a), the same depth of penetration is measured even when the strength is increased. Al pipes are stronger than Ac pipes, which are difficult to expand when the pin penetrates into the cement mortar filled in Al pipes. This is the reason why Al pipe gives the same penetration depth even when the strength is different. Therefore, it is not suitable to use strong materials for pipe because concrete-filled steel pipe structures have high bearing capacity and strong deformation ability [39,40]. In the case of Ac pipe (Figure 15b), the depth of penetration tends to decrease exponentially as the compressive strength increases. When the compressive strength is less than 5 MPa, the depth of penetration shows significant variation. However, the obtained result was different from the previous curve, which was determined in laboratory experiments [35]. The compressive strength range that can be estimated by the proposed method in this study is set from 3 MPa to 15 MPa. A regression curve is set up to obtain the strength estimation equation using the measurement points within this strength range. Here, based on the graph shown in Figure 16, we use the inverse proportionality equation shown in Equation (2) below.

$$fc = \frac{s}{d - t} \tag{2}$$

where,

fc: compressive strength [MPa];

d: corrected penetration depth [mm];

s,t: experimental constant.

The correlation between the compressive strength fc and the penetration depth d is determined from the least squares method as in the following Equations (3) and (4). The coefficients of determination of R^2 are $R^2{}_{Al} = 0.399$ and $R^2{}_{Ac} = 0.903$. It can be confirmed that acrylic pipe shows a higher correlation. Therefore, the result from the acrylic pipe is used to determine early-age compressive strength before demolding.

$$fc = \frac{46.51}{d_{Al} - 2.43} \tag{3}$$

$$fc = \frac{22.02}{d_{Ac} - 5.46} \tag{4}$$

where

d_{Al}: corrected penetration depth of aluminum pipe [mm];

d_{Ac}: corrected penetration depth of acrylic [mm].

Finally, a reference value of the penetration depth is determined with sufficient certainty that the strength required before demolding or preventing early-age freezing has been obtained from the depth of penetration d_i obtained in one penetration test and the corresponding compressive strength fc_i. The corresponding experimental constant s_i in Equation (2) is calculated by the following Equation (5).

$$s_i = (d_i - d) \times fc_i \tag{5}$$

Here, d is the pin penetration depth for the acrylic pipe. s_i is assumed to follow a normal distribution, the mean value s of s_i is calculated by Equation (6), and the standard error SE_{s_i} is calculated by Equation (7) for the acrylic pipes.

$$\bar{s} = \frac{\sum_{i=1}^{n} s_i}{n} \tag{6}$$

$$SE_{s_i} = \frac{\sqrt{\frac{1}{n-1}\sum_{i=1}^{n}(s_i - \bar{s})^2}}{\sqrt{n}} \tag{7}$$

The 99% confidence interval when considering the distribution of the experimental constant s_i is expressed by the following Equations (8) and (9). s_{max} and s_{min} are calculated for the acrylic pipe.

$$s_{max} = \bar{s} + 3SE_{s_i} \tag{8}$$

$$s_{min} = \bar{s} - 3SE_{s_i} \tag{9}$$

Substituting s_{max} and s_{min} into Equation (2) yields the estimated intensity intervals. These equations are expressed in Equations (10) and (11).

$$fc_{max} = \frac{31.24}{d_s - 5.46} \tag{10}$$

$$fc_{min} = \frac{12.81}{d_s - 5.46} \tag{11}$$

Here,

fc_{max}: estimated upper strength of acrylic pipe [MPa];

fc_{min}: estimated lower strength of acrylic pipe [MPa].

These estimated intervals are indicated by the dashed lines in Figure 16, confirming that it is possible to estimate the early-age compressive strength from the depth of penetration obtained. The standard error of SE is 3.07. Considering the respective estimated strength intervals, the penetration depths that ensure the development of the early-age strength required before demolding are shown in Table 5. Figure 16 shows these values for the relationship between penetration depth and compressive strength. If the penetration depth is less than these values, it confirms that the minimum required strength before demolding is achieved.

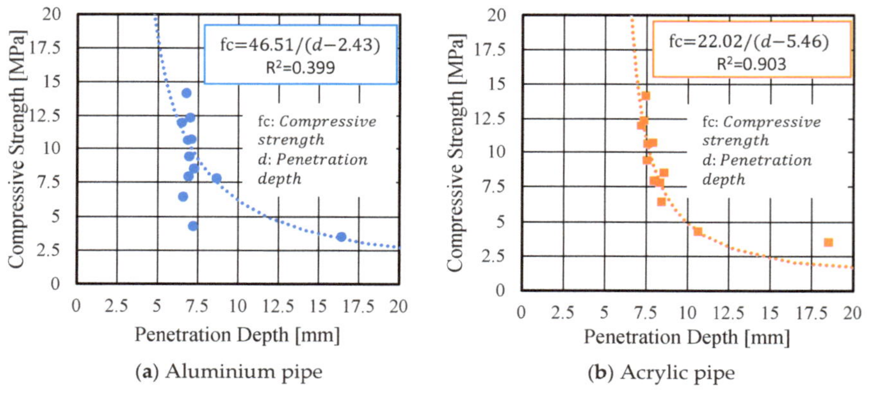

(**a**) Aluminium pipe   (**b**) Acrylic pipe

Figure 15. Relationship between modified penetration depth and compressive strength.

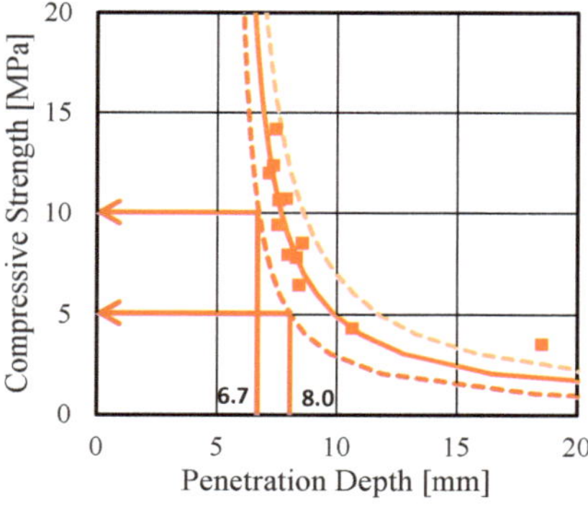

Figure 16. Pin penetration depth to ensure minimum required strength (Ac pipe).

Table 5. Penetration depth to ensure minimum required compressive strength before demolding.

Compressive Strength [MPa]	Penetration Depth [mm] (Acrylic Pipe)
5.0	8.02 (8.0)
10.0	6.74 (6.7)

However, fewer values were obtained at lower strength levels due to higher strength development during testing. Therefore, it is necessary to conduct more experiments to obtain more data at a low strength level in the future, although we were able to propose reference values.

3.4. Surface Condition of Concrete after Demolding

The condition of the surface of the structure after demolding is shown in Figure 17. As shown in these figures, the test hole marks after demolding are small, with acceptable unevenness. It is considered possible to repair these test marks easily without damaging the structural frame. Additionally, when installing finishing materials or insulation, these marks can be covered without any special treatment.

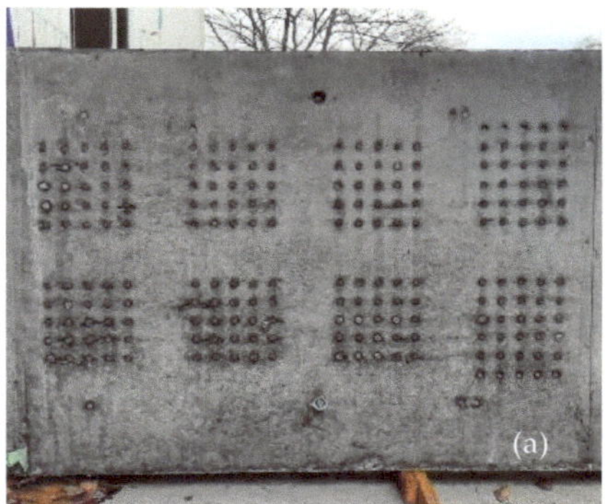

Figure 17. Surface of concrete after demolding (**a**) mock-up specimen (**b**) construction site.

4. Conclusions

In this study, the pin penetration test method was investigated to determine the early-age compressive strength at the actual construction site before demolding. The mock-up specimens were prepared to determine the relationship between penetration depth and compressive strength. The findings of this study are described below.

1. It is confirmed that the pin penetration test method is suitable to measure the early-age compressive strength before demolding at actual construction site.
2. The relationship between pin penetration depth and compressive strength of concrete was determined at actual construction site using the mock-up specimens. The obtained results were different from the existing curves obtained in laboratory experiments.
3. The strength development of specimens was significantly delayed compared to mock-up specimens even when specimens were cured at same ambient air temperatures. Therefore, it confirms that the pin penetration test method is important to determine early-age compressive strength before demolding at actual construction site.
4. The relationship between pin penetration depth and compressive strength gives higher correlation when using the acrylic pipes. However, the compressive strength range that can be estimated by the proposed method in this study was from 3 MPa to 15 MPa.
5. It is confirmed that when the compressive strengths are greater than 5 MPa and 10 MPa, the penetration depths are smaller than 8.0 mm and 6.7 mm, respectively.

Buildings **2024**, *14*, 2099

Author Contributions: Conceptualization, B.N. and T.N.; methodology, B.N. and T.N.; validation, Y.D. and T.N.; formal analysis, B.N. and F.T.; investigation, B.N., F.T., K.K., T.L., A.J. and F.D.; data curation, B.N. and F.T.; writing—original draft preparation, B.N.; writing—review and editing, T.N., Y.D., F.T., K.K., T.L., A.J. and F.D.; visualization, B.N.; supervision, T.N.; project administration, T.N. All authors have read and agreed to the published version of the manuscript.

Funding: This research received no external funding.

Data Availability Statement: The original contributions presented in the study are included in the article, further inquiries can be directed to the corresponding author.

Acknowledgments: We would like to express our gratitude to Sumitomo Mitsui Construction Co., Ltd., for their great support and for allowing us to conduct experiments at their construction site. We also appreciate the support provided by Taihaku Corporation and the Sendai Concrete Testing Center in the preparation and execution of the mock-up specimens.

Conflicts of Interest: The authors declare no conflicts of interest.

References

1. Narantogtokh, B.; Nishiwaki, T.; Pushpalal, D.; Taniguchi, M. Influence of pre-curing period at sub-zero temperature ($-20\,^{\circ}$C) on the compressive strength of concrete. *Cem. Sci. Concr. Technol.* **2022**, *76*, 379–385. [CrossRef]
2. *Japanese Architectural Standard Specification JASS 5 Reinforced Concrete Work*; Architectural Institute of Japan: Tokyo, Japan, 2018. (In Japanese)
3. *Recommendation for Practice of Cold Weather Concreting*; Architectural Institute of Japan: Tokyo, Japan, 2010. (In Japanese)
4. RILEM. *RILEM Recommendations for Concreting in Cold Weather*; VTT Technical Research Centre of Finland: Espoo, Finland, 1988.
5. Revilla-Cuesta, V.; Ortega-Lopez, V.; Faleschini, F.; Espinosa, A.B.; Serrano-Lopez, R. Hammer rebound index as an overall-mechanical-quality indicator of self-compacting concrete containing recycled concrete aggregate. *Constr. Build. Mater.* **2022**, *347*, 128549. [CrossRef]
6. Szilágyi, K.; Borosnyói, A.; Zsigovics, I. Extensive statistical analysis of the variability of concrete rebound hardness based on a large database of 60 years experience. *Constr. Build. Mater.* **2014**, *53*, 333–347. [CrossRef]
7. Kumavat, N.R.; Chandak, N.R.; Patil, I.T. Factors influencing the performance of rebound hammer used for non-destructive testing of concrete members: A review. *Case Stud. Constr. Mater.* **2021**, *14*, e00491. [CrossRef]
8. Odimegwu, T.C.; Amrul Kaish, A.B.M.; Zakaria, I.; Abood, M.M.; Jamil, M.; Ngozi, K.O. Nondestructive determination of strength of concrete incorporating industrial wastes as partial replacement for fine aggregate. *Sensors* **2021**, *21*, 8256. [CrossRef] [PubMed]
9. Mata, R.; Ruiz, R.O.; Nunez, E. Correlation between compressive strength of concrete and ultrasonic pulse velocity: A case of study and a new correlation method. *Constr. Build. Mater.* **2023**, *369*, 130569. [CrossRef]
10. Bogas, J.A.; Gomes, M.G.; Gomes, A. Compressive strength evaluation of structural lightweight concrete by non-destructive ultrasonic pulse velocity method. *Ultrasonics* **2013**, *53*, 962–972. [CrossRef] [PubMed]
11. Andrade, M.; Lopes, A.; Júnior, M.; Cristina, G. Ultrasonic testing on evaluation of concrete residual compressive strength: A review. *Constr. Build. Mater.* **2023**, *373*, 130887. [CrossRef]
12. Moreira, R.; Gondim, L.; Haach, V.G. Monitoring of ultrasonic velocity in concrete specimens during compressive loading-unloading cycles. *Constr. Build. Mater.* **2021**, *302*, 124218. [CrossRef]
13. Liang, M.T.; Wu, J. Theoretical elucidation on the empirical formulae for the ultrasonic testing method for concrete structures. *Cem. Concr. Res.* **2002**, *32*, 1763–1769. [CrossRef]
14. Lencis, U.; Udris, A.; Korjakins, A. Frost influence on the ultrasonic pulse velocity in concrete at early phases of hydration process. *Case Stud. Constr. Mater.* **2021**, *15*, e00614. [CrossRef]
15. Solis-Carcaño, R.; Moreno, E.I. Evaluation of concrete made with crushed limestone aggregate based on ultrasonic pulse velocity. *Constr. Build. Mater.* **2008**, *22*, 1225–1231. [CrossRef]
16. Ali-benyahia, K.; Kenai, S.; Ghrici, M.; Sbartaï, Z. Analysis of the accuracy of in-situ concrete characteristic compressive strength assessment in real structures using destructive and non-destructive testing methods. *Constr. Build. Mater.* **2023**, *366*, 130161. [CrossRef]
17. Hobbs, B.; Kebir, M.T. Non-destructive testing techniques for the forensic engineering investigation of reinforced concrete buildings. *Forensic. Sci. Intern.* **2007**, *167*, 167–172. [CrossRef] [PubMed]
18. Poorarbabi, A.; Ghasemi, M.; Moghaddam, M.A. Concrete compressive strength prediction using non-destructive tests through response surface methodology. *Ain Shams Eng. J.* **2020**, *11*, 939–949. [CrossRef]
19. Helal, J.; Sofi, M.; Mendis, P. Non-destructive testing of concrete: A review of methods. *Elect. J. Struc. Eng.* **2015**, *14*, 97–105. [CrossRef]
20. *JIS A 1155*; Method for Measuring the Degree of Rebound of Concrete. Japanese Standards Association: Tokyo, Japan, 2012. (In Japanese)
21. *ISO 1970-7*; Testing of Concrete—Part 7: Non-Destructive Tests on Hardened Concrete. International Organization for Standardization: Geneva, Switzerland, 2004.

22. Technical Committee on Concrete Strength Estimation; The Society of Materials Science Japan. Guideline for Compressive Strength Estimation of Concrete by Schmidt Hammer (Draft). *Mater. Test.* **1958**, *7*, 426–430. (In Japanese)
23. El-Mir, A.; El-Zahab, S.; Sbartaï, Z.M.; Homsi, F.; Saliba, J.; El-Hassan, H. Machine learning prediction of concrete compressive strength using rebound hammer test. *J. Build. Eng.* **2023**, *64*, 105538. [CrossRef]
24. Breysse, D.; Martínez-Fernández, J.L. Assessing concrete strength with rebound hammer: Review of key issues and ideas for more reliable conclusions. *Mater. Struc.* **2014**, *47*, 1589–1604. [CrossRef]
25. Breysse, D. Nondestructive Evaluation of Concrete Strength: An Historical Review and a New Perspective by Combining NDT Methods. *Const. Build. Mater.* **2012**, *33*, 139–163. [CrossRef]
26. Gunes, B.; Karatosun, S.; Gunes, O. Drilling resistance testing combined with SonReb methods for nondestructive estimation of concrete strength. *Constr. Build. Mater.* **2023**, *362*, 129700. [CrossRef]
27. Al-sabah, S.; Sourav, S.N.A.; Mcnally, C. The post-installed screw pull-out test: Development of a method for assessing in-situ concrete compressive strength. *J. Build. Eng.* **2021**, *33*, 101658. [CrossRef]
28. Maliha, M.; Nishiwaki, T.; Amin, A.F.M.S. A Screening Method for Very-Low-Strength Concrete. *ACI Mater. J.* **2022**, 119. [CrossRef]
29. Nguyen, M.H.; Nakarai, K.; Kubori, Y.; Nishio, S. Validation of simple nondestructive method for evaluation of cover concrete quality. *Constr. Build. Mater.* **2019**, *201*, 430–438. [CrossRef]
30. Shinozaki, T.; Fujii, K.; Kemi, T.; Shirayama, K. Estimation of Concrete Strength in Structures by the BOSS Method. *J. Advanc. Concr. Technol.* **2004**, *2*, 175–185. [CrossRef]
31. *JIS A 1163*; Method of Making and Testing for Compressive Strength BOSS Specimens. Japanese Standards Association: Tokyo, Japan, 2020. (In Japanese)
32. Levent, S.H.; Suleyman, G.; Kamil, K.; Osman, S. A Nondestructive Testing Technique: Nail Penetration Test. *ACI Struct. J.* **2012**, *109*, 245–252. [CrossRef]
33. Herrera-Mesena, C.; Salvadorb, R.P.; Cavalaroc, S.H.P.; Aguadoa, A. Effect of gypsum content in sprayed cementitious matrices: Early age hydration and mechanical properties. *Cem. Concr. Comp.* **2019**, *95*, 81–91. [CrossRef]
34. Maliha, M.; Nishiwaki, T.; Fujiwara, T.; Minemura, T. In-Place Test Method with Penetration Resistance for Low-Strength Concrete. *Proc. Jpn. Concr. Inst.* **2020**, *42*, 1732–1737.
35. Nishiwaki, T.; Takasugi, F.; Narantogtokh, B.; Hara, S.; Maliha, M. A method to estimate the early age compressive strength of concrete before demolding using a pin penetration device. *J. Struct. Constr. Eng.* **2022**, *88*, 18–26. (In Japanese) [CrossRef]
36. Noguchi, T. Quality Control System in Curing Concrete Using Small Integrated Circuit for Monitoring Temperature and Attitude on Formworks. *JACIC Res. Dev. Rep* **2010**, 2010. (In Japanese). Available online: https://www.jacic.or.jp/josei/pdf/2010-16.pdf (accessed on 4 July 2024).
37. *ASTM C1074-11*; Standard Practice for Estimating Concrete Strength by the Maturity Method. ASTM International: West Conshohocken, PA, USA, 2011.
38. Popovics, S. *Strength and Related Properties of Concrete: A Quantitative Approach*; John Wiley & Sons: Hoboken, NJ, USA, 1998; p. 12.
39. Zhou, S.; Sun, Q.; Wu, X. Impact of D/t Ratio on Circular Concrete-Filled High-Strength Steel Tubular Stub Columns under Axial Compression. *Thin-Walled Struct.* **2018**, *132*, 461–474. [CrossRef]
40. Chen, H.; Wu, L.; Jiang, H.; Liu, Y. Seismic Performance of Prefabricated Middle Frame Composed of Special-Shaped Columns with Built-in Lattice Concrete-Filled Circular Steel Pipes. *Structures* **2021**, *34*, 1443–1457. [CrossRef]

 *buildings*

Article

Study on the Bond Performance of Epoxy Resin Concrete with Steel Reinforcement

Peiqi Chen [1,2], Yueqiang Li [1,2], Xiaojie Zhou [1,2], Hao Wang [1,2] and Jie Li [1,2,*]

[1] Tianjin Key Laboratory of Civil Buildings Protection and Reinforcement, Tianjin 300384, China; cpq@tcu.edu.cn (P.C.); zhouxj@tcu.edu.cn (X.Z.)
[2] School of Civil Engineering, Tianjin Chengjian University, Tianjin 300384, China
* Correspondence: lijietcu@163.com

Abstract: Epoxy resin concrete, characterized by its superior mechanical properties, is frequently utilized for structural reinforcement and strengthening. However, its application in structural members remains limited. In this paper, the bond–slip behavior between steel reinforcement and epoxy resin concrete was investigated using a combination of experimental research and finite element analysis, with the objective of providing data support for substantiating the expanded use of epoxy resin concrete in structural members. The research methodology included 18 center-pullout tests and 14 finite element model calculations, focusing on the effects of variables such as epoxy resin concrete strength, steel reinforcement strength, steel reinforcement diameter and protective layer thickness on bond performance. The results reveal that the bond strength between epoxy resin concrete and steel reinforcement significantly surpasses that of ordinary concrete, being approximately 3.23 times higher given the equivalent strength level of the material; the improvement in the strength of both the epoxy resin concrete and steel reinforcement are observed to marginally increase the bond stress. Conversely, an increase in the diameter of the steel reinforcement and a reduction in the thickness of the protective layer of the concrete can lead to diminished bond stress and peak slip. Particularly, when the steel reinforcement strength is below 500 MPa, it tends to reach its yield strength and may even detach during the drawing process, indicating that the yielding of the steel reinforcement occurs before the loss of bond stress. In contrast, for a steel reinforcement strength exceeding 500 MPa, yielding does not precede bond stress loss, resulting in a distinct form of failure described as scraping plough type destruction. Compared to ordinary concrete, the peak of the epoxy resin concrete and steel reinforcement bond stress–slip curve is more pointed, indicating a rapid degradation to maximum bond stress and exhibiting a brittle nature. Overall, these peaks are sharper than those of ordinary concrete, indicating a rapid decline in bond stress post-peak, reflective of its brittle characteristics.

Keywords: epoxy resin concrete; bond performance; bond stress–slip curve; experiment; finite element (method)

Citation: Chen, P.; Li, Y.; Zhou, X.; Wang, H.; Li, J. Study on the Bond Performance of Epoxy Resin Concrete with Steel Reinforcement. *Buildings* **2024**, *14*, 2905. https://doi.org/10.3390/buildings14092905

Academic Editors: Atsushi Suzuki and Dinil Pushpalal

Received: 1 August 2024
Revised: 9 September 2024
Accepted: 10 September 2024
Published: 14 September 2024

1. Introduction

Epoxy resin concrete is a composite material that is hardened and moulded from a mixture of epoxy resin, curing agents, and sand aggregates [1]. Compared with conventional silicate concrete, epoxy resin concrete has excellent corrosion resistance, penetration resistance, and freeze-thaw resistance, as well as high strength and significant early strength [2–6]. Therefore, epoxy resin has been extensively applied in reinforcement projects, including crack repair and road construction [7–11]. Furthermore, research on its application in building materials [11] and structural engineering [12–14] continues to deepen, underscoring the broad development potential of epoxy resin concrete in the construction industry.

The performance of a bond between steel reinforcement and concrete is one of the key influences on the operational performance of concrete structures [15,16]. In these structures, the steel reinforcement is responsible for bearing tensile force, while the bonded concrete disseminates and transfers the loads. However, several factors can compromise this bond, potentially leading to structural failure [17,18]. The application of epoxy resin concrete in engineering structures introduces a bond–slip behavior between it and the steel reinforcement that could significantly diverge from that of ordinary concrete. Investigating this bond performance is of great engineering significance, as it could critically influence structural integrity.

Existing related studies predominantly focus on ordinary concrete and steel reinforcement [19]. However, investigations into the bond–slip behavior of novel concrete types paired with steel reinforcement are less prevalent. With the global development of high-performance concrete [20], research into its application in structural engineering has gradually been carried out. For instance, in 2019, Khaksefidi et al. [21] prepared 60 cubic concrete specimens centered around steel reinforcement with three different bond lengths using Ultra-High-Performance Concrete (UHPC) and high-strength steel rebars in normal as well as two strength types of steel reinforcement. Their tensile test results indicated that several factors, including the concrete strength, ratio of concrete protective layer to diameter, bond length, yield strength of steel reinforcement, and geometry of steel reinforcement, significantly influence the bond performance. Similarly, in the same year, Hu et al. [22] investigated the bond characteristics of high-strength deformed steel reinforcement in UHPC, taking into account variables such as fibre volume content, diameter of the steel reinforcement, embedment length of the steel reinforcement, concrete protective layer, and grade of the steel reinforcement on the bond failure mode. The results highlighted the substantial impact of the concrete protective layer thickness on the bond failure mode, leading to the proposal of theoretical and simplified formulas for calculating the splitting bond strength. Additionally, Liang et al. [23] carried out an experimental study on the bond performance of deformed steel reinforcement with UHPC and analysed the influencing factors of the loading mode, the strength of the UHPC, the steel fibre type and content, the diameter of the steel reinforcement, and the thickness of the protective layer and proposed three bond–slip failure modes, including pullout damage, split damage + pullout damage, and cone failure.

In terms of finite element analysis concerning the bond–slip performance between steel reinforcement and concrete, several significant studies were conducted in recent years. In 2020, Qasem et al. [24] investigated a cohesive zone modeling approach in ABAQUS to explore the interaction between steel reinforcement and concrete, considering factors such as reinforcement diameter, material, and carbon nanotube content in the context of ultra-high-performance concrete. Their analysis focused on the pullout performance of steel reinforcement. In 2023, Zhang et al. [25] used the Coulomb-friction model through ABAQUS to study the effect of waterborne epoxy coating thickness on the bond performance between steel reinforcement and concrete. They adjusted the friction coefficient of the contact surfaces and established a high-fidelity geometric model of ribbed steel reinforcement. Their results showed that the bond stress in waterborne epoxy-coated steel reinforcement is mainly provided by mechanical occlusion, and the coating-induced changes in the shape of the steel reinforcement were the main factors leading to the degradation of the bond performance. Also in 2023, Bai et al. [26] applied a nonlinear spring element in both ANSYS and ABAQUS to model rubberized concrete with varying reinforcement diameters and anchorage lengths. The results demonstrated that an increase in both reinforcement diameter and anchorage length shifts the peak of bond stress and leads to a more inhomogeneous distribution of bond stress. The simulation results are basically consistent with the experimental results, which verifies the feasibility of the method. In 2024, Cui et al. [27] carried out numerical simulations using the cohesive zone model method through ABAQUS, replicating the experimental conditions. Their study verified the bond–slip constitutive models for seawater Sea Sand Alkali Activated Concrete (SSASC) and Fibre Reinforced Plastic (FRP) reinforcement systems. The simulation results obtained from ABAQUS finite element software demonstrated strong consistency with the experimental results.

In summary, the literature reveals a scarcity of research focused on the bond–slip behavior involving new types of concrete and steel reinforcement, particularly between epoxy resin concrete and steel reinforcement. This gap hinders the broader adoption and integration of epoxy resin concrete in structural applications. This paper intends to address this deficiency by employing a dual approach combining experimental methods and finite element analysis to investigate the bond–slip behavior between epoxy resin concrete and steel reinforcement. Specifically, the study will assess how the epoxy resin concrete strength, steel reinforcement strength, steel reinforcement diameter and the thickness of the protective layer influence the bond–slip performance. The results will provide data to support the design and construction of the epoxy resin concrete structure and to promote its application in structural engineering. However, this study has certain limitations. Firstly, the research scope of both the experimental investigations and finite element models is relatively narrow, potentially excluding variables relevant to practical engineering applications, and thus may limit the broad applicability of the results. Secondly, the number of experimental samples is relatively limited, and the simplified assumptions in the finite element models may have a certain impact on the precision of the results. Therefore, it is recommended that future research expand the range of variables and further optimize the model assumptions to enhance the validity and robustness of the conclusions.

2. Experimental Overview

2.1. Specimen Design and Fabrication

According to the "Standard for Testing Methods of Physical and Mechanical Properties of Concrete" (GB/T 50081-2019) [28], this study designed and fabricated 18 center-pullout specimens. The design focused on strength of epoxy resin concrete, bond length, and type of steel reinforcement. Detailed specifications of these specimens, including their primary characteristics, are provided in Table 1. The dimensions for each specimen were 150 mm × 150 mm × 150 mm. The bond length between the epoxy resin concrete and steel reinforcement was controlled by two cut PVC sleeves, illustrated in Figure 1.

Table 1. Basic information of test pieces.

Specimen Number	Cubic Compressive Strength of Epoxy Resin Concrete (MPa)	Bond Length (mm)	Type of Steel Reinforcement
HG2.0-L30	40.3	30	High-strength Reinforced Bar
HG2.0-L40	40.3	40	High-strength Reinforced Bar
HG2.0-L50	40.3	50	High-strength Reinforced Bar
HG3.3-L30	50.5	30	High-strength Reinforced Bar
HG3.3-L40	50.5	40	High-strength Reinforced Bar
HG3.3-L50	50.5	50	High-strength Reinforced Bar
H2.0-L30	40.3	30	Hot Rolled Ribbed Steel Bars (HRB335)
H2.0-L40	40.3	40	Hot Rolled Ribbed Steel Bars (HRB335)
H2.0-L50	40.3	50	Hot Rolled Ribbed Steel Bars (HRB335)
H3.3-L30	50.5	30	Hot Rolled Ribbed Steel Bars (HRB335)
H3.3-L40	50.5	40	Hot Rolled Ribbed Steel Bars (HRB335)
H3.3-L50	50.5	50	Hot Rolled Ribbed Steel Bars (HRB335)
G2.0-L30	40.3	30	Hot Rolled Ribbed Steel Bars (HRB400)
G2.0-L40	40.3	40	Hot Rolled Ribbed Steel Bars (HRB400)
G2.0-L50	40.3	50	Hot Rolled Ribbed Steel Bars (HRB400)
G3.3-L30	50.5	30	Hot Rolled Ribbed Steel Bars (HRB400)
G3.3-L40	50.5	40	Hot Rolled Ribbed Steel Bars (HRB400)
G3.3-L50	50.5	50	Hot Rolled Ribbed Steel Bars (HRB400)

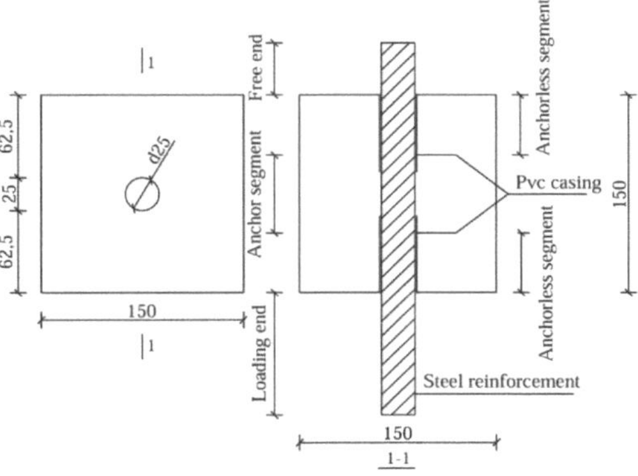

(**a**) Specimen numbers and sizes (All dimensions are reported in mm)

(**b**) Specimen Mould

Figure 1. Specimen design and fabrication.

2.2. Basic Mechanical Properties of Materials

The compositions used in the preparation of epoxy resin concrete for this experiment are detailed in Table 2. The "ring-to-solid ratio" refers to the ratio of the mass of epoxy resin to the mass of the curing agent, while the "glue-to-sand ratio" and "glue-to-stone ratio" represent the mass ratios of epoxy resin to sand and gravel, respectively. For the mechanical properties, the epoxy resin concrete was grouped into two categories: the first group exhibited an average cubic compressive strength of 40.3 MPa, and the average value of the second group was 50.5 MPa. Regarding the steel reinforcement used, the high-strength reinforced bar demonstrated an average yield strength of 1260 MPa and an ultimate strength of 1400 MPa. The yield and ultimate strengths for hot rolled ribbed steel bars (HRB335) were recorded at 335 MPa and 540 MPa, respectively. Similarly, the hot rolled ribbed steel bars (HRB400) had an average yield strength of 400 MPa and an ultimate strength of 540 MPa.

Table 2. Epoxy Resin Concrete Mixing Ratio.

Batch Number	Ring-to-Solid Ratio	Gel-to-Sand Ratio	Glue-to-Stone Ratio	Mixing Ratio (Ring:Solid:Cement:Sand:Stone)
1	2.0	0.25	0.33	1.00:0.50:1.50:4.00:3.03
2	3.3	0.30	0.28	1.00:0.30:1.50:3.33:3.57

2.3. Experiment Setup and Measurement Program

The experimental setup utilized a 2000 kN electro-hydraulic servo universal testing machine; the loading process needs to measure the tensile force at the end of the steel reinforcement, as well as measuring the bond slip between the steel reinforcement and the concrete. The specific setup for the loading and measurement program is shown in Figure 2.

To accurately measure the bond slip between the epoxy resin concrete and the steel reinforcement, displacement gauges were set up at the free end and the loaded end of the steel reinforcement. The bond slip measurement was determined by taking the average of the readings from both displacement gauges, which helps to mitigate any skewing deviations that might occur during the loading process.

It was imperative to maintain continuous and uniform loading throughout the experiment to ensure the reliability of the results. A displacement-controlled loading method was employed, with a loading rate set at 3 mm/min, facilitating the generation of a comprehensive force–displacement curve.

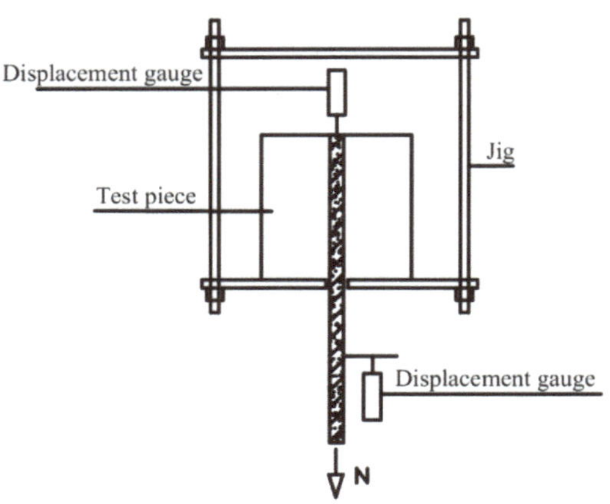

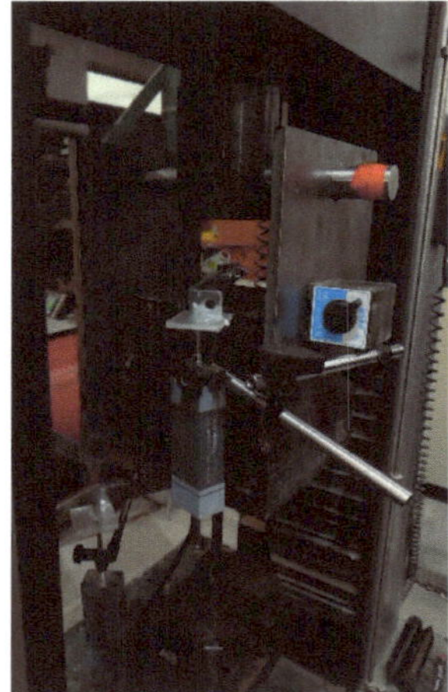

(**a**) Diagram of the experiment equipment　　　　(**b**) Experiment equipment

Figure 2. Experiment equipment.

3. Experiment Results and Analysis

3.1. Experiment Phenomena and Specimen Damage Characteristics

At the end of the experiment, different damage characteristics were observed in the specimens, depending on the type of steel reinforcement used. For specimens that utilized hot rolled ribbed steel bars, specifically groups H and G, no visible damage to the exterior was noted. However, the steel reinforcement was fractured, as shown in Figure 3a. In contrast, the six specimens of group HG, reinforced with High-Strength Reinforced Bar, could be monitored for obvious rebar slippage. Additionally, these specimens exhibited a small number of cracks around the steel reinforcement at the loading end, as detailed in Figure 3b.

The test block of HG group was longitudinally cut along the middle to align with the steel reinforcement. This procedure enabled the observation of the contact surface damage between the ribbed steel reinforcement and the epoxy resin concrete through the cut surface (see Figure 4). Post-experiment analysis revealed distinct friction traces on the contact surface, indicative of the longitudinal pulling out that occurred during the tests (see Figure 5). Importantly, despite the transverse ribs of the steel reinforcement being embedded within the epoxy resin concrete, the contact surface as a whole remained uniformly intact and unbroken (see Figure 6).

According to the above experiment phenomena, all the specimens in the HG group exhibited a scraping plough type destruction pattern. During the experiment, as the external load was progressively increased, a relative slip occurred between the steel reinforcement and epoxy resin concrete. This slip intensified as the load approached the ultimate capacity, leading to the shearing of the concrete situated between the steel reinforcement ribs. At the ultimate load, the superior toughness of the epoxy resin concrete effectively prevented splitting damage. Meanwhile, the concrete interlocked with the steel reinforcement ribs sheared and slid along the cylindrical surface of the outer diameter of the transverse ribs, driven by shear stress, culminating in the scraping plough type destruction.

(**a**) Groups H and G

(**b**) HG group specimens

Figure 3. Overall condition of each group of specimens at the end of the experiment.

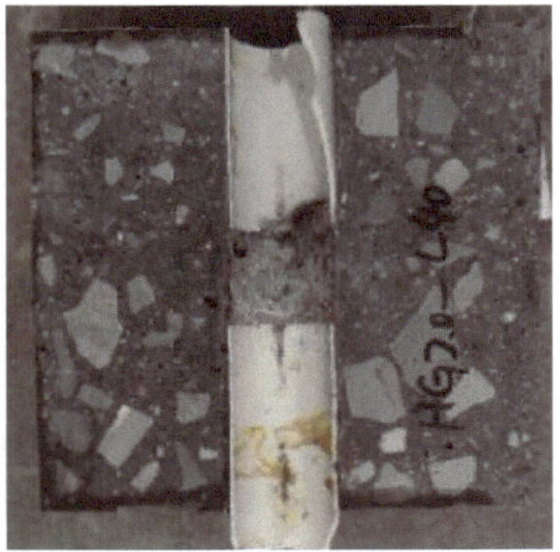

(**a**) Specimen HG2.0−L30

(**b**) Specimen HG2.0−L40

Figure 4. *Cont.*

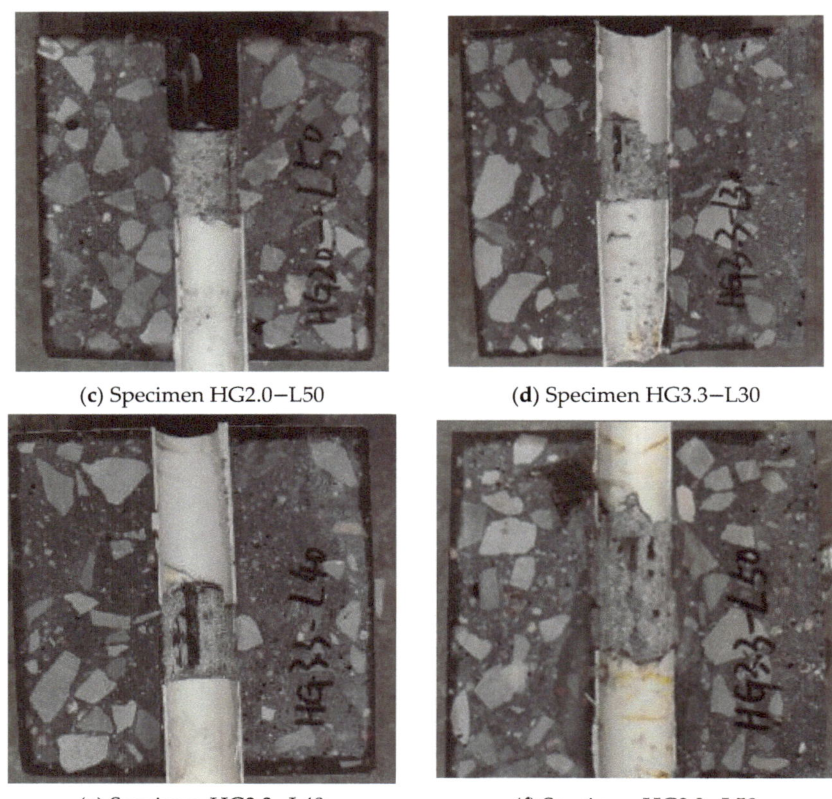

(**c**) Specimen HG2.0−L50

(**d**) Specimen HG3.3−L30

(**e**) Specimen HG3.3−L40

(**f**) Specimen HG3.3−L50

Figure 4. Damage to the contact surface of steel reinforcement with epoxy resin concrete.

(**a**) pre-experiment

(**b**) post-experiment

Figure 5. Comparison of the contact surface condition of steel reinforcement and epoxy resin concrete before and after the experiment.

(**a**) Contact surface situation

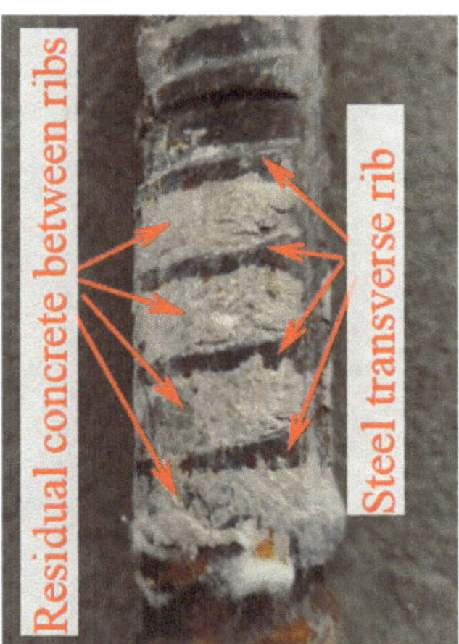

(**b**) Pullout steel reinforcement situation

Figure 6. Condition of contact surface and pullout steel reinforcement after experiment.

3.2. Analysis of Specimen Destruction Process

Due to the high reactivity of the epoxy group in epoxy resin with the active groups (hydroxyl, amino, etc.) on the surface of the substrate, a strong chemical bond is formed. As a result, epoxy resin exhibits strong chemical cemented force with various polar materials, including metal, glass, cement, wood, and plastics.

At the early stage of loading, the bond between the steel reinforcement and the epoxy resin concrete was mainly provided by chemically cemented forces. As the load was applied, these forces diminished upon the onset of relative slip, subsequently allowing the mechanical occlusal forces to become predominant. The high strength and toughness of the epoxy resin concrete prevented it from developing internal diagonal cracks despite the oblique extrusion pressure generated by the drawing steel reinforcement. This pressure was insufficient to penetrate deeply into the epoxy resin concrete, resulting in the dental concrete between the steel bars experiencing a downward longitudinal force due to the oblique extrusion, coupled with a shear stress from the upward support force at its base. Consequently, the bonding force during these stages was largely facilitated by the shear stress acting between the embedded dental concrete and the steel bars, leading to bonding slips predominantly through the shear deformation of the dental concrete, as shown in Figure 7. As the loading continued and the concrete within the intercostal spaces reached its ultimate shear stress, it sheared off at the base, leading to a complete separation of the steel reinforcement from the concrete. Post-ultimate load, although the steel reinforcement, along with the concrete trapped between the ribs, separated from the main body of concrete, frictional resistance at the fracture surfaces prevented an immediate pullout. Instead, the sliding continued to increase gradually until the frictional resistance diminished below the level of the external load. Thus, the ultimate pullout failure of the steel reinforcement from the epoxy resin concrete was primarily due to the loss of mechanical occlusal force.

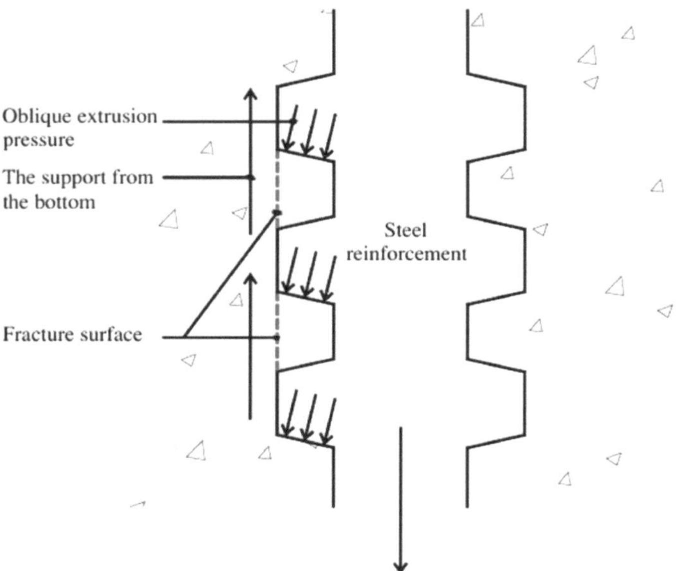

Figure 7. Stress situation around the steel reinforcement.

3.3. Average Bond Stress

According to the "Standard for Testing Methods of Physical and Mechanical Properties of Concrete" (GB/T 50081-2019) [28], the calculation formula of average bond stress is outlined in Equation (1). The test results from these calculations for each specimen are systematically presented in Table 3.

$$\tau = \frac{F}{\pi d l_a} \tag{1}$$

In the formula, F is the external load (N) applied at the loading end; d is the diameter of steel reinforcement (mm); and l_a is the bond length of steel reinforcement (mm).

Table 3. Average bond stress of each specimen.

Specimen Number	Cubic Compressive Strength of Epoxy Resin Concrete (MPa)	Bond Length (mm)	Ultimate Load (kN)	SlipCorresponding ToultiMateload (mm)	Bond Stress (N/mm²)
HG2.0−L30	40.3	30	59.7	5.95	25.3
HG2.0−L40	40.3	40	92.6	9.26	29.5
HG2.0−L50	40.3	50	137.4	9.37	35.0
HG3.3−L30	50.5	30	51.8	5.20	22.0
HG3.3−L40	50.5	40	100.4	7.77	32.0
HG3.3−L50	50.5	50	136.2	8.49	34.7

3.4. Bond Stress–Slip Curve

The whole process curve of bond stress–slip of each specimen is shown in Figure 8, and the bond–slip curve of each specimen is presented in Figure 9.

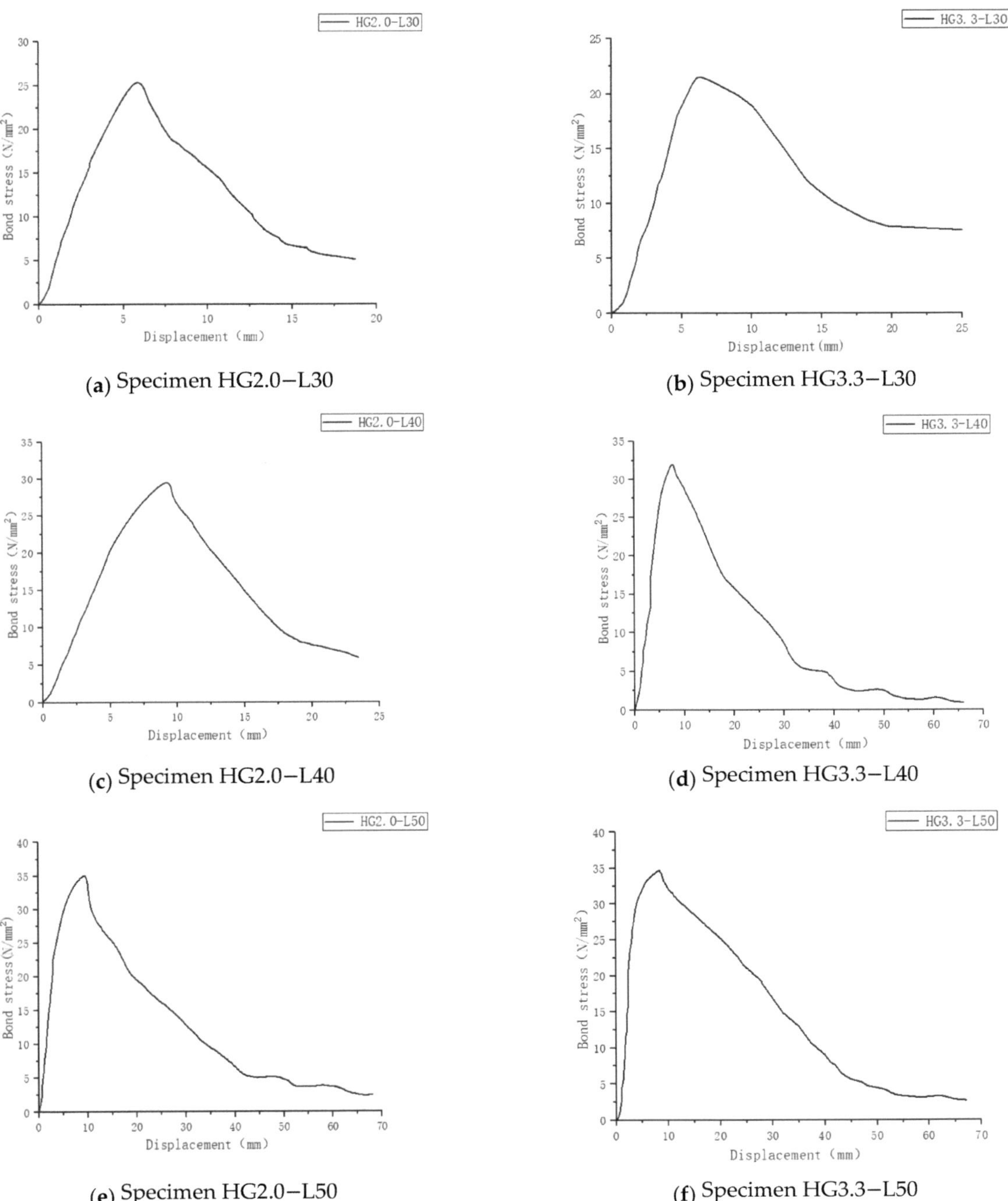

Figure 8. Bond stress–slip curves for each group of specimens.

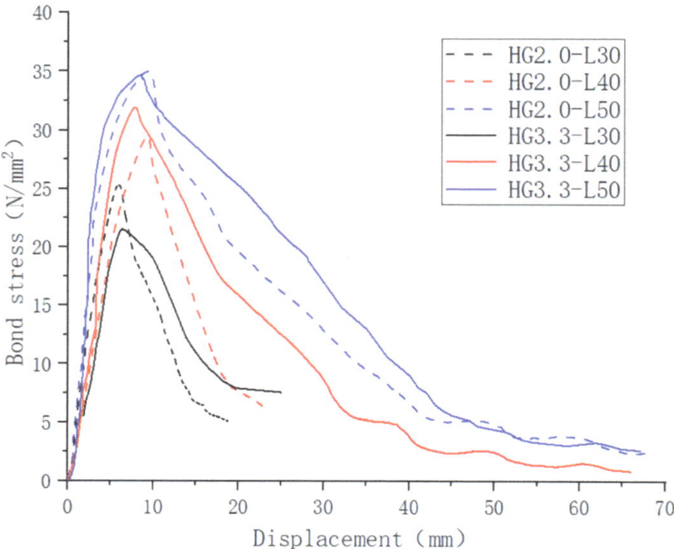

Figure 9. Summary of bond stress–slip curves for each group of specimens.

As illustrated in Figures 8 and 9, the bond stress–slip curves for each specimen display a similar trend, undergoing the following four stages.

(1)   Micro-slip stage: This initial phase of loading is characterized by a very small initial slip between the steel reinforcement and the epoxy resin concrete. During this stage, the bond stress is minimal, predominantly driven by the chemically cemented force. Subsequently, the bond stress rapidly increases, transitioning into the rising curve stage. The micro-slip stage is significantly brief, beginning with a shallow slope from the origin, which then sharpens significantly as the curve steepens. Importantly, specimens with shorter bond lengths exhibit more pronounced micro-slip characteristics.

(2)   Rising stage: As the load increases, the chemically cemented force between the steel reinforcement and the concrete is destroyed, shifting the source of bond stress predominantly to mechanical occlusal forces and frictional resistance. Concurrently, the oblique extrusion pressure exerted by the steel reinforcement on the intercostal epoxy resin concrete intensifies, causing the bond stress–slip curve to ascend nearly linearly. As the stress escalates, micro-cracks begin to develop within the intercostal concrete, and in some instances, crushing occurs. This results in a deceleration of the curve's ascent from approximately 80% of the ultimate bond stress onwards. As the curve approaches the ultimate bond stress, its progression becomes increasingly gradual. Specimens with longer bond lengths exhibit a more pronounced deceleration near the end of the rising section, and the corresponding bond slip at peak stress is greater.

(3)   Declining stage: Upon reaching the peak bond stress, the intercostal concrete typically sustains shear damage, leading to a reduction in the bond stress between the steel reinforcement and concrete. As the slip increases, the damaged area within the intercostal concrete continues to expand, resulting in a significant decrease in mechanical occlusal force.

(4)   Residual stage: Once the mechanical occlusion force is consumed, the steel reinforcement ribs become filled with concrete powder. During this phase, the residual bond stress is solely maintained by the frictional resistance between the concrete broken within the steel reinforcement ribs and the fracture surface. The curve at this stage exhibits a horizontal progression, indicating a stabilization of the residual bond stress.

The steel reinforcement continues to be pulled out until the experiment concludes. Essentially, a larger bond length in the specimen correlates with a reduced residual bond stress between the steel reinforcement and concrete, suggesting that the bond stress has been optimally utilized.

As depicted in Figures 8 and 9, the peak point of the bond stress–slip curve between epoxy resin concrete and steel reinforcement forms a distinctive "sharp point" which is significantly different from ordinary concrete. This "sharp point" phenomenon is primarily caused by two factors: the high strength of epoxy resin concrete and the Poisson effect. First of all, the epoxy resin as a matrix material significantly improves the compressive strength and bond strength of concrete, so that the bond stress between it and the steel reinforcement is also much higher than ordinary concrete. When energy builds up to a critical point, leading to scraping plough type destruction, the intercostal concrete is abruptly sheared. This sudden action causes an instant reduction in both the mechanical biting force and the bond stress between the epoxy resin concrete and steel reinforcement. As a result, the peak point of the curve forms a "sharp point", illustrating the brittle nature of the bond performance. Secondly, the Poisson effect induces transverse contraction of the steel reinforcement under tensile loading, leading to an uneven stress distribution at the bond interface. Specifically, this transverse contraction causes a local stress concentration at the bond interface, disrupting the original uniform stress distribution and significantly increasing the stress in localized regions. These stress concentrations alter the bond failure mode, resulting in the formation of a pronounced "sharp point" in the stress–slip curve at the peak. However, the robust bond strength between epoxy resin concrete and steel reinforcement generally ensures that such characteristics do not lead to detrimental effects under normal engineering conditions.

4. Finite Element Analysis

4.1. Material Model

The finite element model was established and analyzed by ABAQUS software. For the epoxy resin concrete, a plastic damage model was utilized, with the constitutive relationship defined by a compressive stress–strain curve equation, which was fitted by the research group based on experimental data [29]. This is detailed in Equation (2). The compressive stress–strain curve is presented in Figure 10a. For the tensile stress–strain curve of epoxy resin concrete, the same constitutive model as that of ordinary concrete is referenced. The constitutive model for ordinary concrete adopts the plastic damage model, and its uniaxial stress–strain curve is based on the recommended curve in the "Code for design of concrete structures" (GB 50010-2010) [30], as shown in Figure 10b. The constitutive relationship of the steel reinforcement follows the elastic–plastic bilinear model, illustrated in Figure 10c. The key plastic damage parameters for both ordinary concrete and epoxy resin concrete are shown in Table 4.

$$y = \begin{cases} ax + (4.9 - 4.23a)x^2 + (-4.67 + 6.67a)x^3 + (-0.27 - 4.74a)x^4 + (1.07 + 1.27a)x^5, & 0 \le x \le 1 \\ \frac{x}{b(x-1)^2 + x}, & x > 1 \end{cases} \tag{2}$$

where a and b are undetermined parameters, with values ranging from

$$\begin{cases} 0 < a < 1.0 \\ 1.0 < b < 10.0 \end{cases}$$

In this paper, the values for parameters a and b are set at 0.4 and 2.5, respectively.

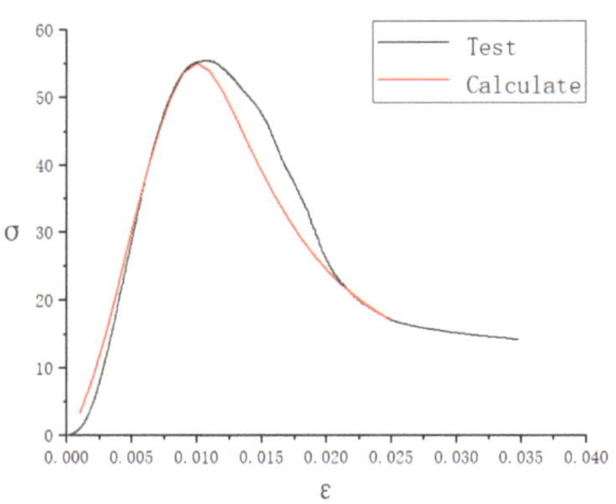

(**a**) Stress–strain curve of epoxy resin concrete under compression

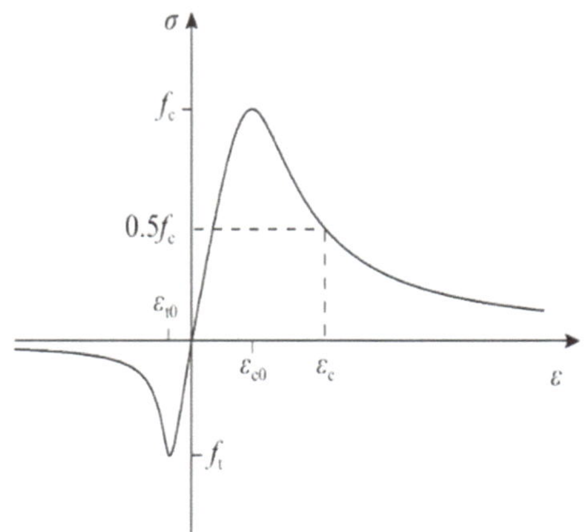

(**b**) Uniaxial stress–strain curve of ordinary concrete

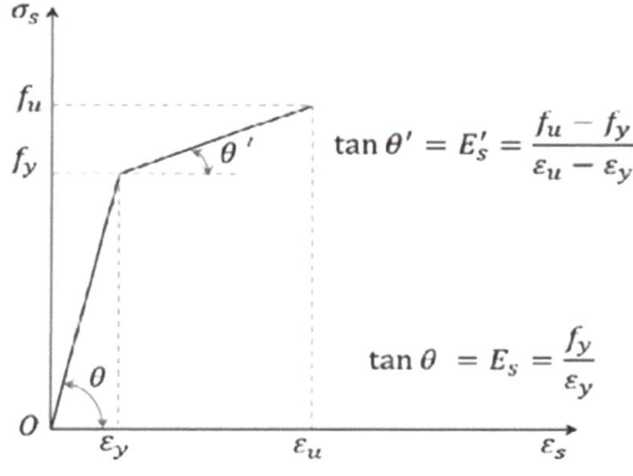

$$\tan\theta' = E_s' = \frac{f_u - f_y}{\varepsilon_u - \varepsilon_y}$$

$$\tan\theta = E_s = \frac{f_y}{\varepsilon_y}$$

(**c**) Reinforcement stress–strain curve

Figure 10. Stress–strain curves of each material.

Table 4. Concrete plastic damage parameters.

Concrete Type	Dilation Angle	Eccentricity	f_{b0}/f_{c0}	K	Viscosity Parameter	Poisson Ratio	Elastic Modulus
Epoxy resin concrete	40°	0.1	1.16	0.6667	0.005	0.3	15,616 Mpa
Ordinary concrete	30°	0.1	1.16	0.6667	0.005	0.3	32,500 Mpa

4.2. Establishment of the Model

Specimens HG2.0−L30 and HG3.3−L40 were selected for finite element modelling and model validation. A separate modeling method is adopted, wherein the epoxy resin concrete, steel reinforcement, and PVC pipe sleeve are individually established, followed by the assembly of these components into the final structure. The cohesive zone model was implemented to simulate the bond–slip between the steel reinforcement and concrete, as shown in Figure 11a. The parameter settings for the cohesion contact were all based on the bond stress–slip curves obtained from the experiments. For example, all the parameters of specimen HG2.0−L30 were sourced from Figure 8a; for the cohesive behavior, the three-way stiffnesses of Knn, Kss, and Ktt were set to 4.25 (the ratio of the peak bond stress to the peak displacement); and for the damage behavior, the three-way stress was set to the peak bond stress of 25.3.

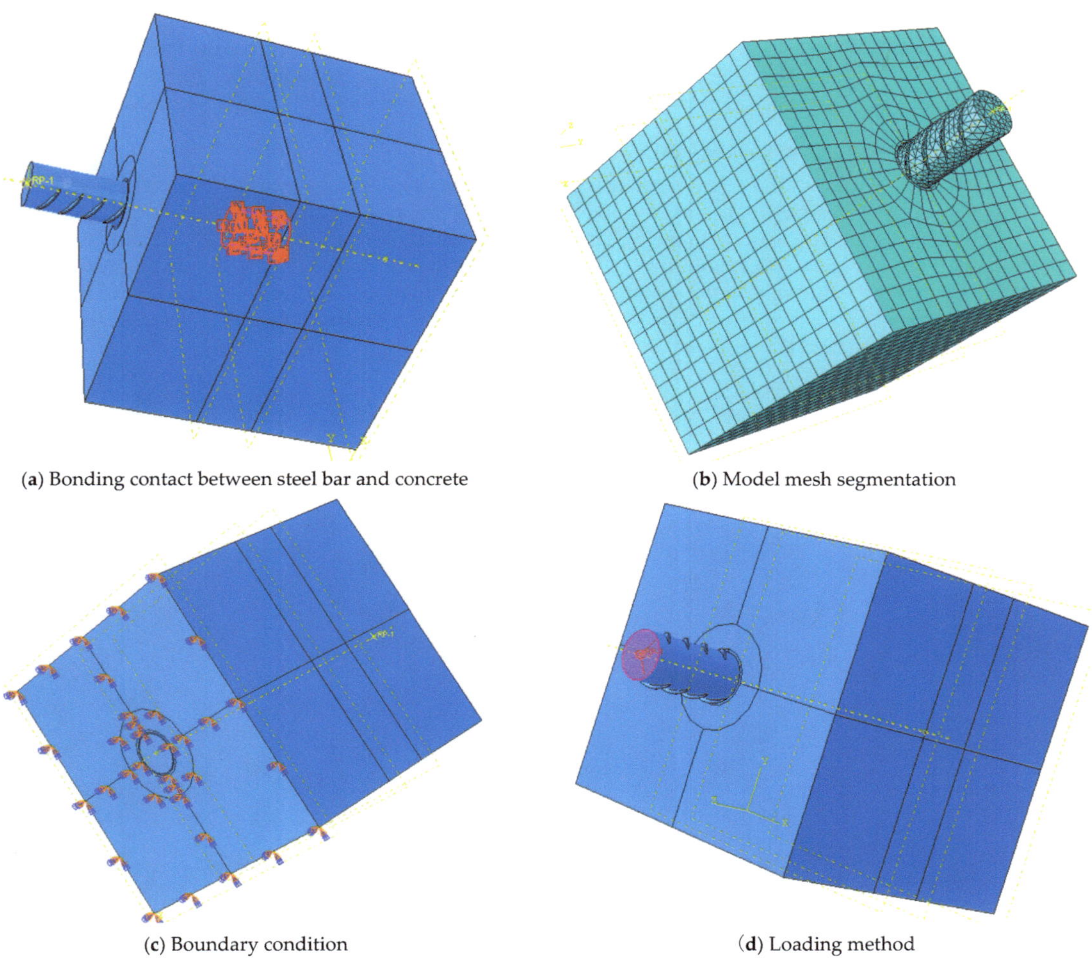

(**a**) Bonding contact between steel bar and concrete (**b**) Model mesh segmentation

(**c**) Boundary condition (**d**) Loading method

Figure 11. Finite element model of pullout specimen.

A separate modelling method was adopted, with C3D8R cells for the epoxy resin concrete and PVC casing, and C3D4 cells for the steel reinforcement. The PVC casing model adopts swept grid technology for grid division, with a unit size of 4 mm. The steel reinforcement model is meshed using free meshing technology, also with an element size of 4 mm. Due to the high irregularity of epoxy resin concrete around the ribbed steel reinforcement, region segmentation is first applied to simplify the subsequent mesh generation process. In the segmented regular area, structured grid technology is employed with a unit size of 10 mm; for the irregular areas, free mesh technology is applied with a unit size of 4 mm. This process is illustrated in Figure 11b.

The boundary conditions of the model are consistent with those employed in the experiment. Specifically, completely fixed constraints are applied to the bottom surface of specimen to ensure that all nodes on the bottom surface are completely restricted from displacement in all three spatial directions (x, y, z), as well as rotation around any axis (x, y, z). This prevents any form of displacement or rotation, as illustrated in Figure 11c. For the loading method, a reference point (RP-1) is established at the midpoint of the steel bar end, and RP−1 is coupled with the cross-section of the steel reinforcement end. Displacement loading is applied at the reference point along the Z direction, as shown in Figure 11d. A static general analysis step is used, and displacement loading is applied at a reference point based on the force–displacement curve obtained from the experiments. The total applied displacement is 20 mm. An amplitude function is defined within the amplitude module to achieve a uniform loading rate until a displacement of 20 mm is reached.

4.3. Model Verification

The distribution of concrete compression damage factors for specimens HG2.0−L30 and HG3.3−L40 is shown in Figure 12. From this figure, it is apparent that the primary damage within the model occurs in the bond region between the steel reinforcement and the epoxy resin concrete. Predominantly, the epoxy resin concrete retains its integrity throughout the loading process. This simulation result aligns well with the observations made during the experimental phase, demonstrating consistency between the simulated and experimental phenomena.

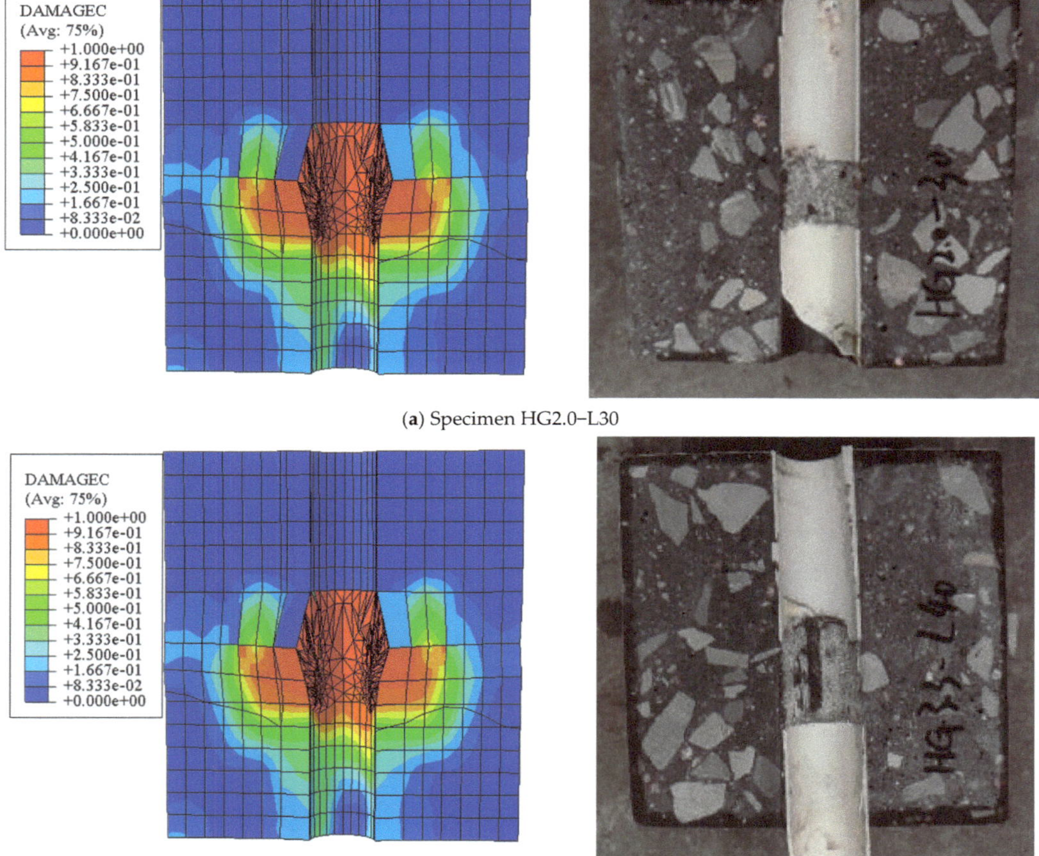

(**a**) Specimen HG2.0–L30

(**b**) Specimen HG3.3–L40

Figure 12. Distribution of damage factor of each specimen.

The comparison between the experimental and simulated bond–slip curves for specimens HG2.0−L30 and HG3.3−L40 is presented in Figure 13. Although there are some deviations between the two curves, they are basically aligned in terms of the overall trend. The initial slope of the simulated curve is steeper compared to the experimental curve, and the slip at the peak stress point is smaller in the simulation. This discrepancy is primarily attributed to the finite element model's exclusion of certain unfavorable factors, such as residual stress and the compression effect between the testing apparatus. Despite these differences, the finite element model developed in this study successfully approximates the bond–slip behavior between epoxy resin concrete and steel reinforcement, providing a generally reliable representation of the experimental outcomes.

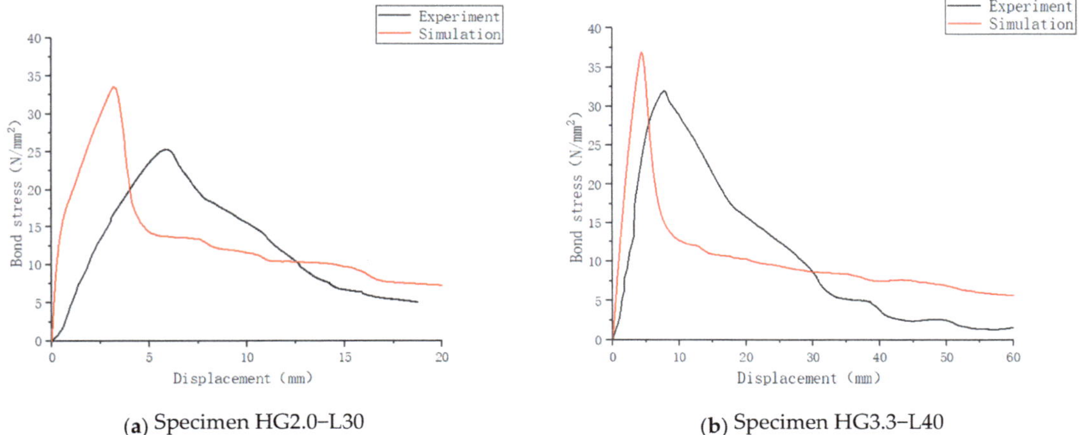

(**a**) Specimen HG2.0–L30

(**b**) Specimen HG3.3–L40

Figure 13. Comparison of simulation and experimental results of bond stress–slip curve.

4.4. Analysis of Influencing Factors of Bond–Slip Performance between Epoxy Resin Concrete and Steel Reinforcement

4.4.1. Effect of Concrete Type and Strength

Based on the original specimen HG2.0−L30, this study varied the type and strength of the concrete while keeping the parameters for steel reinforcement strength and bond length constant. Four groups of finite element models were established, with their respective model numbers and parameter settings detailed in Table 5. Among them, the model C−40 adopts ordinary concrete, and the other four groups of models adopt epoxy resin concrete at different strength levels. The compressive strength and stress–strain curves for each group's epoxy resin concrete cubes were obtained through the preliminary experiment, and the corresponding properties for ordinary concrete were derived in accordance with standard specifications.

Table 5. Basic information of specimens with different epoxy resin concrete strengths.

Specimen Number	Concrete Cube Compressive Strength (MPa)	Bond Length (mm)
HG2.0−L30	40.3	30
EPC−44.5	44.5	30
EPC−45.3	45.3	30
EPC−57.3	57.3	30
C−40	40	30

Figure 14 illustrates the stress field diagram of steel reinforcement at the peak load for four groups of epoxy resin concrete specimens and the compressive damage factor distribution of concrete in the final state. The stress distribution in the steel reinforcement is predominantly concentrated in the bond region across all groups, with minimal variation among them. Remarkably, damage within the concrete also occurs in the bond region between the steel reinforcement and the concrete, and the higher the strength grade of the concrete, the more confined the damage range. Figure 15 displays the bond stress–slip curves for each specimen. The peak bond stresses of specimens HG2.0−L30, EPC−44.5, EPC−45.3, and EPC−57.3 are 33.6 N/mm^2, 35.7 N/mm^2, 35.4 N/mm^2, and 35.9 N/mm^2, respectively. The results indicate minimal variation among these groups, suggesting that changes in the strength of epoxy resin concrete have a negligible impact on bond stress. On the other hand, the peak bond stress of specimen C−40 is only 10.4 N/mm^2, which is approximately 3.23 times less than that of specimen HG2.0−L30. Regarding the slip at peak bond stress, the displacement is greater in epoxy resin concrete specimens compared to ordinary concrete, attributed to the superior toughness of the epoxy resin concrete. This analysis confirms that the bonding performance of epoxy resin concrete to steel reinforcement markedly surpasses that of ordinary concrete.

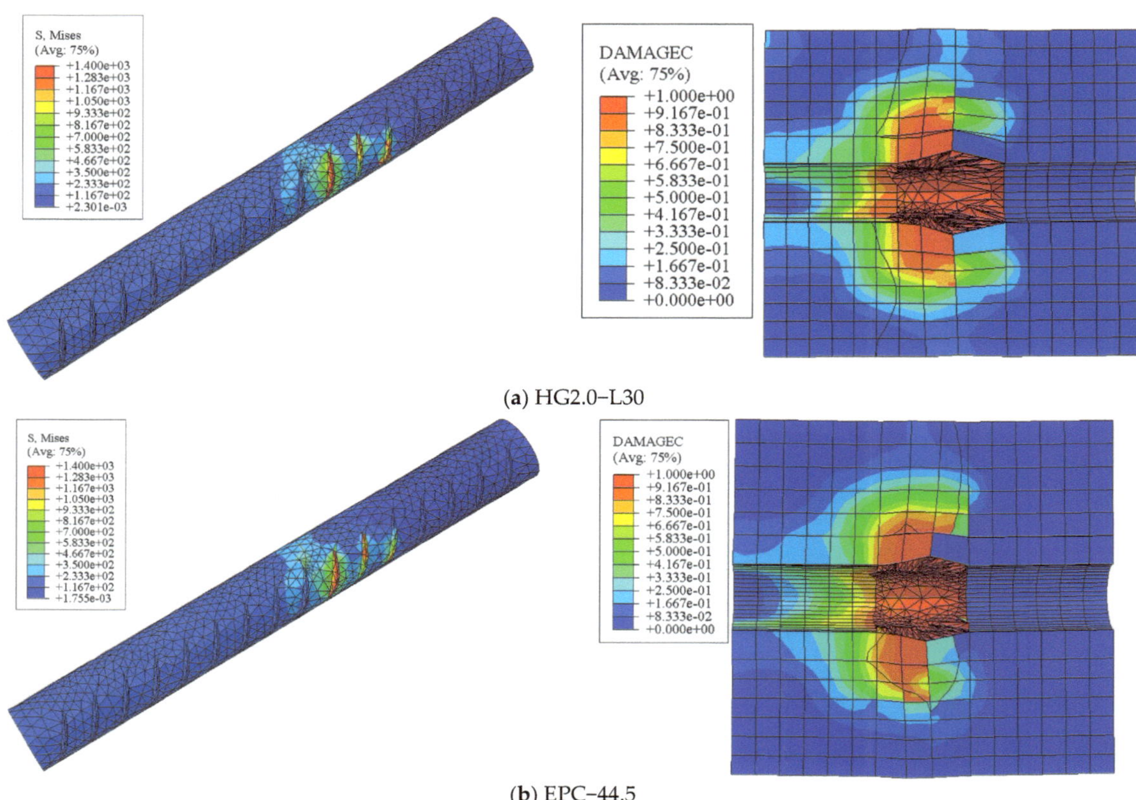

(**a**) HG2.0−L30

(**b**) EPC−44.5

Figure 14. *Cont.*

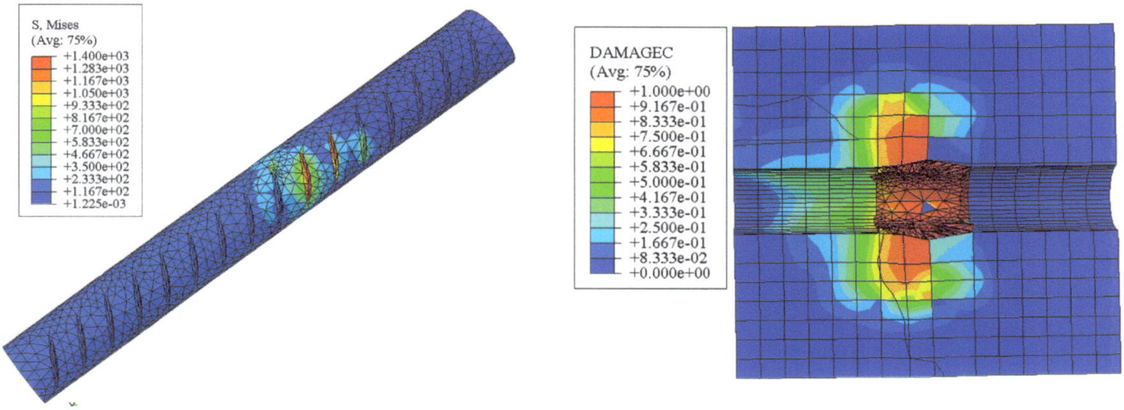

(**c**) EPC–45.3

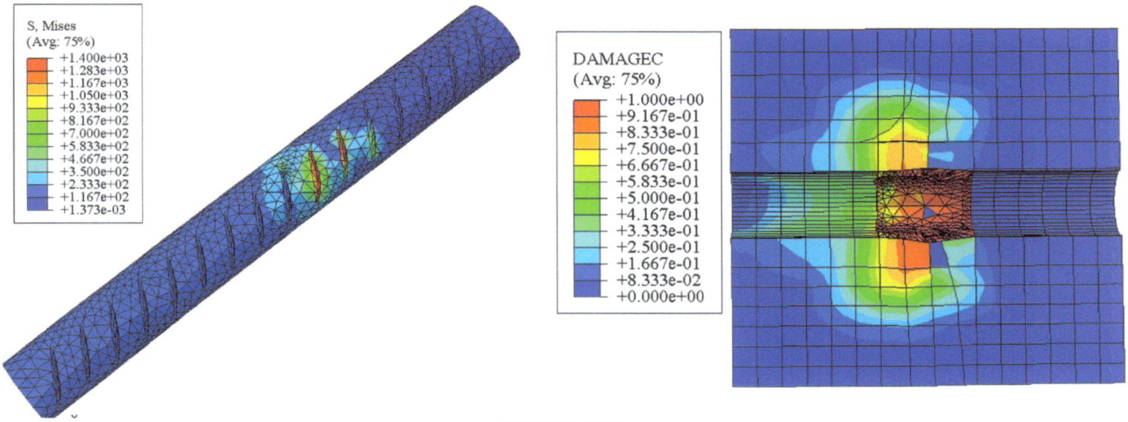

(**d**) EPC–57.3

Figure 14. Steel reinforcement stress field diagram and concrete damage factor distribution diagram of specimens with different epoxy resin concrete strengths.

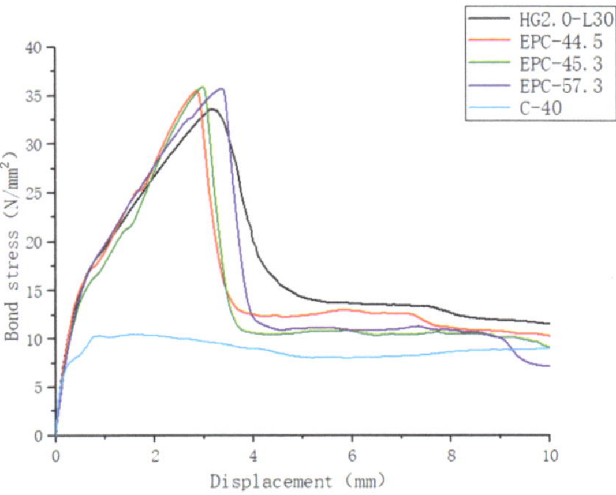

Figure 15. Bond stress–slip curves of specimens with different concrete types and strengths.

4.4.2. Effect of Steel Reinforcement Type and Strength

Building on the basis of specimen HG2.0−L30, this analysis maintained consistent parameters for epoxy resin concrete strength and bond length while varying the type and strength of steel reinforcement. Six sets of finite element models were created to explore these variations. The model numbers and specific parameter settings for each are provided in Table 6. For these models, aside from specimen HG2.0−L30, the strength settings for the steel reinforcement for all other specimens were determined in accordance with established specifications.

Table 6. Basic information of specimens with different steel reinforcement types and strengths.

Specimen Number	Type of Steel Reinforcement	Yield Strength of Steel Reinforcement (MPa)	Ultimate Strength of Steel Reinforcement (MPa)	Bond Length (mm)
HG2.0−L30	High-strength Reinforced Bar	1260	1400	30
F−1080	High-strength Reinforced Bar	1080	1230	30
F−930	High-strength Reinforced Bar	930	1080	30
F−785	High-strength Reinforced Bar	785	980	30
HRB500	Hot Rolled Ribbed Steel Bars	500	630	30
HRB400	Hot Rolled Ribbed Steel Bars	400	540	30
HRB335	Hot Rolled Ribbed Steel Bars	335	490	30

Figure 16 presents the stress field diagram of steel reinforcement at peak load and the compressive damage factor distribution of concrete in the final state for each group of specimens. It is observed that the maximum values of steel reinforcement stresses in the steel reinforcement are predominantly localized in the bond region, especially at the transverse ribs. Overall, the relative stresses (i.e., the ratio of the actual stress to the yield stress) at the loading end of the steel reinforcement and within the bonded region increase as the strength of the steel reinforcement decreases, indicating increasingly effective utilization of the steel reinforcement's strength. At the peak load, the maximum stress in the bonded section of High-Strength Reinforced Bar remains below the yield stress, indicating that the steel reinforcement does not yield prior to the loss of bond stress. This scenario points to a potential scraping plough type destruction, as shown in Figures 16a–d and 17a–d. In contrast, with ordinary hot rolled ribbed steel bars, the maximum stress in the bonded section often approaches or reaches the yield stress. This indicates that yielding of the steel reinforcement occurs before any significant loss of bond stress, potentially leading to a pull-off scenario, see Figures 16e–g and 17e–g. Throughout the pullout process, the epoxy resin concrete maintains its integrity, with damage primarily concentrated in the bond area. The stress emanates from the bond area and diminishes as it spreads to the surrounding regions. Crucially, in the models where the steel reinforcement strength is below 500 MPa, the damage to the epoxy resin concrete is minimized because the steel reinforcement yields before significant bond stress loss occurs, as illustrated in Figure 16f,g.

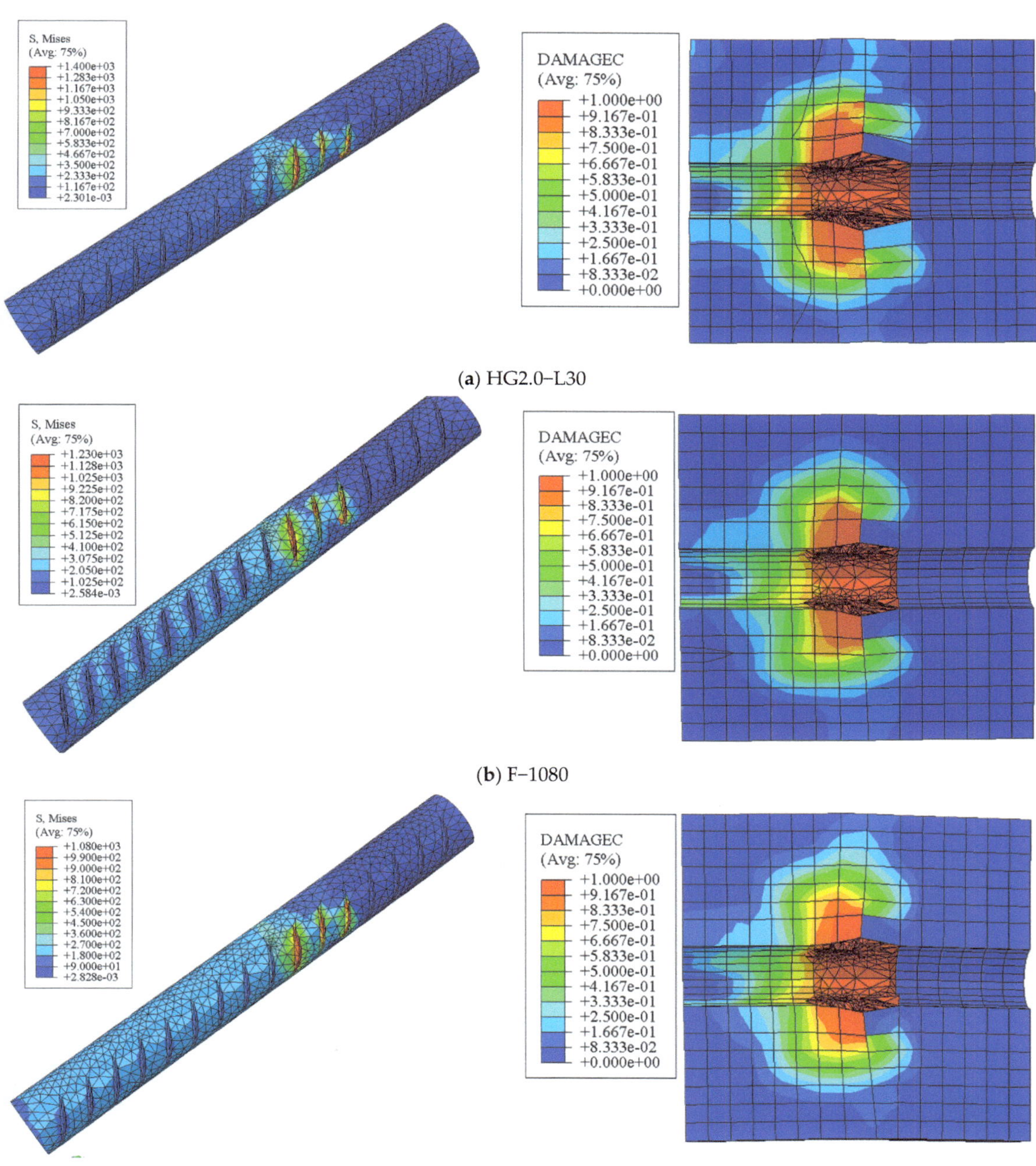

(**a**) HG2.0–L30

(**b**) F–1080

(**c**) F–930

Figure 16. *Cont.*

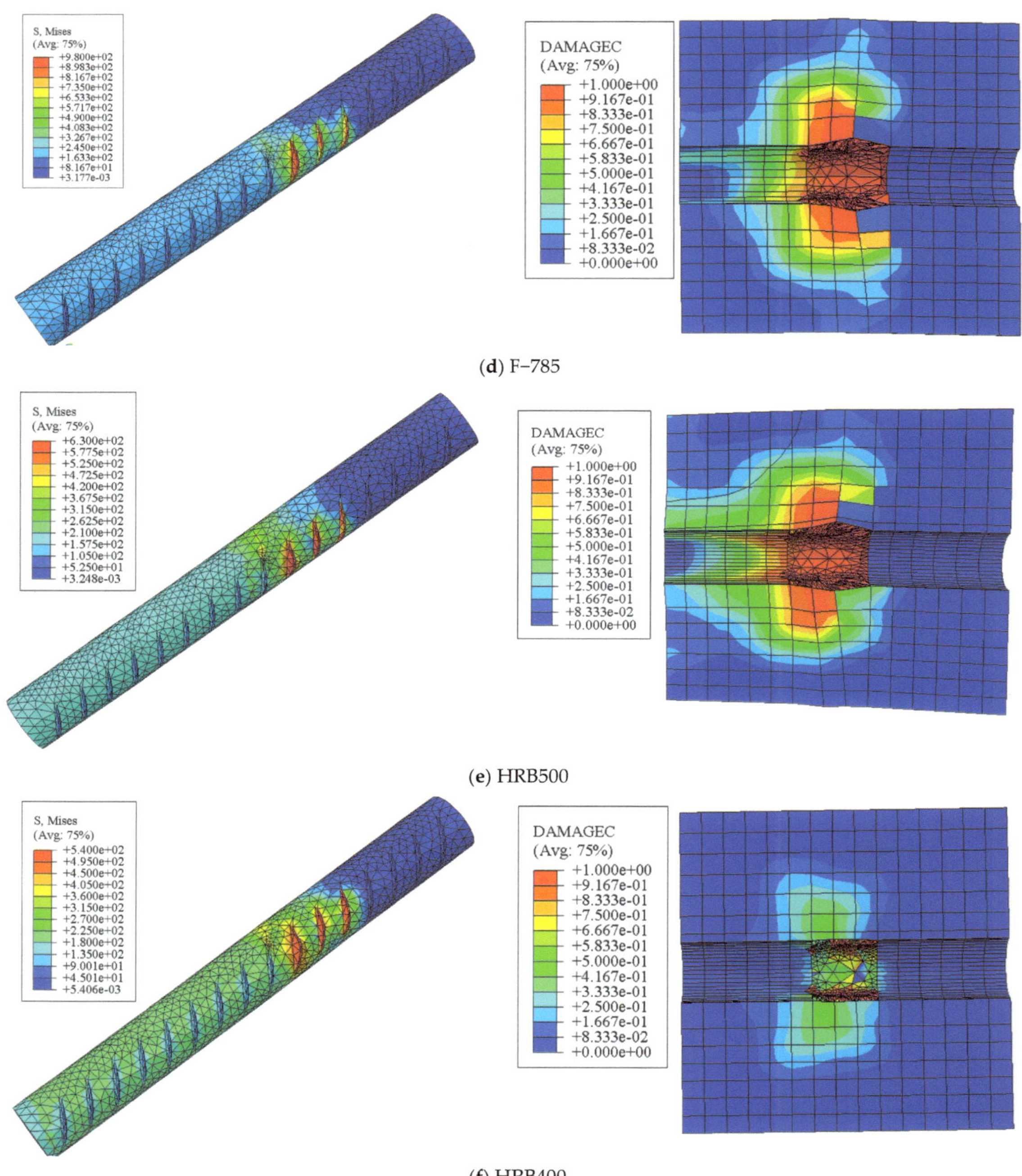

(**d**) F–785

(**e**) HRB500

(**f**) HRB400

Figure 16. *Cont.*

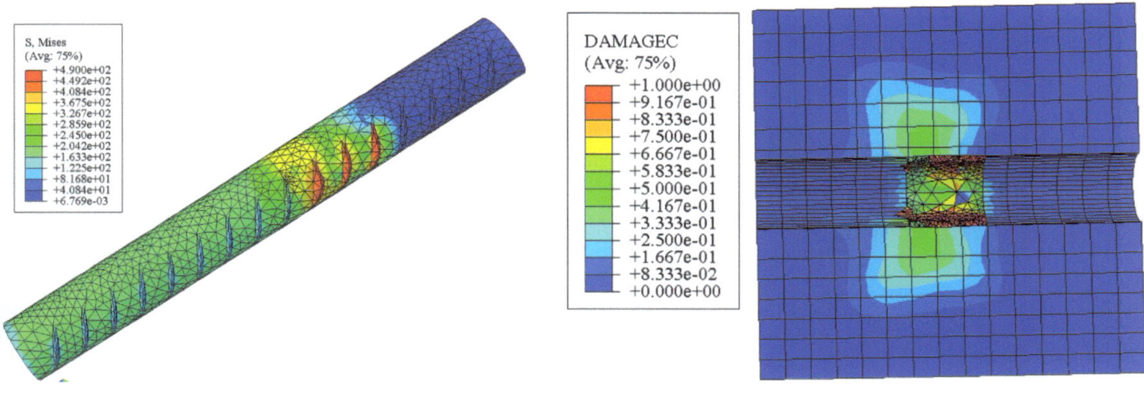

(g) HRB335

Figure 16. Steel reinforcement stress field diagram and concrete compression damage factor distribution of specimens with different steel reinforcement types and strengths.

Figure 17 presents the stress field diagram of steel reinforcement bonding area at peak load. It can be seen that the stresses on the surface of the steel reinforcement are significantly higher in plots in Figure 17 e–g compared to plots in Figure 17a–e. This discrepancy is primarily due to the fact that, for steel reinforcement with a strength class below 500 MPa, the bond stresses remain intact when the steel reinforcement yields or is pulled off. This facilitates stress concentration, resulting in higher localized stress values. In contrast, for steel reinforcement with a strength class exceeding 500 MPa, the bond stress between the reinforcement and concrete is lost before the reinforcement yields, leading to minor overall slippage and reduced stress concentration. Therefore, the stresses in plots in Figure 17e–g are noticeably higher than those in plots in Figure 17a–e.

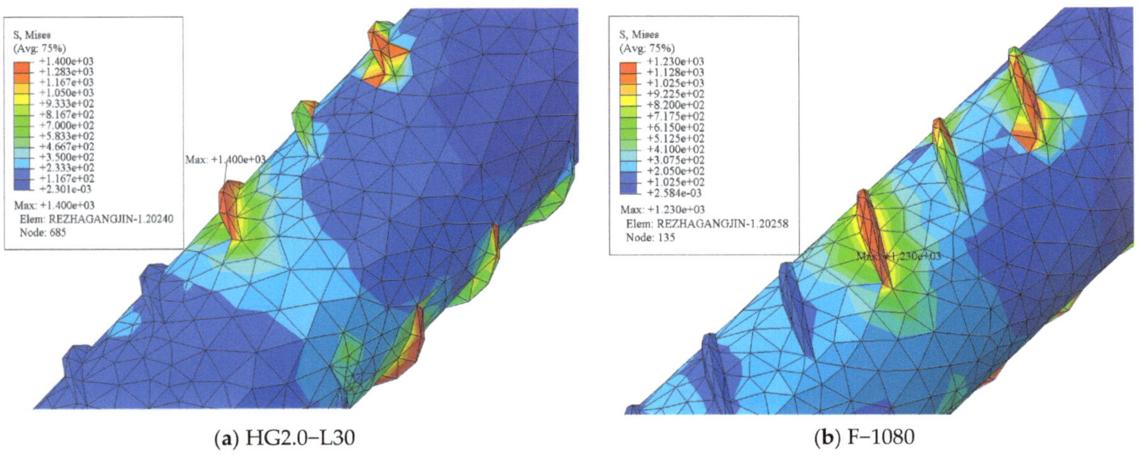

(a) HG2.0–L30　　　　　　　　　　　　**(b)** F–1080

Figure 17. *Cont.*

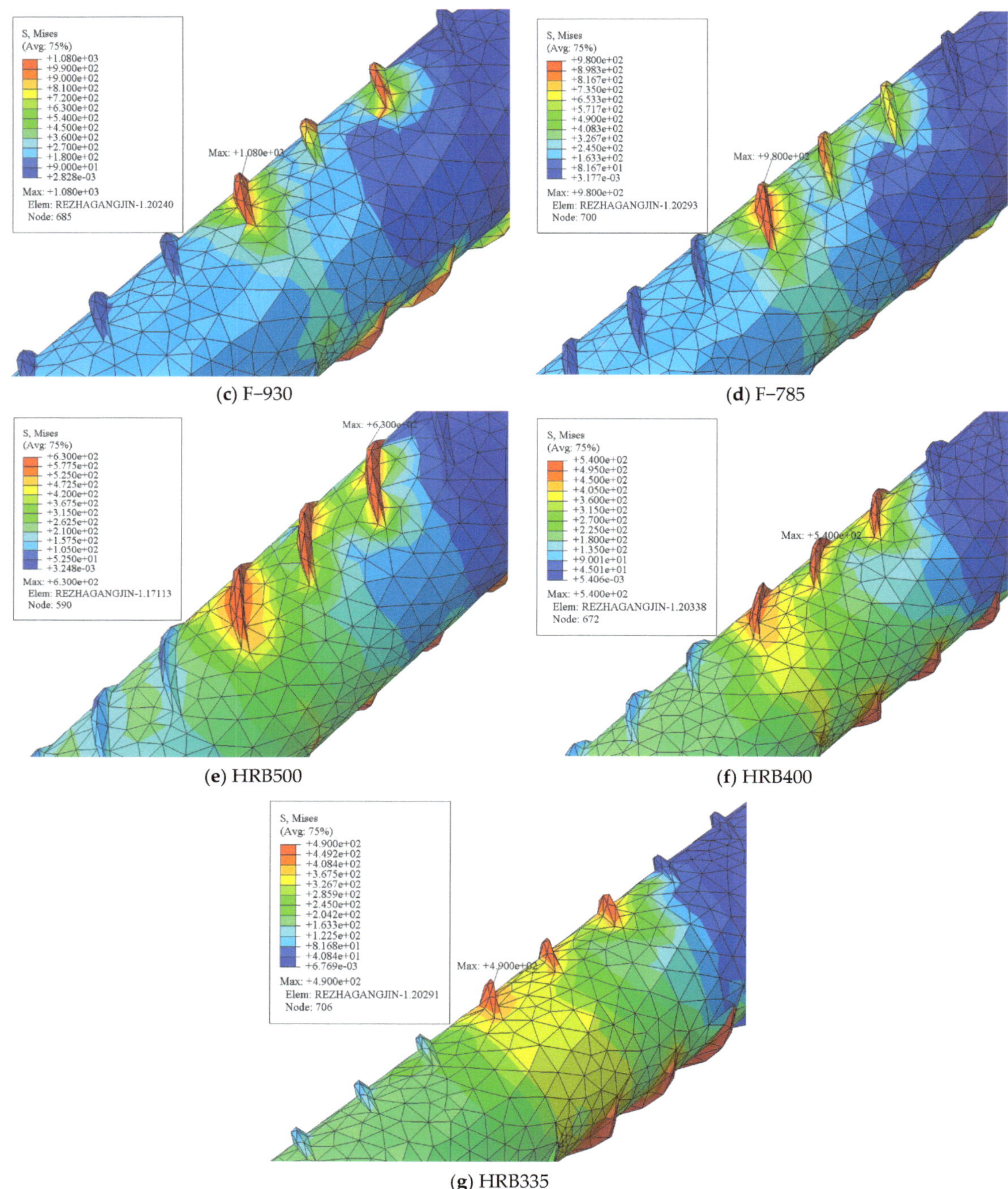

Figure 17. Stress field diagram of steel reinforcement bonding area at peak load.

Figure 18 presents the bond stress–slip curves for each specimen, revealing that the bond stress increases with increasing steel reinforcement strength, but the effect is not significant. The figure demonstrates the bond stress–slip curves of low-strength and high-strength reinforced bars differ in both shape and curvature. This variation arises from the following factors: the slope and curvature of the bond stress–slip curves for high-strength reinforced bars are smaller due to the gradual loss of bond stress and occurrence of slight slip. In contrast, low-strength reinforced bars exhibit better bonding performance before yielding, resulting in smaller slip and a steeper slope; when the steel reinforcement approach or reaches its yield strength, slip increases rapidly, causing the curve to bend towards the horizontal axis until the maximum stress is reached. Therefore, the bond stress–slip curves of low-strength reinforced bars display distinct shapes and curvatures compared to those of high-strength reinforced bars.

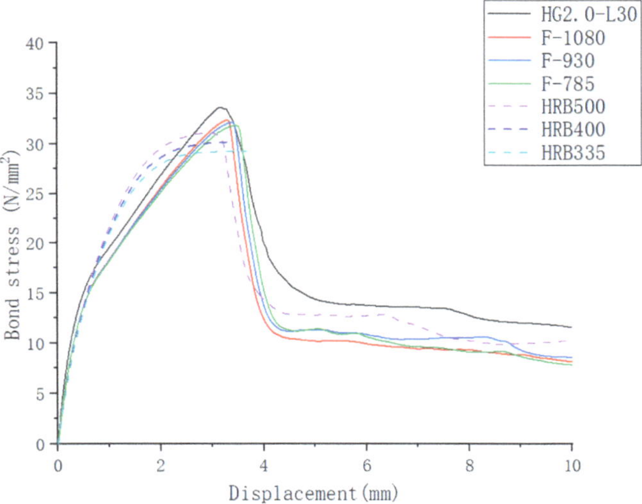

Figure 18. Bond stress–slip curves of specimens with different steel reinforcement types and strength.

4.4.3. Effect of Steel Reinforcement Diameter

Building on the basis of specimen HG2.0−L30, this analysis held constant the parameters epoxy resin concrete strength. To assess the impact of steel reinforcement diameter on bond–slip performance, the diameter was varied, resulting in the creation of two additional finite element models. The model numbers and specific parameter settings for these variations are detailed in Table 7.

Table 7. Basic information of specimens with different steel reinforcement diameters.

Specimen Number	Steel Reinforcement Diameter (mm)	Yield Strength of Steel Reinforcement (MPa)	Ultimate Strength of Steel Reinforcement (MPa)	Rib Spacing of Steel Reinforcement (mm)	Bond Length (mm)
HG2.0−L30	25	1260	1400	11.8	30
R−32	32	1260	1400	11.8	30
R−40	40	1260	1400	11.8	30

Figure 19 illustrates the stress field diagram of steel reinforcement at peak load of each specimen, alongside the compressive damage factor distribution of concrete in the final state. It is observed that the maximum stress within the steel reinforcement is mainly concentrated in the bond region. Interestingly, the stress distribution decreases with the increase of the steel reinforcement diameter. However, difference in the diameter of the High-Strength Reinforced Bar has no significant effect on the damage pattern in the bond region.

Figure 20 presents the bond–slip curves of each specimen. At the beginning of loading, the curves for all specimens align closely, indicating that the initial stiffness remains consistent across the variations. As the diameter of the steel reinforcement increases, both the peak stress and peak slip exhibit a decrease, and the descending section of the curves becomes steeper. This suggests that the degradation of mechanical occlusion force accelerates post-peak. Specifically, the maximum bond stresses for the specimens with steel reinforcement diameters of 25 mm, 32 mm, and 40 mm were 33.6 N/mm², 31.6 N/mm², and 29.0 N/mm², respectively. This trend demonstrates that, with constant rib spacing, the bond stress decreases in correlation with the increase in the steel reinforcement diameter.

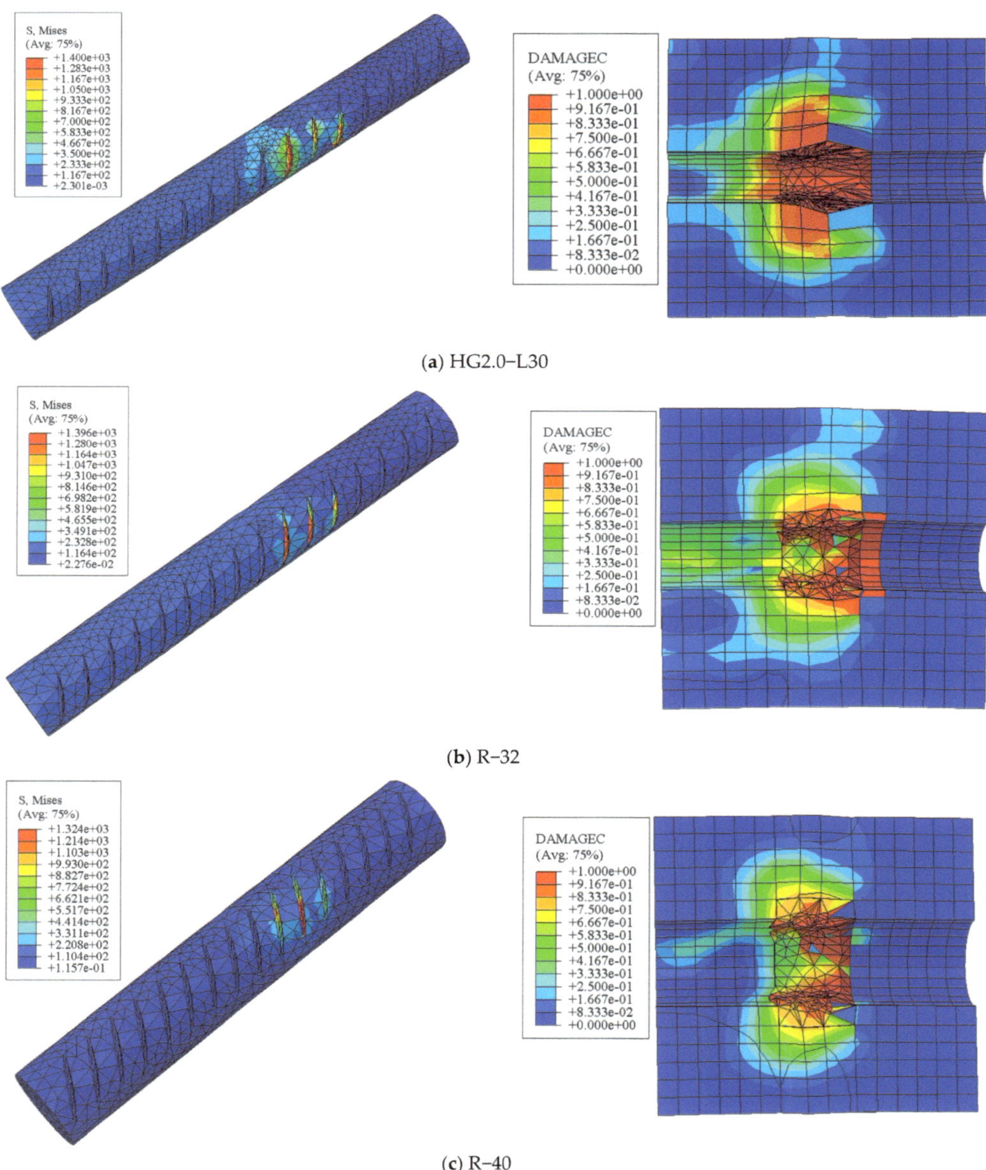

(**a**) HG2.0–L30

(**b**) R–32

(**c**) R–40

Figure 19. Steel reinforcement stress field diagram and concrete compressive damage factor distribution of specimens with different steel reinforcement diameters.

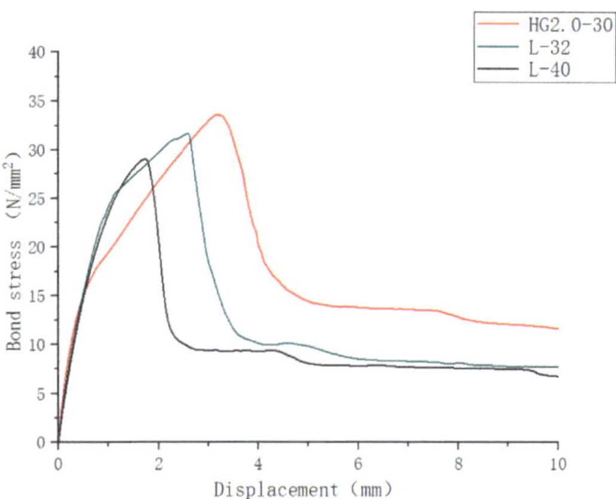

Figure 20. Bond stress–slip curves of specimens with different steel reinforcement diameters.

4.4.4. The Influence of the Thickness of Steel Reinforcement Protective Layer

On the basis of specimen HG2.0−L30, this study maintained constant the parameters for epoxy resin concrete strength, steel reinforcement strength, and bond length. To explore the effects of varying the thickness of the protective layer on steel reinforcement, two additional finite element models were developed. The specific model numbers and parameter settings for these new configurations are detailed in Table 8. The setups for these finite element models are visually represented in Figure 21.

Table 8. Basic information of specimens with different thicknesses of steel reinforcement protective layers.

Specimen Number	Thickness of Steel Reinforcement Cover (mm)	Yield Strength of Steel Reinforcement (MPa)	Ultimate Strength of Steel Reinforcement (MPa)	Rib Spacing of Steel Reinforcement (mm)	Bond Length (mm)
HG2.0−L30	62.5	1260	1400	11.8	30
R−32	37.5	1260	1400	11.8	30
R−40	12.5	1260	1400	11.8	30

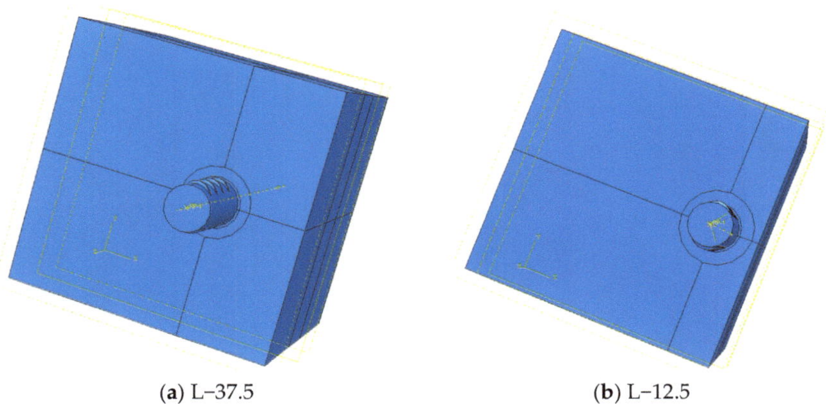

(**a**) L–37.5   (**b**) L–12.5

Figure 21. Finite element model of specimens with different thicknesses of steel reinforcement protective layers.

Figure 22 presents the stress field diagrams of steel reinforcement at peak load and the compressive damage factor distribution of concrete in the final state for each specimen with different thicknesses of protective layer. Observations indicate that damage in all three specimens is concentrated in the bond section. Particularly, the extent of damage area diminishes with the decrease of the protective layer thickness of the steel reinforcement. The maximum stress within the steel reinforcement is concentrated at the transverse rib of the bond section, and this stress value decreases with a reduction in the protective layer thickness.

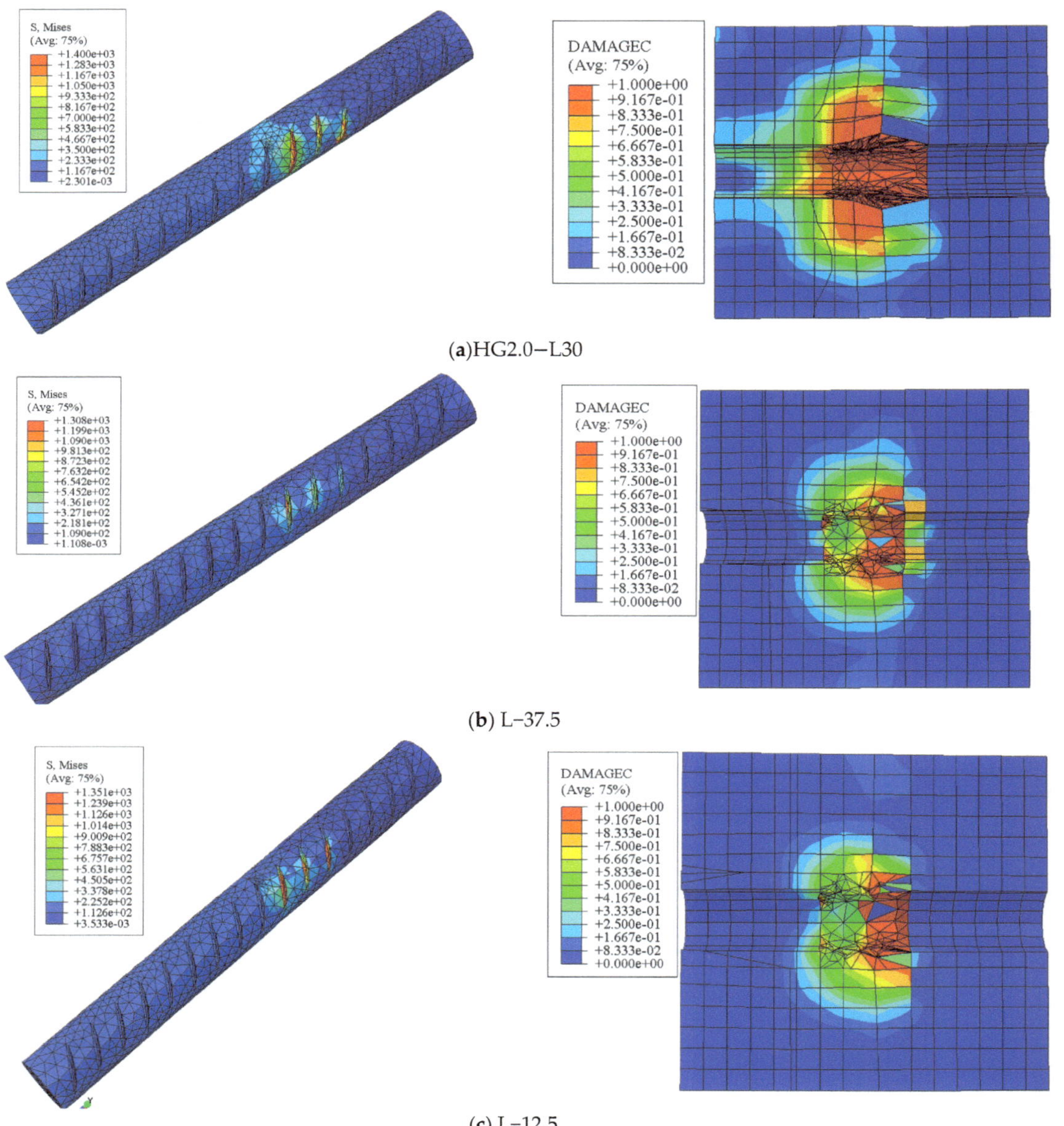

(**a**)HG2.0−L30

(**b**) L−37.5

(**c**) L−12.5

Figure 22. Steel reinforcement stress field diagram and concrete compressive damage factor distribution of specimens with different thicknesses of steel reinforcement protective layers.

Figure 23 shows the bond–slip curves of each specimen. It can be seen that the changes in the thickness of the protective layer of the steel reinforcement have a significant effect on the bond stress. Specifically, the bond stresses for the specimens with protective layer thicknesses of 62.5 mm, 37.5 mm, and 12.5 mm were 33.6 N/mm^2, 26.9 N/mm^2, and 16.7 N/mm^2, respectively; i.e., the bond stress decreases significantly as the thickness of the protective layer is reduced. Additionally, the peak slip is reduced. For the conventional protective layer thickness of 30–40 mm, the peak bond slip between the epoxy resin concrete and the deformed steel reinforcement is approximately 1 mm.

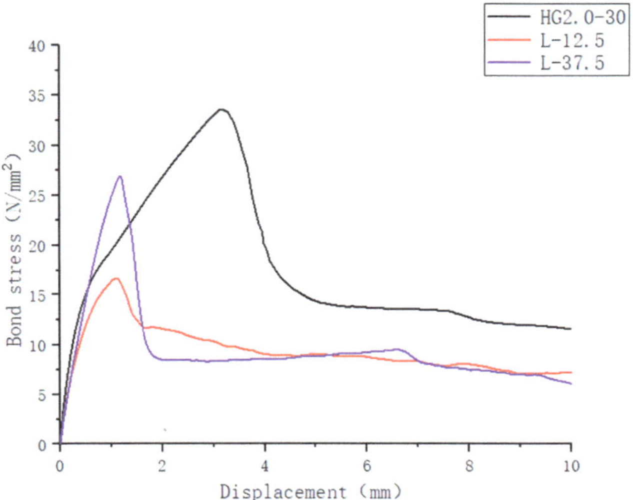

Figure 23. Bond stress–slip curves of specimens with different thicknesses of steel reinforcement protective layers.

5. Conclusions

In this paper, the bonding mechanism between epoxy resin concrete and steel reinforcement is investigated through a pullout test and finite-element simulation. It also explores the influence of different factors on the bond–slip performance between the two materials, to provide data support for the practical application of epoxy resin concrete in structural engineering. The primary conclusions are as follows:

(1) Due to the inherent adhesive property of the epoxy resin material, the bond strength between epoxy resin concrete and steel reinforcement significantly surpasses that of ordinary concrete, and it is approximately 3.23 times higher at an equivalent material strength level.

(2) An increase in both epoxy resin concrete strength and steel reinforcement strength can slightly increase the bond stress; the effect remains relatively subtle. Conversely, increasing the diameter of the steel reinforcement and reducing the thickness of the protective layer predominantly diminishes the bond stress, and the peak bond slip is reduced.

(3) Given the substantial bond strength between epoxy resin concrete and steel reinforcement, when the strength of the steel reinforcement is below 500 MPa, due to the high bond strength between epoxy resin concrete and steel reinforcement, when the strength of the steel bar is below 500 MPa, the bond between the steel reinforcement and epoxy resin concrete remains strong throughout the drawing process. In such cases, the steel reinforcement reaches its the yield strength or even fractures, meaning that yielding occurs before any significant loss of bond strength. In contrast, for high-strength reinforced bar with a strength exceeding 500 MPa, the bond strength between

the steel reinforcement and the epoxy resin concrete gradually diminishes during the drawing process, resulting in micro-slip without the steel reinforcement reaching yield. This leads to a failure mode characterized by scraping- or ploughing-type damage. Consequently, when ordinary steel reinforcement is used in epoxy resin concrete members, the risk of bond anchorage failure is largely eliminated.

(4) Due to the high strength of epoxy resin concrete and the localized stress concentration induced by the Poisson effect in the steel reinforcement, compared with ordinary concrete, the peak of the bond stress–slip curve for the epoxy resin concrete and steel reinforcement is notably more pointed. This sharp peak indicates that the bond stress reaches its maximum value of degradation faster, demonstrating a more brittle behavior.

Author Contributions: Conceptualization, P.C.; Investigation and Methodology, X.Z.; Software, Y.L. and H.W.; Writing—original draft, Y.L.; Writing—review & editing, P.C. and X.Z.; Project administration, P.C. and J.L.; Resources and Supervision, J.L. All authors have read and agreed to the published version of the manuscript.

Funding: This research was funded by the National Natural Science Foundation of China, grant number 51678389.

Data Availability Statement: Dataset available on request from the authors.

Conflicts of Interest: The authors declare no conflict of interest.

References

1. Beeldens, A.; Monteny, J.; Vincke, E.; De Belie, N.; Van Gemert, D.; Taerwe, L.; Verstraete, W. Resistance to biogenic sulphuric acid corrosion of polymer-modified mortars. *Cem. Concr. Compos.* **2001**, *23*, 47–56. [CrossRef]
2. Qian, Y.; Li, Z.; Jin, Y.; Wang, R. Experimental Study on Axial Tension Members of a New Epoxy Resin Concrete. *Sci. Adv. Mater.* **2021**, *13*, 2005–2015. [CrossRef]
3. Liu, K.F.; Li, Y.Z. Experimental Study on Mechanical Properties of Epoxy Resin Concrete. *Key Eng. Mater.* **2015**, *629*, 518–521. [CrossRef]
4. Diaconescu, R.M.; Barbuta, M.; Harja, M. Prediction of properties of polymer concrete composite with tire rubber using neural networks. *Mater. Sci. Eng. B* **2013**, *178*, 1259–1267. [CrossRef]
5. Haddad, H.; Al Kobaisi, M. Influence of moisture content on the thermal and mechanical properties and curing behavior of polymeric matrix and polymer concrete composite. *Mater. Des.* **2013**, *49*, 850–856. [CrossRef]
6. El-Hawary, M.M.; Abdel-Fattah, H. Temperature effect on the mechanical behavior of resin concrete. *Constr. Build. Mater.* **2000**, *14*, 317–323. [CrossRef]
7. El-Hawary, M.; Al-Khaiat, H.; Fereig, S. Effect of Sea Water on Epoxy-repaired Concrete. *Cem. 538 Concr. Compos* **1998**, *20*, 41–52. [CrossRef]
8. Yu, T. *The Experimental Study of Repairing Concrete Cracks with Epoxy Resin Grouting*; Beijing University of Technology: Beijing, China, 2016. (In Chinese)
9. Zhang, W.; Guan, X.; Ren, J.; Gu, X. Experimental study on chloride permeability of concrete surface-treated with epoxy resin. *J. Build. Mater.* **2008**, *11*, 339–344.
10. Xiang, Q.; Xiao, F. Applications of epoxy materials in pavement engineering. *Constr. Build. Mater.* **2020**, *235*, 117529. [CrossRef]
11. Rahman, M.M.; Akhtarul Islam, M. Application of epoxy resins in building materials: Progress 546 and prospects. *Polym. Bulletin.* **2022**, *79*, 1949–1975. [CrossRef]
12. El-Mandouh, M.A.; Abd El-Maula, A.S. Shear strength of epoxy-modified reinforced concrete beams. *Innov. Infrastruct. Solut.* **2021**, *6*, 105. [CrossRef]
13. Jin, Y. *Research of Epoxy Resin Concrete (Mortar) Modified and Application*; Jilin Jianzhu University: Jilin, China, 2010.
14. Li, X. *Study on the Behavior of the Axial Compression Members of the New Epoxy Resin Concrete*; Jilin Jianzhu University: Jilin, China, 2019.
15. Mousavi, S.S.; Mousavi Ajarostaghi, S.S.; Bhojaraju, C. A critical review of the effect of concrete composition on rebar-concrete interface (RCI) bond strength: A case study of nanoparticles. *SN Appl. Sci.* **2020**, *2*, 893. [CrossRef]
16. Mo, K.H.; Goh, S.H.; Alengaram, U.J.; Visintin, P.; Jumaat, M.Z. Mechanical, toughness, bond and durability-related properties of lightweight concrete reinforced with steel fibres. *Mater. Struct.* **2017**, *50*, 46. [CrossRef]
17. Shunmuga Vembu, P.R.; Ammasi, A.K. A comprehensive review on the factors affecting bond strength in concrete. *Buildings* **2023**, *13*, 577. [CrossRef]
18. Jathar, R. Study of Bond Strength of Fibre Reinforced Concrete. *Int. J. Res. Appl. Sci. Eng. Technol.* **2018**, *6*, 2371–2380. [CrossRef]

19. Kang, S.B.; Wang, S.; Long, X.; Wang, D.D.; Wang, C.Y. Investigation of dynamic bond-slip behaviour of reinforcing bars in concrete. *Constr. Build. Mater.* **2020**, *262*, 120824. [CrossRef]
20. Azmee, N.M.; Shafiq, N. Ultra-high performance concrete: From fundamental to applications. *Case Stud. Constr. Mater.* **2018**, *9*, e00197. [CrossRef]
21. Khaksefidi, S.; Ghalehnovi, M.; De Brito, J. Bond behaviour of high-strength steel rebars in normal (NSC) and ultra-high performance concrete (UHPC). *J. Build. Eng.* **2021**, *33*, 101592. [CrossRef]
22. Hu, A.; Liang, X.; Shi, Q. Bond characteristics between high-strength bars and ultrahigh-performance concrete. *J. Mater. Civ. Eng.* **2020**, *32*, 04019323. [CrossRef]
23. Liang, R.; Huang, Y.; Xu, Z. Experimental and analytical investigation of bond behavior of deformed steel bar and ultra-high performance concrete. *Buildings* **2022**, *12*, 460. [CrossRef]
24. Qasem, A.; Sallam, Y.S.; Eldien, H.H.; Ahangarn, B.H. Bond-slip behavior between ultra-high-performance concrete and carbon fiber reinforced polymer bars using a pull-out test and numerical modelling. *Constr. Build. Mater.* **2020**, *260*, 119857. [CrossRef]
25. Zhang, W.; Liu, Z.; Liu, Q.; Zhang, W. Investigation of bond behavior between waterborne epoxy-coated steel reinforcing bars (WECR) and concrete. *J. Build. Eng.* **2023**, *76*, 107328. [CrossRef]
26. Bai, Y.; Zhang, D.; Shen, K.; Yan, Y. Experimental study and numerical simulation on the bonding properties of rubber concrete and deformed steel bar. *Constr. Build. Mater.* **2023**, *383*, 131416. [CrossRef]
27. Cui, Y.; Qu, S.; Tekle, B.H.; Ai, W.; Liu, M.; Xu, N.; Zhang, Y.; Zhang, P.; Leonovich, S.; Sun, J.; et al. Experimental and finite element study of bond behavior between seawater sea-sand alkali activated concrete and FRP bars. *Constr. Build. Mater.* **2024**, *424*, 135919. [CrossRef]
28. *GB/T 50081-2019*; Standard for Testing Methods of Physical and Mechanical Properties of Concrete. National's Standard of the Peoples' Republic of China: Beijing, China, 2019.
29. Chen, P.; Xu, S.; Zhou, X.; Xu, D. An Experimental Study on Flexural-Shear Behavior of Composite Beams in Precast Frame Structures with Post-Cast Epoxy Resin Concrete. *Buildings* **2023**, *13*, 3137. [CrossRef]
30. *GB 50010-2010*; Code for Design of Concrete Structures. China Architecture & Building Press: Beijing, China, 2015. (In Chinese)

Article

Performance Assessment of Wood-Based Composite Materials Subjected to High Temperatures

Ruxandra Irina Erbașu *, Andrei-Dan Sabău *, Daniela Țăpuși and Ioana Teodorescu

Department of Civil, Urban and Technological Engineering, Faculty of Civil, Industrial and Agricultural Buildings, Technical University of Civil Engineering Bucharest, 020396 Bucharest, Romania; daniela.tapusi@utcb.ro (D.Ț.); ioana.teodorescu@utcb.ro (I.T.)
* Correspondence: ruxandra.erbasu@utcb.ro (R.I.E.); andreisabau228@gmail.com (A.-D.S.)

Abstract: This paper is based on research placed within the broader framework of the growing environmental impact requirements of building materials. Given this context, wood-based composite materials have emerged as a promising and innovative solution for structural elements. The current work aims to define a system for testing the mechanical behavior of glued laminated timber elements when exposed to high temperatures, in the neighborhood of the pyrolytic decomposition of materials. These tests monitor the transient behavior of the composite material and characterize the parameters involved in the thermo-mechanical analysis of elements constructed using this type of engineered wood product. The tests are used for the calibration of the material models involved in the numerical analysis and for the analysis of potential prototypes, considering the transient thermal load and heat propagation through the materials. By taking such tests, benchmark models and laboratory procedures are defined that can be used in the future to evaluate different materials, existing or new, and material combinations used to construct such a composite.

Keywords: glulam; engineered wood; fire resistance; finite element analysis

Citation: Erbașu, R.I.; Sabău, A.-D.; Țăpuși, D.; Teodorescu, I. Performance Assessment of Wood-Based Composite Materials Subjected to High Temperatures. *Buildings* **2024**, *14*, 3177. https://doi.org/10.3390/buildings14103177

Academic Editors: Atsushi Suzuki and Dinil Pushpalal

Received: 4 September 2024
Revised: 26 September 2024
Accepted: 1 October 2024
Published: 5 October 2024

1. Introduction

Mass timber, including glued laminated timber (glulam), has become an increasingly popular and widely used building material for several important reasons, such as its design versatility, dimensional stability, excellent strength-to-weight ratio, sustainability and aesthetic appearance. These benefits, along with improvements in manufacturing processes and a growing recognition of sustainable building practices, have contributed to the success and adoption of glulam structures.

The behavior of mass timber elements when subjected to fire and high temperatures is of great importance in their design and construction processes. Whenever these materials are subjected to intense heat, their mechanical properties can deteriorate significantly [1]. The thermal stability of the adhesives used in the construction of mass timber structural elements is known to have a great impact on the overall performance of the element [2]. On the other hand, the numerical modeling of such a behavior is a non-trivial task and involves advanced techniques as well as computationally intensive models, as shown in [3,4]. This is why designing mass timber structures to withstand fire loads safely is an expensive and time-consuming process that can rarely account for the type of adhesive used in the fabrication of the element.

The global behavior of glulam elements is greatly influenced by the behavior of the glued connections between the glued lamella forming the cross-section, especially when subjected to high temperatures. The production process itself (i.e., hot pressing) and the type of adhesive used bring about a complex combination of physical and chemical changes in wood [5,6] that influence the material behavior in terms of bearing capacity and mechanical properties. The glued interface is governed by the combination of normal and tangential stresses. Prior testing programs [7] have shown that the shear strength of this bonded interface strongly influences the global behavior of the composite material. This shear strength is obviously dependent on the working temperature of the adhesive. As such, the present work will further study the shear behavior of the glued joint when subjected to high temperatures.

The objective of this paper is to establish a laboratory mechanical testing protocol for glulam specimens subjected to high temperatures, which will yield data that are useful for the numerical modeling of the bonded interface of glulam elements using the finite element method. The numerical models obtained herein will be calibrated using the test data, thus obtaining an experimentally confirmed model for the bonded interface that can be used in further evaluations of prototypes and the benchmarking of the performance of newly developed adhesives.

2. Materials and Methods

2.1. Test Specimens

The wood used in the construction of the test samples is a soft wood of class GL24, as described in Table 1, which was chosen due to its very common application in the industry. The specimens were produced in the laboratory using a Melamine–Urea–Formaldehyde (MUF) adhesive.

Table 1. Strength and stiffness properties in N/mm^2 for GL24 class wood according to [8].

Property	Symbol	GL24c
Flexural Strength	$f_{m,g,k}$	24
Tensile Strength	$f_{t,0,g,k}$	17
	$f_{t,90,g,k}$	0.5
Compressive Strength	$f_{c,0,g,k}$	21.5
	$f_{c,90,g,k}$	2.5
Shear strength (shear and twist)	$f_{v,g,k}$	3.5
Shear Strength Normal to the Wood Fibers	$f_{r,g,k}$	1.2
Modulus of Elasticity	$E_{0,g,med}$	11,000
	$E_{0,g,05}$	9100
	$E_{90,g,med}$	300
	$E_{90,g,05}$	250
Shear Modulus	$G_{g,med}$	650
	$G_{g,05}$	540
Shear Modulus Normal to the Wood Fibers	$G_{g,med}$	65
	$G_{g,05}$	54
Density	$\rho_{g,k}\left[\frac{kg}{m^3}\right]$	365
	$\rho_{g,med}\left[\frac{kg}{m^3}\right]$	400

The test samples consist of two lamellas of size $60 \times 60 \times 10$ mm glued together by means of MUF adhesive. The dimensions of the samples were given by the capacity of the horizontal load sensor. The individual elements and the test samples were kept in a temperature and humidity-controlled environment during fabrication and curing, as well as until the mechanical tests were performed. Inside the samples, a nickel–chromium alloy heating element was placed in order to heat the sample directly at the bond interface. Due to the fact that NiCr alloys have a very high melting point, contact with the copper wire connectors was achieved via mechanical pressed sleeves.

The pattern used for the placement of the heating element, as shown in Figure 1b, was designed to ensure a uniform temperature distribution inside the bond interface. The continuity of the heating element was checked after curing by measuring the resistance of the wire. A total of 40 thermally active samples were produced for the testing campaign, 7 of which had the heating element shorted or interrupted. These were used in order to benchmark the unheated sample that contained the nickel–chromium alloy wire against glued samples lacking the heating element.

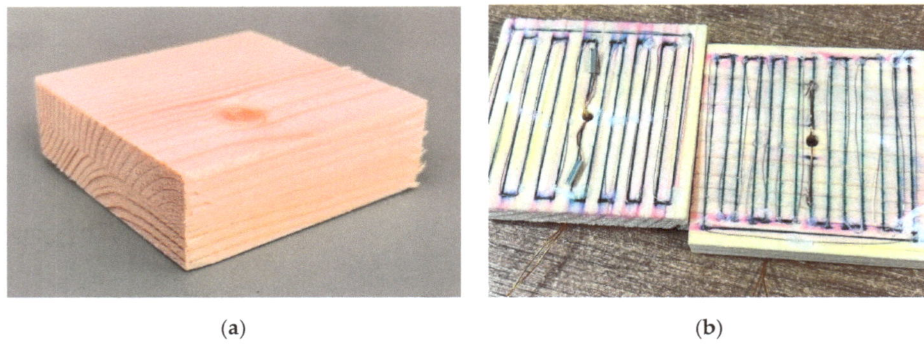

(a)   (b)

Figure 1. Wood sample construction: (**a**) The final two-piece glulam sample. (**b**) The nickel–chromium alloy wire heating element distribution inside the sample.

2.2. Test Equipment

The main difficulty in performing coupled thermo-mechanical tests for the characterization of the bonded interface comes from the fact that the heating element and the temperature sensor need to be as close as possible to the mechanical failure point of the sample. This is why the present study employs a direct shear apparatus in which the failure plane is imposed. A direct shear box had to be adapted in order to accommodate the wiring necessary for the heating element and temperature sensor (inserted in each sample as close as possible to the shearing plane) without damaging the circuits, as shown in Figure 2a.

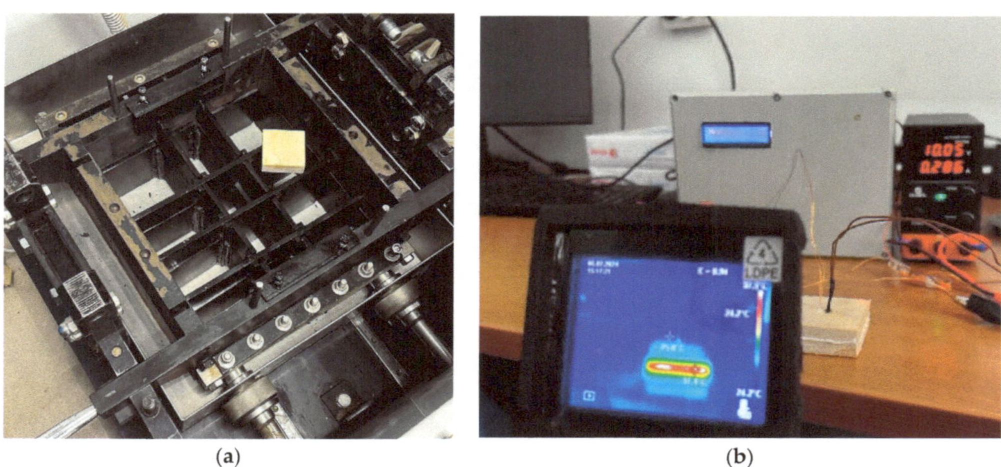

(a)   (b)

Figure 2. Thermo-mechanical testing equipment: (**a**) direct shear box and (**b**) temperature control and measurement equipment checked by a thermal camera.

The ends of the heating element pass through a hole at the top of the specimen, together with the connections to the resistive temperature sensor, as shown in Figure 2b. The temperature is set by switching a relay on or off whenever the measured temperature drifts from the target by more than 1 degree. The current for the heating element is taken from a programmable current source, with the voltage ranging between 10 V and 30 V, depending on the target temperature. A programmable current source was chosen so that it is able to provide a high enough voltage in order to keep the sample temperature constant throughout the test.

2.3. Testing Procedure

In order to compensate for the heatsink effect of the shear box, the samples were first preheated to the target temperature. After this step, the sample was placed inside the shearing box, a vertical stress was applied, and the temperature was stabilized before starting the mechanical test. Previous contributions [9] tested the behavior of wood at pure shear loading, where the interaction consisted only of mechanical coupling with no aid of confinement. The shearing box can be used to study the effect of various normal stresses on the shearing plane, too. The shearing force is sampled every 10 s, and the test is performed at a constant displacement rate of 1 mm/min. The maximum horizontal displacement was set to 20 mm. For this particular sample size, the maximum shear force encountered was below 30 kN, while the capacity of the shear force sensor is up to 44 kN. The test specimens should be large enough to allow just a negligible influence of the local material variation (such as the obliquity of fibers and the presence of small knots); therefore, the loading capacity of the shearing force sensor must be large enough to accommodate the test with reasonable headroom. The glulam sample was placed inside the shearing box with the wood fibers parallel to the shearing direction as they were loaded in a girder subjected to bending.

The testing temperatures were chosen starting from ambient values up to the ones expected inside a normal glulam cross-section subjected to fire, as shown in [10]. Thus, the discrete values were as follows: 20°, 40°, 60°, 70°, 80°, 100°, and 120°.

For each temperature, a number of at least 3 samples were tested.

2.4. Checking Sample Preparation Bias

The first step in the testing campaign was to evaluate the sensitivity of the samples to the fabrication process and the presence of the nickel–chromium alloy wire. To that end, 20 industrially made samples were first cut off an existing industrially manufactured beam and sheared at room temperature. These samples were divided into 5 sample groups that were sheared at different vertical stresses in order to check whether or not frictional behavior developed at failure. The axial forces used ranged from 50 kPa up to 600 kPa, and the results clearly exclude any kind of frictional behavior, as shown in Figure 3.

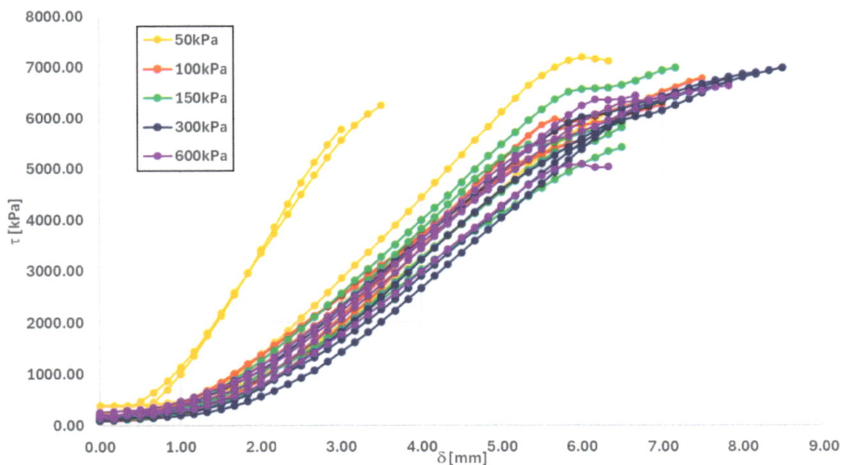

Figure 3. Shear mobilization curves at different vertical stresses for the industrially made samples.

The shear strength results indicate the fact that the fabrication process is an important factor in the final results obtained; however, they remain inside comparable domains in terms of mean and standard deviation.

It was noticed that every sample actually failed along the surface situated in the wood, immediately adjacent to the glued surface, showing the proper behavior of the adhesive. In order to check this assumption, some samples made of plain wood were also tested, and the compatibility of these results was observed. Another bias test also involved the shearing of the samples prepared in the laboratory and that those were instrumented but that had not been subjected to any heat load (Figure 4).

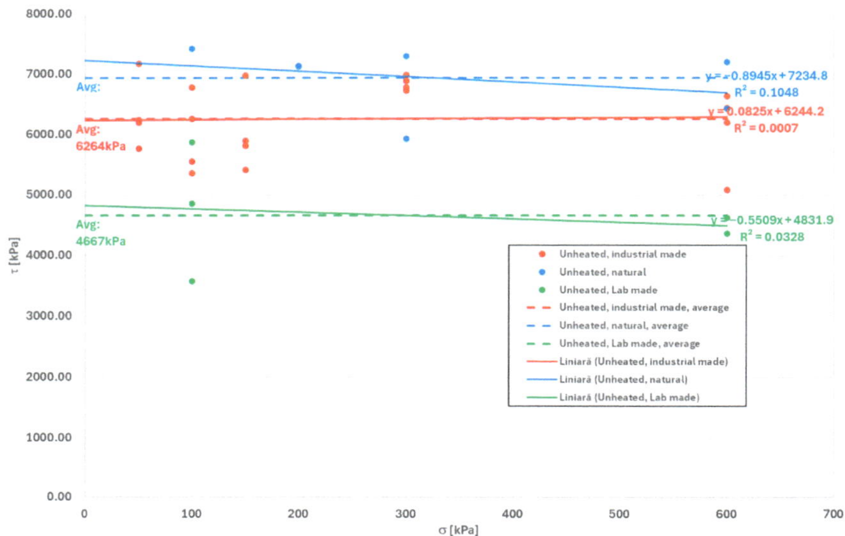

Figure 4. Comparison between the room temperature behavior of natural wood, industrially made glulam, and laboratory-made instrumented samples.

The distribution of the values is noticeable, yet within statistically acceptable margins. The shearing strength values of the natural wood and the industrially made samples are close and intertwined, with the actual averages being only governed by the number of samples. The behavior of the laboratory-made and instrumented samples is within an acceptable range, yet the data scattering is higher due to the smaller number of tests performed when trying to avoid testing additional electrically functional specimens.

3. Results

This section presents the mobilization curves during shearing at each temperature (Figure 5). The represented data start from the point of contact and about 5% mobilization of strength to avoid depicting a large initial plateau. This is necessary to close the tolerance gap between the sample and the shear box. This tolerance also refers to the slight obliquity of the sample, which is negligeable with respect to the overall behavior.

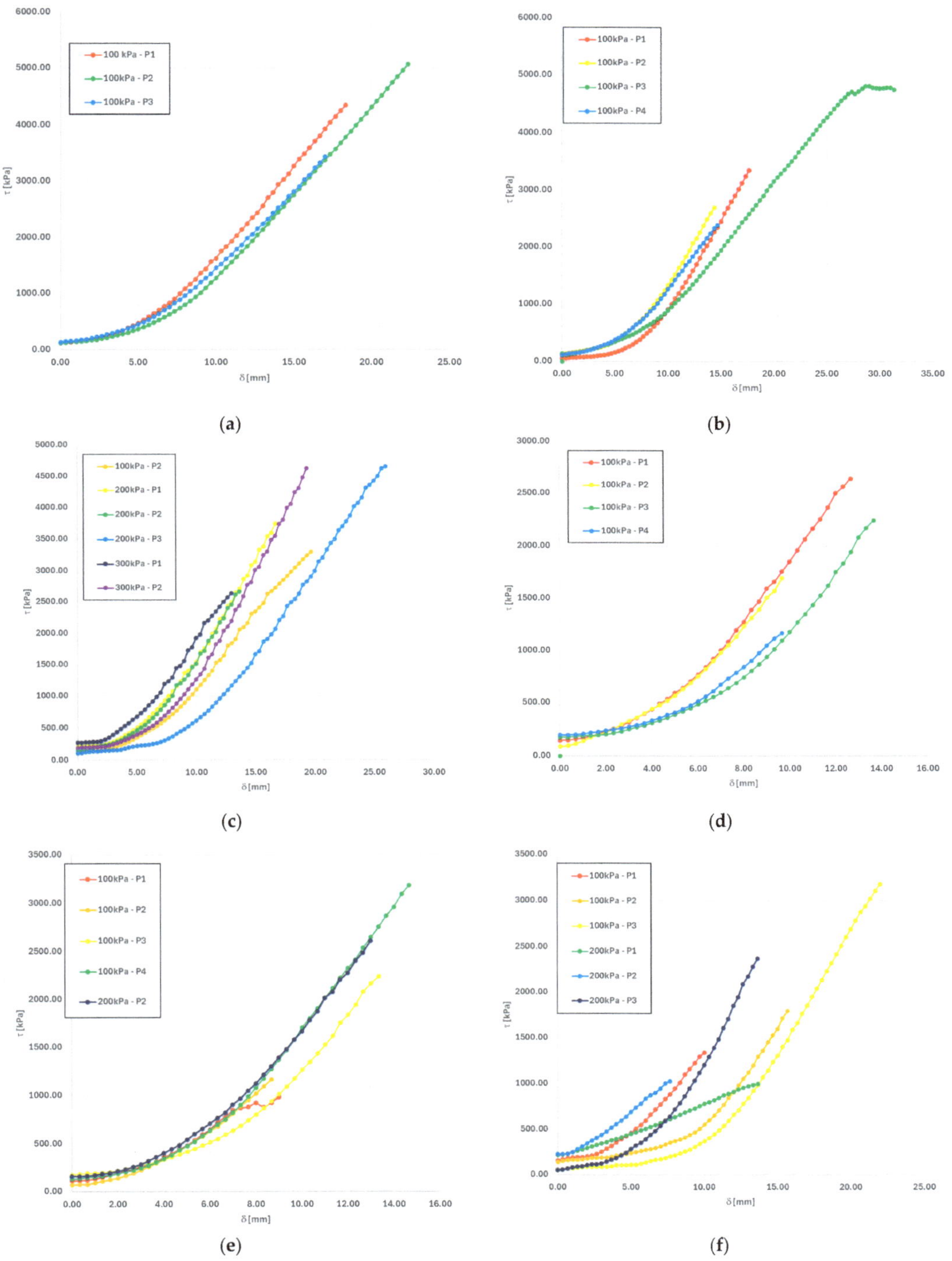

Figure 5. Shear mobilization curves for the thermo-mechanical tests at the following temperatures: (**a**) 40°, (**b**) 60°, (**c**) 70°, (**d**) 80°, (**e**) 100°, and (**f**) 120° Celsius.

It may be noted that for except very few cases, all failures occurred in a brittle fashion, with the post-failure domain being virtually inexistent (after the peak value recorded on the graph, the next force reading was virtually zero, and thus, not represented on the mobilization curves).

In order to plot the ultimate shearing stress variation with temperature, the peak values are given in Table 2.

Table 2. Peak values for the shear strength for each test specimen.

T [°C]	Stress [kPa]	Sample 1	Sample 2	Sample 3	Sample 4	Sample 5	Sample 6
20	s	100	100	100	600	600	-
	t	5880.56	4866.67	3583.333	4627.78	4375.00	-
40	s	100	100	100	-	-	-
	t	4347.22	5066.67	3433.333	-	-	-
60	s	100	100	100	100	-	-
	t	3347.22	2700.00	4816.667	2394.44	-	-
70	s	100	200	200	200	300	300
	t	3302.78	3741.667	2663.89	4661.11	2636.11	4630.56
80	s	100	100	100	100	-	-
	t	2644.44	1697.22	2241.67	1163.89	-	-
100	s	100	100	100	100	200	-
	t	925.00	1169.44	2241.667	3188.89	3238.89	-
120	s	100	100	100	100	200	200
	t	1338.89	1794.44	3180.56	997.22	1025.00	2369.44

When plotting the obtained data (Figure 6), consistent scattering was observed, as showcased by the error bars, which indicate a linear variation in shearing strength with temperature, having an R-squared value of 0.863. This linearity is maintained despite the fact that starting with about 60–70°, the failure mode no longer passes through the wood but gradually switches to the debonding of the glued lamellae and, ultimately, temperature softening of the glue.

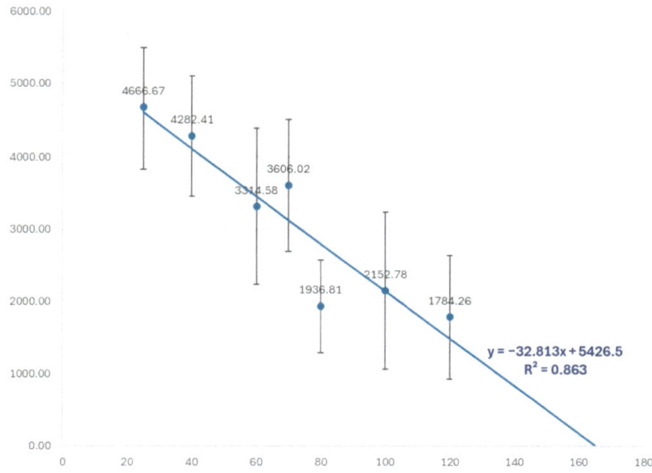

Figure 6. Ultimate shearing strength against temperature.

The experimental tests performed on MUF adhesive glued samples have shown that the results may fall within 3 stages corresponding to 3 main identified failure modes as shown in Figures 7–9.

Figure 7. Stage I of tests in which failure occurs due to wood shearing parallel to the fibers.

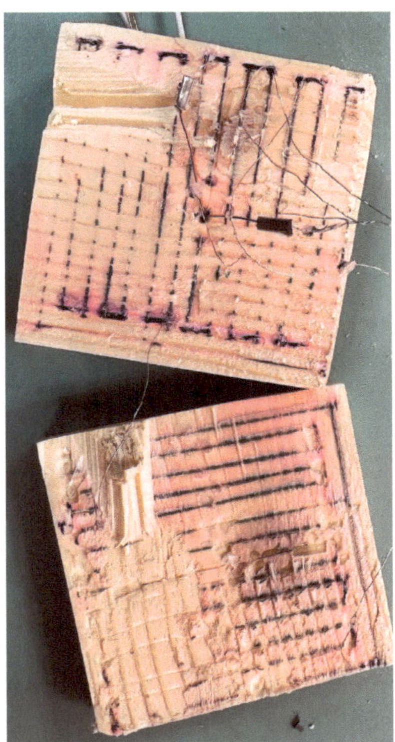

Figure 8. Stage II of tests in which failure occurs due to debonding at the interface.

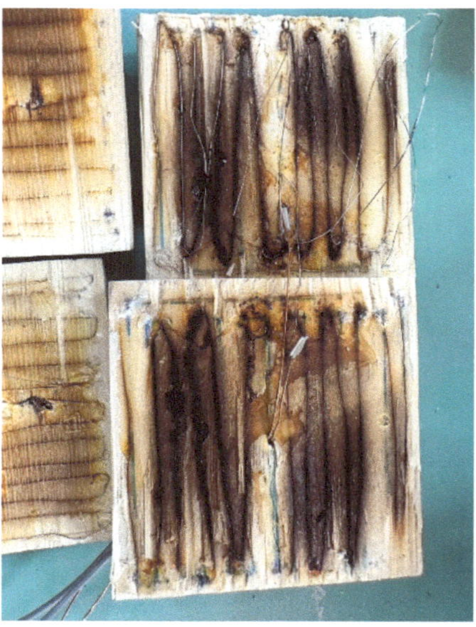

Figure 9. Stage III of tests in which failure occurs due to the thermal decomposition of the materials.

4. Numerical Modeling

In order to implement a finite element model, Figure 10, that captures the behavior of the bonded interface of the composite material tested, transient coupled temperature–displacement analysis is required. The finite element model presented herein was developed using the software ABAQUS 2024 since it implements coupled temperature–displacement elements and the contact models that it implements enable the modeling of the relevant behavior. The material parameters for wood are presented in Table 1. The interface was modeled as having a contact interaction type defined by a cohesive traction–separation behavior, as characterized by the average slope of the mobilization curves presented in Figure 5 for the elastic domain, while post-elasticity is described using a temperature-dependent damage model, with the criterion set at the nominal shear force value for each temperature determined during the experiments, as shown in Figure 6.

The applied loadings for the transient analysis were defined as time-dependent functions of amplitude, as shown in Figure 11. The temperature increased to its target value, in accordance with the test procedure after the initial stabilization null step. After the temperature distribution was stable at the interface, the model was loaded at a constant displacement rate of 1 mm/min, as in the testing procedure.

The outer boundary conditions imposed on the model reflect the heatsink effect of the actual steel shearing box. This was considered to be an ideal dissipator at a constant temperature of 20 °C. The mechanical boundary conditions are simple surface supports. In the case of the lower half of the sample, the mobile part in this experiment, the supports were displaced during transient analysis by a translation that ramps linearly over the analysis time, exactly as during the actual experiment, with the ramp amplitude function set up so as to reflect the constant shear rate imposed by the shear box.

The geometry reflects the actual shape of the test sample. The orientation of the local axes was chosen to obey an orthotropic material orientation.

The shear mobilization of the adhesive should match the quasi-linear relationship obtained during the tests; thus, the finite element implementation was chosen as linear, fitting the slope of the shear mobilization curve. The behavior of the adhesive was chosen to linearly ramp down after reaching peak strength in order to ensure proper convergence

(Figure 12). This was carried out because the delamination phenomenon was still captured with numerical stability ensured.

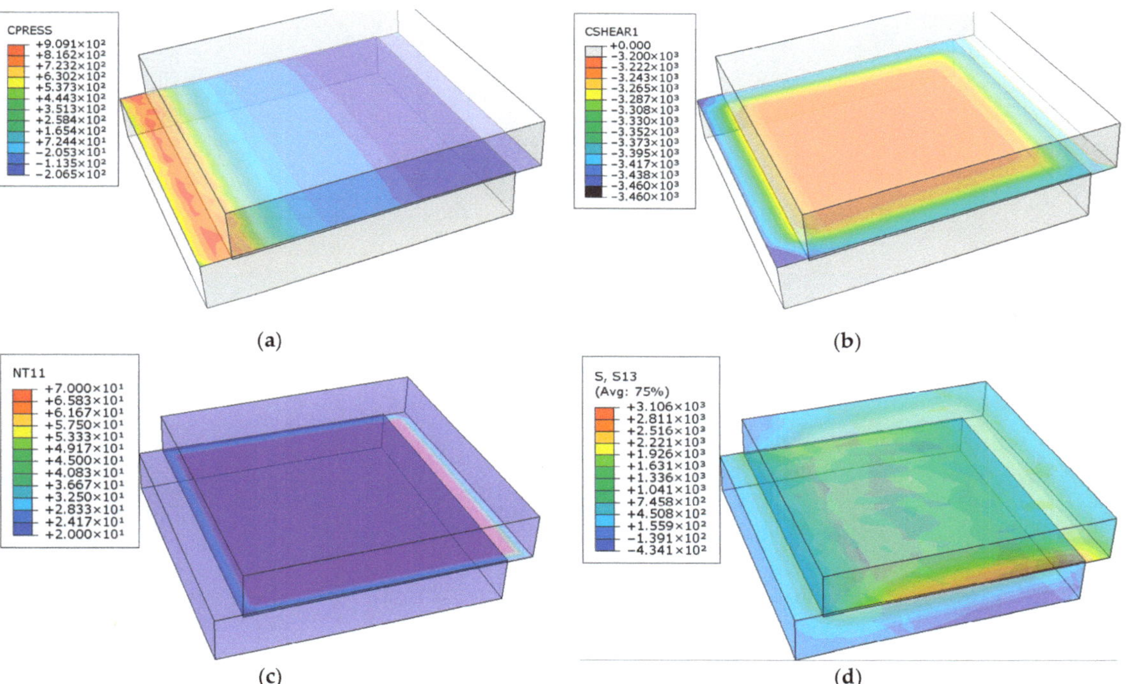

Figure 10. FEM model results at 70 °C and time of bonded interface failure: (**a**) normal contact pressure [kPa], (**b**) contact shear stress [kPa], (**c**) nodal temperatures [°C], and (**d**) internal shear stress [kPa].

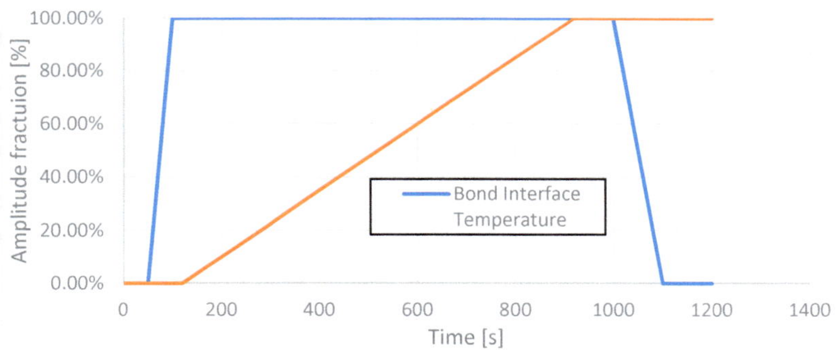

Figure 11. Amplitude variation for the loads applied in the numerical model of the test specimen.

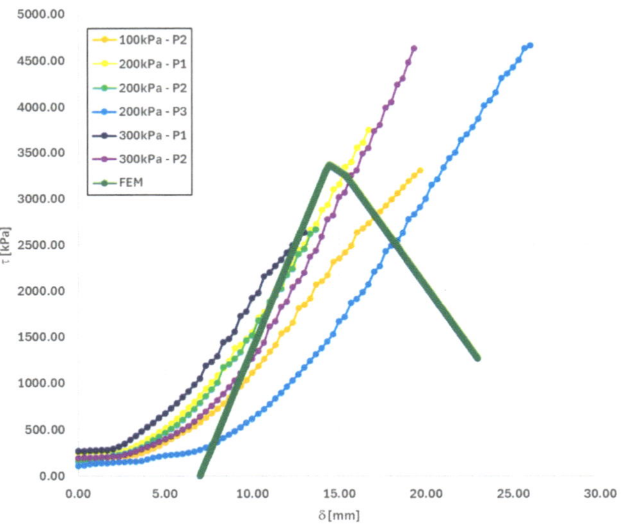

Figure 12. Comparison of FEM modeled behavior for 70 °C with respect to the test data.

5. Discussion

The experimental campaign shows that for the MUF adhesive tested, three main failure modes are identified, as shown in Figures 7–9. The manner of failure remains brittle for the three modes; however, it is clear that for low temperatures, the point of failure develops inside the wood; for medium temperatures, the point of failure occurs due to adhesive debonding; and for high temperatures, failure appears due to the adhesive softening. Unfortunately, the temperatures to be expected inside a glulam cross-section during a fire, as documented by [10], are higher than the thresholds found in the current research regarding the changes in the bonding characteristics. This means that the overall behavior of the element subjected to fire will not only be affected by the mass loss inherent to pyrolysis but also by delamination due to coupled thermo-mechanical loads.

It is obvious that new adhesives need to be developed to partially mitigate the delamination effect. However, each new solution must fulfill the minimum criterion of structural element connections; namely, in the case of mechanical ultimate loads, the failure has to occur inside the connected parts, not inside the connection itself. In the case of an MUF adhesive, this holds true for service temperatures yet loses validity as temperatures increase. This behavior should be known (Figure 6) and accounted for by structural designers. The compounds resulting from pyrolysis were not within the scope of the present work; however, such compounds should also be considered for new bonding agents, both from the point of view of the environment as well as from the point of view of toxic substance exposure during fires. Currently, the presence of MUF adhesive renders the reuse and recycling of glulam difficult.

When the experimental campaign started, the best option was studied in terms of subjecting the glued samples to controlled temperature loads. One solution was to isolate the shearing box inside a controlled heating chamber. The advantage of this method was the possibility of testing industrially prepared samples cut out from existing structural members; however, the insulation of the sample from actuators and sensors was a real challenge, along with the size of the equipment itself. Heating just the shearing box was still a challenge regarding heating the actuators and the sensors; therefore, the best solution was narrowed down to embedding the heating elements inside the specimen.

An assessment based on a limited number of samples indicates that once the glued elements subjected to high temperatures are cooled back to their service temperature, the shearing strength reverts to a nominal one; however, this research path was not followed since it deviated from the initial set scope of the work, with the application of a variable number of heating–cooling cycles. This study may be significant for elements such as roofing purlins and for those exposed to direct sun exposure in areas with hot climates.

The shearing box was modified to accommodate for the sample size and, most importantly, the applied axial stress was switched from a hydraulic solution to a pneumatic solution in order to achieve a more consistent vertical load value. The thermal camera was an important tool when checking the thermal features of the system and the sample.

If the compounds resulting from pyrolysis are not known, we strongly recommend performing this kind of experiment in a well-ventilated environment.

Testing the fire behavior at actual scale in the case of structural glulam elements is rather difficult, especially due to the precautions to be taken against fire hazards. This means that such a method should only be employed after numerical analysis for the confirmation of results.

In common engineering practices, the numerical analysis of glulam elements subjected to fire rarely accounts for the heat degradation of bond mechanical strengths. This is because, normally, the geometry of the numerical model should account for the bond interface by means of solid elements in order to ensure continuity. However, for the dimensions of the thin bond interface, a large number of elements should be used, or compromises should be made in terms of the accuracy of the results. These effects are especially pronounced when damage models and post-elastic behaviors are required, such as in the case studied in the present work. The transient nature of fire loads also means that approaching this problem using solid continuum elements will lead to excessively long analysis times. This is why the approach chosen here when modeling this phenomenon was achieved by means of a cohesive interface, where the actual shear test results can be directly linked to the governing law of the interface. The main components of the interface behavior are the elastic traction–separation model, given by the slope of the mobilization curve, as can be seen in Figure 11, and the damage model that governs the post-elastic behavior. Both of these functions are clearly temperature-dependent, and the values chosen for bond interface definition should be, as in the present case, backed by experimental data if the contact definition is to be useful for the further prototyping structural elements, such as beams and columns subjected to fire load. Defining damage as a total loss of strength after reaching the nominal shear resistance for the adhesive may lead to issues in terms of solution convergence; thus, steps need to be taken to mitigate this phenomenon in future developments.

If some future adhesive will exhibit a nonlinear temperature–strength behavior, some measures that will aid convergence in the transient analysis are first to decrease the analysis time step size. Another way of accommodating nonlinearity in thermo-mechanical behavior is to define a maximum temperature change over the step. This is especially useful if some noisy fire load curves are to be used.

The member prototypes should be loaded in the transient case by design-imposed fire loads, such as those defined by [11], as well as real fire load curves that contain the flashover and post-flashover phases if a proper evaluation of the structural element is to be conducted.

6. Conclusions

Wooden structures have two main issues: the inhomogeneity inherent to natural material, which is successfully mitigated by engineered wood products and, secondly, their sensitivity to fire. Theoretically, the latter should be dealt with by specific code provisions that impose constructive measures. These provisions mainly focus on the wood itself and less on the glued joints in the composite material. This paper aims to establish procedures that allow structural designers to analyze the complex behavior of bonded interfaces subjected to thermo-mechanical loads.

Some limitations of the procedure presented in this paper are related to the influence of the sample fabrication process and the presence of instrumentation inside the sample on the shear strength, as shown in Table 3. This, however, seems to be only an offset of the mean values, and they were found to be inside a reasonable value domain. Further research must be carried out in order to quantify the magnitude of this offset.

Table 3. Shear strength mean and standard deviation for the 3 modes of fabrication.

Fabrication Method	Mean Shear Strength [kPa]	Standard Deviation for Shear Strength [kPa]
Plain wood	6940.87	505.5724
Industrially made	6264.03	599.7707
Laboratory made	4666.67	744.81

Usually, in relevant codes, thermal loading is defined as an over-conservative time–temperature function [11]; this does not consider the fact that fires are finite events. Furthermore, they do not characterize the cooling phase and do not capture the real temperature distribution within the cross-section of elements. The results indicate that the thermal degradation of mechanical properties may appear close to service temperatures, especially in hot regions, as in the context of climate change for elements exposed to the environment.

The results indicate several key performance indices for developing future adhesives, such as better mechanical properties than the wood itself at service temperature as well as maintaining said characteristics at higher temperatures. Another key element is avoiding decomposition into potentially dangerous compounds, both during fire events and the post-utilization phase. Even if some key requirements are not fully met, the numerical methods presented showcase the possibility of finding special use cases where the exposure of the structural elements is less severe or where these elements are not critical for global structural integrity. Such a case should be applied for a potential adhesive with lower mechanical performance yet good recycling capacity.

Since the combination of wood types with certain adhesives is relatively limited, catalogs may be compiled for bonding properties to be used by structural engineers in the numerical modeling of glulam structures.

Supplementary Materials: The following supporting information can be downloaded at: https://www.mdpi.com/article/10.3390/buildings14103177/s1, Supplementary File S1.

Author Contributions: Conceptualization, R.I.E.; methodology, R.I.E. and A.-D.S.; data curation, A.-D.S.; writing—original draft preparation, I.T., writing—review and editing, R.I.E., A.-D.S. and D.T.; visualization, A.-D.S.; supervision, R.I.E. and D.T.; numerical modeling, A.-D.S., project administration, R.I.E., funding acquisition, R.I.E. All authors have read and agreed to the published version of the manuscript.

Funding: This research was funded by the Technical University of Civil Engineering Bucharest, Romania [grant no. GnaC$_{2023}^{ARUT}$-UTCB-25].

Data Availability Statement: The original contributions presented in the study are included in the article and Supplementary Materials, further inquiries can be directed to the corresponding authors.

Acknowledgments: The authors wish to acknowledge the contribution of Manole-Stelian Șerbulea (Department of Geotechnical and Foundation Engineering) from the Technical University of Civil Engineering Bucharest for their assistance in the development of the testing equipment and the following volunteer students for helping in the laborious work of preparing the instrumented samples (in alphabetical order): Alexandru Flangea, Luca Georgescu, Matei-Stelian Șerbulea, Noa Shraiter (Faculty of Engineering in Foreign Languages) and Denis Simota (Faculty of Railways, Roads and Bridges) of the Technical University of Civil Engineering Bucharest. Also, we aknowledge the support of Stefan Castravete of Caelynx Europe and of Dana Erbașu, MCHEM student, Faculty of Chemistry, University of Southampton, UK, for their theoretical support.

Conflicts of Interest: The authors declare no conflicts of interest. The funders had no role in the design of the study; in the collection, analyses, or interpretation of data; in the writing of the manuscript; or in the decision to publish the results.

References

1. Sinha, A.; Gupta, R.; Nairn, J. Effect of heat on the mechanical properties of wood and wood composites. In Proceedings of the 11th World Conference on Timber Engineering 2010 (WCTE 2010), Trentino, Italy, 20–24 June 2010; pp. 661–668.
2. Zhang, R.; Dai, H.; Smith, G.D. Investigation of the high temperature performance of a polyurethane adhesive used for structural wood composites. *Int. J. Adhes. Adhes.* **2022**, *116*, 102882. [CrossRef]
3. Fonseca, E.M.M.; Gomes, C. FEM Analysis of 3D Timber Connections Subjected to Fire: The Effect of Using Different Densities of Wood Combined with Stee. *Fire* **2023**, *6*, 193. [CrossRef]
4. Pereira, D.; Fonsec, E.M.M.; Osório, M. Computational Analysis for the Evaluation of Fire Resistance in Constructive Wooden Elements with Protection. *Appl. Sci.* **2024**, *14*, 1477. [CrossRef]
5. Wei, P.; Rao, X.; Yang, J.; Guo, Y.; Chen, H.; Zhang, Y.; Chen, S.; Deng, X.; Wang, Z. Hot Pressing of Wood-Based Composites: A Review. *For. Prod. J.* **2016**, *66*, 419–427. [CrossRef]
6. Liu, J.; Kong, Y.; Wang, F.; Wu, J.; Tang, Z.; Chen, Z.; Lu, W.; Liu, W. Effects of Moisture Content on Lap-shear, Bending, and Tensile Strength of Lap-jointed and Finger-Jointed Southern Pine using Phenol Resorcinol Formaldehyde and Melamine Urea Formaldehyde. *Bioresources* **2020**, *15*, 3534–3544. [CrossRef]
7. Dhima, D.; Audebert, M.; Racher, P.; Bouchair, A.; Taazount, M. Shear tests of glulam at elevated temperatures. *Fire Mater.* **2014**, *38*, 827–842. [CrossRef]
8. EN 14080:2013; Timber Structures—Glued Laminated Timber and Glued Solid Timber—Requirements. CEN: Brussels, Belgium, 2013.
9. Ng, A.L.Y.; Lau, H.H.; Fang, Z.; Roy, K.; Raftery, G.M.; Lim, B.P.J. The behavior of cold-formed steel and Belian hardwood self-tapping screw connections. In Proceedings of the Cold-Formed Steel Research Consortium Colloquium, Online, 17–19 October 2022.
10. Björn, K.; Per, L. Strength properties of wood adhesives after exposure to fire. In Proceedings of the Wood Adhesives 2005, San Diego, CA, USA, 2–4 November 2005.
11. EN 1991-1-2; Actions on Structures—Part 1–2: General Actions—Actions on Structures Exposed to Fire. CEN: Brussels, Belgium, 1991.

Article

Risk Assessment of Overturning of Freestanding Non-Structural Building Contents in Buckling-Restrained Braced Frames

Atsushi Suzuki [1,*], Susumu Ohno [2] and Yoshihiro Kimura [1]

[1] Graduate School of Engineering, Tohoku University, Sendai 980-8579, Japan; kimura@tohoku.ac.jp
[2] International Research Institute of Disaster Science, Tohoku University, Sendai 980-8577, Japan; susumu.ohno.e2@tohoku.ac.jp
* Correspondence: atsushi.suzuki.c2@tohoku.ac.jp; Tel.: +81-22-795-7876

Abstract: The increasing demand in structural engineering now extends beyond collapse prevention to encompass business continuity planning (BCP). In response, energy dissipation devices have garnered significant attention for building response control. Among these, buckling-restrained braces (BRBs) are particularly favored due to their stable hysteretic behavior and well-established design provisions. However, BCP also necessitates the prevention of furniture overturning—an area that remains quantitatively underexplored in the context of buckling-restrained braced frames (BRBFs). Addressing this gap, this research designs BRBFs using various design criteria and performs incremental dynamic analysis (IDA) with artificially generated seismic waves. The results are compared with previously developed fragility curves for furniture overturning under different BRB design conditions. The findings demonstrate that the fragility of furniture overturning can be mitigated by a natural frequency shift, which alters the threshold of critical peak floor acceleration. These results, combined with hazard curves obtained from various locations across Japan, quantify the mean annual frequency of furniture overturning. The study reveals that increased floor acceleration in stiffer BRBFs can lead to a 3.8-fold higher risk of furniture overturning compared to frames without BRBs. This heightened risk also arises from the greater hazards at shorter natural periods due to stricter response reduction demands. The probabilistic risk analysis, which integrates fragility and hazard assessments, provides deeper insights into the evaluation of BCP.

Keywords: fragility curve; furniture overturning; incremental dynamic analysis; probabilistic risk assessment

Citation: Suzuki, A.; Ohno, S.; Kimura, Y. Risk Assessment of Overturning of Freestanding Non-Structural Building Contents in Buckling-Restrained Braced Frames. *Buildings* **2024**, *14*, 3195. https://doi.org/10.3390/buildings14103195

Academic Editor: Muxuan Tao

Received: 8 September 2024
Revised: 1 October 2024
Accepted: 5 October 2024
Published: 8 October 2024

1. Introduction

In the face of significant earthquakes occurring globally, appropriate seismic retrofitting is essential. One effective method for collapse prevention is the implementation of seismic retrofitting or strengthening by buckling-restrained braces (BRBs). A BRB consists of a steel core encased in concrete, which is further confined by a steel tube. Typically, BRBs are installed in diagonal, V, or chevron (inverted-V) configurations, as depicted in Figure 1.

Since the invention of the BRB by Kimura et al. [1], extensive research has been conducted to evaluate BRBs at the component level [2–4]. These studies have demonstrated that BRBs possess exceptional ductility and energy-dissipation capacity, highlighting their ability to withstand strong earthquakes.

Currently, the design procedure is outlined in the prevailing design specifications [5,6]. The method for calculating the required amount of BRBs was investigated by Kasai et al. [7], and the proposed method is now incorporated into the guidelines of the Architectural Institute of Japan (AIJ) [5]. The necessary design considerations for the main frame have been extensively studied at both the member and frame levels [8–31].

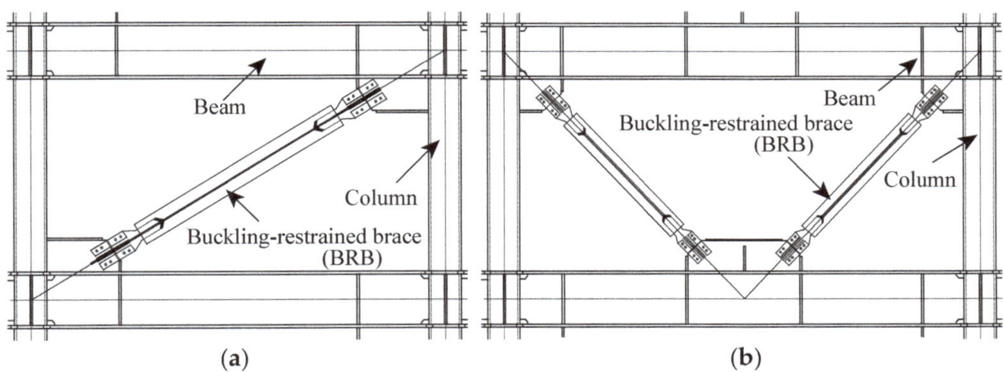

Figure 1. Configuration of Buckling-Restrained Braced Frame (BRBF): (**a**) diagonal configuration; (**b**) V configuration.

Thanks to these major advancements, buildings equipped with BRBs can withstand more significant earthquakes. As illustrated in Figure 2a,b, buckling-restrained braced frames (BRBFs) effectively prevent collapse even during large-scale seismic events, which could otherwise lead to the collapse of moment-resisting frames (MRFs). In recent years, the structural engineering field has increasingly emphasized the importance of ensuring business continuity after a major earthquake, aligning with the concept of Business Continuity Planning (BCP). However, previous research has predominantly focused on collapse prevention [32–35]. Faced with this need, structural health monitoring and damage detection methods have been emerging and have recently advanced [36–51]. While they effectively contribute to resilience, avoiding any damage that impacts BCP is the best option. Figure 2c illustrates the interior of a room following the 2022 Fukushima-Oki earthquake. The building was equipped with an energy dissipation device. Although the building remained intact and did not collapse, furniture toppled over, leading to a suspension of business operations during the repair work. Particularly after the Great East Japan Earthquake (GEJE) in 2011, researchers have shown increased interest in the damage to non-structural components, such as ceiling failures [52,53].

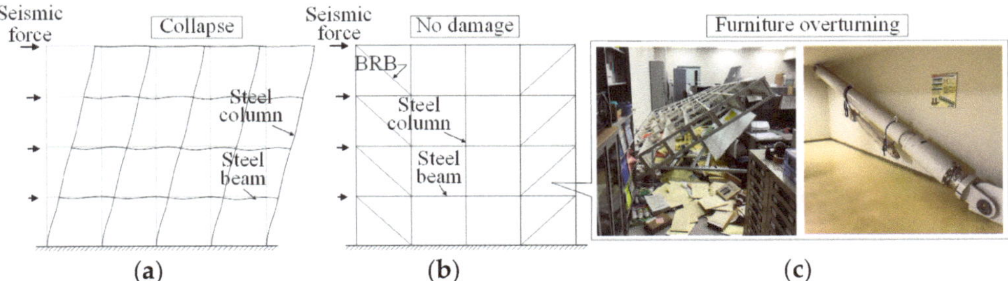

Figure 2. Concept of this research: (**a**) MRF; (**b**) BRBF; (**c**) furniture overturning after the 2022 Fukushima-Oki earthquake.

Regarding furniture overturning, a fundamental study was conducted by Kaneko [54,55]. The essential characteristics of overturning ratios of rigid bodies subjected to large input motions were examined through analytical studies. The results revealed that overturning ratios are influenced by the sizes and shapes of rigid bodies, as well as the levels and predominant frequencies of the input motions. Based on these studies, a general fragility curve was proposed to describe the relationship between overturning ratios and the amplification of input motions.

Subsequent studies have advanced the understanding of the fragility of rocking bodies by considering deformation, nonlinearity, and sliding [56–61]. These studies typically derived the moment equilibrium around the center of rotation using the moment of inertia and applied horizontal forces. Additionally, deformation is modeled using a spring and dashpot system at the corners of the rocking bodies to better reflect realistic conditions.

Subsequently, Saito et al. [62] conducted a seismic risk evaluation study considering furniture overturning in mid- and low-rise office buildings. Saito et al. [62] focused on mid- and low-rise steel frame office buildings. The study used the ratio of Q_u/Q_{un} (the horizontal load-bearing capacity, Q_u, to the required horizontal yield strength, Q_{un}) as an analytical parameter. Seismic response analysis was used to evaluate response values, and building damage was assessed through Seismic Risk Analysis. Additionally, the impact of furniture within the building was analyzed by determining the overturning ratio. The study demonstrated that by increasing Q_u/Q_{un}, the accuracy of seismic safety analysis for the building can be enhanced.

The probabilistic risk assessment (PRA) framework is well-suited for quantifying the effectiveness of seismic retrofitting in preventing furniture overturning. In a foundational study, Ishikawa et al. [63] proposed a procedure for seismic risk evaluation in buildings incorporating the results of probabilistic seismic hazard assessment (PSHA), structural fragility analysis, and economic loss estimation. The study produced a seismic risk curve that illustrates the relationship between financial loss and its annual exceedance probability with case studies validating the proposed methodology. This approach is widely applied globally [64–67].

PRA typically involves incremental dynamic analysis (IDA), comprehensively outlined by Vamvatsikos and Cornell [68]. IDA is a parametric method developed in various forms to estimate structural performance under seismic loads. It entails applying scaled ground motion records to a structural model across multiple intensity levels, producing response curves parameterized by intensity. Their study established unified terminology, introduced suitable algorithms, and examined the properties of IDA curves for both single- and multi-degree-of-freedom systems. Additionally, it discussed methods for summarizing multi-record IDA results, comparing IDA with conventional static pushover analysis, and evaluating the yield reduction R-factor.

Eads et al. [69] investigated various aspects of calculating the mean annual frequency of collapse and introduced an efficient approach for estimating the sideways collapse risk of structures in seismic regions. The deaggregation of the mean annual frequency of collapse showed that this risk is generally dominated by earthquake ground motion intensities in the lower half of the collapse fragility curve. Their study also quantified the uncertainty in both the collapse fragility curve and the mean annual frequency of collapse based on the number of ground motions used in the analysis. The proposed method significantly reduced the computational effort and the uncertainty associated with these estimates.

Deylami and Mahdavipour [70] critiqued a significant drawback of BRBFs, namely the low post-yield stiffness of steel cores, which results in large residual drifts in a story after earthquakes. Their study examined the seismic demands of low- and mid-rise BRBFs and Dual-BRBFs using Probabilistic Seismic Demand Analysis (PSDA). By comparing demand hazard curves, they concluded that employing BRBFs as a dual system could significantly reduce residual drift demands, improving the resilience of such structures after earthquakes with lower repair costs. Additionally, the study used two nonlinear models—a deteriorating model and a non-deteriorating model—to examine the impact of degradation on Dual-BRBFs' seismic demands. Their analysis revealed residual deformations are more susceptible to degradation than maximum deformations. However, deterioration is mitigated by the significant stiffness of BRBs, which keeps MRFs within a low range of nonlinearity.

The aforementioned studies primarily concentrated on structural collapse or damage related to structural safety. However, as previously noted, the recent societal demand increasingly emphasizes the importance of maintaining business continuity even after major earthquakes. Despite this, approaches to BCP based on PRA remain limited. In response to this, the present study specifically focuses on applying PRA to the issue of furniture overturning in BRBFs. Additionally, the effectiveness of retrofitting with BRBs for BCP is evaluated analytically. Therefore, building collapse is not assessed in this study, as advancements in BRBF design have already enhanced the structural capacity to prevent collapse effectively.

In this study, the model structure is derived from the design provisions of the AIJ and retrofitted with BRBs according to various design criteria. Realistic seismic waves are generated, reflecting the underlying ground structure and potential fault fracture scenarios. Using these seismic waves, IDA is conducted to establish fragility curves for furniture overturning. Finally, fragility and hazard curves for various locations in Japan are compared, and the mean annual frequency of furniture overturning is computed.

The findings of this research contribute to advancing seismic design by incorporating considerations for BCP. Moreover, the study highlights the importance of adequately securing furniture, even in buildings retrofitted with BRBs. In practical applications, engineers can quantify the risk of furniture overturning and assess its impact on BCP, providing a reasonable criterion for discussions with building owners or tenants. As such, the methodology presented in this research offers a platform for developing more sophisticated design strategies that account for business contingency.

2. Model Structure for Seismic Retrofitting Using Buckling-Restrained Braces

2.1. Outline of MRF Model Structure

This study considers the seismic upgrading of a four-story Japanese office building. The structure in question is a representative model of a steel moment-resisting frame (MRF) as specified in the provisions of the AIJ [5].

The structure represents a typical office building designed in accordance with Japanese regulations, featuring a regular floor plan. Consequently, torsional response is not a significant concern. This regularity makes the building suitable for studying the relationship between retrofitting with BRBs and the risk of furniture overturning.

A drawing of the structure is presented in Figure 3. The building's floor plan spans 25.6 m in the x direction and 19.2 m in the y direction. The weight distribution for each story is shown in Figure 3. The columns are square hollow section (SHS) tubes, with widths ranging from 350 to 450 mm and wall thicknesses between 16 and 22 mm.

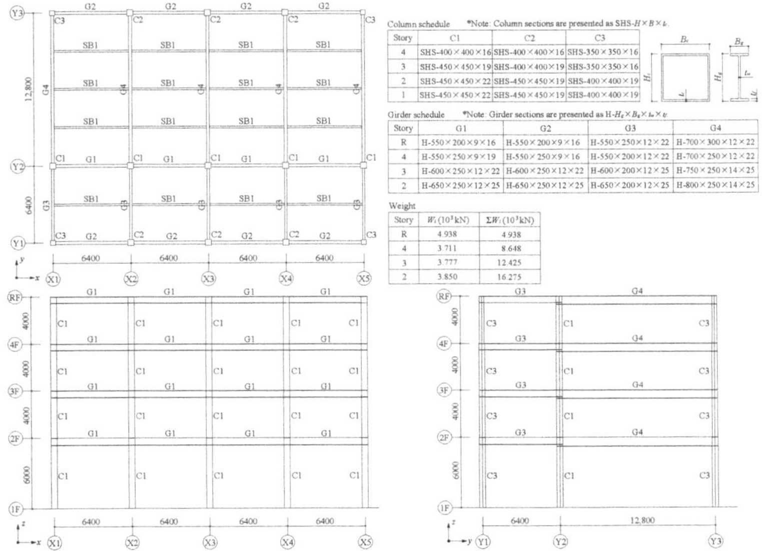

Column schedule   *Note: Column sections are presented as SHS-$H \times B \times t$.

Story	C1	C2	C3
4	SHS-400×400×16	SHS-400×400×16	SHS-350×350×16
3	SHS-450×450×19	SHS-400×400×19	SHS-350×350×16
2	SHS-450×450×22	SHS-450×450×19	SHS-400×400×19
1	SHS-450×450×22	SHS-450×450×19	SHS-400×400×19

Girder schedule   *Note: Girder sections are presented as H-$H_g \times B_g \times t_w \times t_f$.

Story	G1	G2	G3	G4
R	H-550×200×9×16	H-550×200×9×16	H-550×250×12×22	H-700×300×12×22
4	H-550×250×9×19	H-550×250×9×16	H-550×200×12×22	H-700×250×12×22
3	H-600×250×12×22	H-600×250×12×22	H-600×200×12×25	H-750×250×14×25
2	H-650×250×12×25	H-650×250×12×25	H-650×200×12×25	H-800×250×14×25

Weight

Story	W_i (10^3 kN)	ΣW_i (10^3 kN)
R	4.938	4.938
4	3.711	8.648
3	3.777	12.425
2	3.850	16.275

Figure 3. Drawing and member schedule of model structure.

The main girders consist of wide-flange sections with depths ranging from 550 to 800 mm and flange widths of 200 to 300 mm, with plate thicknesses between 9 and 25 mm. All girder-column connections are designed as rigid moment connections. The steel used is SN490, per Japanese standards [71], with a yield stress of 325 N/mm^2.

2.2. Equivalent Linearization-Based Design of a Buckling-Restrained Brace

For this study, the BRBs are designed based on the principles of equivalent linearization, a method introduced initially by Kasai et al. [7] and currently codified in the Japanese design specifications [5]. Figure 4 illustrates the concept of response reduction of BRBFs. Respective figures indicate the idealized displacement, velocity, and acceleration spectrum. The design primarily aims to reduce the displacement response. This stems from the enhanced stiffness using the added BRBs. Also, the enlargement of the damping factor reduces the response spectrum; thereby, the displacement response further decreases, as presented in Figure 4a.

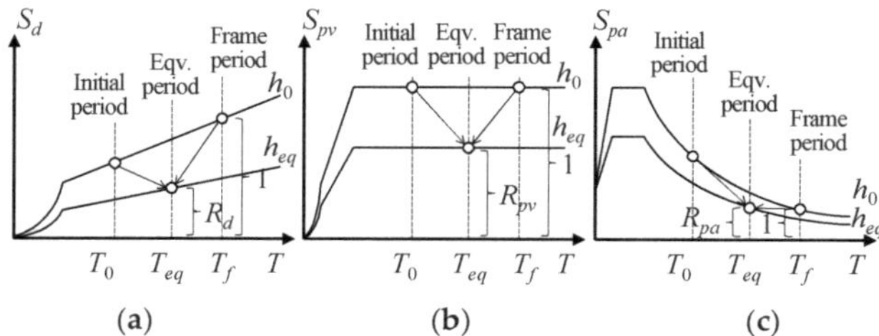

Figure 4. Idealized response spectra: (**a**) displacement; (**b**) velocity; (**c**) acceleration.

Using mathematical expressions for the effective vibration period and damping ratio of a building equipped with BRBs, along with idealized seismic response spectra, the peak seismic responses of the system and its local components can be expressed as continuous functions of structural and seismic parameters. These relationships are depicted in "performance curves".

Figure 5 illustrates the concept of equivalent linearization for a BRB integrated with an elastic frame, represented as an equivalent single-degree-of-freedom (SDOF) system. Figure 5a shows the displacement response spectrum used in this study, where the vertical axis represents the displacement response, denoted as S_e/ω^2, and the horizontal axis represents the natural period, T. The design spectrum used in this study is derived from Eurocode-8 [72], with the ground type assumed to be Type C and the reference ground acceleration, a_g, set at 3.0 m/s^2. Additionally, importance factors γ_I of 1.0, 1.2, and 1.4, corresponding to importance classes II, III, and IV, respectively, are applied in the BRB design.

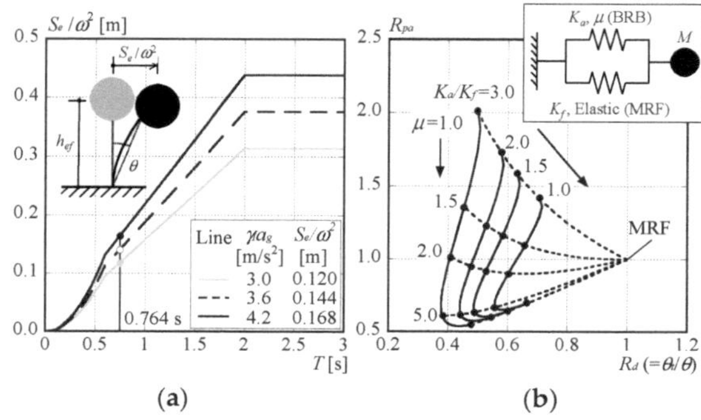

Figure 5. Calibration of BRB specification: (**a**) displacement response spectrum; (**b**) performance curve.

Five scenarios are considered for seismic upgrading:

1. $\theta_t = 1/120$ rad with $\gamma_1 a_g = 3.0$ m/s² peak ground acceleration (PGA),
2. $\theta_t = 1/150$ rad with $\gamma_1 a_g = 3.0$ m/s² PGA,
3. $\theta_t = 1/200$ rad with $\gamma_1 a_g = 3.0$ m/s² PGA,
4. $\theta_t = 1/200$ rad with $\gamma_1 a_g = 3.6$ m/s² PGA, and
5. $\theta_t = 1/200$ rad with $\gamma_1 a_g = 4.2$ m/s² PGA.

Here, θ_t represents the target story drift.

Figure 5b illustrates the performance curve, which consists of the displacement reduction ratio (R_d) and the force (or pseudo-acceleration) reduction ratio (R_{pa}), both expressed relative to the frame's responses without BRBs. In equivalent linearization, the target story drift θ_t is initially determined as a sufficiently small value to prevent structural damage. The displacement response can be computed using the equivalent SDOF system and response spectra. This response is converted into story drift θ by dividing by the effective height, h_{ef}, of the equivalent SDOF. The displacement reduction ratio, R_d, is calculated as $R_d = \theta_t / \theta$.

The curve depicted in Figure 5b theoretically captures the influence of the stiffness ratio K_a / K_f between the added component (a damper and brace combined in series) and the frame, as well as the ductility demand μ of the added component. These curves incorporate these critical factors and provide the necessary K_a / K_f and μ values to achieve the target drift or displacement reduction ratio R_d. The required yield force of the damper can also be derived from the R_d and force reduction ratio R_a values. The results of this study suggest that the design ductility of the BRB is 4.0, as the performance curve identifies this value as the optimal response reduction point, indicated by the convex portion in Figure 5b.

Once the necessary K_a / K_f value is determined for the SDOF system, it can be translated into the requirements for dampers in a multi-story system under the following assumptions:

1. The equivalent damping factor is consistent between the SDOF and multi-story systems.
2. The inter-story drift is uniform across all stories and equals the target story drift for the seismic force, calibrated according to the A_i distribution.
3. The ductility of the BRB is uniform across all stories.

Here, the A_i distribution represents the vertical seismic force distribution guided by the AIJ [73]. The necessary stiffness of individual stories K_{ai} is computed using the following equations to meet these criteria.

$$K_{ai} = \frac{Q_i}{h_i} \frac{\sum_{i=1}^{N}(K_{fi}h_i{}^2)}{\sum_{i=1}^{N}(Q_i h_i)} \left(\mu + \frac{K_a}{K_f} \right) - \mu \cdot K_{fi} \tag{1}$$

In this context, Q_i represents the story shear force, h_i denotes the height of the i-th floor, K_{fi} signifies the story stiffness of the i-th floor, and μ denotes the ductility of the BRBs at the target story drift. Additionally, the required yield axial displacement u_{ayi} is determined based on the design ductility μ, the story height h_i, and the target story drift θ_t.

$$u_{ayi} = \frac{\theta_t}{\mu} \cdot h_i \tag{2}$$

Ultimately, the yield axial forces of the respective BRBs, F_{ay}, can be computed based on the axial stiffness K_{ai} and the yield axial displacement u_{ay}. The formulation for F_{ay} is provided below.

$$F_{ayi} = K_{ai} \cdot u_{ayi} \tag{3}$$

The calibrated specifications for the BRBs are detailed in Table 1. These BRBs are positioned in bay Y2, which is situated at the center of the frame. The design-based shear ratios are 0.47, 0.57, and 0.66 for seismic accelerations of $\gamma_1 a_g = 3.0$ m/s², 3.6 m/s², and 4.2 m/s², respectively.

Table 1. Required capacities of BRBs (unit: kN/mm for K_{fi} and K_{ai}, kN for F_{ai}).

| Θ_t | | 1/120 rad | | 1/150 rad | | 1/200 rad | | 1/200 rad | | 1/200 rad | |
| $\gamma_I a_g$ | | 3.0 m/s^2 | | 3.0 m/^{2} | | 3.0 m/s^2 | | 3.6 m/s^2 | | 4.2 m/s^2 | |
Story	K_{fi}	K_{ai}	F_{ai}	K_{ai}	F_{ai}	K_{ai}	F_{ai}	K_{ai}	F_{ai}	K_{ai}	F_{ai}
R	194.5	—	—	—	—	—	—	90.00	450.2	249.2	1246
4	249.2	—	—	2.800	18.60	186.2	936.9	378.8	1906	631.1	3175
3	323.1	—	—	32.20	216.1	275.2	1385	530.4	2669	864.6	4350
2	211.8	198.5	2253	270.2	2556	475.2	3371	690.5	4898	972.4	6898

3. Outline of Nonlinear Response History Analyses and Fragility Assessment of Furniture Overturning

3.1. Outline of Nonlinear Response History Analyses

A three-dimensional model was constructed, as detailed in Figure 6. Nonlinear response history analyses (NRHA) were conducted using ABAQUS ver. 2021, a finite element analysis (FEA) software package. For specific settings and element definitions, readers are referred to the ABAQUS manual [74]. The ABAQUS/Standard module is employed for the analyses described. The model utilizes a three-dimensional configuration with encastre constraints applied at the column base. As this research focuses on behavior along the *x*-axis, out-of-plane (OOP) deformation is constrained by the boundary conditions.

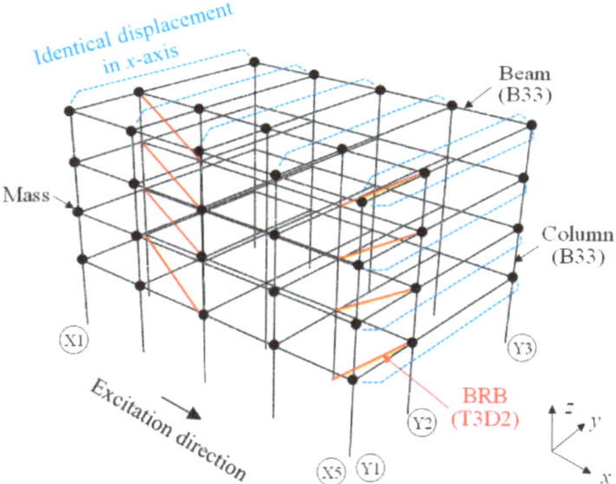

Figure 6. FEA model of model structure.

Beams and columns are represented using beam elements (B31), with distributed plasticity modeling adopted. Member yielding is identified based on von Mises stress, considering the sum of stress from both axial force and flexural moment for yield detection. The yield strength is determined to be 325 N/mm^2. Elastic-perfectly plastic (EPP) hysteresis is applied to each member, with column and beam intersections defined as rigid bodies.

The flexural stiffness of girders is doubled for beams with concrete slabs on both sides and increased by 1.5 times for beams with a half slab in order to model the influence of the concrete slab. The contributions of the concrete slabs to flexural stiffness are calculated according to Japanese design guidelines [75]. The contribution of the concrete slab is influenced by the shear connector performance [76–112]. Also, structural members can originate the buckling in a huge deformation, as summarized in the previous studies [113–139]. However, this research assumes idealized conditions to examine the risk of furniture overturning specifically. Additionally, Rayleigh damping is utilized for each structural member, with primary and secondary damping ratios set at 0.02.

The established model was verified by comparing the natural period and load-displacement relationship with those provided in the AIJ provision [5].

3.2. Fragility Assessment of Furniture Overturning

The fragility of furniture overturning is assessed statistically. Although various approaches exist for evaluating fragility [56–61], this research adopts the methodology proposed by Kaneko [54]. Kaneko et al. [54] integrated a theoretical framework with post-earthquake inspections of overturned bodies following several large seismic events. The resulting trend is represented by a continuous function, making it straightforward to apply when determining the threshold acceleration. The application of alternative evaluation methods will be explored in future research.

Based on a previous study [54], the probability of furniture overturning is determined using the following equations. The acceleration that results in a 50% rate of furniture overturning is calculated using Equation (6).

$$R\left(A_f\right) = \alpha \cdot \varphi \left(\frac{lnA_f - lnA_{R50}}{\zeta} \right) \tag{4}$$

$$A_{R50} = \begin{cases} \frac{B}{H} \cdot g \left(1 + \frac{B}{H}\right) \left(F_f \leq F_b\right) \\ 10\frac{B}{\sqrt{H}} \left(1 + \frac{B}{H}\right)^{2.5} \cdot 2\pi F_f \left(F_f > F_b\right) \end{cases} \tag{5}$$

$$F_f = \frac{V_f}{2\pi D_f} \tag{6}$$

where φ represents the log-normal distribution with a mean of lnA_{R50} and a standard deviation of ζ (where $\zeta = 0.50$). A_f denotes the peak floor acceleration (PFA) response, V_f the peak floor response velocity, D_f the peak floor response displacement, α is the coefficient representing the influence of slip, B is the depth of the furniture, H is the height of the furniture, and g is the gravitational acceleration. In this research, the constants in Equation (5) are determined as $\alpha = 0.8$ and $\zeta = 0.50$.

This study examines three types of furniture with dimensions specified in Table 2. The depth of furniture B remains constant while the height H is varied to modify the dimensions. The respective cases are categorized as "low", "middle", and "high". The furniture is assumed to be not anchored to the walls or floors. The frequency of furniture F_b varies depending on the height of the furniture.

$$F_b = \frac{15.6}{\sqrt{H}} \left(1 + \frac{B}{H}\right)^{-1.5} \tag{7}$$

Table 2. Furniture profiles.

Type	B [mm]	H [mm]	F_b [Hz]	
I (low)	45.0	120	0.883	
II (medium)	45.0	150	0.859	
III (tall)	45.0	200	0.814	

The fragility curves calculated using the above equations are presented in Figure 7. In Figure 7, the taller furniture is more likely to overturn at lower floor accelerations.

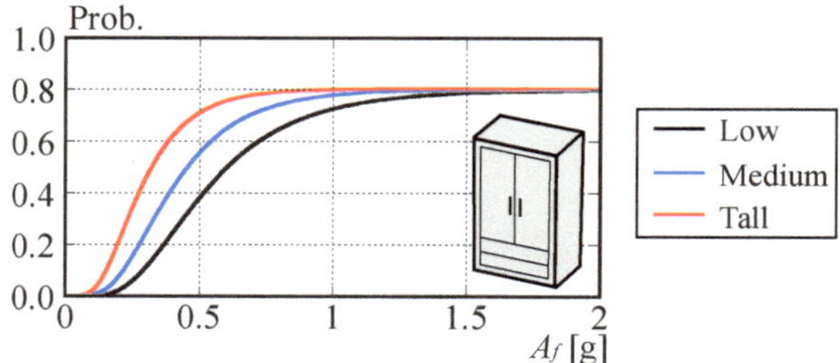

Figure 7. Fragility curve of furniture overturning.

4. Computation of Artificial Seismic Waves Reflecting Ground Characteristics

The seismic waves are calculated considering the rupture process of faults and the ground structure. Three locations in central Sendai were selected, focusing on the Nagamachi-Rifu Fault (M7.5) located directly beneath the city. Figure 8 illustrates the surface projection of the characterized source model from the National Seismic Hazard Map for Japan [140] and the locations of the two strong motion generation areas [141]. In this analysis, two cases with different rupture initiation points are evaluated. The respective locations are marked by ★.

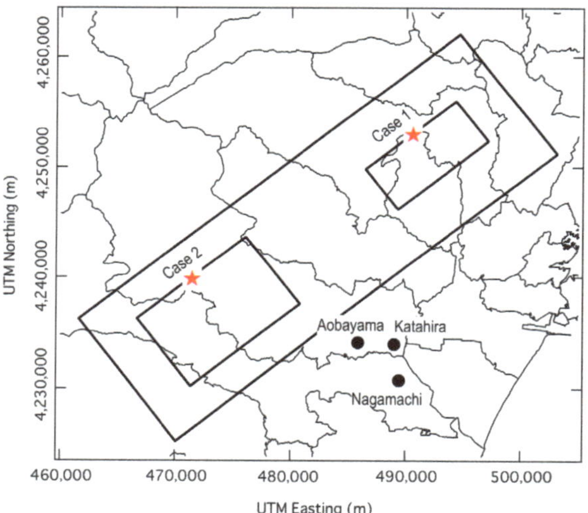

Figure 8. Earthquake scenarios concerned.

The evaluation of seismic waveforms follows the methodology outlined in the National Seismic Hazard Map for Japan [140]. First, the waveforms at the engineering bedrock level are predicted using the stochastic waveform synthesis method [142]. The component waveforms are artificially generated using the envelope shape from Boore [143] and random phase. The Q value of the propagation path is given by the following equation [144]:

$$Q = 110 f^{0.7} \tag{8}$$

where f represents the frequency. The amplification rate of the deep ground from the seismic bedrock to the engineering bedrock is derived from the 1D structure directly beneath each site, using the subsurface structure model applied in earthquake damage estimation for Sendai [145]. Ultimately, the amplification rate is calibrated under the condition of vertical incidence of S-waves. The response of the surface ground layer above the engineering bedrock is approximated using a modified R-O model [146], which accounts for strain-dependent rigidity and attenuation as used in the earthquake damage simulation for Sendai [145]. Surface acceleration waveforms are then obtained through a fully nonlinear time-domain analysis of total stress, using the seismic motion at the engineering bedrock as the input.

The coordinates of the three locations, along with the PGA and PGV of the calculated waveforms at the surface, are shown in Table 3. Figure 9 presents the acceleration response spectra with a damping factor of 2%. All evaluation points are located in central Sendai City, near the upper edge of the strong motion generation area on the southern side. The differences observed between the evaluation points are primarily due to variations in ground conditions.

Table 3. Profile of seismic waves (unit: m/s^2 for PGA and m/s for PGV).

Site	Longitude	Latitude	Case	Direction	PGA	PGV
			Case 1	NS	7.77	0.64
				EW	6.55	0.58
Aobayama	38°15′18″ N	140°50′18″ E	Case2	NS	10.1	0.65
				EW	11.3	1.64
			Case 1	NS	4.27	0.51
				EW	4.48	0.43
Katahira	38°15′14″ N	140°52′26″ E	Case2	NS	6.23	0.48
				EW	6.42	1.21
			Case 1	NS	5.63	0.54
				EW	5.25	0.52
Nagamachi	38°13′26″ N	140°52′49″ E	Case2	NS	6.71	0.73
				EW	5.89	1.32

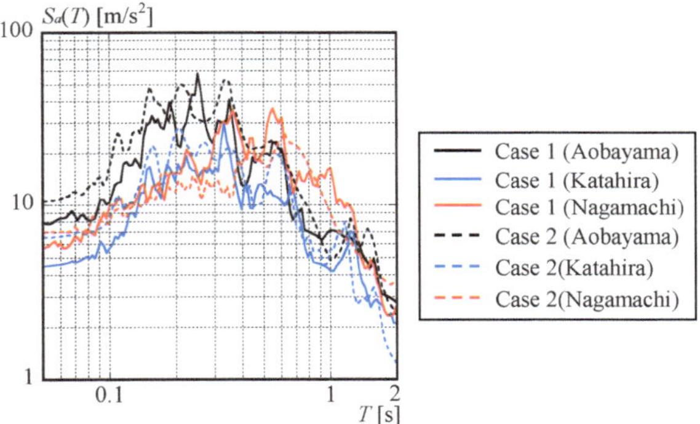

Figure 9. Acceleration response spectra ($h = 0.02$).

5. Contribution of Seismic Retrofitting to Structural Response Reduction Based on Incremental Dynamic Analysis

The input is standardized based on the acceleration response of the Single-Degree-of-Freedom (SDOF) system with a damping factor of 0.02. The natural periods of the structure for the MRF and BRBFs are derived from eigenvalue analysis. The intervals range from 0.1 G to 3.0 G with a step size of 0.1 G, resulting in a total of 2160 cases. The seismic waves utilized in the IDA are the 12 waves generated in the preceding chapter.

Figure 10 presents the peak story drift for the MRF and BRBFs. The horizontal axis represents $S_a(T)$, and the vertical axis shows the peak story drift. Since the story drift becomes the most significant in the first story, the floor displacement of the second story δ_2 is extracted in Figure 10. The median and $16^{th}/84^{th}$ percentile lines are depicted by bold black lines. Overall, the story drift decreases with more stringent design criteria, and building collapse can be prevented by retrofitting with BRBFs.

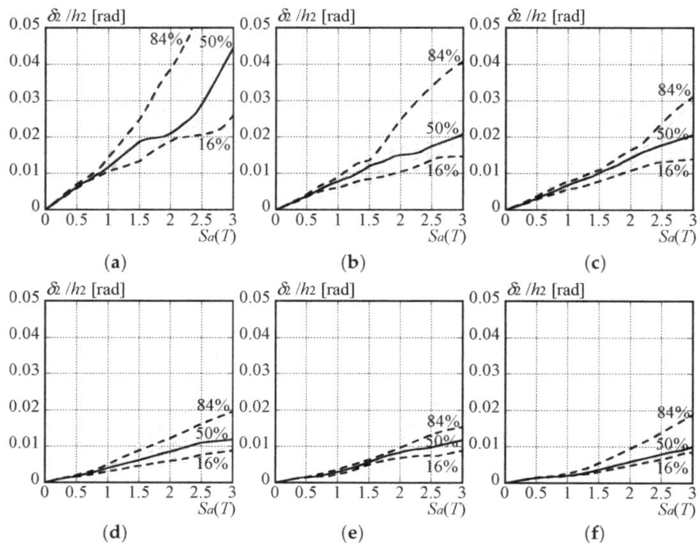

Figure 10. Maximum inter-story drift obtained from IDA: (**a**) MRF; (**b**) $\theta_t = 1/120$ rad; (**c**) $\theta_t = 1/150$ rad; (**d**) $\theta_t = 1/200$ rad ($\gamma_I a_g = 3.0$ m/s^2); (**e**) $\theta_t = 1/200$ rad ($\gamma_I a_g = 3.6$ m/s^2); (**f**) $\theta_t = 1/200$ rad ($\gamma_I a_g = 4.2$ m/s^2).

Figure 11 summarizes the PFA. According to previous studies [54], furniture overturning is associated with PFA. Unlike story drift, PFA cannot be mitigated by more stringent BRB design criteria (see Figure 4c). Increased BRB stiffness, intended to reduce displacement, may result in higher acceleration responses.

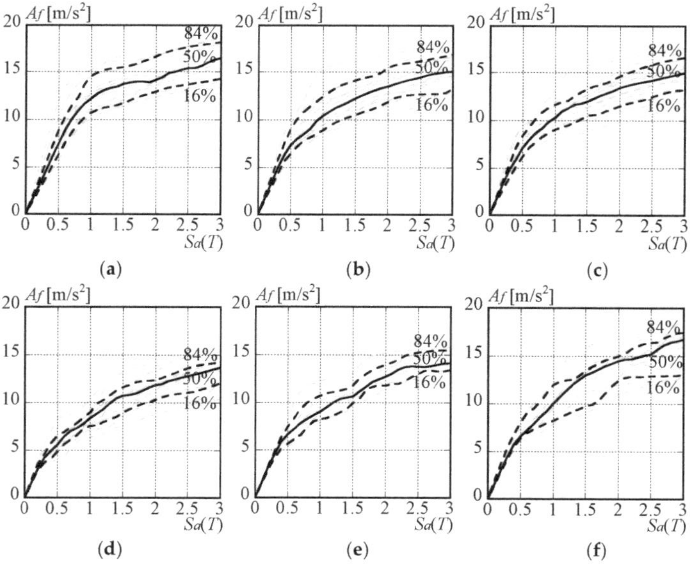

Figure 11. Peak floor acceleration obtained from IDA: (**a**) MRF; (**b**) $\theta_t = 1/120$ rad; (**c**) $\theta_t = 1/150$ rad; (**d**) $\theta_t = 1/200$ rad ($\gamma_I a_g = 3.0$ m/s^2); (**e**) $\theta_t = 1/200$ rad ($\gamma_I a_g = 3.6$ m/s^2); (**f**) $\theta_t = 1/200$ rad ($\gamma_I a_g = 4.2$ m/s^2).

Additionally, the occurrence of furniture overturning is related to the building's equivalent frequency, as considered by Equation (7). The A_{R50} differs depending on the balance of the unique frequency of furniture F_b and the characteristic frequency of the building, thereby altering the fragility curve. The threshold acceleration for 50% of furniture overturning is shown in Figure 12. The furniture dimension is low. The median and 16th/84th percentile lines are also depicted in the figure. The equivalent frequency is calculated based on the building's response, precisely the peak floor velocity V_f and peak floor displacement D_f, with the equation provided in Equation (7). Figure 12 demonstrates that the threshold acceleration increases with more stringent BRB design criteria primarily due to smaller peak floor displacements.

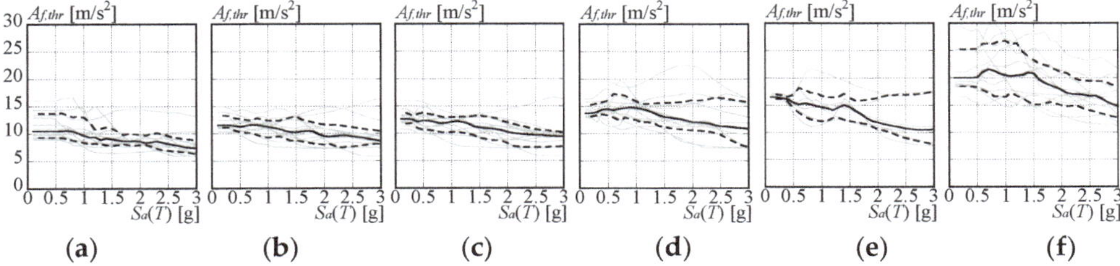

Figure 12. Threshold acceleration causing 50% of furniture overturning (low): (**a**) MRF; (**b**) $\theta_t = 1/120$ rad; (**c**) $\theta_t = 1/150$ rad; (**d**) $\theta_t = 1/200$ rad ($\gamma_I a_g = 3.0$ m/s^2); (**e**) $\theta_t = 1/200$ rad ($\gamma_I a_g = 3.6$ m/s^2); (**f**) $\theta_t = 1/200$ rad ($\gamma_I a_g = 4.2$ m/s^2).

Figure 13 illustrates the calculation flow for deriving the fragility curve of furniture overturning. As summarized in Figure 11, the PFA is determined for each story. Simultaneously, the threshold acceleration for a 50% probability of furniture overturning is calculated using the equation provided by Kaneko [54]. The exceedance of this threshold is then evaluated for each ground motion. Finally, the exceedance ratio is computed for the respective spectral accelerations. The fragility curve is obtained by fitting a cumulative distribution function to a log-normal distribution.

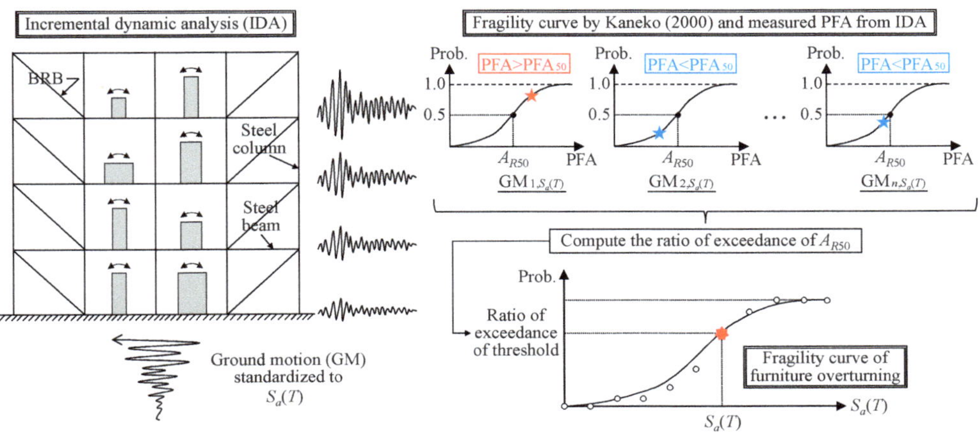

Figure 13. Calculation flow of fragility curve of furniture overturning.

Figure 14 illustrates the probability of furniture overturning. The horizontal axis represents the intensity of the input wave, and the vertical axis denotes the probability of exceeding 50% furniture overturning. The plots suggest that the likelihood of overturning decreases with more stringent design criteria. Although PFA can increase with greater BRB stiffness and capacity, the threshold acceleration also rises with more stringent BRB design criteria. Consequently, the fragility curve for $\theta_t = 1/200$ rad ($\gamma_I a_g = 4.2$ m/s^2) becomes the smallest in Figure 14.

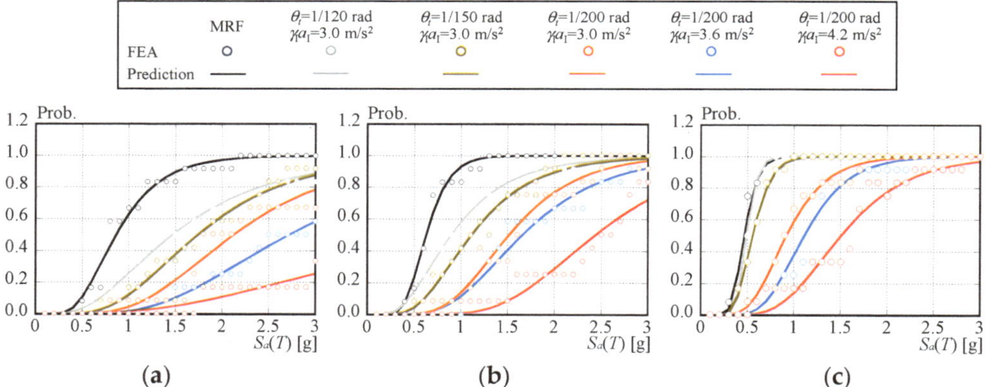

Figure 14. Fragility curve of furniture overturning: (**a**) low; (**b**) medium; (**c**) tall.

6. Evaluation of Mean Annual Frequency of Furniture Overturning

To assess the effectiveness of seismic retrofitting in business continuity planning, the mean annual frequency of furniture overturning is determined based on the framework depicted in Figure 15.

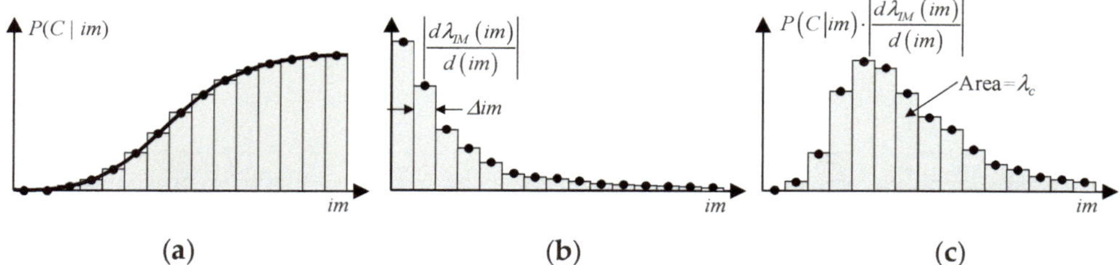

Figure 15. Computation concept of mean annual frequency: (**a**) fragility curve; (**b**) derivative of hazard curve; (**c**) mean annual frequency.

This study utilizes the mean annual frequency λ_c for risk evaluation. Calculating λ_c requires two key elements: the seismic hazard curve, which represents the mean annual frequency of exceeding ground motion intensities at a specific site, and the fragility curve for furniture overturning. Ground motion intensity is quantified using an intensity measure (*IM*), such as $S_a(T)$, the spectral acceleration at the structure's fundamental period.

The value of λ_c is obtained by integrating the furniture fragility curve with the site-specific seismic hazard curve using the following equation:

$$\lambda_c = \int_0^\infty P(C|im) \cdot |d\lambda_{IM}(im)| \tag{9}$$

where $P(C|IM)$ represents the probability of furniture overturning under an earthquake with a given ground motion intensity level *IM*, and λ_{IM} denotes the mean annual frequency of exceeding the ground motion intensity *IM*. By multiplying and dividing the right-hand side of Equation (10) by $d(IM)$, the expression for calculating λ_c can be reformulated as the following:

$$\lambda_c = \int_0^\infty P(C|im)\cdot\left|\frac{d\lambda_{IM}(im)}{d(im)}\right|d(im) \tag{10}$$

where $d\lambda_{IM}(IM)/d(IM)$ represents the slope of the seismic hazard curve at the site. A closed-form solution for the integral in Equation (11) is usually unavailable, so the integral is typically evaluated through numerical integration. This involves computing the product of the collapse probability conditioned on IM, the slope of the seismic hazard curve at specific IM levels, multiplying by the increment in $IM(\Delta IM)$, and summing the results across all IM levels. This procedure is expressed in Equation (12) and visually depicted in Figure 15.

$$\lambda_c = \sum_{i=1}^\infty P(C|im_i)\cdot\left|\frac{d\lambda_{IM}(im_i)}{d(im)}\right|\cdot\Delta im \tag{11}$$

6.1. Relationship between λ_c and Probability of Collapse

λ_c denotes the mean annual frequency of furniture overturning. Assuming that earthquake occurrences follow a Poisson process over time, the probability of at least one overturning event within a period of t years can be determined using the following equation:

$$P_c(in\ t\ years) = 1 - exp(-\lambda_c t) \tag{12}$$

Given that λ_c is typically a small value for most buildings, the annual probability of overturning is approximately equal to λ_c, expressed as follows:

$$P_c(in\ 1\ years) \cong \lambda_c \tag{13}$$

6.2. Deaggregation of λ_c

λ_c alone does not reveal which ground motion intensities contribute most to the collapse risk. To address this, a deaggregation of λ_c provides a valuable method for identifying the relative contributions of various ground motion intensities to the overall collapse risk. This process corresponds to the deaggregation by magnitude, distance, and epsilon (ε) commonly used in PSHA to determine the primary sources contributing to the hazard. The parameter ε quantifies the deviation, in terms of logarithmic standard deviations, between the observed ground spectral acceleration and the spectral acceleration predicted by an attenuation relationship at an arbitrary period.

A point on the λ_c deaggregation curve is retrieved as the product of the overturning probability at a ground motion intensity and the slope of the hazard curve as an intensity function. As indicated in Equation (12), the contribution of a specific (narrow) range of ground motion intensities to λ_c is calculated by multiplying the collapse probability at the midpoint intensity of the range by the slope of the hazard curve at the intensity and by the width of the intensity, ΔIM. As illustrated in Figure 15, the total area under the deaggregation curve equals λ_c. Ground motion intensities with higher values on the deaggregation curve represent more remarkable contributions to λ_c.

Figure 15 demonstrates that ground motion intensities associated with high overturning probabilities do not always contribute significantly to λ_c, as these intensities occur less frequently than those in the lower half of the overturning fragility curve. Consequently, the most significant contribution to λ_c generally comes from intensities within the lower half of the fragility curve, where the collapse probabilities may be smaller. Still, the seismic hazard curve's slope is typically steeper compared to that for higher intensities.

6.3. Hazard Curve Deaggregation of λ_c

As one of the most earthquake-prone regions on the planet, Japan is chosen as the case study. The PSHA requires a hazard curve in spectral form. The National Research Institute for Earth Science and Disaster Resilience (NIED) provides the Japan Seismic

Hazard Information (J-SHIS) system [147,148]. J-SHIS has developed a database of hazard curves based on the response spectrum for exceedance probabilities corresponding to a 50-year return period. These hazard curves are constructed by combining various scenarios involving subduction zones and near-fault earthquakes.

The damping factor provided by J-SHIS is set at 5%, which is generally applicable to reinforced concrete (RC) structures. However, the typical damping factor for steel structures is 2%, as used in the simulations conducted in this research. Therefore, the spectral response data from J-SHIS must be adjusted to reflect the corresponding damping factor. For this conversion, the following equation is provided by AIJ [5].

$$D_h = \sqrt{\frac{1 + 25h_0}{1 + 25h_{eq}}} \tag{14}$$

where h_0 is the initial damping factor, and h_{eq} is the equivalent damping factor. Additionally, the probability for a return period of T_c is converted to the mean annual frequency using the following equation.

$$\lambda_c = -\frac{ln(1 - P_c)}{T_c} \tag{15}$$

where P_c is the probability of exceedance and T_c designates the time interval where the probability P_c is calculated [149].

Additionally, the hazard curve is provided for specific natural periods: 0.1 s, 0.2 s, 0.3 s, 0.5 s, 1.0 s, 2.0 s, 3.0 s, and 5.0 s. To derive the hazard curve corresponding to the natural periods of the MRF and BRBF systems, the hazard curves are linearly interpolated in log-log space. Since the hazard curves in J-SHIS are derived from empirical equations using arbitrary constants, theoretical interpolation is not possible. Instead, previous studies have customarily interpolated the hazard curve in log-log space [69,150]. Therefore, the interpolation method used in this study follows earlier investigations [69,150].

The target locations are listed in Table 4, which includes sites selected from various regions across Japan. The hazard curves retrieved from J-SHIS are summarized in Figure 16. The frequency generally becomes greater in a shorter period. The discrepancy varies depending on the region, and Aichi exhibits a relatively huge gap among the natural period.

Table 4. Target locations for risk assessment.

Location	J-SHIS Mesh Code	Longitude	Latitude	Elevation
Hokkaido	6441427814	43.0615 N	141.3547 E	21 m
Miyagi	5740362921	38.2677 N	140.8703 E	46 m
Tokyo	5339452532	35.6885 N	139.6922 E	38 m
Ishikawa	5436657223	36.5615 N	136.6578 E	27 m
Aichi	5236671243	35.1823 N	136.9078 E	10 m
Hyogo	5235012543	34.6906 N	135.1953 E	6 m
Hiroshima	5132436612	34.3844 N	132.4547 E	2 m
Kumamoto	4930156623	32.8031 N	130.7078 E	13 m

Figure 17a illustrates the mean annual frequency for low-height furniture. Overall, the MRF demonstrates the highest mean annual frequency except in Aichi. The greatest frequency is observed in the case of $\theta_t = 1/200$ rad ($\gamma_1 a_g = 3.6$ m/s^2). The hazard curve in Figure 16 shows that Aichi has a higher probability of occurrence in a shorter natural period. Due to the seismic retrofitting principle using BRB, the natural period shortens. As Aichi's seismic motion characteristics cause more significant acceleration responses in buildings with shorter natural periods, the computed mean annual frequency can exceed that of buildings without seismic retrofitting.

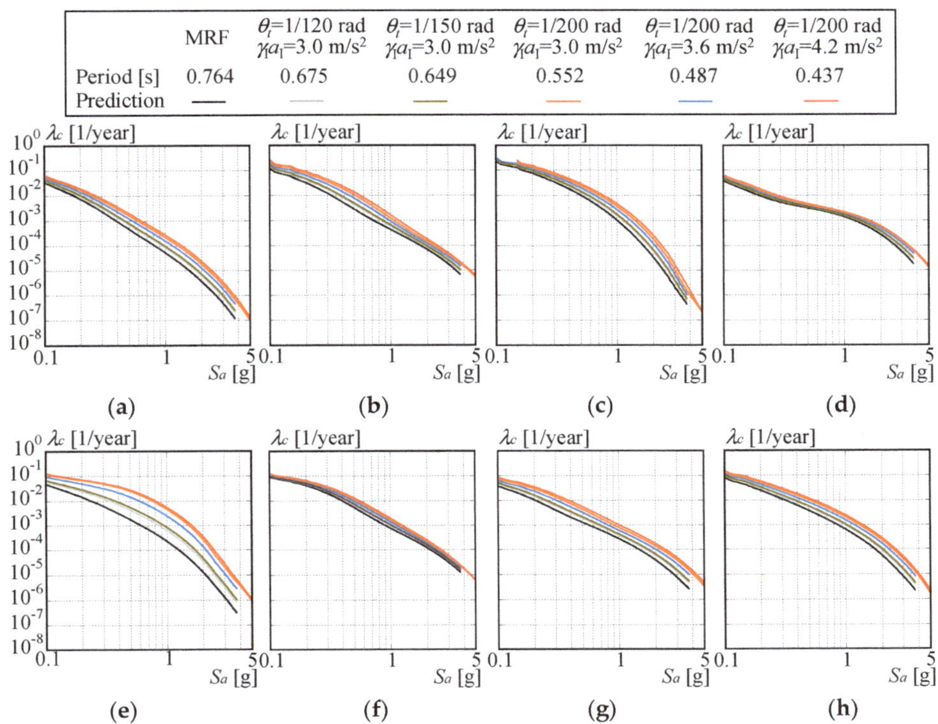

Figure 16. Hazard curves for locations across Japan: (**a**) Hokkaido; (**b**) Miyagi; (**c**) Tokyo; (**d**) Ishikawa;
(**e**) Aichi; (**f**) Hyogo; (**g**) Hiroshima; (**h**) Kumamoto.

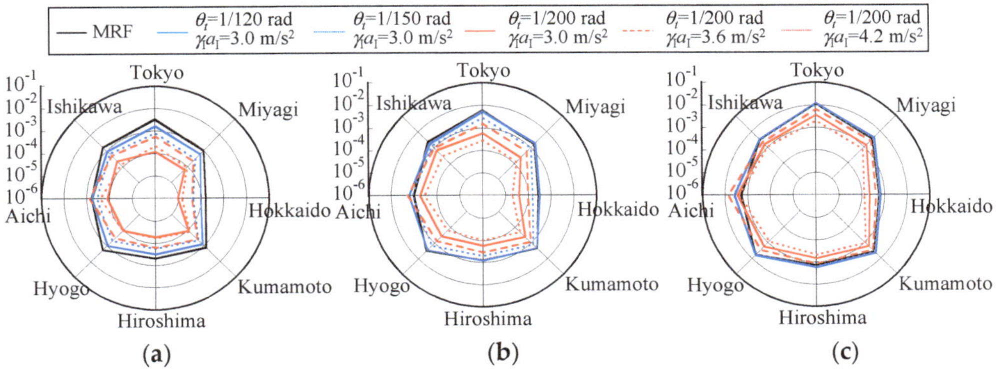

Figure 17. Mean annual frequency of furniture overturning: (**a**) low; (**b**) medium; (**c**) tall.

Additionally, the risk of furniture overturning at $\theta_t = 1/200$ rad ($\gamma_I a_g = 3.0$ m/s^2) and
$\theta_t = 1/200$ rad ($\gamma_I a_g = 4.2$ m/s^2) is nearly identical in most cases. As shown in Figure 11, the
acceleration response increases in $\theta_t = 1/200$ rad ($\gamma_I a_g = 4.2$ m/s^2) due to greater stiffness,
while the threshold acceleration also increases. Consequently, the probability of furniture
overturning is lowest in $\theta_t = 1/200$ rad ($\gamma_I a_g = 4.2$ m/s^2) according to the fragility curve in
Figure 14. Meanwhile, the probability of exceedance in acceleration response tends to be
more pronounced at shorter natural periods. As previously mentioned, the mean annual
frequency is calculated as the product of the fragility curve and the derivative of the hazard

curve. Thus, the effectiveness of seismic retrofitting with BRB depends on the balance between the reduction in fragility and the resonance with the seismic wave.

Figure 17b presents the case for medium-height furniture, where the risk of overturning is nearly the same for MRF and $\theta_t = 1/120$ rad. According to the fragility curve in Figure 17b, fragility is reduced through seismic retrofitting. However, the spectral response increases due to the shorter natural period, resulting in nearly identical computed mean annual frequencies. The lowest frequency is observed in $\theta_t = 1/200$ rad ($\gamma_I a_g = 4.2$ m/s^2) across all locations, with the second lowest in $\theta_t = 1/200$ rad ($\gamma_I a_g = 3.0$ m/s^2). The second-most stringent design criterion of $\theta_t = 1/200$ rad ($\gamma_I a_g = 3.6$ m/s^2) exceed the aforementioned two cases and is highest in Aichi.

Figure 17c displays the mean annual frequency for tall furniture. The difference between retrofitting cases is less pronounced compared to low and medium-height furniture. The magnitude relation remains consistent with Figure 17b. However, the lowest mean annual frequency is achieved by MRF, owing to a more significant hazard curve at shorter natural periods. The mean annual frequency reaches 3.8 times greater in $\theta_t = 1/200$ rad ($\gamma_I a_g = 3.6$ m/s^2) compared to the MRF.

6.4. Discussion

Seismic retrofitting with BRBs effectively reduces displacement response and prevents building collapse. The design procedure has become straightforward due to the use of equivalent linearization. This method begins with determining the target story drift, making displacement response the primary concern. However, the risk of furniture overturning can increase as the building's natural period shortens due to the added stiffness from the BRBs. In particular, the mean annual frequency increases when the spectral response at shorter periods becomes more significant. The proposed method quantifies this risk, enabling engineers to consider BCP when designing BRBFs for seismic retrofitting.

This study assumes that the furniture is not anchored. The fragility of furniture overturning can be mitigated through proper anchorage. Even with the installation of seismic damping devices, careful attention is required to maintain business continuity.

7. Conclusions

This study applies probabilistic risk assessment (PRA) to the issue of furniture overturning. Buckling-restrained braced frames (BRBFs) are designed using various design criteria based on the moment-resisting frame (MRF) provisions. Incremental dynamic analysis (IDA) is performed using artificial seismic waves that reflect realistic ground conditions. The resulting fragility curve for furniture overturning is then compared with hazard curves across Japan. The findings are summarized as follows:

(1) Peak story drift is effectively mitigated with more stringent BRB design criteria. However, when the target story drift is reduced, peak floor acceleration can increase in BRBFs compared to MRFs.

(2) The fragility of furniture overturning decreases with more stringent design criteria due to an increased critical acceleration threshold.

(3) The mean annual frequency of acceleration response spectra generally increases with shorter natural periods. Consequently, the mean annual frequency of furniture overturning can be higher in BRBFs compared to MRFs.

(4) The risk of furniture overturning is influenced by the balance between the building's performance and the specific hazard characteristics at the location. Particular attention should be given to the shape of the hazard curve, especially in cases where short-period vibrations are prominent.

(5) The calculation flow outlined in this research provides engineers and researchers with a clear methodology to quantify the risk of furniture overturning in a precise and explicit manner.

The model structure used in this research is a four-story building. In future research, it will be essential to assess the potential risk of furniture overturning in high-rise structures to extend safety warnings to the broader community. Future studies will explore the application of this methodology to buildings with varying aspect ratios and address additional business continuity planning (BCP) concerns, such as roof collapses. A more comprehensive investigation will be presented in subsequent papers.

Author Contributions: Conceptualization, A.S.; methodology, S.O.; software, A.S.; validation, S.O. and Y.K.; formal analysis, A.S.; investigation, A.S. and S.O.; resources, S.O., and Y.K.; data curation, A.S.; writing—original draft preparation, A.S.; writing—review and editing, S.O. and Y.K.; visualization, A.S.; supervision, Y.K.; project administration, A.S.; funding acquisition, A.S. All authors have read and agreed to the published version of the manuscript.

Funding: This research was funded by JSPS KAKENHI (Grant Number 24H00342) (Principal Investigator: Dr. Yoshihiro Kimura), JSPS KAKENHI (Grant Number 23K13392) (Principal Investigator: Dr. Atsushi Suzuki), and The Japan Iron and Steel Federation (Principal Investigator: Dr. Atsushi Suzuki). We express our deepest gratitude for their sincere support.

Data Availability Statement: The raw/processed data necessary to reproduce these findings cannot be shared at this time because the data also form part of an ongoing study.

Conflicts of Interest: The authors declare that they have no known competing financial interests or personal relationships that could have appeared to influence the work reported in this paper.

References

1. Kimura, K.; Yoshioka, K.; Takeda, T.; Furuya, N.; Takemoto, Y. An experimental study of braces encased in steel tube and mortar. In Proceedings of the Architectural Institute of Japan Annual Meeting, Hokkaido, Japan, 15–17 September 1976; pp. 1041–1042.
2. Takeuchi, T.; Hajjar, J.F.; Matsui, R.; Nishimoto, K.; Aiken, I.D. Local buckling restraint condition for core plates in buckling restrained braces. *J. Constr. Steel Res.* **2010**, *66*, 139–149. [CrossRef]
3. Takeuchi, T.; Ida, M.; Yamada, S.; Suzuki, K. Estimation of cumulative deformation capacity of buckling restrained braces. *J. Struct. Eng.* **2008**, *134*, 822–831. [CrossRef]
4. Takeuchi, T.; Ozaki, H.; Matsui, R.; Sutoh, F. Out-of-plane stability of buckling-restrained braces including moment transfer capacity. *Earthq. Eng. Struct. Dyn.* **2014**, *43*, 851–869. [CrossRef]
5. Architectural Instutite of Japan (AIJ). *Recommended Provisions for Seismic Damping Systems Applied to Steel Structures*; Maruzen Publishing Co., Ltd.: Tokyo, Japan, 2014. (In Japanese)
6. American Instutite of Steel Construction (AISC). *Seismic Provisions for Structural Steel Buildings*; American Instutite of Steel Construction (AISC): Chicago, IL, USA, 2016.
7. Kasai, K.; Fu, Y.; Watanabe, A. Passive control systems for seismic damage mitigation. *J. Struct. Eng.* **1998**, *124*, 501–512. [CrossRef]
8. Kimura, Y.; Suzuki, A.; Kasai, K. Estimation of plastic deformation capacity for H-shaped beams on local buckling under compressive and tensile forces. *J. Struct. Constr. Eng. (Trans. AIJ)* **2016**, *81*, 2133–2142. (In Japanese) [CrossRef]
9. Suzuki, A.; Kimura, Y.; Kasai, K. Plastic deformation capacity of h-shaped beams collapse with combined buckling under reversed axial forces. *J. Struct. Constr. Eng. (Trans. AIJ)* **2018**, *83*, 297–307. (In Japanese) [CrossRef]
10. Suzuki, A.; Kimura, Y.; Kasai, K. Rotation capacity of I-shaped beams under alternating axial forces based on buckling-mode transitions. *J. Struct. Eng.* **2020**, *146*, 04020089. [CrossRef]
11. Kimura, Y.; Fujak, M.S.; Suzuki, A. Elastic local buckling strength of I-beam cantilevers subjected to bending moment and shear force based on flange–web interaction. *Thin-Walled Struct.* **2021**, *162*, 107633. [CrossRef]
12. Kimura, Y.; Sato, Y.; Suzuki, A. Effect of fork restraint of column on lateral buckling behavior for H-shaped steel beams with continuous braces under flexural moment gradient. *J. Struct. Constr. Eng. (Trans. AIJ)* **2022**, *87*, 316–327. (In Japanese) [CrossRef]
13. Fujak, M.S.; Kimura, Y.; Suzuki, A. Estimation of elastoplastic local buckling capacities and novel classification of I-beams based on beam's elastic local buckling strength. *Structures* **2022**, *39*, 765–781. [CrossRef]
14. Suzuki, A.; Kimura, Y. Rotation capacity of I-shaped beam failed by local buckling in buckling-restrained braced frames with rigid beam-column connections. *J. Struct. Eng.* **2023**, *149*, 04022243. [CrossRef]
15. Fujak, M.S.; Suzuki, A.; Kimura, Y. Estimation of ultimate capacities of moment-resisting frame's subassemblies with mid-story pin connection based on elastoplastic local buckling. *Structures* **2023**, *48*, 410–426. [CrossRef]
16. Suzuki, A.; Kimura, Y.; Matsuda, Y.; Kasai, K. Rotation capacity of I-shaped beams with concrete slab in buckling-restrained braced frames. *J. Struct. Eng.* **2024**, *150*, 04023204. [CrossRef]
17. Suzuki, A.; Kimura, Y. Rotation capacity of I-beams under cyclic loading with different kinematic/isotropic hardening characteristics. *J. Constr. Steel Res.* **2024**, *223*, 109007. [CrossRef]
18. Aiken, I.D.; Mahin, S.A.; Uriz, P. Large-scale testing of buckling restrained braced frames. In *Proceedings of the Japan Passive Control Symposium*; Tokyo Institute of Technology: Tokyo, Japan, 2002.
19. Berman, J.W.; Bruneau, M. Cyclic testing of a buckling restrained braced frame with unconstrained gusset connections. *J. Struct. Eng.* **2009**, *135*, 1499–1510. [CrossRef]

20. Fahnestock, L.A.; Sause, R.; Ricles, J.M. Seismic response anad performance of buckling-restrained braced frames. *J. Struct. Eng.* **2007**, *133*, 1195–1204. [CrossRef]
21. Hsiao, P.C.; Lehman, D.E.; Roeder, C.W. Improved analytical model for special concentrically braced frames. *J. Constr. Steel Res.* **2012**, *73*, 80–94. [CrossRef]
22. Jia, M.; Lu, D.; Guo, L.; Sub, L. Experimental research and cyclic behavior of buckling-restrained braced composite frame. *J. Constr. Steel Res.* **2014**, *95*, 90–105. [CrossRef]
23. Lin, P.C.; Tsai, K.C.; Wang, K.J.; Yu, Y.J.; Wei, C.Y.; Wu, A.C.; Tsai, C.Y.; Lin, C.H.; Chen, J.C.; Schellenberg, A.H.; et al. Seismic design and hybrid tests of a full-scale three-story buckling-restrained braced frame using welded end connections and thin profile. *Earthq. Eng. Struct. Dyn.* **2012**, *41*, 1001–1020. [CrossRef]
24. Mcmanus, P.S.; Maxmahon, A.; Puckett, J. Buckling restrained braced frame with all-bolted gusset connections. *Eng. J.* **2013**, *50*, 89–116. [CrossRef]
25. Palmer, K.D.; Christopulos, A.S.; Lehman, D.E.; Roeder, C.W. Experimental evaluation of cyclically loaded, large-scale, planar and 3-D buckling restrained braced frames. *J. Constr. Steel Res.* **2014**, *101*, 415–425. [CrossRef]
26. Qu, Z.; Xie, J.; Cao, Y.; Li, W.; Wang, T. Effect of strain rate on the hysteretic behavior of buckling-restrained braces. *J. Struct. Eng.* **2020**, *146*, 06019003. [CrossRef]
27. Tsai, K.; Hsiao, K.P.; Wang, K.; Weng, Y.; Lin, M.; Lin, K.; Chen, C.; Lai, J.; Lin, S. Pseudo-dynamic tests of a full-scale CFT/BRB frame—Part I: Specimen design, experiment and analysis. *Earthq. Eng. Struct. Dyn.* **2008**, *37*, 1081–1098. [CrossRef]
28. Tsai, K.; Hsiao, P. Pseudo-dynamic test of a full scale CFT/BRB frame–Part II: Seismic performance of buckling-restrained braces and connections. *Earthq. Eng. Struct. Dyn.* **2008**, *37*, 1099–1115. [CrossRef]
29. Ozkilic, Y.O.; Bozkurt, M.B.; Topkaya, C. Evaluation of seismic response factors for BRBFs using FEMA P695 methodology. *J. Constr. Steel Res.* **2018**, *151*, 41–57. [CrossRef]
30. Ozkilic, Y.O.; Bozkurt, M.B.; Topkaya, C. Mid-spliced end-plated replaceable links for eccentrically braced frames. *Eng. Struct.* **2021**, *237*, 112225. [CrossRef]
31. Ozkilic, Y.O.; Bozkurt, M.B. Numerical validation on novel replaceable reduced beam section connections for moment-resisting frames. *Structures* **2023**, *50*, 63–79. [CrossRef]
32. Lignos, D.G.; Krawinkler, H. Development and utilization of structural component databases for performance-based earthquake engineering. *J. Struct. Eng.* **2013**, *139*, 1382–1394. [CrossRef]
33. Chenouda, M.; Ayoub, A. Probabilistic collapse analysis of degrading Multi Degree of Freedom structures under earthquake excitation. *Eng. Struct.* **2009**, *31*, 2909–2921. [CrossRef]
34. Scozzese, F.; Tubaldi, E.; Dall'Asta, A. Assessment of the effectiveness of multiple-stripe analysis by using a stochastic earthquake input model. *Bull. Earthq. Eng.* **2020**, *18*, 3167–3203. [CrossRef]
35. Bradley, C.R.; Fahnestock, L.A.; Hine, E.M.; Sizemore, J.G. Full-scale cyclilc testing of low-ductility concentrically braced frames. *J. Struct. Eng.* **2017**, *143*, 04017029. [CrossRef]
36. Chida, H.; Takahashi, N. Study on image diagnosis of timber houses damaged by earthquake using deep learning. *Jpn. Archit. Rev.* **2021**, *4*, 420–430. [CrossRef]
37. Chang, P.C.; Flatau, A.; Liu, S.C. Review paper: Health monitoring of civil infrastructure. *Struct. Health Monit.* **2003**, *2*, 257–267. [CrossRef]
38. Carden, E.P.; Fanning, P. Vibration based condition monitoring: A review. *Struct. Health Monit.* **2004**, *3*, 355–377. [CrossRef]
39. Ji, X.; Fenves, G.L.; Kajiwara, K.; Nakashima, M. Seismic damage detection of a full-scale shaking table test structure. *J. Struct. Eng.* **2011**, *137*, 14–21. [CrossRef]
40. Okada, K.; Shiroishi, R.; Morii, Y.; Sagawa, R. Method of structural health monitoring after earthquake using limited accelerometer—Case study of large-scale shaking table test on six story RC building. *Concr. J.* **2017**, *55*, 138–145. (In Japanese) [CrossRef]
41. Gislason, G.P.; Mei, Q.; Gul, M. Rapid and automated damage detection in buildings through ARMAX analysis of wind induced vibrations. *Front. Built Environ.* **2019**, *5*, 16. [CrossRef]
42. Alampalli, S.; Fu, G.; Dillon, E.W. Signal versus noise in damage detection by experimental modal analysis. *J. Struct. Eng.* **1997**, *123*, 237–245. [CrossRef]
43. Ju, M.; Dou, Z.; Li, J.; Qiu, X.; Shen, B.; Zhang, S.; Yao, F.; Gong, W.; Wang, K. Piezoelectric materials and sensors for structural health monitoring: Fundamental aspects, current status, future perspectives. *Sensors* **2023**, *23*, 543. [CrossRef]
44. Kishiki, S.; Iwasaki, Y. Evaluation of residual strength based on local buckling deformation of steel column—Quick inspection method for steel structures based on the visible damage Part 3. *J. Struct. Constr. Eng. (Trans. AIJ)* **2017**, *82*, 735–743. (In Japanese) [CrossRef]
45. Gong, F.; Maekawa, K. Multi-scale simulation of freeze-thaw damage to RC column and its restoring force characteristics. *Eng. Struct.* **2018**, *156*, 522–536. [CrossRef]
46. Gong, F.; Wang, Z.; Xia, J.; Maekawa, K. Coupled thermos-hydro-mechanical analysis of reinforced concrete beams under the effect of frost damage and sustained load. *Struct. Concr.* **2021**, *22*, 3430–3445. [CrossRef]
47. Shimoi, N.; Nishida, T.; Obata, A.; Nakasho, K.; Cuadra, C. Comparison of displacement measurements in an exposed type column base using piezoelectric vibration sensors and piezoelectric limit sensors. *Akita Prefect. Univ. Web J. B* **2019**, *6*, 27–36.

48. Aabid, A.; Parveez, B.; Raheman, M.A.; Ibrahim, Y.E.; Anjum, A.; Hrairi, M.; Parveen, N.; Zayan, J.M. A review of piezoelectric material-based structural control and health monitoring techniques for engineering structures: Challenges and opportunities. *Actuators* **2021**, *10*, 101. [CrossRef]
49. Harada, T.; Yokoyama, K.; Tanabe, K. Study of crack detection of civil infrastructure using PVDF film sensor. *J. Struct. Eng.* **2013**, *59A*, 47–55.
50. Shimoi, N.; Cuadra, C.; Madokoro, H.; Nakasho, K. Comparison of displacement measurements and simulation on fillet weld of steel column base. *Int. J. Mech. Eng. Appl.* **2020**, *8*, 111–117. [CrossRef]
51. Suzuki, A.; Liao, W.; Shibata, D.; Yoshino, Y.; Kimura, Y.; Shimoi, N. Structural damage detection technique of secondary building components using piezoelectric sensors. *Buildings* **2023**, *13*, 2368. [CrossRef]
52. Matsumoto, Y.; Yamada, S.; Iyama, J.; Koyama, T.; Kishiki, S.; Shimada, Y.; Asada, H.; Ikenaga, M. Damage to steel educational facilities in the 2011 East Japan Earthquake: Part 1 Outline of the reconnaissance and damage to major structural components. In Proceedings of the 15th World Conference on Earthquake Engineering, Lisbon, Portugal, 24–28 September 2012.
53. Suzuki, A.; Fujita, T.; Kimura, Y. Identifying damage mechanisms of gymnasium structure damaged by the 2011 Tohoku earthquake based on biaxial excitation. *Structures* **2022**, *35*, 1321–1338. [CrossRef]
54. Kaneko, M.; Hayashi, Y. Proposal of a curve to describe overturning ratios of rigid bodies. *J. Struct. Constr. Eng. (Trans. AIJ)* **2000**, *536*, 55–62. (In Japanese) [CrossRef]
55. Kaneko, M. Method to estimate overturning ratios of furniture during earthquakes. *J. Struct. Constr. Eng. (Trans. AIJ)* **2002**, *551*, 61–68. (In Japanese) [CrossRef]
56. Zhang, J.; Makris, N. Rocking response of free-standing blocks under cycloidal pulses. *J. Eng. Mech.* **2001**, *127*, 473–483. [CrossRef]
57. Kounadis, A.N. Parametric study in rocking instability of a rigid block under harmonic ground pulse: A unified approach. *Soil Dyn. Earthq. Eng.* **2013**, *45*, 125–143. [CrossRef]
58. Vassiliou, M.F.; Mackie, K.R.; Stojadinovic, B. Dynamic response analysis of solitary flexible rocking bodies: Modeling and behavior under pulse-like ground excitation. *Earthq. Eng. Struct. Dyn.* **2014**, *43*, 1463–1481. [CrossRef]
59. Fragiadakis, M.; Diamantopoulos, S. Fragility and risk assessment of freestanding building contents. *Earthq. Eng. Struct. Dyn.* **2020**, *49*, 1028–1048. [CrossRef]
60. Avgenakis, E.; Psycharis, I. An integrated macroelement formulation for the dynamic response of inelastic deformable rocking bodies. *Earthq. Eng. Struct. Dyn.* **2020**, *49*, 1072–1094. [CrossRef]
61. Egidio, A.D.; Leo, A.M. Improvement of the dynamic and seismic response of non-structural rocking bodies through the ability to change their geometrical configuration. *Eng. Struct.* **2023**, *275*, 115231. [CrossRef]
62. Saito, K.; Higuchi, T.; Nakazawa, S. Study of seismic risk evaluation consideration of overturning furniture for middle and low-rise buildings such as office buildings. *Steel Constr. Eng.* **2016**, *23*, 77–88. (In Japanese)
63. Ishikawa, Y.; Takeda, M.; Okumura, T.; Hayashi, Y.; Kanegawa, S. Procedure for seismic risk evaluation of buildings. *AIJ J. Technol. Des.* **2000**, *6*, 275–278. (In Japanese) [CrossRef]
64. Elkady, A.; Lignos, S.G. Effect of gravity framing on the overstrength and collapse capacity of steel frame buildings with perimeter special moment frames. *Earthq. Eng. Struct. Dyn.* **2015**, *44*, 1289–1307. [CrossRef]
65. Jisr, H.E.; Lignos, D.G. Fragility assessment of beam-slab connections for informing earthquake-induced repairs in composite-steel moment resisting frames. *Front. Built Environ.* **2021**, *7*, 691553.
66. Jisr, H.E.; Kohrangi, M.; Lignos, D.G. Proposed nonlinear macro-model for seismic risk assessment of composite-steel moment resisting frames. *Earthq. Eng. Struct. Dyn.* **2022**, *51*, 1180–1200. [CrossRef]
67. Paranonessso, M.; Lignos, D.G. Seismic design and performance of steel concentrically braced frames buildings with dissipative floor connectors. *Earthq. Eng. Struct. Dyn.* **2022**, *51*, 3505–3525. [CrossRef]
68. Vamvatsikos, D.; Cornell, C.A. Incremental dynamic analysis. *Earthq. Eng. Struct. Dyn.* **2002**, *31*, 491–514. [CrossRef]
69. Eads, L.; Miranda, E.; Krawinkler, H.; Lignos, D. An efficient method for estimating the collapse risk of structures in seismic regions. *Earthq. Eng. Struct. Dyn.* **2013**, *42*, 25–41. [CrossRef]
70. Deylami, A.; Mahdavipour, M.A. Probabilistic seismic demand assessment of residual drift for buckling-restrained braced frames as a dual system. *Struct. Saf.* **2016**, *58*, 31–39. [CrossRef]
71. Architectural Instutite of Japan (AIJ). *AIJ Standard for Allowable Stress Design of Steel Structures*; Maruzen Publishing Co., Ltd.: Tokyo, Japan, 2019. (In Japanese)
72. European Committee for Standardization (CEN). Design of Structures for Earthquake Resistance—Part I: General Rules, Seismic Actions and Rules for Buildings. Eurocode 8. 2003. CEN. Available online: https://www.confinedmasonry.org/wp-content/uploads/2009/09/Eurocode-8-1-Earthquakes-general.pdf (accessed on 4 October 2024). Eurocode 8, CEN..
73. Architectural Instutite of Japan (AIJ). *Recommendations for Loads on Buildings*; Maruzen Publishing Co., Ltd.: Tokyo, Japan, 2015. (In Japanese)
74. Dassault Systems. *Simulia User Assistance 2024*; Dassault Systems: Vélizy-Villacoublay, France, 2024.
75. Architectural Instutite of Japan (AIJ). *Design Recommendations for Composite Constructions*; Maruzen Publishing Co., Ltd.: Tokyo, Japan, 2010. (In Japanese)
76. Suzuki, A.; Kimura, Y. Cyclic behavior of component model of composite beam subjected to fully reversed cyclic loading. *J. Struct. Eng.* **2019**, *145*, 04019015. [CrossRef]

77. Suzuki, A.; Abe, K.; Kimura, Y. Restraint performance of stud connection during lateral-torsional buckling under synchronized in-plane displacement and out-of-plane rotation. *J. Struct. Eng.* **2020**, *146*, 04020029. [CrossRef]
78. Suzuki, A.; Abe, K.; Suzuki, K.; Kimura, Y. Cyclic behavior of perfobond shear connectors subjected to fully reversed cyclic loading. *J. Struct. Eng.* **2021**, *147*, 04020355. [CrossRef]
79. Suzuki, A.; Suzuki, K.; Kimura, Y. Ultimate shear strength of perfobond shear connectors subjected to fully reversed cyclic loading. *Eng. Struct.* **2021**, *248*, 113240. [CrossRef]
80. Suzuki, A.; Kimura, Y. Mechanical performance of stud connection in steel–concrete composite beam under reversed stress. *Eng. Struct.* **2021**, *249*, 113338. [CrossRef]
81. Suzuki, A.; Hiraga, K.; Kimura, Y. Cyclic behavior of steel-concrete composite dowel by clothoid-shaped shear connectors under fully reversed cyclic stress. *J. Adv. Concr. Technol.* **2023**, *21*, 76–91. [CrossRef]
82. Suzuki, A.; Hiraga, K.; Kimura, Y. Mechanical performance of puzzle-shaped shear connectors subjected to fully reversed cyclic stress. *J. Struct. Eng.* **2023**, *149*, 04023087. [CrossRef]
83. Suzuki, A.; Hiraga, K.; Kimura, Y.; Takahashi, J.; Kikuchi, K. Evaluation of superiority mechanical shear connectors based on flexural stiffness of composite beam. *AIJ J. Technol. Des.* **2024**, *30*, 154–159. (In Japanese) [CrossRef]
84. Ollgaard, J.G.; Slutter, R.G.; Fisher, J.W. Shear strength of stud connectors in lightweight and normal weight concrete. *AISC Eng. J.* **1971**, *71*, 55–64. [CrossRef]
85. Hawkins, N.M.; Mitchell, D. Seismic response of composite shear connections. *J. Struct. Eng.* **1984**, *110*, 2120–2136. [CrossRef]
86. Oehlers, D.J. Deterioration in strength of stud connectors in composite bridge beams. *J. Struct. Eng.* **1990**, *116*, 3417–3431 [CrossRef]
87. Bursi, O.S.; Gramola, G. Behaviour of headed stud shear connectors under low-cycle high amplitude displacements. *Mater. Struct.* **1999**, *32*, 290–297. [CrossRef]
88. Lin, W.; Yoda, T.; Taniguchi, N. Fatigue tests on straight steel-concrete composite beams subjected to hogging moment. *J. Constr. Steel Res.* **2013**, *80*, 42–56. [CrossRef]
89. Lin, W.; Yoda, T.; Taniguchi, N.; Kasano, H.; He, J. Mechanical performance of steel-concrete composite beams subjected to a hogging moment. *J. Struct. Eng.* **2014**, *140*, 04013031. [CrossRef]
90. Song, A.; Wan, S.; Jiang, Z.; Xu, J. Residual deflection analysis in negative moment regions of steel-concrete composite beams under fatigue loading. *Constr. Build. Mater.* **2018**, *158*, 50–60. [CrossRef]
91. Japan Society of Civil Engineers (JSCE). *Standard Specification for Hybrid Structures–2014*; Maruzen Publishing Co., Ltd.: Tokyo, Japan, 2015.
92. Oguejiofor, E.C.; Hosain, M.U. Behavior of perfobond rib shear connectors in composite beams. *Can. J. Civ. Eng.* **1992**, *19*, 1–10. [CrossRef]
93. Oguejiofor, E.C.; Hosain, M.U. Numerical analysis of pushout specimens with perfobond rib connectors. *Comput. Struct.* **1997**, *62*, 617–624. [CrossRef]
94. Kim, S.H.; Kim, K.S.; Park, S.; Ahn, J.H.; Lee, M.K. Y-type perfobond rib shear connectors subjected to fatigue loading on highway bridges. *J. Constr. Steel Res.* **2016**, *122*, 445–454. [CrossRef]
95. Kim, S.H.; Kim, K.S.; Han, O.; Park, J.S. Influence of transverse rebar on shear behavior of Y-type perfobond rib shear connection. *Constr. Build. Mater.* **2018**, *180*, 254–264. [CrossRef]
96. Kim, S.H.; Batbold, T.; Shah, S.H.A.; Yoon, S.; Han, O. Development of shear resistance formula for the Y-type perfobond rib shear connector considering probabilistic characteristics. *Appl. Sci.* **2021**, *11*, 3877. [CrossRef]
97. Tanaka, T.; Sakai, J.; Kawano, A. Development of new mechanical shear connector using burring steel plate. *J. Struct. Constr. Eng. (Trans. AIJ)* **2013**, *78*, 2237–2245. [CrossRef]
98. Tanaka, T.; Sakai, J.; Kawano, A. Experimental study on elastic-plastic flexural behavior of composite beam by burring shear connector and perfobond-rib shear connector. *Steel Constr. Eng.* **2014**, *21*, 111–123.
99. Tanaka, T.; Norimatsu, K.; Sakai, J.; Kawano, A. Development of innovative shear connector in steel-concrete joint and establishment of rational design method. *Proc. Jpn. Concr. Inst.* **2013**, *35*, 1237–1242.
100. Lorenc, W.; Kozuch, M.; Rowinski, S. The behaviour of puzzle-shaped composite dowels—Part I: Experimental study. *J. Constr. Steel Res.* **2014**, *101*, 482–499. [CrossRef]
101. Lorenc, W.; Kozuch, M.; Rowinski, S. The behaviour of puzzle-shaped composite dowels—Part II: Theoretical investigations. *J. Constr. Steel Res.* **2014**, *101*, 500–518. [CrossRef]
102. Kopp, M.; Wolters, K.; Classen, M.; Hegger, J.; Gundel, M.; Gallwoszus, J.; Heinemeyer, S.; Feldmann, M. Composite dowels as shear connectors for composite beams—Background to the design concept for static loading. *J. Constr. Steel Res.* **2018**, *147*, 488–503. [CrossRef]
103. Lorenc, W. Concrete failure of composite dowels under cyclic loading during full-scale tests of beams for the "Wierna Rzeka" bridge. *Eng. Struct.* **2020**, *209*, 110199. [CrossRef]
104. Classen, M.; Hegger, J. Shear-slip behaviour and ductility of composite dowel connectors with pry-out failure. *Eng. Struct.* **2017**, *150*, 428–437. [CrossRef]
105. Lorenc, W.; Kurz, W.; Seidl, G. Hybrid steel-concrete sections for bridges: Definition and basis for design. *Eng. Struct.* **2022**, *270*, 114902. [CrossRef]

106. Christou, G.; Hegger, J.; Classen, M. Fatigue of clothoid shaped rib shear connectors. *J. Constr. Steel Res.* **2020**, *171*, 106133. [CrossRef]
107. Classen, M.; Gallwoszus, J. Concrete fatigue in composite dowels. *Struct. Concr.* **2016**, *17*, 63–73. [CrossRef]
108. Classen, M.; Gallwoszus, J.; Stark, A. Anchorage of composite dowels in UHPC under fatigue loading. *Struct. Concr.* **2016**, *17*, 183–193. [CrossRef]
109. Kim, H.Y.; Jeong, Y.J. Experimental investigation on behaviour of steel-concrete composite bridge decks with perfobond ribs. *J. Constr. Steel Res.* **2006**, *62*, 463–471. [CrossRef]
110. Christou, G.; Classen, M.; Wolters, K.; Broschart, Y. Fatigue of composite constructions with composite dowels. In Proceedings of the 14th Nordic Steel Construction Conference, Copenhagen, Denmark, 18–20 September 2019; pp. 277–282.
111. Kopp, M.; Christou, G.; Stark, A.; Hegger, J.; Feldmann, M. Integrated slab system for the steel and composite construction. *Bauingenieur* **2018**, *87*, 136–148.
112. Japan Society of Steel Construction (JSSC). *Guideline of Standard Push-Out Tests of Headed Stud and Current Situation of Research on Stud Shear Connectors*; Japan Society of Steel Construction (JSSC): Tokyo, Japan, 1996.
113. Ahmed, S.E. Ultimate Shear Resistance of Plate Girders Part 2—Hoglund Theory. *Int. J. Civ. Environ. Eng.* **2013**, *7*, 918–926.
114. Alinia, M.M.; Shakiba, M.; Habashi, H.R. Shear Failure Characteristics of Steel Plate Girders. *Thin-Walled Struct.* **2009**, *47*, 1498–1506. [CrossRef]
115. Anthimos, A.; Marius, M.; Victor, G. Prediction of available rotation capacity and ductility of wide-flange beams part 2 applications. *J. Constr. Steel Res.* **2012**, *68*, 176–191.
116. Basler, K. Strength of plate girders in shear. *J. Struct. Div.* **1961**, *87*, 151–180. [CrossRef]
117. Cheng, X.; Chen, Y.; Pan, L. Experimental study of steel beam-columns composed of slender H-sections under cyclic bending. *J. Constr. Steel Res.* **2013**, *88*, 279–288. [CrossRef]
118. Elkady, A.; Lignos, D.G. Full-scale testing of deep wide-flange steel columns under multiaxis cyclic loading. *J. Struct. Eng.* **2018**, *144*, 2018. [CrossRef]
119. Gioncu, V.; Marius, M.; Anthimos, A. Prediction of available rotation capacity and ductility of wide-flange beams part 1 DUCTROT-M computer program. *J. Constr. Steel Res.* **2012**, *69*, 8–19. [CrossRef]
120. Glassman, D.J.; Garlock, M.E. A compression model for ultimate postbuckling shear strength. *Thin-Walled Struct.* **2016**, *102*, 258–272. [CrossRef]
121. Gulen, O.; Harris, J.; Uang, C. Observations from cyclic tests on deep, wide-flange beam-columns. *Eng. J.* **2017**, *54*, 45–59.
122. Hanna, M.T. Failure loads of web panels loaded in pure shear. *J. Constr. Steel Res.* **2015**, *105*, 39–48. [CrossRef]
123. Ikarashi, K.; Suekuni, R.; Shinohara, T.; Wang, T. Evaluation of plastic deformation capacity of H-shaped steel beams with newly proposed limitation value of plate slenderness. *J. Struct. Constr. Eng. (Trans. AIJ)* **2011**, *76*, 1865–1872. (In Japanese) [CrossRef]
124. Kasai, K.; Popov, E.P. General behavior of WF steel shear link beams. *J. Struct. Eng.* **1986**, *112*, 362–382. [CrossRef]
125. Kasai, K.; Popov, E.P. Cyclic Web Buckling Control for Shear Link Beams. *J. Struct. Eng.* **1986**, *112*, 505–523. [CrossRef]
126. Kato, B. Rotation capacity of H-section members as determined by local buckling. *J. Constr. Steel Res.* **1989**, *13*, 95–109. [CrossRef]
127. Lee, C.S.; Davidson, J.S.; Yoo, C.H. Shear buckling coefficients of plate girder web panels. *Comput. Struct.* **1996**, *59*, 789–795. [CrossRef]
128. Lee, C.D.; Lee, D.; Yoo, C.H. Flexure and Shear Interaction in Steel I-girders. *J. Struct. Eng.* **2013**, *139*, 1882–1894. [CrossRef]
129. MacRae, G.A. The Seismic Response of Steel Frames. Ph.D. Thesis, University of Canterbury, Christchurch, New Zealand, 1999.
130. Miguel, A.; Luis, M.; Jose, M.C. Evaluation of the rotation capacity limits of steel members defined in EC8-3. *J. Constr. Steel Res.* **2017**, *135*, 11–29.
131. Nakashima, M.; Takanashi, K.; Kato, H. Test of steel beam-columns subject to sidesway. *J. Struct. Eng.* **1990**, *116*, 2516–2531. [CrossRef]
132. Nowell, J.D.; Uang, C. Cyclic behavior of steel wide-flange columns subjected to large drift. *J. Struct. Eng.* **2008**, *134*, 1334–1342. [CrossRef]
133. Shahabedding, T.; Schafer, B.W. Role of local slenderness in the rotation capacity of structural steel members. *J. Constr. Steel Res.* **2014**, *95*, 32–43.
134. Shokouhian, M.; Shi, Y. Classification of I-section flexural members based on member ductility. *J. Constr. Steel Res.* **2014**, *95*, 198–210. [CrossRef]
135. Suzuki, T.; Ikarashi, K.; Tsuneki, Y. A study of collapse mode and plastic deformation capacity of H-shaped steel beams under shear bending. *J. Struct. Constr. Eng. (Trans. AIJ)* **2001**, *547*, 185–191. (In Japanese) [CrossRef]
136. Torsten, H. Simply supported long thin plate I-girders without web stiffeners subjected to distributed transverse load. *IABSE Rep. Work. Comm.* **1971**, 85–97.
137. Torsten, H. Shear buckling resistance of steel and aluminium plate girders. *Thin-Walled Struct.* **1997**, *29*, 13–30.
138. Wael, F.R. Local buckling of welded steel I-beams considering flange–web interaction. *Thin-Walled Struct.* **2015**, *97*, 241–249.
139. Yiyi, C.; Xin, C.; Nethercot, N.D.A. An overview study on cross-section classification of steel H-sections. *J. Constr. Steel Res.* **2013**, *80*, 386–393.
140. The Headquarters for Earthquake Research Promotion. National Seismic Hazard Map for Japan. Available online: https://www.jishin.go.jp/evaluation/seismic_hazard_map/ (accessed on 5 September 2024).

141. Japan Seismic Hazard Information Station (J-SHIS). Available online: https://www.j-shis.bosai.go.jp/ (accessed on 5 September 2024).
142. Kamae, K.; Irikura, K. Prediction of site-specific strong ground motion using semi-empirical methods. In Proceeding of the 10th World Conference on Earthquake Engineering, Madrid, Spain, 19–24 July 1992; pp. 801–806.
143. Boore, D.M. Stochastic simulation of high-frequency ground motions based on seismological models of the radiated spectra. *Bull. Seismol. Soc. Am.* **1983**, *73*, 1865–1984.
144. Satoh, T.; Kawase, H.; Sato, T. Statistical spectral model of earthquakes in the eastern Tohoku district, Japan, based on the surface and borehole records observed in Sendai. *Bull. Seismol. Soc. Am.* **1997**, *87*, 446–462. [CrossRef]
145. Sendai City. A distribution map of seismic intensity in JMA in the expected off Miyagi earthquake. *Investig. Rep. Earthq. Damage Estim.* **2002**.
146. Ohsaki, Y.; Hara, A.; Kiyota, Y. Stress-strain model for soils for seismic analysis. In Proceedings of the 5th Japan Earthquake Engineering Symposium, Tokyo, Japan, 28–30 November 1978; pp. 697–704.
147. Fujiwara, H.; Morikawa, N.; Maeda, T.; Iwaki, A.; Senna, S.; Kawai, S.; Azuma, H.; Xiansheng, K.H.; Imoto, M.; Wakamatsu, K. et al. *Improved Hazard Assessment after the 2011 Great East Japan Earthquake (Part 2)*; National Research Institute for Earth Science and Disaster Resilience: Tsukuba, Japan, 2023. (In Japanese)
148. Subcommittee on Strong Motion Evaluation of the Earthquake Research Committee. *Seismic Hazard of Response Spectra (Prototype Version)*; National Research Institute for Earth Science and Disaster Resilience: Tsukuba, Japan. Available online: https://www.jishin.go.jp/evaluation/seismic_hazard_map/sh_response_spectrum/ (accessed on 4 October 2024). (In Japanese).
149. Bantilas, K.E.; Naoum, M.C.; Kavvadias, I.E.; Karayannis, C.G.; Elenas, A. Structural pounding effect on the seismic performance of a multistorey reinforced concrete frame structure. *Infrastructures* **2023**, *8*, 122. [CrossRef]
150. Garcia, J.R.; Miranda, E. *Performance-Based Assessment of Existing Structures Accounting for Residual Displacements*; Stanford University: Stanford, CA, USA, 2005.

Article

Numerical Methods for Topological Optimization of Wooden Structural Elements

Daniela Țăpuși [1,*], **Andrei-Dan Sabău** [1,*], **Adrian-Alexandru Savu** [2], **Ruxandra-Irina Erbașu** [1] **and Ioana Teodorescu** [1]

1 Department of Civil, Urban and Technological Engineering, Faculty of Civil, Industrial and Agricultural Buildings, Technical University of Civil Engineering Bucharest, 020396 Bucharest, Romania; ruxandra.erbasu@utcb.ro (R.-I.E.); ioana.teodorescu@utcb.ro (I.T.)
2 Department of Structural Mechanics, Faculty of Civil, Industrial and Agricultural Buildings, Technical University of Civil Engineering Bucharest, 020396 Bucharest, Romania; adrian.a.savu@utcb.ro
* Correspondence: daniela.tapusi@utcb.ro (D.Ț.); andreisabau228@gmail.com (A.-D.S.)

Abstract: Timber represents a building material that aligns with the environmental demands on the impact of the construction sector on climate change. The most common engineering solution for modern timber buildings with large spans is glued laminate timber (glulam). This project proposes a tool for a topological optimized geometry generator of structural elements made of glulam that can be used for building a database of topologically optimized glulam beams. In turn, this can be further used to train machine learning models that can embed the topologically optimized geometry and structural behavior information. Topological optimization tasks usually require a large number of iterations in order to reach the design goals. Therefore, embedding this information into machine learning models for structural elements belonging to the same topological groups will result in a faster design process since certain aspects regarding structural behavior such as strength and stiffness can be quickly estimated using Artificial Intelligence techniques. Topologically optimized geometry propositions could be obtained by employing generative machine learning model techniques which can propose geometries that are closer to the topologically optimized results using FEM and as such present a starting point for the design analysis in a reduced amount of time.

Keywords: glulam; topological optimization; finite element method; machine learning; artificial neural network

Citation: Țăpuși, D.; Sabău, A.-D.; Savu, A.-A.; Erbașu, R.-I.; Teodorescu, I. Numerical Methods for Topological Optimization of Wooden Structural Elements. *Buildings* **2024**, *14*, 3672. https://doi.org/10.3390/buildings14113672

Academic Editors: Atsushi Suzuki and Dinil Pushpalal

Received: 29 September 2024
Revised: 5 November 2024
Accepted: 15 November 2024
Published: 18 November 2024

1. Introduction

The construction industry contributes to approximately 40–50% of greenhouse gas emissions. This figure includes both the energy consumed during building operations on site and the energy used in the manufacturing of construction materials and products. Two main strategies have been thoroughly investigated to mitigate the environmental impact of building elements: structural optimization by minimizing the elements' section and the use of low-carbon materials [1–4]. This is the reason why wood and wood products such as glued laminated timber are valued and have gained success as a construction material nowadays.

Glued laminated timber, known as "glulam" or GLT, is a composite material made of overlapping layers of wood bonded with synthetic resins and pressed into the desired shape. This type of product is one of the most popular options in civil engineering—while the ability to absorb carbon from the environment makes it a sustainable material. Additionally, the good mechanical properties of the material along with its low self-weight make it a great seismic performance solution [5,6].

The production of GLT elements represents a major advantage in terms of reducing workmanship on site, through the quality of the machined elements and the obtaining of complex geometries with various sizes that can be customized to meet individual specifications and requirements. The design and execution of civil structures is, however still dependent on inertia in terms of the geometric dimensioning of structures, which favors constant rectangular sections, easy to size.

Another significant advantage of wood is its strength-to-weight ratio when compared to materials like concrete and steel, making it an excellent choice [1]. Reducing the weight of the element diminishes the dead loads along with the size of elements in the structure.

Recent studies have shown that topological optimization can lead to better dimensions of the element able to support the forces applied. According to existing studies in the field, the characteristics of the model resulting from different types of analysis in the Finite Element Method [5,7], simple regression on metamodels [8], tests on models [4] or by Monte-Carlo simulation [9] may produce material reductions of over 10% [10].

The topological optimization is used in a multitude of fields such as aerospace field (for application in the manufacturing process) [11,12]; robotics (in order to reduce the weight of the product, while maintaining the mechanical properties) [13,14]; medical devices (for the shape of the element), the energy and battery life [15–17], cases where different scenarios were considered with the variation of geometric dimensions and structural properties [18], architecture [19,20]; and civil engineering which was the focus in this current research with the findings presented below.

Optimization model examples are presented in [21,22] where structural analysis and dimensioning constraints defined by Eurocode standards are used in order to create a model of profiles that can reduce the cost of a building with elements made of glued laminated timber and steel.

Moreover, the study of [23] demonstrates that optimization techniques reduce the mass of elements up to 30% creating high-performance with low-weight design and reduction of deflection by 15–20% [8]. The variation in geometric dimensions of elements, materials, curved elements with large span, height and volume of buildings creates new designs where the elements can be used in a rational manner, reducing the manufacturing process and the cross sections while maintaining the resistance [24–28].

At the same time, during the current research campaigns, topological optimization can be carried out using AI that gives new perspectives and a multitude of possible scenarios [29]. The methods mentioned above improve cost efficiency and optimize the mechanical performance of structural elements in GLT. However, they derive from analyses that require financial resources, whether in laboratory experiments or through time-consuming programs. This happens considering that the operation research processing on models using the Finite Element Method (FEM) involves systems solving inequations having as input data the results of the FEM and the geometry variation until the objective function is reached.

This is the current scientific context, where the premises exist for the use of machine learning techniques in order to embed geometry and structural behavior data obtained by topological optimization and to reduce the computation requirements for the deployment of such a solution [30]. Techniques such as deep neural networks [31] or generative adversarial networks [32] have already proven useful for this kind of application. The part this paper addresses is the training data for such an algorithm. The success of any development in this direction depends on the quality of the training data; for example, the present project proposes some domain limits for the exploration of topologically optimized solutions and presents an algorithm for computing this kind of optimization task, using the Dassault ABAQUS 2023 software package.

For the experimental campaign, GLT elements were used in the experiments and tested as well as their combination to ascertain the contact properties.

In terms of the wood species used, spruce was the best solution considering that it is a common species used for the realization of glued laminated timber elements according to [8].

The GLT elements are formed by joining together layers of wood material in the form of lamellas with the thickness $n_l = 2$ cm with a structural epoxy resin-based adhesive.

The adhesive used for processing the glued laminated timber elements, Melamine-urea-formaldehyde (MUF), is one of the most commonly used adhesives in the wood industry. A representation of a glued laminated timber element can be seen in Figure 1 where the lamellas can be identified. However, the adhesive is not visible in the glued laminated timber section being a part of the element. The element is good quality which means that poor glue bonding between the layers is not encountered so the section is evenly created.

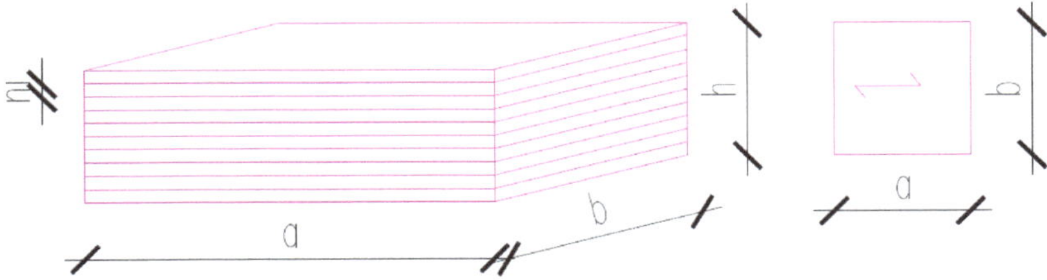

Figure 1. Representation of a physical model of a glued laminated timber element with a section representing the position of the fibers in the element; a = depth; b = width; h = height.

Given the natural composition of wooden material in the form of annual growth rings, the position of the fibers is essential when using the material in a structure. The mechanical properties are different along the axis of the element and for maximum utilization of it they have to be parallel to the longitudinal axis [33–35].

The laboratory determinations were carried out in two stages: in the first one, the strength parameters of glued laminated timber were directly determined, and in the second one, the shear strength parameters on the glulam material were determined. The tests revealed values much higher than the shear strength of wood material, which makes it so that the FEM modeling of the glulam-type element can be achieved either in the form of homogeneous material or using "tie"-type contact elements. The experimental campaign was conducted in the laboratories of the Technical University of Civil Engineering Bucharest.

First of all, the fundamental characteristics of wood in the form of glued laminated timber can be seen in Table 1 with some dimensions of elements, the types of forces applied on the elements (traction and compression parallel with the fibers and bending) and the average value of the strength results. Ten samples for each test were performed.

Secondly, 20 samples of glued laminated timber elements with dimensions 60 × 60 mm × 20 mm were inserted into a shear box where force was applied on them until breaking point. In Figure 2a, the shearing box can be observed, with specific dimensions in which the wood sample can be inserted perfectly with the force being applied on the center of the piece following [1]. After the breaking point, the sample is retrieved and in Figure 2b,c the breaking pattern is visible. The wooden piece has been broken completely.

(**a**) (**b**) (**c**)

Figure 2. Shear tests in the laboratory: (**a**) The force applied on the wooden sample in the shear box. (**b**) The wooden sample after the application of the force. (**c**) The broken sample.

Table 1. Results of laboratory tests on wooden samples according to [36].

Test Number	Test Specimen Sample Size (w × h − L)	Average Strength Value
Traction parallel to the fibers		$f_t = 93.61 \ \text{N/mm}^2$
Compression parallel to the fibers		$f_c = 40.85 \ \text{N/mm}^2$
Bending		$f_{inc} = 79.58 \ \text{N/mm}^2$

2. Test Equipment

The experimental campaign was carried out following some test steps that were needed in order to verify the correctness of the results considering the correlation between the Mohr–Coulomb parameters and the unitary forces.

In order to verify the mechanical parameters obtained, elasto-plastic modeling was performed in explicit dynamics formulations in order to be able to follow the evolution of the deformations over time. The geometry of the element takes into consideration the restrictions of the model in the experimental campaign due to the dimensions of the shear box [8].

The model created consisted of replicating the sample used in the direct shearing machine with the parallelepiped sample dimensions mentioned above. The shear plan was determined midway through the test, based on the boundary conditions (Figure 3). The load was applied by a certain imposed deformation, with constant speed. The color map represents the sensitivity of parts of the wooden elements when force is applied.

The deformation model using the parameters defined beforehand can be identified in Figure 3a. After the identification of the plastic zones in the sample, the deformation of the proposed glued laminated timber sample can be seen Figure 3b,c.

The model with the same parameters and characterizations as the wooden sample subjected to deformations by the Finite Element Method can be compared directly with the samples subjected to loadings in the shearing box. It can be seen in the two images in Figure 3c (deformation of the model) and Figure 3d (the real wooden sample) that the pattern follows the same distribution of the deflection.

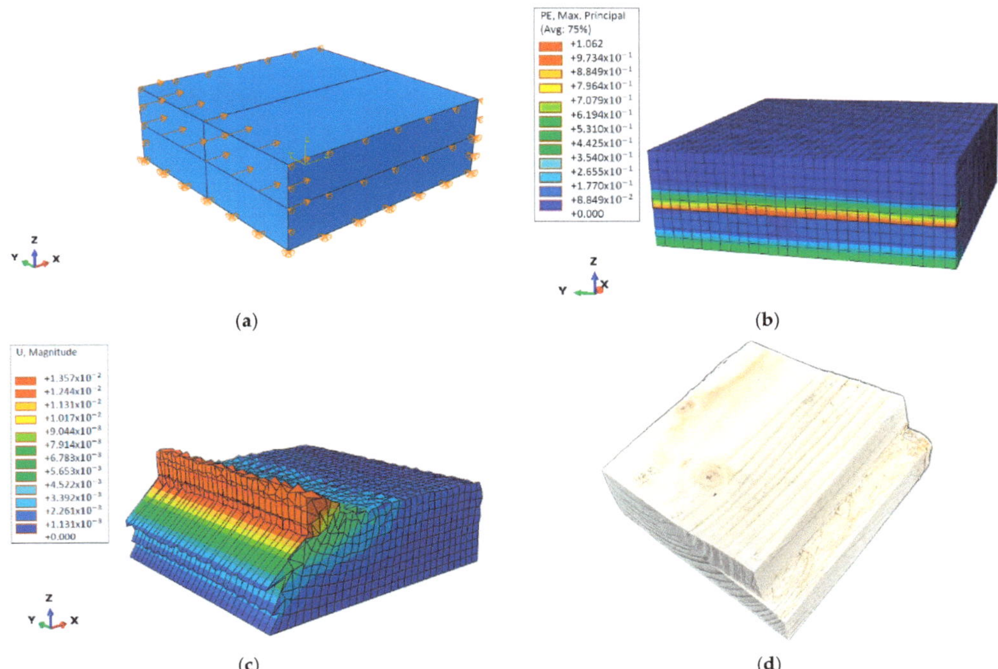

Figure 3. Deformation model using FEM: (**a**) The model created. (**b**) Formation of the plastic zones of tangents in the sample. (**c**) Deformation of the model [6]. (**d**) The real model sample of GLT.

Strain compatibility was monitored by recording the strain history at a point belonging to the failure surface (Figure 4). Comparing these results with those obtained from the laboratory determination, it can be seen that the mobilization of deformations occurs in a similar way.

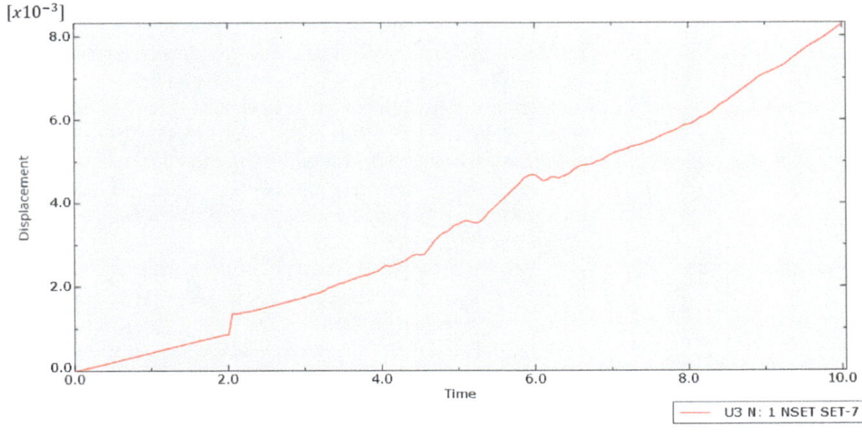

Figure 4. Shear mobilization curves at different vertical stresses for the industrially made samples [36].

3. Numerical Modeling and Methods

Following the steps described above, a database was created with optimized models obtained by generating, following Monte-Carlo simulation principle multiple models that are created by using scripts for the automation of FEM runs. The general characterization of the samples considered for the FEM model can be seen in Figure 5a. The dimensions of the elements are 10,000 mm × 160 mm × 40 mm with two fixed supports of 600 mm × 160 mm × 40 mm. The number of wooden samples used is 23. The discretization of the model can be analyzed in Figure 5b.

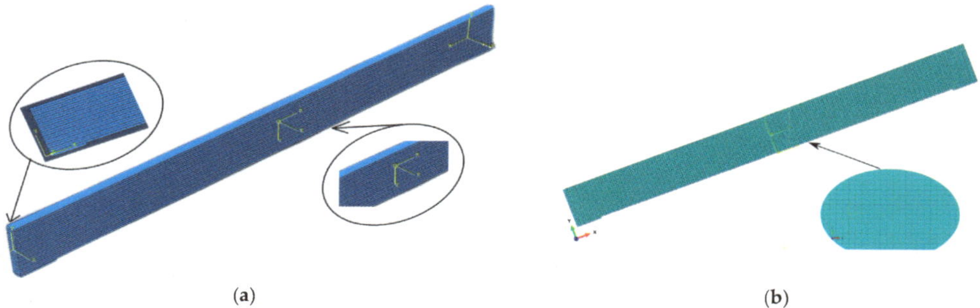

(a) (b)

Figure 5. (**a**) Three-dimensional compute assembly. (**b**) Discretization of the model.

For the optimization, the topological optimization method was chosen, blocking the areas of application of the conditions on the contour and the area of application of the uniform pressure. In the optimization process, the finite element density was allowed to vary between 1 and 0.001 with a maximum variation per analysis cycle of 0.25. Three design response functions have been defined, namely "Energy Stiffness Measure", "Volume" and "Signed Von Mises Stress". The objective of the optimization was to minimize the design response values for "Volume". The optimization task initially applied to the model was to limit the volume variation below 45%. Unfortunately, the software applied the 45% material reduction directly to the model and then reshaped it in order to withstand the increased interior stresses, so it was not a correct approach. Due to this shortcoming, in the final models, the optimization task was to have displacements under the allowable limits specified in the current norms (beam span/200). This instructed the software to gradually remove material until the allowable displacement is reached, and then remodel it to create a more uniform distribution of stresses.

The software allows the application of several types of geometric constraints to the model in order to control how the material is removed (see Figure 6). All of them were tested in the early models, but, since some of them did not show a significant optimization or were not pertinent to the tested model, only three were chosen for the final analysis: "Planar symmetry", "Rotational symmetry", "Point symmetry". The "Planar symmetry" constraint forces the optimized model to be symmetric about a specified plane, the "Rotational symmetry" constraint forces the optimized model to be symmetric about a specified axis and the "Point symmetry" constraint forces the optimized model to be symmetric about a specified point.

Based on the laboratory tests carried out within the current project, a Monte Carlo analysis was carried out using a proprietary script in which the material parameters were varied using the mean and the standard deviation resulting from the laboratory tests. For each model to which different geometric constraints were applied, the script was applied to run 30 models with different mechanical parameters. The results were collected in a database compared and contrasted.

In order to incorporate as close as possible the geometric results obtained by Monte-Carlo simulation, it was initially proposed to use a generative machine learning model (diffuser) which, starting from the synthetic data of each beam (opening, spans, loads), can generate topologically optimized glued laminated timber elements in elevation. The first attempts resulted in a lack of alignment of the machine learning model.

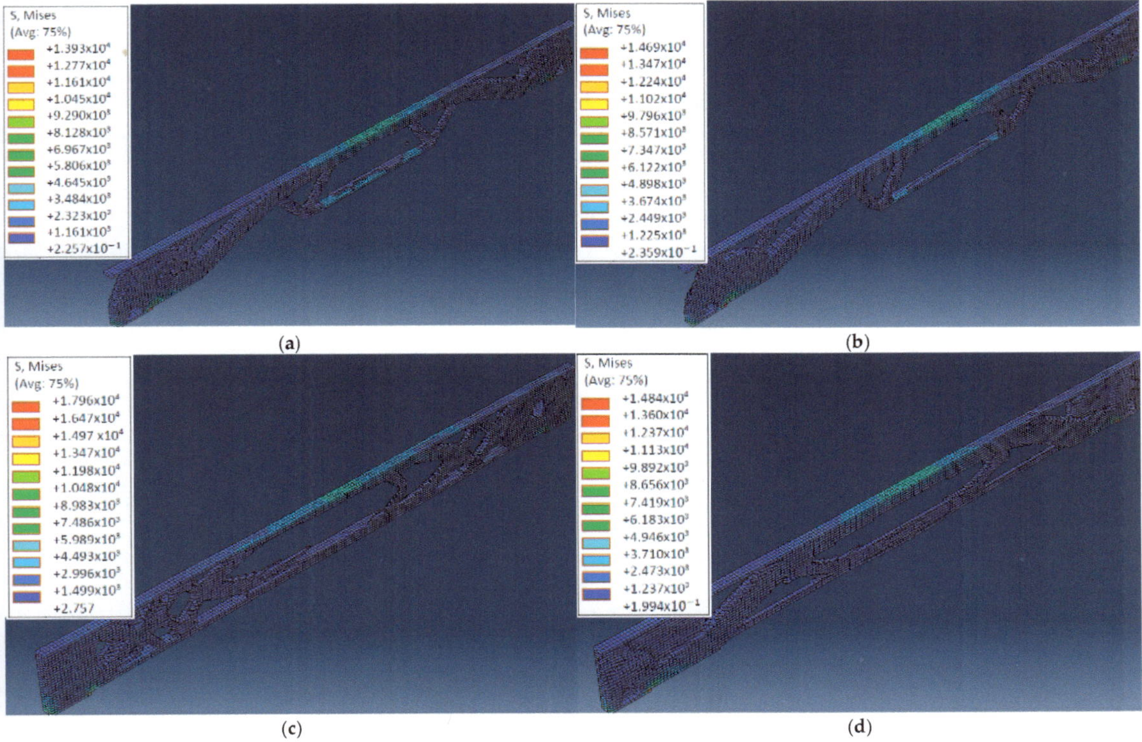

Figure 6. Geometric restriction results for the topology optimization tasks: (**a**) Condition Demold. (**b**) Milling condition. (**c**) Planar symmetry condition. (**d**) Condition Symm Point [36].

Noting the lack of alignment of the generative model, another machine learning method took the place of the current one, namely an artificial neural network (ANN) that would embed the results of topological optimizations and be able to indicate, based on the data of the problem, a measure of the stiffness of a topologically optimized beam in order to be able to identify whether or not such optimization would produce satisfactory results from the design point of view.

4. Results

For the characterization of the model, it was preferred to generate all possible dimensional combinations of beams (Table 2), respectively, to vary the dimensions as presented below:

Table 2. Dimensional combination of beams.

Type of Element	Dimensions	Pitch
Beam length (L)	from 4 to 8 m	10 cm
Beam width (W)	From 65 cm to 1.65 m	5 cm
Beam span	from 3 to 5 m	50 cm
Thickness of a plank	4 cm	-
Length of the support area	20 cm, 30 cm and 40 cm	-

The height of the section was established through pre-dimensioning rules.

Following the combinations in the model resulted in 6526 geometrically distinct patterns. Running these models required over 24,000 h/core, which was executed by outsourcing the service.

Each of these patterns were statically analyzed in the Abaqus 2023 program with the following optimization strategies:

- Planar symmetry with respect to the vertical plane perpendicular to the longitudinal axis of the beam in the center of gravity;
- Planar symmetry with respect to the horizontal median plane;
- Polar symmetry with respect to the center of gravity of the beam.

Each optimizer was run with 100 iterations and the final iteration was processed.

Due to the large space occupied by the results of a turnover, only the useful synthetic information presented in the Annexes was saved.

In order to test the feasibility of embedding the topological optimization information into a machine learning model, the first step is to create a regressive model and to test how it responds to the input parameters and objectives. The approach chosen was to develop a deep neural network in order to see if it is possible to predict useful information such as a deflection for a beam in the case of the topologically optimized elements, without having to run the topological optimization. This represents a tried and tested technique for civil engineering applications as shown by [31]. The architecture of the deep neural network is described in Figure 7.

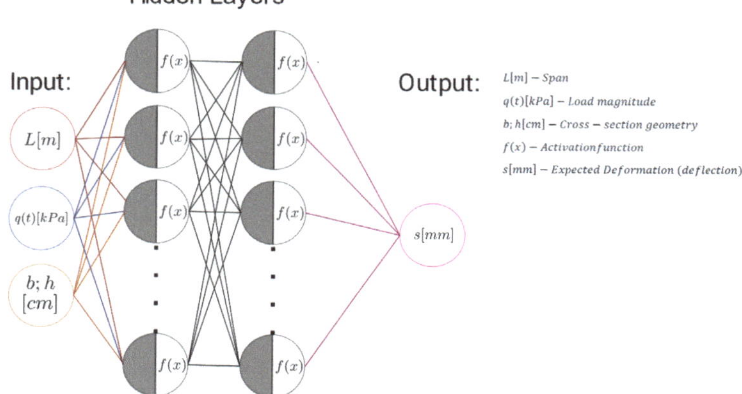

Figure 7. Neural network architecture.

A total of 6526 topologically optimized models/datasets (pairs of synthetic parameters—displacements) was divided into two parts: 6388 models for ANN training and 138 models for validating the result. The ANN optimization algorithm was the SGD (Stochastic Gradient Descent) algorithm with the hyper parameters $l_r = 0.001$ and Momentum = 0.9.

In order to preprocess the input and output data (key–value pairs), the normalization constants presented in Table 3 were used.

Table 3. Normalization parameters for the proposed ANN.

Parameter	Minimum Value	Δ
Beam length (L)	4.0	2.2
Beam span	3.0	2.0
Beam width (W)	0.065	0.1
Optimization constraint	0	2
Deflection	0.019020382	0.029697631

The ANN model thus obtained shows a good alignment both from the point of view of predictions for training data and from the point of view of predictions for the set dedicated to validation as shown in Figure 8 where the prediction of the ANN model is represented on the horizontal axis, and the actual value is displayed on the vertical axis. Figure 8 also showcases some edge-cases where the model tends to underpredict the actual deflection; this is why there are more points above the 100% alignment line, which can be improved by further developing the ANN architecture. After finalizing the training, a benchmark was run over the training dataset (6388 models) in order to benchmark how much faster the neural network is in predicting the deflection than running the FEM optimization task using the timeit Python (3.13.0) library. The results obtained indicate that this information can be obtained in 1.4 milliseconds using the ANN (best result out of 10,000 runs).

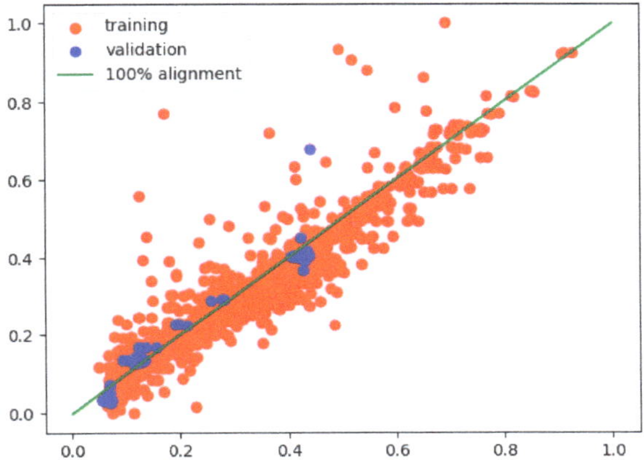

Figure 8. Alignment of trained ANN to input data, and validation data.

Figure 9 shows the evolution of the objective function value over the training epochs. In this case, the objective function value was considered to be the sum of the Mean Squared Loss value for each data point, over one training epoch.

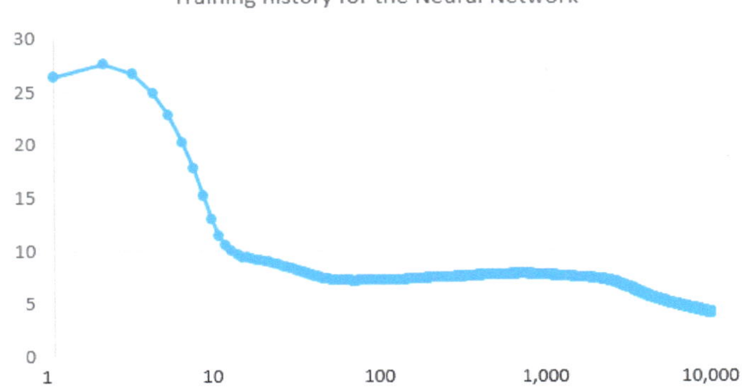

Figure 9. Training history for the neural network.

5. Conclusions

The current project aims to start the development of an expert software tool, based on Artificial Intelligence, which will provide optimized geometric solutions for glulam elements.

The formation of the database for training the Artificial Intelligence model is carried out based on numerical modeling using FEM analysis of a number of glued laminated wood samples and laboratory testing to confirm the correctness of the material model. The training of the AI model involves the use of bent glulam beams characterized by a series of variable parameters, obtaining a large number of topologically optimized models/datasets (pairs of synthetic parameters—displacements). These models were used for ANN training and for the validation of the result.

A deep neural network model is proposed herein with the goal of testing whether or not a useful alignment can be obtained from machine learning techniques for this kind of problem and this kind of optimization task.

The ANN network obtained can indicate based on the problem data a measure of the stiffness of a topologically optimized beam in order to be able to identify whether such optimization will produce satisfactory results from a design point of view. The resulting neural network can be used as it is for the task mentioned in the present work, being properly sized as to avoid overfitting issues. Further iteration on this neural network will be developed in order to improve the architecture in terms of size and activation function for the task previously mentioned. The models proposed for this paper follow certain parameters chosen beforehand which means that further improvements can be added for future models and experimental campaigns. These parameters are related to the properties of the wood material (type of wood, density, modulus of elasticity), dimensions and then the discretization of the model. The results obtained herein indicate that in the case of beam models, larger cross-section sizes and spans will benefit even more from the topological optimization. Also, the topological optimization task can be improved to take into account the lamellar structure of glulam and optimize the element such that the final geometry could rather be obtained by placing the correctly cut lamella inside the element without having to do a CNC machining step for the element.

Future developments will need to account for larger structural element sizes (cross-sections and span) and also take into account different loading conditions such as unevenly distributed loads or point loads. Also, it is clear that a great step in developing machine learning applications for this kind of task is to deploy a generative adversarial network architecture that can eventually propose geometries that can then be optimized in fewer steps.

Author Contributions: Conceptualization, D.Ț.; methodology, D.Ț. and A.-D.S.; data curation, A.-A.S. and A.-D.S.; writing—original draft preparation, I.T., writing—review and editing D.Ț., A.-D.S and R.-I.E.; visualization, A.-D.S.; supervision D.Ț. and R.-I.E.; numerical modelling A.-A.S. and A.-D.S.; project administration D.Ț., funding acquisition D.Ț. All authors have read and agreed to the published version of the manuscript.

Funding: This research was funded by the Technical University of Civil Engineering Bucharest [Grant No. $GnaC^{ARUT}_{2023}$-UTCB-28].

Data Availability Statement: The processed data is included in the paper and the raw data is available on demand from the corresponding authors.

Conflicts of Interest: The authors declare no conflicts of interest. The funders had no role in the design of the study; in the collection, analyses, or interpretation of data; in the writing of the manuscript; or in the decision to publish the results.

References

1. Mayencourt, P.; Mueller, C. Structural Optimization of Cross-laminated Timber Panels in One-way Bending. *Structures* **2019**, *18*, 48–59. [CrossRef]
2. Mayencourt, P.; Mueller, C. Structural Optimization and Digital Fabrication of Timber Beams. *IABSE Symp. Rep.* **2017**, *108*, 153–154. [CrossRef]
3. Skaržauskaitė, M.; Aleksandraviciute-Sviaziene, A.; Kovaite, K.; Gulevičiūtė, G.; Skarzauskiene, A. Innovating Civil Engineering: Strategies for Fostering Stakeholder Engagement in the Circular Economy. *Eur. Conf. Innov. Entrep.* **2024**, *19*, 416–424. [CrossRef]
4. Lobos Calquin, D.; Mata, R.; Correa, C.; Nuñez, E.; Bustamante, G.; Caicedo, N.; Blanco Fernandez, D.; Díaz, M.; Pulgar-Rubilar, P.; Roa, L. Implementation of Building Information Modeling Technologies in Wood Construction: A Review of the State of the Art from a Multidisciplinary Approach. *Buildings* **2024**, *14*, 584. [CrossRef]
5. De Vito, A.F., Jr.; Vicente, W.M.; Xie, Y.M. Topology optimization applied to the core of structural engineered wood product. *Structures* **2023**, *48*, 1567–1575. [CrossRef]

6. Du, H.; Xu, R.; Liu, P.; Yuan, S.; Wei, Y. Experimental and theoretical evaluation of inclined screws in glued laminated bamboo and timber-concrete composite beams. *J. Build. Eng.* **2024**, *91*, 109579. [CrossRef]

7. Jaaranen, J.; Fink, G. A finite element simulation approach for glued-laminated timber beams using continuum-damage model and sequentially linear analysis. *Eng. Struct.* **2024**, *304*, 117679. [CrossRef]

8. Pech, S.; Kandler, G.; Lukacevic, M.; Füss, J. Metamodel assisted optimization of glued laminated timber beams by using metaheuristic algorithms. *Eng. Appl. Artif. Intell.* **2019**, *79*, 129–141. [CrossRef]

9. Christoforo, A.; de Moraes, M.; Fraga, I.; Pereira Junior, W.; Lahr, F. Computational Intelligence Applied in Optimal Design of Wooden Plane Trusses. *Eng. Agríc.* **2022**, *42*, e20210123. [CrossRef]

10. Vicente, W.; Picelli, R.; Pavanello, R.; Xie, Y. Topology Optimization of Periodic Structures for Coupled Acoustic-Structure Systems. In Proceedings of the VII European Congress on Computational Methods in Applied Sciences and Engineering, Crete, Greece, 5–10 June 2016.

11. Boado Cuartero, B.; Pérez-Álvarez, J.; Roibas, E. Material Characterization of High-Performance Polymers for Additive Manufacturing (AM) in Aerospace Mechanical Design. *Aerospace* **2024**, *11*, 748. [CrossRef]

12. Zheng, H. A review on the topology optimization of the fiber-reinforced composite structures. *Aerosp. Technic Technol.* **2021**, 54–72. [CrossRef]

13. Anciferov, S.; Karachevceva, A.; Sychyov, E.; Litvishko, A. Topological optimization of design elements of a robotic cell. *Bull. Belgorod State Technol. Univ. Named After V. G. Shukhov* **2023**, *8*, 93–102. [CrossRef]

14. Ding, Y.-M.; Wang, Y.-C.; Zhang, S.-X.; Yan, Z. Exploring the topological sector optimization on quantum computers. *Phys. Rev. Appl.* **2024**, *22*, 034031. [CrossRef]

15. Méndez, D.; Garcia Cena, C.; Bedolla-Martinez, D.; González, A. Innovative Metaheuristic Optimization Approach with a Bi-Triad for Rehabilitation Exoskeletons. *Sensors* **2024**, *24*, 2231. [CrossRef] [PubMed]

16. Osakpolor, O.; Jimu, H.; Odion, D. Optimization Techniques for Maximizing Energy Harvested in Solar-Powered Wearable Medical Devices. *Int. J. Wearable Device* **2024**, *171*.

17. Peto, M.; García-Ávila, J.; Rodriguez, C.; Siller, H.; Silva, J.; Ramirez-Cedillo, E. Review on structural optimization techniques for additively manufactured implantable medical devices. *Front. Mech. Eng.* **2024**. [CrossRef]

18. Lin, K.; He, Y.; Yang, Y.; Xiong, L. From Topology Optimization to Complex Digital Architecture: A New Methodology for Architectural Morphology Generation. *Adv. Civil Eng.* **2021**, *2021*, 1–13. [CrossRef]

19. Korus, K.; Salamak, M.; Jasiński, M. Optimization of geometric parameters of arch bridges using visual programming FEM components and genetic algorithm. *Eng. Struct.* **2021**, *241*, 112465. [CrossRef]

20. Milner, R. A study of the strength of glued laminated timber. *Aust. J. Struct. Eng.* **2018**, *16*, 256–265. [CrossRef]

21. Kravanja, S.; Žula, T. Optimization of a single-storey timber building structure. *Int. J. Comput. Methods Exp. Meas.* **2021**, *9*, 126–140. [CrossRef]

22. Simón-Portela, M.; Villar-García, J.R.; Vidal-López, P.; Rodríguez-Robles, D. Enhancing Sustainable Construction: Optimization Tool for Glulam Roof Structures According to Eurocode 5. *Sustainability* **2024**, *16*, 3514. [CrossRef]

23. Ochieng, D. Kiu Publication Extension. Introduction to Lightweight Structures: A Review and Analysis of Topological Optimization Methods and Applications. *Res. Invent. J. Eng. Phys. Sci.* **2024**, *3*, 9–15.

24. Fedchikov, V. Topological optimization methods in the design of metal structures of buildings. *E3S Web Conf.* **2024**, *533*, 02023. [CrossRef]

25. Li, H.; Gao, L.; Li, H.; Li, X.; Tong, H. Full-scale topology optimization for fiber-reinforced structures with continuous fiber paths. *Comput. Methods Appl. Mech. Eng.* **2021**, *377*, 113668. [CrossRef]

26. Wellershoff, F.; Baudisch, R.; Posavec, M. Design optimization of glued-laminated timber freeform strcutrues with multi-objective constraints. In Proceedings of the IASS Annual Symposium 2017 Interfaces: Architecture, Engineering, Science, Hamburg, Germany, 25–28 September 2017.

27. Simón-Portela, M.; Villar-García, J.R.; Rodríguez-Robles, D.; Vidal-López, P. Optimization of Glulam Regular Double-Tapered Beams for Agroforestry Constructions. *Appl. Sci.* **2023**, *13*, 5731. [CrossRef]

28. Stanić, A.; Hudobivnik, B.; Brank, B. Economic-design optimization of cross laminated timber plates with ribs. *Compos. Struct.* **2016**, *154*, 527–537. [CrossRef]

29. Wenhao, R. Prediction and optimization of civil engineering material properties based on artificial intelligence. *Appl. Comput. Eng.* **2024**, *51*, 298–312. [CrossRef]

30. Taye, M.M. Understanding of Machine Learning with Deep Learning: Architectures, Workflow, Applications and Future Directions. *Computers* **2023**, *12*, 91. [CrossRef]

31. Park, H.I.; Cho, C. Neural Network Model for Predicting the Resistance of Driven Piles. *Mar. Georesour. Geotechnol.* **2010**, *28*, 324–344. [CrossRef]

32. Nauata, N.; Hosseini, S.; Chang, K.-H.; Chu, H.; Cheng, C.-Y.; Furukawa, Y. House-GAN++: Generative Adversarial Layout Refinement Network towards Intelligent Computational Agent for Professional Architect. In Proceedings of the IEEE/CVF Conference on Computer Vision and Pattern Recognition, Nashville, TN, USA, 20–25 June 2021; pp. 13632–13641.

33. Livas, C.; Ekevad, M.; Öhman, M. Experimental analysis of passively and actively reinforced glued-laminated timber with focus on ductility. *Wood Mater. Sci. Eng.* **2020**, *17*, 129–137. [CrossRef]

34. *EN 14080:2013*; Timber Structures—Glued Laminated Timber and Glued Solid Timber—Requirements. European Committee for Standardization (CEN): Brussels, Belgium, 2013.
35. Gong, Y.; Chen, X.; Ren, H.; Liu, B.; Zhang, H.; Wang, Y. Theoretical and experimental studies on the bending properties of glued laminated timber manufactured with Chinese fir. *Structures* **2024**, *68*, 107149. [CrossRef]
36. Tapusi, D.; Andronic, A.; Tufan, N.; Teodorescu, I.; Erbasu, R. Development of a generator of glued laminated timber element sections using artificial intelligence. In Proceedings of the XXIVth International Multidisciplinary Scientific GeoConference Surveying, Geology and Mining, Ecology and Management, SGEM 2024, Albena, Bulgaria, 29 June–8 July 2024.

Case Report

Apparent Influence of Anhydrite in High-Calcium Fly Ash on Compressive Strength of Concrete

Dinil Pushpalal [1,*] and Hiroo Kashima [2]

[1]   Graduate School of International Cultural Studies, Tohoku University, Sendai 980-8576, Japan
[2]   Central Research Institute, Shin-Etsu Industry Co., Ltd., Saitama 355-0071, Japan; gdstk545@yahoo.co.jp
*   Correspondence: dinil.pushpalal.b4@tohoku.ac.jp

Abstract: This case study investigates five fly ashes with high CaO and SO_3 levels in their chemical composition and compares the apparent influence of the presence and absence of anhydrite on compressive strength. Another distinguishing feature of the above ashes is that they, more or less, naturally contain anhydrite. Two different series of mixed proportions were adopted. Series 1 is designed to understand the maximum possible replacement level of fly ash. Series 2 is designed to understand the effect of anhydrite on compressive strength development. The mineral composition and glass phase of fly ashes were determined by X-ray diffraction Rietveld analysis. As a result of this study, we have found that concrete containing anhydrite-rich fly ash exhibits a higher strength than concrete containing anhydrite-free fly ash at all ages. The compressive strength increases with an increasing fly ash replacement ratio when anhydrite-rich ash is used, but strength decreases when the replacement level exceeds a certain point. The optimal amount of anhydrite was 2 ± 0.5 kg/m^3 of concrete, excluding the anhydrite contained in cement.

Keywords: anhydrite; Baganuur; compressive strength; insoluble residue; lignite; mineral composition; Mongolia; ready-mixed concrete; Shivee Ovoo

Citation: Pushpalal, D.; Kashima, H. Apparent Influence of Anhydrite in High-Calcium Fly Ash on Compressive Strength of Concrete. *Buildings* 2024, 14, 2899. https://doi.org/10.3390/buildings14092899

Academic Editors: Grzegorz Ludwik Golewski, Rajai Zuheir Al-Rousan, Ciro Del Vecchio and Daiyu Wang

Received: 8 July 2024
Revised: 7 September 2024
Accepted: 12 September 2024
Published: 13 September 2024

1. Introduction

It is well known that fly ash has versatile capabilities to make durable, affordable, tough, and environmentally friendly (DATE) concrete. However, the chemical composition of fly ash is highly variable in countries other than Japan, and most of them contain less SiO_2 and more SO_3 and CaO than Japanese fly ash [1,2]. In this context, anhydrite became apparent as one of the key ingredients. Anhydrite-rich fly ash is an unavoidable byproduct of lignite-burning coal power plants, disregarding the method of burning coal. According to the World Energy Outlook 2018, there are 23.07 trillion tons of coal reserves in terms of resources, of which 20% is lignite [3]. Lignite is a very soft coal that contains water up to 70% by weight and it pollutes the air higher than other coals. With a carbon content of as low as 25 to 35%, lignite is the lowest rank of coal because of its low heat content compared to other coals such as anthracite [4]. Another unwelcome property of lignite coal is that it generates 10–50% of ash [4]. Burning lignite produces ash containing high amounts of Ca- and S-bearing minerals.

Several authors have reported the presence of anhydrite in ash resulting from burning lignite. McCarthy et al. [5] investigated 26 fly ash (FA) samples from North Dakota and argued that if lignite or sub-bituminous coal is burned, the formation of anhydrite is not astonishing, because of high SO_3 and CaO contained in the ash. They showed that free lime in the ash is acting as a built-in "scrubber" for SO_3. Examining Cretaceous lignite from the Hailar Basin, Inner Mongolia, China, Jia and coauthors [6] reported that burning temperature depends on the mineral phases of ash. Quartz and anhydrite with a high melting point dominated the high-temperature ashes burnt at 815 °C, because of naturally available sulfate minerals such as gypsum dehydrate at high temperatures, resulting in these changing into bassanite and anhydrite.

However, opinions on the effect of anhydrite in fly ash on the fresh and hardened properties of concrete are divided into those for and against. Ravina and Mehta [7] showed the impact of replacing up to 50% of cement with ASTM Class C fly ash and F fly ash on the compressive strength of lean concrete mixtures. They obtained a targeted compressive strength of 14 MPa at an earlier age with mixtures that incorporated CaO- and SO_3-rich Class C fly ash than with CaO- and SO_3-poor Class F fly ash. However, the delay of setting times was longer in the concrete mixtures with ASTM Class C fly ash than those with ASTM Class F fly ash [8]. This delay is attributed to the sulfate content of the fly ash that is found to be on the surface of the fly ash particles. Zhang and coauthors [9] reported that concrete made with fly ash containing a high content of CaO with SO_3 has shown high compressive strength, even at early ages.

Burning lignite is not the only way that anhydrite-rich ash can be produced. Fluidized Bed Combustion (FBC) generally produces ashes richer in anhydrite because FBC boilers use ground limestone as a sulfur-absorbing material, and several authors have studied the properties of concrete that incorporates anhydrite-rich ashes [10–13]. Comparing two Circulating Fluidized Bed Combustion (CFBC) ashes, Shen and coauthors [14] reported that the compressive strength of mortar specimens containing high-anhydrite CFBC ash is higher than low-anhydrite CFBC ash. Furthermore, they have insisted that CFBC ash can be efficiently used without harm to the volume stability of cement paste if the proportions are properly designed.

Poon and coauthors [15] have investigated the role of anhydrite in the activation of ASTM Class F fly ash mortar by adding anhydrite to mortar. Their study found that early and later age strengths were increased. A mortar system that incorporated fly ash up to 55% has been activated by adding 10% anhydrite, and the existence of an optimum quantity of anhydrite for obtaining the highest strength was discovered. However, the optimum amount of anhydrite, which should be contained in a unit volume of mortar or concrete, has not been quantified.

Fly ash replacement of up to 70% of the total cementitious content of concrete was investigated in this study. This research aimed to clarify the influence of the anhydrite phase contained naturally in fly ash on the compressive strength of concrete. Furthermore, an attempt has been made to quantify the optimum amount of anhydrite, which should be contained in a unit volume of concrete to obtain a high compressive strength.

2. Experimental Procedures

2.1. Materials

2.1.1. Fly Ash

Five types of fly ashes, which were collected from Ulaanbaatar City 4th coal power plant (PP4) in different seasons, were used. These ashes are a byproduct of burning lignite. FA-0, FA-1, and FA-2 derived from Shivee Ovoo coal [16] and FA-3 and FA-4 from Baganuur coal. Shivee Ovoo ashes contain a considerably high volume of anhydrite due to sulfur-bearing minerals naturally contained in coal. According to ASTM C 618 [17], these fly ashes are designated as Class C. The chemical composition of the fly ashes measured by an X-ray fluorescence spectrometer is given in Table 1. Ashes derived from Shivee Ovoo coal (FA-0, FA-1, and FA-2) show a higher SO_3 content than ashes from Baganuur coal, as also previously confirmed by several authors [18–22]. It should be noted that there are contradictory specifications on SO_3 content in fly ash. Chinese (GBT 1596:2018) and European (BS EN 450-1:2012) standards recommended that less than 3 wt%. ASTM C 618:2017 standards recommended less than 5 wt%, while there is no limit set by Japanese (JIS A6201:2015) and Korean (KSL 5405:2018) standards [3].

Table 1. Chemical composition of fly ashes.

Type of Fly Ash	Chemical Composition (wt%)									
	SiO_2	Al_2O_3	Fe_2O_3	CaO	SO_3	MgO	Na_2O	K_2O	TiO_2	MnO
FA-0	30.2	10.7	5.8	33.7	8.2	6.4	1.3	0.7	0.5	1.0
FA-1	43.4	12.4	7.1	23.8	4.2	4.6	0.9	1.1	0.8	0.5
FA-2	38.1	11.7	7.3	27.9	5.2	5.2	1.2	0.9	0.6	0.6
FA-3	47.2	14.2	12.0	18.6	1.2	2.2	0.6	1.1	0.6	0.3
FA-4	53.1	14.4	12.8	13.2	1.2	1.6	0.4	1.4	0.6	0.4

2.1.2. Other Materials

Class 42.5 Portland cement (specific gravity: 3.00, Blaine specific surface area: 3260 cm^2/g) available in the market was used. The chemical composition and mineral composition are given in Tables 2 and 3, respectively. Sand (specific gravity: 2.60) sieved through 5 mm mesh was used. Gravel was 20 mm maximum in size, and the specific gravity was 2.65. A polycarboxylate ether-based high-range water-reducing admixture was used to achieve appropriate workability for the concrete mixtures.

Table 2. Chemical composition of Portland cement (wt%).

SiO_2	Al_2O_3	Fe_2O_3	CaO	MgO	K_2O	Na_2O	SO_3	Insoluble Residue	Loss on Ignition	Total
19.69	5.63	3.88	59.13	1.2	1.15	0.2	2.19	3.74	3.19	100

Table 3. Mineral composition of Portland cement (wt%).

C_3S	C_2S	C_3A	C_4AF	Gypsum	Free-CaO	Total
41.75	25.20	8.36	11.81	4.00	1.45	92.75

2.2. Testing Procedures

2.2.1. X-ray Diffraction

An X-ray diffractometer (PHILIPS X'Pert MPD) was used. Fly ash was pulverized in an alumina mortar until grains could no longer be felt between the fingers and used as the sample. Powder X-ray diffraction was performed under the following conditions: target CuKα, tube voltage 45 kV, tube current 40 mA, scanning range 2θ = 10–60°, and step width 0.05.

Rietveld analysis of fly ash was performed on fly ashes FA-1, FA-2, FA-3, and FA-4 with 20% corundum as an internal standard. High Score Plus was used as the Rietveld analysis software. The minerals quartz, hematite, magnetite, lime, akermanite, merwinite, and anhydrite were identified.

2.2.2. Insoluble Residues and Other Properties

The insoluble residue (IR) of fly ash and cement was determined following JIS R 5202 [23]. This method is a conventional method in which the insoluble residue in cement is obtained by treating the sample with a dilute hydrochloric acid solution so that the precipitation of soluble SiO_2 is minimal. The residue from this treatment is treated again with a boiling solution of Na_2CO_3 to re-dissolve traces of SiO_2 which may have precipitated. The residue is determined gravimetrically after ignition.

Loss on ignition was measured following JIS A 6201 [24]. Specific gravity was measured per JIS R 5201 [25]. A laser diffraction particle size distribution analyzer (SHIMADZU, SALD-7000) was used to measure the specific surface area and average particle size. Free CaO content in cement was measured according to procedures recommended by the Japan Cement Association [26].

2.2.3. Specimen Preparation and Testing

Two different series of mixed proportions were adopted, and they are tabulated in Tables 4 and 5. Series 1 is designed to understand the maximum possible replacement level of fly ash. Series 2 is designed to understand the effect of anhydrite on compressive strength development. All concrete mixtures were made to obtain a slump of 210 ± 20 mm by using the water-reducing admixture, considering easy pumpability in construction sites as required by RMC. Mixtures were mixed in a single-axis horizontal mixer according to the procedures explained below. First, the binder and aggregates were added to the mixer and mixed for 30 s. Next, water which was premixed with water-reducing admixture (SP) was added and mixed for 90 s. The slump was measured in accordance with Japanese Industrial Standard A 1101 [27], and 100 $\varphi \times 200$ mm specimens were molded. The specimens were demolded after two days of curing in a moist room at $20 \pm 2°$ and then cured in water at 20 ± 2 °C. All specimens were continuously water-cured until the time of strength testing. Compressive strength tests were carried out according to the Japanese Industrial Standard A 1108 [28] at 3, 7, 28, and 91 days.

Table 4. Mixed proportions of concrete (Series 1).

Mix Index	Type of FA	W/B (%)	FA/B (%)	Unit Weight (kg/m^3)					
				W	C	FA	S	G	SP
Control	-	35	0	136	389	0	994	961	6.8
FA-0-30	FA-0	35	30	133	266	114	971	938	6.8
FA-0-40	FA-0	35	40	132	226	151	964	931	6.8
FA-0-50	FA-0	35	50	131	187	187	956	924	6.8
FA-0-60	FA-0	35	60	130	148	223	949	917	6.8
FA-0-70	FA-0	35	70	129	111	258	942	910	6.8

W/B: water—binder (cement plus fly ash), FA/B: fly ash—binder (cement plus fly ash), W: water, C: cement, FA: fly ash, S: sand, G: gravel, SP: superplasticizer.

Table 5. Mixed proportions of concrete (Series 2).

Mix Index	Type of FA	W/B (%)	FA/B (%)	Unit Weight (kg/m^3)					
				W	C	FA	S	G	SP
Control	-	46	0	164	356	0	1001	878	5.0
FA-1-10	FA-1	46	10	164	320	36	999	878	5.0
FA-1-20	FA-1	46	20	163	284	71	996	875	4.9
FA-1-40	FA-1	46	40	162	212	141	991	871	4.9
FA-2-10	FA-2	46	10	164	320	36	999	878	5.5
FA-2-20	FA-2	46	20	164	284	71	997	876	5.2
FA-3-10	FA-3	46	10	164	320	36	999	878	5.5
FA-3-20	FA-3	46	20	163	284	71	996	876	5.2
FA-3-40	FA-3	46	40	163	212	141	992	871	5.1
FA-4-10	FA-4	46	10	164	320	36	999	878	5.7
FA-4-20	FA-4	46	20	163	284	71	997	876	5.7

W/B: water—binder (cement plus fly ash), FA/B: fly ash—binder (cement plus fly ash), W: water, C: cement, FA: fly ash, S: sand, G: gravel, SP: superplasticizer.

3. Results and Discussion

3.1. Mineral Composition and Other Properties of Fly Ashes

Figure 1A shows the X-ray diffraction pattern of FA-0. Figure 1B represents the X-ray diffraction patterns of fly ashes FA-1, FA-2, FA-3, and FA-4. These XRD patterns show that the main constituents of FA-0, FA-1, and FA-2 are quartz (SiO_2), hematite (Fe_2O_3), lime (CaO), and anhydrite ($CaSO_4$); however, akermanite ($Ca_2Mg(Si_2O_7)$) and merwinite ($Ca_3Mg(SiO_4)_2$) are also traced. On the other hand, quartz (SiO_2), hematite (Fe_2O_3), and magnetite (Fe_3O_4) are available in FA-3 and FA-4; however, anhydrite does not exist. The mineral composition and glass phase of fly ashes determined by XRD Rietveld analysis are shown in Table 6. Supporting the XRD patterns, Rietveld analysis proves the availability of anhydrite in FA-1 and FA-2.

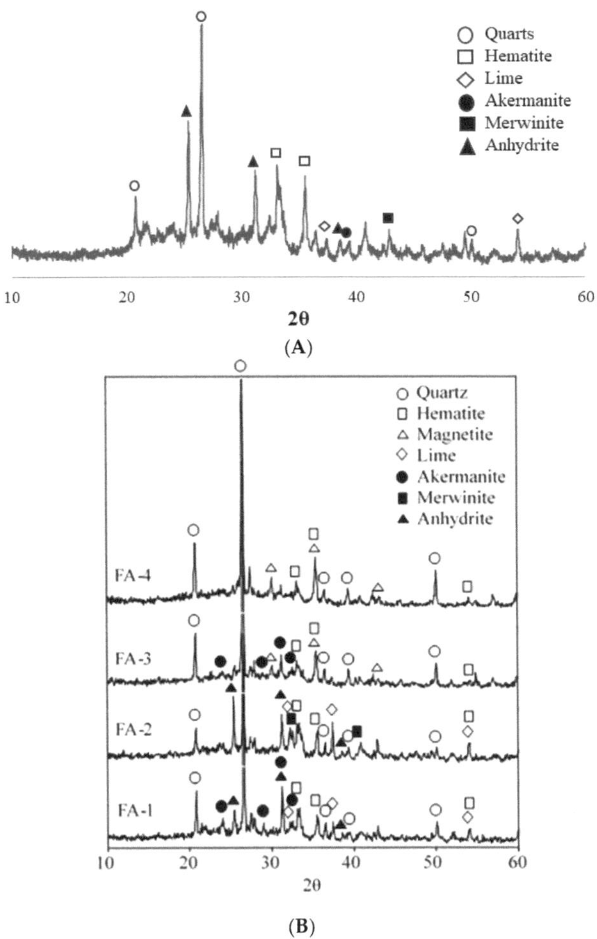

Figure 1. (**A**) X-ray diffraction pattern of fly ash FA-0. (**B**) X-ray diffraction patterns of fly ashes FA-1, FA-2, FA-3, and FA-4.

Table 6. Mineral composition and other properties of fly ashes FA-1, FA-2, FA-3, and FA-4.

Measured Item	Type of Fly Ash			
	FA-1	FA-2	FA-3	FA-4
Quartz (wt%)	13.4	8.4	18.4	21.2
Hematite (wt%)	4.5	3.6	1.7	1.6
Magnetite (wt%)	-	-	1.8	3.3
Lime (wt%)	0.7	1.9	-	-
Akermanite (wt%)	8.7	-	6.0	-
Merwinite (wt%)	-	6.9	-	-
Anhydrite (wt%)	2.8	5.7	-	-
Glass Phase (wt%)	67.1	72.7	70.6	73.4
Insoluble residue (wt%)	48.6	38.6	56.6	64.6
Loss on ignition (wt%)	0.8	0.8	1.5	0.5
Specific gravity	2.5	2.6	2.5	2.5
Blaine specific surface area (cm^2/g)	3746	5946	5770	7241
Average particle size (μm)	35.9	18.4	19.8	8.1

The percentage of the insoluble residue of fly ashes FA-1, FA-2, FA-3, and FA-4 is tabulated in Table 6. IR does not take part in the cementing action. Therefore, international standards limit IR to less than 1.5% in the case of cement. Kiattikomol remarked that IR is a measure of the adulteration of cement, largely coming from impurities [29]. IR in anhydrite-rich FA-2 is extremely low compared to other fly ashes expecting its high cementing action. Hanehara et al. [30] have shown that IR relates to the reaction ratio of fly ash, and this phenomenon has been confirmed by other authors too [31,32]. Figure 2 shows the relationship between CaO content (refer Table 1) and insoluble residue (refer Table 6), and it proves a strong correlation between the two. Though there is no similar correlation, we have found in the previous literature that studies made by the Greek Public Power Corporation remarked that "when insoluble residue is high the CaO content is low [33]." Furthermore, Goswami [34] has found that IR is higher in low-calcium FAs than high-calcium FAs, supporting our findings.

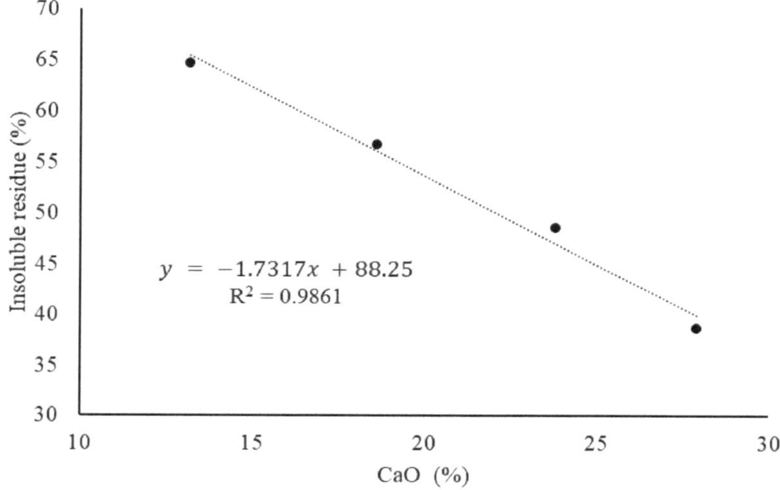

Figure 2. The relationship between CaO content and insoluble residue.

3.2. Compressive Strength

Figure 3 shows the time-dependent compressive strength of mixed proportions of Series 1. Fly ash replacement up to 60% of total cementitious content shows strength nearly equal to the control at later ages beyond 14 days. A large decrement in strength is seen when the fly ash content increases beyond 60%. While the early strength of all fly ash mixtures is lower than the control, 30 and 40 percent fly ash-incorporated mixtures have shown slightly higher strength development after 7 days, overpassing the control. A similar tendency has been confirmed by Zhang and coauthors [9], in which 50% replacement of high-calcium fly ash has shown higher later-age strength than the control mix proportion without fly ash. We attributed this tendency to being a result of the high CaO and SO_3 contents of FA-0. In order to confirm this hypothesis, we conducted the experiments in Series 2, using two clearly different sets of Class C fly ashes, in which FA-1 and FA-2 represent the set of ashes, which contains a considerable amount of anhydrite, as measured by XRD Rietveld analysis, while FA-3 and FA-4 have shown no anhydrite (see Table 6).

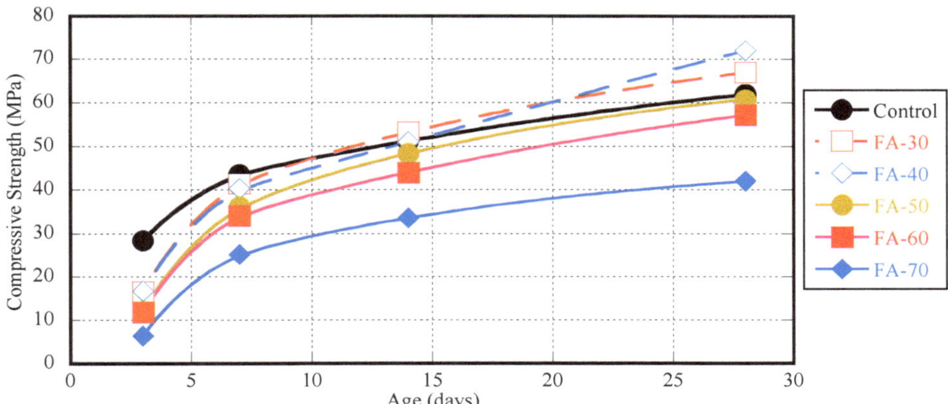

Figure 3. Compressive strength vs. age for control concrete and FA-0-incorporated fly ash concretes.

Figures 4–6 show the compressive strength of all eleven mixtures in Series 2. At all ages, FA-1 and FA-2 fly ash-incorporated concrete with replacement levels of 10% and 20% showed a higher strength compared to the control. The 40% replacement by FA-1 showed a lower strength up to the 28th day; however, it began to surpass the control after 28 days. On the contrary, FA-3 and FA-4 (i.e., both with no anhydrite) show a lower strength than the control, especially at early ages for all replacement ratios. Besides, FA-1- and FA-2-incorporated concrete exhibit higher strength than concretes with FA-3 and FA-4 at all ages. The existence of anhydrite in FA-1 and FA-2 might contribute to this achievement. Similar results have been obtained by Poon and coauthors [15] by adding 10% anhydrite to mortar with 35% fly ash replacement. When they compare gypsum and anhydrite in terms of an equivalent SO_3 content, the latter is more productive in increasing the early-age strength but less productive in increasing the later-age strength than gypsum. However, anhydrite is more beneficial in increasing the strength at all ages if a comparison is made in terms of an equal amount of addition. Therefore, anhydrite has shown itself to be advantageous over gypsum. Zhang and coauthors [9] used high-CaO and -SO_3 fly ash to make concrete and obtained a slightly higher early strength than low-CaO and -SO_3 fly ash and argued that the formation of ettringite might increase the early strength. Enders has suggested that anhydrite particles contaminated with aluminum are an easily accessible elemental source for the formation of the first ettringite during the hydration reaction of lignite fly ashes [35]. It is interesting in this context to note that these conclusions are in good agreement with what we found.

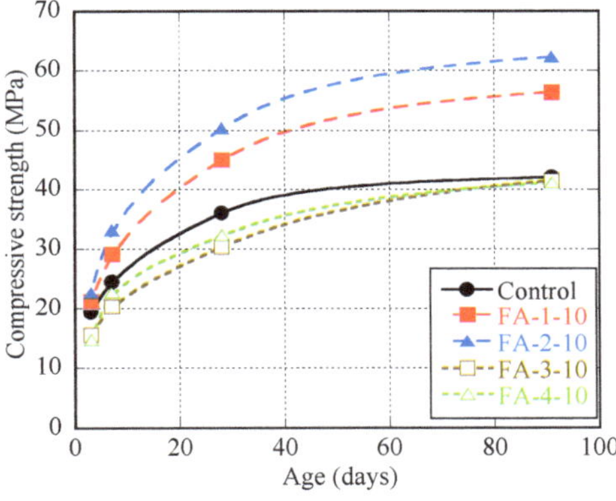

Figure 4. Compressive strength vs. age for control concrete and fly ash concretes at 10% replacement.

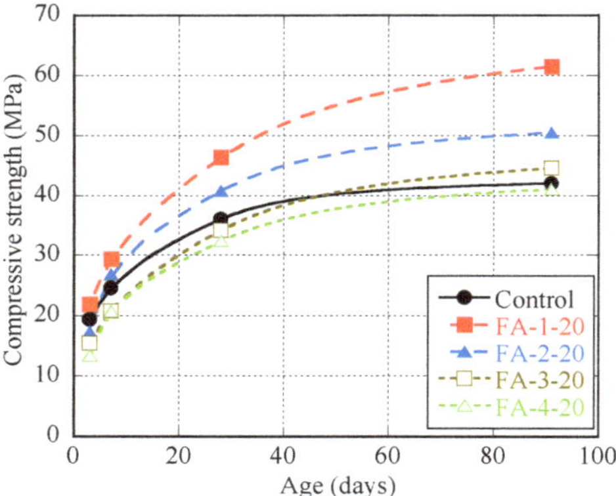

Figure 5. Compressive strength vs. age for control concrete and fly ash concretes at 20% replacement.

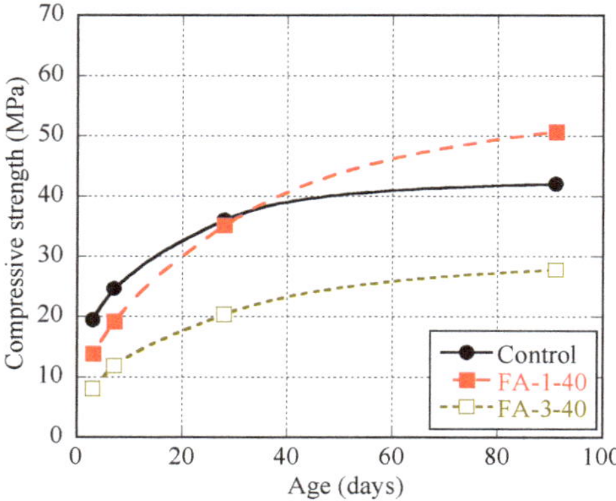

Figure 6. Compressive strength vs. age for control concrete and fly ash concretes at 40% replacement.

3.3. Optimum Amount of Anhydrite

Figures 7 and 8 show the relationship between anhydrite content, fly ash replacement ratio, and compressive strength of FA-1- and FA-2-incorporated concrete. Anhydrite quantity in fly ash concrete per 1 m^3 was calculated according to the following Equation (1):

$$W_{Anhydrite} = FA \times (\%Anhydrite) \tag{1}$$

where $W_{Anhydrite}$: quantity of anhydrite in fly ash concrete (kg/m^3), FA: quantity of fly ash in concrete (kg/m^3), and $\%Anhydrite$: weight percent of anhydrite in fly ash (wt%). Figures 7 and 8 show the availability of an optimum level of anhydrite to be contained in concrete to obtain the highest strength. As exhibited by the curves, we arbitrate that the optimum amount lies within 1.5 to 2.5 kg/m^3 of concrete. In support of this observation, Poon and coauthors [15] have also reported the existence of an optimum level of anhydrite, proving this fact by performing compressive strength tests of mortars cured at elevated temperatures. Zhou et al. [36] have found that the pozzolanic activity of anhydrite-rich CFBC is higher than that of pulverized coal combustion fly ash (PCA), which is poor in anhydrite. The activity index corresponding to the seventh-day compressive strength of CFBC can reach 103.76% when the mixing amount is 25%, while that of PCA is only 93.36%. However, the pozzolanic activity of CFBC decreases with increasing the mixing amount, the same as in our study.

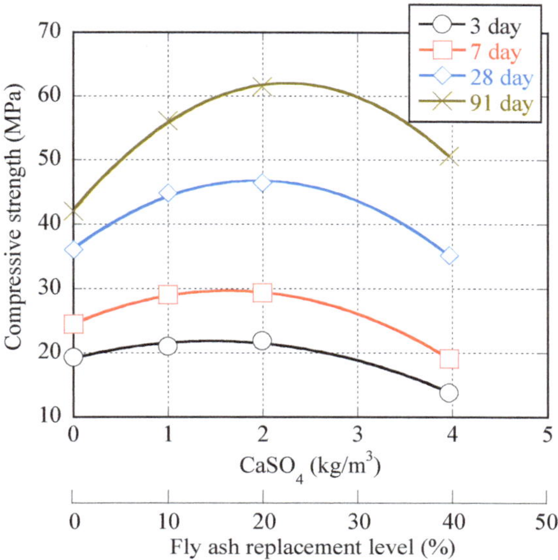

Figure 7. The relationship between anhydrite content, fly ash replacement ratio, and compressive strength for FA-1-incorporated concrete.

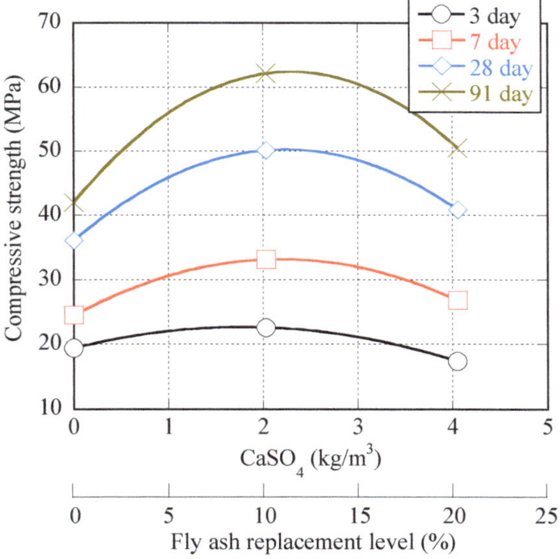

Figure 8. The relationship between anhydrite content, fly ash replacement ratio, and compressive strength for FA-2-incorporated concrete.

4. Conclusions

Two series of concrete incorporated with five high-calcium fly ashes were investigated to understand the influence of anhydrite in fly ash on the compressive strength of concrete. The main conclusions can be drawn as follows:

1. It is possible to achieve a compressive strength nearly equal to the control at a late age beyond 14 days if fly ash containing a high amount of CaO and SO_3 is used, up to a replacement level of 60%.
2. The replacement levels of up to 20% of fly ash give higher strength than control concrete at all ages except the third day when fly ash containing anhydrite is used.
3. When the CaO content is high, insoluble residue is low, showing high cementitious properties.
4. The compressive strength increases with an increasing fly ash replacement ratio when anhydrite-rich ash is used, but strength decreases when the replacement level exceeds a certain point. The optimal amount of anhydrite was $2 \pm 0.5 \ \mathrm{kg/m^3}$ of concrete.
5. This case study shows that anhydrite-rich fly ash is stronger than concrete-containing anhydrite-free fly ash even at early ages. However, this phenomenon is apparent. The factors behind this observation are still under investigation but may include anhydrite being an indirect indicator of other influential factors for producing high performances.

Author Contributions: D.P. and H.K. contributed equally to this work in all areas. All authors have read and agreed to the published version of the manuscript.

Funding: Part of this work was funded by JSPS KAKENHI Grant Number JP 25303004.

Data Availability Statement: Data are contained within the article.

Acknowledgments: The authors would like to thank Ochirbat Batmunkh (Premium Concrete LLC, Ulaanbaatar, Mongolia) for providing research support for this article.

Conflicts of Interest: Author Hiroo Kashima was employed by the company Shin-Etsu Industry Co., Ltd. The remaining author declares that the research was conducted in the absence of any commercial or financial relationships that could be construed as a potential conflict of interest.

References

1. Pushpalal, D.; Danzandorj, S.; Narantogtokh, B.; Nishiwaki, T.; Sashka, U.; Erdenebat, S.; Sambuu, C. Freezing-Thawing Durability and Chemical Characteristics of Air-Entrained Sustainable Concrete Incorporated High-Calcium Fly Ash in High-Volume. *J. Build. Eng.* **2024**, *89*, 109307. [CrossRef]
2. Yamamoto, T. Formulation optimization and proposal of Assessed Pozzolanic-activity Index (API) method for rapid evaluation of pozzolanic activity of fly ash. *Constr. Build. Mater.* **2024**, *432*, 136394. [CrossRef]
3. Japan Fly Ash Association (Ed.) *Coal Ash Handbook*; Japan Fly Ash Association: Tokyo, Japan, 2021. (In Japanese)
4. Bowen, B.H.; Irwin, M.W. *Indiana Center for Coal Technology Research COAL CHARACTERISTICS CCTR Basic Facts File # 8*; Purdue University: West Lafayette, IN, USA, 2008; pp. 1–28.
5. McCarthy, G.J.; Swanson, K.D.; Keller, L.P.; Blatter, W.C. Mineralogy of Western Fly Ash. *Cem. Concr. Res.* **1984**, *14*, 471–478. [CrossRef]
6. Jia, R.; Liu, J.; Han, Q.; Zhao, S.; Shang, N.; Tang, P.; Zhang, Y. Mineral Matter Transition in Lignite during Ashing Process: A Case Study of Early Cretaceous Lignite from the Hailar Basin, Inner Mongolia, China. *Fuel* **2022**, *328*, 125252. [CrossRef]
7. Ravina, D.; Mehta, P.K. Compressive Strength of Low Cement/High Fly Ash Concrete. *Cem. Concr. Res.* **1988**, *18*, 571–583. [CrossRef]
8. Ravina, D.; Mehta, P.K. Properties of Fresh Concrete Containing Large Amounts of Fly Ash. *Cem. Concr. Res.* **1986**, *16*, 227–238. [CrossRef]
9. Zhang, Y.; Sun, W.; Shang, L. Mechanical Properties of High Performance Concrete Made with High Calcium High Sulfate Fly Ash. *Cem. Concr. Res.* **1997**, *27*, 1093–1098. [CrossRef]
10. Zahedi, M.; Jafari, K.; Rajabipour, F. Properties and Durability of Concrete Containing Fluidized Bed Combustion (FBC) Fly Ash. *Constr. Build. Mater.* **2020**, *258*, 119663. [CrossRef]
11. He, P.; Zhang, X.; Chen, H.; Zhang, Y. Waste-to-Resource Strategies for the Use of Circulating Fluidized Bed Fly Ash in Construction Materials: A Mini Review. *Powder Technol.* **2021**, *393*, 773–785. [CrossRef]
12. Hanisková, D.; Bartoníčková, E.; Koplík, J.; Opravil, T. The Ash from Fluidized Bed Combustion as a Donor of Sulfates to the Portland Clinker. *Procedia Eng.* **2016**, *151*, 394–401. [CrossRef]
13. Sebök, T.; Šimoník, J.; Kulísek, K. The Compressive Strength of Samples Containing Fly Ash with High Content of Calcium Sulfate and Calcium Oxide. *Cem. Concr. Res.* **2001**, *31*, 1101–1107. [CrossRef]
14. Shen, Y.; Qian, J.; Zhang, Z. Investigations of Anhydrite in CFBC Fly Ash as Cement Retarders. *Constr. Build. Mater.* **2013**, *40*, 672–678. [CrossRef]

15. Poon, C.S.; Kou, S.C.; Lam, L.; Lin, Z.S. Activation of Fly Ash/Cement Systems Using Calcium Sulfate Anhydrite (CaSO$_4$). *Cem. Concr. Res.* **2001**, *31*, 873–881. [CrossRef]
16. Avid, B.; Purevsuren, B.; Born, M.; Dugarjav, J.; Davaajav, Y.; Tuvshinjargal, A. Pyrolysis and TG Analysis of Shivee Ovoo Coal from Mongolia. *J. Therm. Anal. Calorim.* **2002**, *68*, 877–885. [CrossRef]
17. *ASTM C618-17a; Standard Specification for Coal Fly Ash and Raw or Calcined Natural Pozzolan for Use in Concrete.* ASTM International: West Conshohocken, PA, USA, 2017.
18. Temuujin, J.; Minjigmaa, A.; Davaabal, B.; Bayarzul, U.; Ankhtuya, A.; Jadambaa, T.; Mackenzie, K.J.D. Utilization of Radioactive High-Calcium Mongolian Flyash for the Preparation of Alkali-Activated Geopolymers for Safe Use as Construction Materials. *Ceram. Int.* **2014**, *40*, 16475–16483. [CrossRef]
19. Kashima, H. Study on Effective Utilization of Fly Ash in Mongolia: Focusing on Technology Development and Local Adaptability. Doctoral Thesis, Tohoku University, Sendai, Japan, 2016. (In Japanese)
20. Suzuki, A.; Pushpalal, D.; Kashima, H. An Appraisal of Compressive Strength of Concrete Incorporated with Chemically Different Fly Ashes. *Open Civ. Eng. J.* **2020**, *14*, 188–199. [CrossRef]
21. Pushpalal, D.; Danzandorj, S.; Bayarjavkhlan, N.; Nishiwaki, T.; Yamamoto, K. Compressive Strength Development and Durability Properties of High-Calcium Fly Ash Incorporated Concrete in Extremely Cold Weather. *Constr. Build. Mater.* **2022**, *316*, 125801. [CrossRef]
22. Alyeksandr, A.; Bai, Z.; Bai, J.; Janchig, N.; Barnasan, P.; Feng, Z.; Hou, R.; He, C. Thermal Behavior of Mongolian Low-Rank Coals during Pyrolysis. *Carbon Resour. Convers.* **2021**, *4*, 19–27. [CrossRef]
23. *JIS R5202; Methods for Chemical Analysis of Cement.* Japanese Standards Association: Tokyo, Japan, 2018. (In Japanese)
24. *JIS A6201; Fly Ash for Concrete (Standard Specification).* Japanese Industrial Standards Committee: Tokyo, Japan, 2015. (In Japanese)
25. *JIS R5201; Physical Testing Methods for Cement.* Japanese Standards Association: Tokyo, Japan, 2015. (In Japanese)
26. *JCAS I-01:1997; Method for Quantifying Free Calcium Oxide.* Japan Cement Association: Tokyo, Japan, 1997. (In Japanese)
27. *JIS A1101; Method of Test for Slump of Concrete (Standard Specification).* Japanese Standards Association: Tokyo, Japan, 2020. (In Japanese)
28. *JIS A1108;* JIS A 1108 Method of Test for Compressive Strength of Concrete. Japanese Standards Association: Tokyo, Japan, 2018. (In Japanese)
29. Kiattikomol, K.; Jaturapitakkul, C.; Tangpagasit, J. Effect of Insoluble Residue on Properties of Portland Cement. *Cem. Concr. Res.* **2000**, *30*, 1209–1214. [CrossRef]
30. Hanehara, S.; Tomosawa, F.; Kobayakawa, M.; Hwang, K.R. Effects of Water/Powder Ratio, Mixing Ratio of Fly Ash, and Curing Temperature on Pozzolanic Reaction of Fly Ash in Cement Paste. *Cem. Concr. Res.* **2001**, *31*, 31–39. [CrossRef]
31. Ji, X.; Takasu, K.; Suyama, H.; Koyamada, H. The Effects of Curing Temperature on CH-Based Fly Ash Composites. *Materials* **2023**, *16*, 2645. [CrossRef] [PubMed]
32. Taniguchi, M.; Sagawa, T.; Katsura, O. *Research on Reaction Rate of Fly Ash;* Japan Concrete Institute: Tokyo, Japan, 2007; pp. 189–194. (In Japanese)
33. Papayianni, J. Use of a High-Calcium Fly Ash in Blended Type Cement Production. *Cem. Concr. Compos.* **1993**, *15*, 231–235. [CrossRef]
34. Goswami, A.P. Determining Physico-Chemical Parameters for High Strength Ambient Cured Fly Ash-Based Alkali-Activated Cements. *Ceram. Int.* **2021**, *47*, 29109–29119. [CrossRef]
35. Enders, M. Microanalytical Characterization (AEM) of Glassy Spheres and Anhydrite from a High-Calcium Lignite Fly Ash from Germany. *Cem. Concr. Res.* **1995**, *25*, 1369–1377. [CrossRef]
36. Zhou, M.; Chen, P.; Chen, X.; Ge, X.; Wang, Y. Study on Hydration Characteristics of Circulating Fluidized Bed Combustion Fly Ash (CFBCA). *Constr. Build. Mater.* **2020**, *251*, 118993. [CrossRef]

MDPI AG

Grosspeteranlage 5

4052 Basel

Switzerland

Tel.: +41 61 683 77 34

Buildings Editorial Office

E-mail: buildings@mdpi.com

www.mdpi.com/journal/buildings